Correlations, Coherence, and Order

Correlations, Coherence, and Order

Edited by

Diana V. Shopova and
Dimo I. Uzunov

Bulgarian Academy of Sciences
Sofia, Bulgaria

Springer Science+Business Media, LLC

Proceedings of the First Pamporovo Winter Workshop on Cooperative Phenomena in Condensed Matter, held March 7 – 15, 1998, in Pamporovo, Bulgaria

ISBN 978-1-4613-7142-7 ISBN 978-1-4615-4727-3 (eBook)
DOI 10.1007/978-1-4615-4727-3

Originally published by Kluwer Academic / Plenum Publishers, New York in 1999
Softcover reprint of the hardcover 1st edition 1999

10 9 8 7 6 5 4 3 2 1

A C.I.P. record for this book is available from the Library of Congress.

PREFACE

This volume contains a collection of review articles that are extended versions of invited lectures given at the *First Pamporovo Winter Workshop on Cooperative Phenomena in Condensed Matter* held in villa "Orlitza" (7th–15th March 1998, Pamporovo Ski Resort, Bulgaria). Selected research works reported at the Workshop have been published in the *Journal of Physical Studies* — a new International Journal for research papers in experimental and theoretical physics (Lviv University, Lviv, Ukraine).

These reviews are supposed to be status reports and present new insights gained from the rapidly developing research of outstanding problems in condensed matter physics such as structural properties and phase transitions in fullerene crystals, superconductivity of strongly interacting electrons in copper oxides, spin polarized Fermi liquids, chaotic vortex filaments in superfluid turbulent Helium-II, desorption induced by electronic transitions in ionic compounds, fluctuation phenomena in superconductors, and quantum critical phenomena in low dimensional magnets and quantum liquids. We have set the material according to the alphabetic order of authors' names although the high temperature superconductivity seems to be the hard kernel in condensed matter physics.

The authors have taken care to present the recent advances in their research in a form which is readable and useful not only to experts in the respective field, but also to young scientists. That is why the lectures include a comprehensive introduction to the matter and also an extended discussion of methodical details.

We hope that the lectures collected in this volume will be helpful and stimulating to scholars and students interested in the developments of different topics in condensed matter physics and are very much indebted to authors for their collaboration.

The Workshop has been organized with the kind support of the Bulgarian Council of Ministers and the President of the Bulgarian Academy of Sciences. It could not have taken place without invaluable help and financial support from the following institutions: the Austrian Institute for East and South East Europe, the Joint Institute of Nuclear Research (Dubna, Russia), the 3M Corporation, and the INTAS program which ensured the travel and local expenses of about twenty Russian scientists. The Workshop has been practically supported, including a substantial funding, also by the President of Pamporovo Tourist Corporation.

One of us (D.I.U.) thanks for the hospitality of the Department of Theoretical Physics of Salerno University (Italy), where a part of this edition has been done with the kind technical assistance of Mrs. C. D'Apolito, and Mr. V. Di Marino. A technical help by Mrs. Svetlana Savova and Mr. Ivaylo P. Takov is also gratefully acknowledged.

Diana V. Shopova,
Dimo I. Uzunov

CONTENTS

STRUCTURAL PECULIARITIES IN FULLERENE CRYSTALS

V. L. Aksenov and V. S. Shakhmatov

Frank Laboratory of Neutron Physics,
Joint Institute for Nuclear Research,
141980 Dubna, Moscow region, Russia

Fullerene molecules are the new form of carbon in Nature discovered in 1985 in investigations stimulated by interest in certain unexplained features of the absorption and emission spectra of interstellar matter. During the last years many chemists and physicists have found themselves involved in the new area of research lying on the intersection of quantum chemistry, mathematics, architecture, space physics and space chemistry. Fullerene-based compounds exhibit many interesting physical properties and are called the materials of the 21st century.

The review is devoted to a symmetry theory of structural transformations in subsystems of C_{60} molecules and intercalated metals. Structural properties of compounds based on fullerene molecules are discussed. A Landau phenomenological theory of the phase transition to a polymer-like phase in AC_{60} (A=K,Rb) crystals is developed. The theory describes adequately spontaneous crystalline cell deformations at phase transition and predicts partial ordering of alkali metal atoms over possible positions in the octahedral environment of C_{60} molecules.

The local symmetry of the orientational states of C_{60} molecule in crystals is investigated. It is shown that in different crystals different orientational phase transitions are connected with different orientational orbits. The model of orientational phase transitions based on a sequence of orientational states of different symmetry is proposed. It is discovered that the local symmetry and the symmetry of internal vibrations of C_{60} molecule increase as the spatial symmetry of the crystal decreases at phase transition. This effect has a general character and may be observed following orientational phase transitions of the order-disorder type with a wave vector at the boundary of the Brillouin zone. Some experimental data on structural peculiarities in fullerene crystals are discussed.

1. INTRODUCTION

During the last 10–15 years many very interesting discoveries have been made in the field of condensed matter physics. Among them the most important are: Bose con-

Correlations, Coherence, and Order. Edited by Shopova and Uzunov
Kluwer Academic/Plenum Publishers, 1999

densation, high temperature superconductors, and fullerenes. All three have a quite different history of discovery. The effect of Bose condensation was predicted theoretically in the beginning of this century. Since then scientists had long striven for its experimental observation and finally, in 1995, they succeeded.

Superconductivity was discovered in 1911. The microscopic theory of the phenomenon was constructed in 1957. The theory formalism gives the limit for superconducting temperature in metals in the vicinity of 30 K. Numerous attempts to overcome this limit failed. As a result, in the early 80's just few enthusiasts dared to continue searching for materials with the high ($T_c > 30$ K) superconducting temperature. In 1986 the effort of J. G. Bednorz and K. A. Müller was successful.

In 1985, H. W. Kroto et al.[1] announced the existence of the new form of carbon–stable giant molecules C_{60}, called fullerenes. This was not expected. Unlike above mentioned discoveries it did not complete a long period of work but opened the door to the new area of research. As a result, during the last five years many chemists and physicists have found themselves involved in this new area. They try to do physics and find applications related with the newly discovered material.

In this lecture, we discuss the structural properties of compounds based on fullerene molecules. Knowledge of the crystalline structure of new materials is the basis for conducting investigations and gaining the understanding of any physical properties. To begin, we discuss in short the history of the discovery of the new form of carbon.

Carbon is a unique element in Nature which gives rise to many chains, circles and compounds with carbon-carbon bonds being the basis for live and synthetic organic chemistry forms. Actually, carbon chemistry bases on two natural forms: diamond, the hardest, and graphite, a very soft form of substance in Nature. Diamond is organized in the form of 3–dimensional ($3D$) nets of carbon atoms connected via simple single bonds. On the other hand, graphite is built of chemically nonconnected $2D$ planes, each of which represents itself the nets of strong coupled 6-atom aromatic rings. Graphite is the most stable form of carbon.

Initially, the work of Kroto and colleagues was stimulated by the interest in certain unexplained features of absorption and emission spectra of interstellar matter. During the experiments to understand the mechanisms of the formation of long-chain carbon molecules in the interstellar space and circumstellar shells, graphite was vaporized by laser irradiation and, as a result, produced a remarkably stable cluster consisting of sixty carbon atoms.

To answer to the question what kind of C_{60} atom structure gives rise to superstable species, the authors assumed it to be a truncated icosahedron, the polygon with 60 vertices and 32 faces, 12 of which are pentagonal and 20 are hexagonal.

C_{60} is special, because of all structures made of pentagons and hexagons that curl around and close, it is the only one that does that so smoothly that all atoms have equal curvatures. This comes from mathematics.

Sixty is the most factorable of all integers. That is why Babylonians used it as a basis of their number system and we continue to divide circles in 360 degrees, have 60 minutes in an hour and 60 seconds in a minute. For reasons that so far seem obscure and are probably connected with a high factorability, sixty is also the maximum finite number of ways you can rotate an object around the central point in a $3D$ space so that when the rotating is finished the object looks exactly the same as before. Such an object has the icosahedron symmetry, the highest finite point group with 60 proper rotational symmetry elements. This object is commonly encountered as a football. The C_{60} molecule obtained by placing a carbon atom in every vertex of this structure has all valences satisfied by two single and one double bonds, has many resonance structures and appears to be aromatic.

The title of paper[1] is: "C_{60}: Buckminsterfullerene". Kroto et al.[1] used the ideas of Buckminster Fuller, the architect who in 1895 proposed to use truncated icosahedral structures for constructing cupola buildings.

In his work, B. Fuller used the ideas of famous mathematician Leonardo Euler who in 1758 developed a theory of polygons and advanced the theorem: $N_{vertices} + N_{faces} - N_{edges} = 2$. For $N_{vertices} = 60$, one has a truncated icosahedron as a football.

The possibility of the existence of stable C_{60}, C_{70} as well as clusters was demonstrated by the E. Osawa[2] and D. Bochvar groups[3] in the early 70's by quantum chemistry calculations. They predicted such carbon structures to be chemically stable due to closed electron shells and aromaticity.

The last point in this story was put by W. Kratschmer et al.[4], who proposed a method for the synthesis of solid C_{60}–fullerites. Following this, the Nobel Prize in chemistry was given to Robert Curl, Harold Kroto and Richard Smaley in 1996. On the intersection of quantum chemistry, mathematics, architecture, space physics and space chemistry the new era of various physical and chemical experiments with novel materials has begun.

2. FULLERITES AND FULLERIDES

At room temperature, pure C_{60} (fullerite) has the crystalline face-centered cubic (fcc) structure (*Fm3m* symmetry) but the molecules are orientationally disordered due to fast quasi random rotation. On the X-ray time scale, all four molecules of a cubic cell are structurally equivalent. The lattice parameter a=14.16 Å includes the Van der Waals molecular diameter $d \sim 7$ Å.

Cooling below $T_c = 260$ K, we observe the phase transition to a simple cubic lattice (*Pa3* symmetry). At $T_s = 90$ K, another phase transition connected with the freezing of molecule rotations – the orientational glass transition, is observed.

Investigations of fullerenes intercalated with metals, fullerides, gave rise to a new field of physics. The following isolated phases are documented: A_xC_{60}, where A=Na, K, Rb, Cs; $x = 1, 2, 3, 4, 6$. Special interest in studies of the physical properties of fullerene crystals has been excited by the report[5] on observation of superconductivity at T_c =10 K in fullerite specimens treated in potassium vapor (K_3C_{60}). This phenomenon is attributed to the process of intercalation of potassium atoms into interstitial sites of the crystalline lattice of fullerites as is the case in graphite. Further games played to vary the type of intercalants (alkali metals) resulted in the variation of T_c from 10 K for Na_2CsC_{60} to 31 K for Rb_2CsC_{60}. The appearance of superconductivity in fullerene crystals is, in a certain sense, very interesting and exotic. At the same time, the mechanism of superconductivity is classical but has some peculiarities in the conductivity band and the vibrational modes (see, for example, Ref.[6]).

Another class is fullerides with the AC_{60} stoicheometry phases[7] which demonstrate a number of interesting characteristics. At high temperatures (above 380 K) AC_{60} adopts a fcc rocksalt structure. Below 380 K, AC_{60} has been reported to have two phases, a stable orthorhombic polymer (*Pnnm*)[8] and a metastable orthorhombic dimer phase ($P2_1/a$)[9].

The polymer phase consists of linear chains of C_{60} molecules and exists when the sample is slowly cooled. It is stable and conducting. On the other hand, the dimer phase is formed when the sample is cooled fast from the high temperature fcc phase and is found to be an insulating phase. The stable polymer phase demonstrates a number of interesting physical properties. It is magnetic and at low temperatures, it exhibits

Table 1. Phase transitions from the phase with the wave vector (001).

IR	τ_1	τ_2	τ_3	τ_4	τ_5	τ_6	τ_7
SG	D_{4h}^1	D_{4h}^4	D_{4h}^6	D_{4h}^7	D_{4h}^{14}	D_{4h}^{15}	D_{4h}^9
IR	τ_8	τ_9	τ_9	τ_9	τ_{10}	τ_{10}	τ_{10}
SG	D_{4h}^{12}	D_{2h}^{12}	D_{2h}^{18}	C_{2h}^5	D_{2h}^{13}	D_{2h}^{17}	C_{2h}^2

instabilities wave. It is shown[10] that there are several metastable phases (one of them has a simple cubic structure with an unknown space group) at fast cooling to T=150 K for CsC_{60} and T=125 K for RbC_{60}.

The results of X-ray (synchrotron) diffraction[8] are an evidence in favor of a model where one molecular 2-fold axis is aligned with the 9.13 Å a axis. One more result is that a strong deformation of C_{60} molecules in the a-direction is observed. This leads to the appearance of the bonding structure. The partial π-character of the inter-cage bonds in polymeric chains supports the conclusion that this material is a one-dimensional metal. The important issue of structure determination is connected with the positions of metal atoms.

This short review of experimental results shows that fullerites C_{60} and especially, fullerides AC_{60} have many interesting structural peculiarities connected with structural transformations. In the next Section, we construct the Landau phenomenological theory of these phase transitions based on the symmetry analysis.

3. SYMMETRY THEORY

We start with discussion of the fulleride crystals AC_{60} (A=K, Rb). From experimental investigations it is known that the phase transition from the high temperature fcc phase of $Fm3m$ (O_h^5) symmetry to the low temperature polymer-like phase of $Pnnm$ (D_{2h}^{12}) orthorhombic symmetry can be observed.

The orthorhombic phase is characterized by shorter distances between the nearest C_{60} molecules along one of the three crystalline directions in which the $Pnnm$ phase lattice translational vectors are directed. This difference is of the order of 8% and is the result of the existence of a specific double bond between two nearest C_{60} molecules. The inter-arrangement of crystalline cells in two above mentioned phases is shown in Fig. 1.

According to the experimental data the $Fm3m \rightarrow Pnnm$ phase transition leads to the doubling of the primitive cell volume in the $Fm3m$ high temperature phase. Hence, this phase transition is connected with one of the following wave vectors on the Brillouin zone boundary: (100), (010) or (001). These wave vectors form a single wave vector star, $\mathbf{k}_{10}$, in the $Fm3m$ phase.[11]

To be specific, let us consider a phase transition with the wave vector (001), i.e., the phase transition to one of three ray domains (see Table 1). In the one-dimensional irreducible representation (IR), $\tau_1,...,\tau_8$, the order parameter has one component and in two-dimensional IR, τ_9 and τ_{10}, the following combinations of the order parameter components are possible: $C_1 \neq 0, C_2=0$ (or $C_1=0, C_2 \neq 0$), $C_1 = C_2 \neq 0$ and $C_1 \neq C_2 \neq 0$, and the corresponding space groups (SG) of low symmetry phases are given in Table 1.

As is seen from Table 1, the low symmetry phase of the $Pnnm(D_{2h}^{12})$ space group observed in experiments[8] may arise as a result of the $C_1 \neq 0$, $C_2=0$ (or $C_1=0$, $C_2 \neq 0$) type condensation for the order parameter of the τ_9 symmetry. Note that the order parameter of the τ_9 symmetry[12] and some other type of condensation, namely the

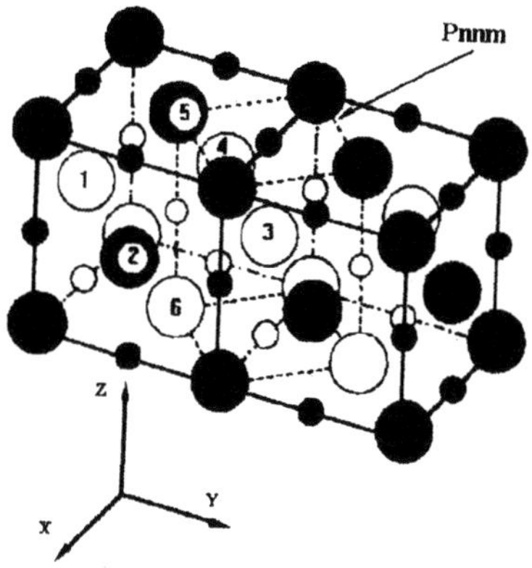

Figure 1. The inter-arrangement of crystalline cells in the *Fm3m* and *Pnnm* phases. Larger clean circles refer to C_{60} molecules and smaller ones – to metal atoms. Black circles show metal atoms and C_{60} molecules situated on the 'visible' faces of two face-centered cubic cells in the *Fm3m* phase. The *Pnnm* phase crystalline cell (for the phase transition with the wave vector (001)) is shown by the dashed lines. The C_{60} molecules creating the octahedral environment for the metal atom are denoted by numbers 1,...,6.

$C_1 = C_2 = C_3 = C_4 = C_5 = C_6 \neq 0$ type, lead to the symmetry change in the C_{60} crystal. Thus, the phase transition in the C_{60} crystal is connected with all three rays of the $\mathbf{k}_{10}$ wave vector star and the phase transition in AC_{60} goes along one of three rays of the same vector star. The phenomenological theory of the phase transition in C_{60} crystals is developed in a number of papers; see, for example Refs.[12-14] The free energy expansion can be written as

$$F_1 = F_c + F_{ce} + F_e. \tag{1}$$

The F_c term describes the orientational degrees of freedom of C_{60} molecules:

$$\begin{aligned} F_c &= A\sum_i C_i^2 + B(C_1C_3C_6 + C_1C_4C_5 + C_2C_3C_5 + C_2C_4C_6) \\ &+ D_1(\sum_i C_i^2)^2 + D_2\sum_i C_i^4 + ..., \end{aligned} \tag{2}$$

where $A = \alpha(T - T_c)$, $\alpha > 0, B, D_i$ are the phenomenological constants of the phase transition theory, $\{C_i\}$ with $i = 1, ..., 6$ is the six-fold order parameter. The C_1 and C_2 components have the wave vector (001); C_3, C_4 – (100) and C_5, C_6 – (010). Note, that we can rewrite the cubic terms in (2) using the unitary transformation in the form $C_1'C_3'C_5' + C_2'C_4'C_6'$, from Ref.[13]

For cubic symmetry crystals, the elastic energy has the form[15]:

$$F_e = \lambda_{11}(e_1^2 + e_2^2 + e_3^2) + \lambda_{12}(e_1e_2 + e_1e_3 + e_2e_3) + \lambda_{44}(e_4^2 + e_5^2 + e_6^2), \tag{3}$$

where λ_{ij} are the elastic constants and e_i are the spontaneous strain tensor components, all in Voigt notation. The interaction between the primary order parameter components $\{C_i\}$ and the strain tensor components $\{e_i\}$ can be written as follows[14]:

$$F_{ce} = \gamma(C_1^2 + C_2^2 + C_3^2 + C_4^2 + C_5^2 + C_6^2)(e_1 + e_2 + e_3). \tag{4}$$

It should be noted that the above interaction (4) is sufficient for the description of spontaneous strains of crystalline cell of pristine C_{60} compound following phase transition with $C_1 = C_2 = C_3 = C_4 = C_5 = C_6 \neq 0$ condensation. However, it is completely insufficient for the description of the phase transition in AC_{60} with $C_1 \neq 0$, $C_2 = 0$ (or $C_2 \neq 0$, $C_1 = 0$), $C_3 = C_4 = C_5 = C_6 = 0$ condensation. In the latter case, additional interactions between the order parameter components and spontaneous strain tensor components, which transform according to the same irreducible representation, take place.[16]

So the two quantities, $\sqrt{3}(C_3^2 + C_4^2 - C_5^2 - C_6^2)$ and $2C_1^2 + 2C_2^2 - C_3^2 - C_4^2 - C_5^2 - C_6^2$, transform according to the two–dimensional E_g irreducible representation with the wave vector **k**=0:

$$E_g \sim \begin{array}{l} \sqrt{3}(C_3^2 + C_4^2 - C_5^2 - C_6^2) \\ 2C_1^2 + 2C_2^2 - C_3^2 - C_4^2 - C_5^2 - C_6^2, \end{array} \qquad (5)$$

and three quantities, $C_1^2 - C_2^2$, $C_3^2 - C_4^2$, $C_5^2 - C_6^2$, transform according to the F_{2g} irreducible representation with the wave vector **k**=0 as well:

$$F_{2g} \sim \begin{array}{l} C_1^2 - C_2^2 \\ C_3^2 - C_4^2 \\ C_5^2 - C_6^2. \end{array} \qquad (6)$$

In a similar manner, the two quantities, $\sqrt{3}(e_1 - e_2)$, $e_1 + e_2 - 2e_3$, transform according to the E_g irreducible representation and three quantities, e_6, e_4, e_5, transform according to the F_{2g} irreducible representation. Note that the quantities e_6, e_4 and e_5 can be rewritten in full notation as e_{xy}, e_{yz} and e_{xz}, respectively. Thus, one obtains the following additional interactions:

$$\begin{aligned} F'_{ce} = \ & \gamma'_1 \, [3(C_3^2 + C_4^2 - C_5^2 - C_6^2)(e_1 - e_2) \\ & +(2C_1^2 + 2C_2^2 - C_3^2 - C_4^2 - C_5^2 - C_6^2)(e_1 + e_2 - 2e_3)] \\ & +\gamma'_2 \, [(C_1^2 - C_2^2)e_6 + (C_3^2 - C_4^2)e_4 + (C_5^2 - C_6^2)e_5]. \end{aligned} \qquad (7)$$

Note that the above mentioned unitary transformation of order parameter components $\{C_i\}$ will lead to substitution of only $C_1^2 - C_2^2$, $C_3^2 - C_4^2$, $C_5^2 - C_6^2$, which transform according to the F_{2g} irreducible representation by C_1C_2, C_3C_4, C_5C_6.

Interactions (4) and (7) allow us to give a complete description of spontaneous crystal cell strains following phase transition with $C_1 \neq 0$, $C_2 = C_3 = C_4 = C_5 = C_6 = 0$ condensation of order parameter components.

In addition to a different type of order parameter condensation, phase transition in AC_{60} fullerides differs from the phase transition in C_{60} by the presence of a subsystem of alkali metal atoms in AC_{60}. This subsystem of alkali metal atoms has a number of degrees of freedom which can have the E_g, F_{2g} (wave vector **k**=0) or τ_9 ($\{\mathbf{k}\}_{10}$ wave vector star) symmetries.

Let $\{\theta_i\}$ with $i = 1, 2$, $\{\eta_i\}$ with $i = 1, 2, 3$ and $\{\xi_i\}$ with $i = 1, ..., 6$ denote the secondary order parameters with degrees of freedom of E_g, F_{2g} and τ_9 symmetries, respectively. In analogy with interactions (7), one can write

$$\begin{aligned} F_{c\theta,c\eta} = \ & \beta_1 \, [\sqrt{3}(C_3^2 + C_4^2 - C_5^2 - C_6^2)\theta_1 \\ & +(2C_1^2 + 2C_2^2 - C_3^2 - C_4^2 - C_5^2 - C_6^2)\theta_2] \\ & +\beta_2 \, [(C_1^2 - C_2^2)\eta_1 + (C_3^2 - C_4^2)\eta_2 + (C_5^2 - C_6^2)\eta_3]. \end{aligned} \qquad (8)$$

The secondary order parameter $\{\xi_i\}$ and the primary order parameter $\{C_i\}$ have the same symmetry and therefore, interact linearly with each other:

$$F_{c\xi} = \beta_3(C_1\xi_1 + ... + C_6\xi_6). \tag{9}$$

Adding quadratic terms to all of introduced and defined above secondary order parameters:

$$F_{\theta,\eta,\xi} = \delta_1(\theta_1^2 + \theta_2^2) + \delta_2(\eta_1^2 + \eta_2^2 + \eta_3^2) + \delta_3(\xi_1^2 + ... + \xi_6^2), \tag{10}$$

we obtain the free energy expansion which describes the phase transition in AC_{60} crystals as:

$$F = F_1 + F'_{ce} + F_{c\xi} + F_{c\theta,c\eta} + F_{\theta,\eta,\xi}. \tag{11}$$

The primary order parameter $\{C_i\}$ describes the orientational ordering of the rotational axes in C_{60} molecules. The secondary order parameters $\{\theta_i\}$, $\{\eta_i\}$ and $\{\xi_i\}$ describe changes in a metal atom subsystem.

If the primary order parameter condensation is of the type $C_1 \neq 0$, $C_2 = C_3 = C_4 = C_5 = C_6 = 0$, interactions (4) and (7)-(10) will lead to the condensation of the secondary order parameters $\theta_1 \neq 0$, $\eta_2 \neq 0$ and $\xi_1 \neq 0$ as well as of different spontaneous strain tensor components.

Let us investigate strains in crystalline cells. Analogously to the phase transition in C_{60}, the interaction (4) leads to the appearance of non-zero components $e_1 = e_2 = e_3 \neq 0$ which do not affect the cubic symmetry of the crystal lattice. Then, as is seen from (7), additional deformations of the type $e_1 = e_2 = -2e_3 \neq 0$ and $e_6 \neq 0$ arise. The first deformation reduces the symmetry of the crystal lattice to tetragonal and the second, $e_6 \neq 0$, establishes the orthorhombic symmetry of the crystal lattice as is expected for the primary order parameter condensation leading to the phase transition $Fm3m$ (O_h^5) $\rightarrow Pnnm$ (D_{2h}^{12}). Note that it is the shear deformation component $e_6 \neq 0$ that leads to strong deformation of the stress type in the crystal cell in the [110] or [$\bar{1}$10] directions and as a result, to shorter distances between the nearest C_{60} molecules along these crystal directions.

It is necessary to keep in mind that the symmetry analysis was carried out for the wave vector (001). For the wave vector (100), there are shorter distances between nearest C_{60} molecules along the directions [011] or [0$\bar{1}$1] and for the (010) vector, along the directions [101] or [$\bar{1}$01]. So, free energy expansion (11) describes the experimental data, namely, the symmetry of the low temperature phase and the crystal cell strains.

4. SUBSYSTEM OF ALKALI METAL ATOMS

In the further we discuss changes in the subsystem of alkali metal atoms connected with the phase transition.[16,17] At present, there exist no experimental data on the behavior of a subsystem of metal atoms at phase transition in AC_{60}. Therefore, all conclusions made about structural changes following the phase transition are just model predictions which need to be verified in further diffraction experiments.

It is known from the experimental data[8] that a metal atom is in the octahedral environment of C_{60} molecules (see Fig. 2a). The smaller value of the alkali atom radius in comparison with the size of an octahedron composed of large C_{60} molecules together with large anisotropic Debye-Waller factors for metal atoms[8] allow us to assume that the true position of an alkali atom is different from the central position shown in Fig. 2a. In the general case[11], in the $Fm3m$ phase the following positions are possible for

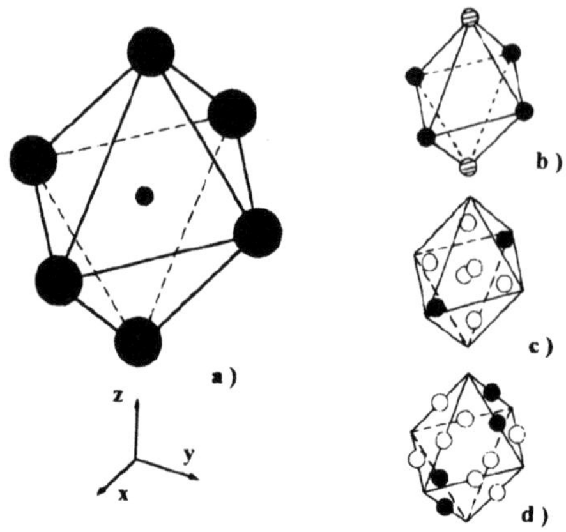

Figure 2. Possible types of positions for metal atoms (smaller circles) in the octahedral environment of C_{60} molecules (larger circles): a) The central position **1b**; b), c) and d) are the **6e**, **8f** and **12i** positions, respectively.

the octahedral environment: the central position **1b** with the (111) coordinate and the local symmetry O_h; the six-fold position **6e** with the characteristic coordinate (00z) and the C_{4v} symmetry as well as the **8f** position, (xxx), C_{3v}; **12i**, (xx1), C_{2v}; **24j**, (xy0), C_s and **24k**, (xxz), C_s.

In the position **1b** (Fig. 2a) a metal atom only oscillates relative to its equilibrium position. The displacements that have the wave vector **k**=0 transform in accordance with the three dimensional F_{1u} irreducible representation and therefore, cannot be related to the secondary order parameters of E_g or F_{2g} symmetries. Analogous displacements but with the wave vector (001) transform in accordance with the τ_4 and τ_{10} irreducible representations and also cannot be described by the $\xi_1 \neq 0$ order parameter of symmetry τ_9. As a consequence, a metal atom in the octahedral center position experiences no structural changes at phase transition.

If a metal atom occupies a noncentral position, partial ordering over the positions indicated in Figs. 2b, c), and d) takes place following phase transition. For the position **6e**, the single secondary order parameter $\theta_2 \neq 0$ of the E_g symmetry describes the partial ordering of metal atoms either on the horizontal plane (dark circles) or over two positions situated above or below the plane (shaded circles). For the positions **8f** and **12i** (Fig. 2c, and d) the secondary order parameters can be written as $\eta_1 \neq 0$, $\xi_1 \neq 0$ and $\theta_2 \neq 0$, $\eta_1 \neq 0$, $\xi_1 \neq 0$, respectively. The occupied positions are shown by black circles. The lower symmetry positions, 24j and 24k, arise from the positions **12i** and **8f**, respectively. The ordering of metal atoms over these positions is described by three secondary order parameters.

It can be assumed that if the temperature further decreases, complete ordering of metal atoms can take place and the phase transition leads to a lower crystal symmetry fully determined by the type of positions. For example, one of three possible variants of complete ordering of metal atoms for the position **8f** with the wave vector **k**=0 leads to the space group C_{2v}^7.

5. THE SYMMETRY OF THE ORIENTATIONAL STATES OF C_{60} MOLECULE

To determine the type of orientational states, let us investigate the orientations of C_{60} molecules of icosahedric symmetry occupying a position in the crystal lattice

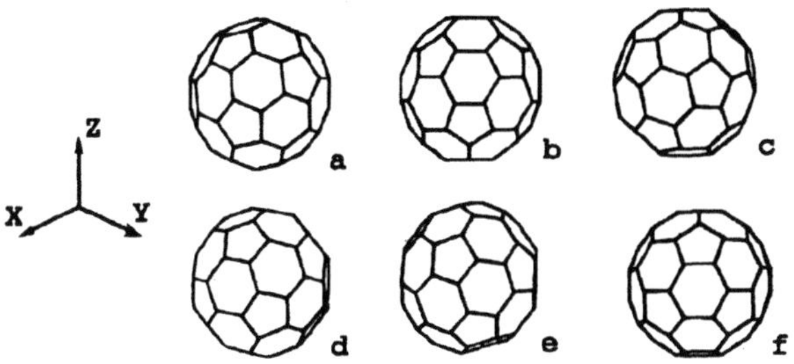

Figure 3. The initial orientations of a C_{60} molecule for different orientational bases. a), b), c), d), e), and f) are one of the orientational states of a C_{60} molecule for the 2-, 8-, 12(a)-, 12(b)-, 24-, or the 6-fold bases.

of cubic symmetry. The molecules C_{60} have the point symmetry group Y_h which, as is shown in Table 2, has 6 five-fold symmetry axes (C_5), 10 three-fold symmetry axes (C_3), 15 two-fold (C_2) symmetry axes, and the inversion (I). In the crystal lattice, C_{60} molecules occupy the positions with local symmetry O_h. The point group O_h has 3 four-fold symmetry axes (C_4), 4 three-fold axes, 6 two-fold, and inversion.

A comparison of the sets of symmetry axes for the groups O_h and Y_h shows that to obtain one of the four-fold symmetry axes of the cubic crystal, the C_{60} molecule must occupy two discrete orientational states, if a two-fold symmetry axis of the C_{60} molecule is directed along one of three $\langle 100 \rangle$ directions, or four discrete states, if the C_{60} molecule has any other orientation. If in a crystal the two-fold symmetry axis of the C_{60} molecule is oriented along one of three $\langle 100 \rangle$ directions, the C_{60} molecule can be rotated about this two-fold symmetry axis so that its three-fold symmetry axes would be oriented along the $\langle 111 \rangle$ crystal directions. This provides most symmetric occupation by C_{60} molecule of a position in a cubic crystal lattice (see Fig. 3a) and results in two so-called "standard" orientations interrelated by 90^0 rotation about the axis coinciding with any direction of the $\langle 100 \rangle$ type. The C_{60} molecule in the crystal lattice of the phase $Fm3m$ which occupies a position with the symmetry O_h and has the standard orientation has a local symmetry $T_h = O_h \cap Y_h$[18]. This is realized in A_3C_{60} crystals, where the C_{60} molecules are statically disordered over two "standard" orientations.[19]

In what follows, we shall discuss the situation when only one of three-fold symmetry axes of a C_{60} molecule coincides with one of $\langle 111 \rangle$ directions in a cubic crystal (see Fig. 3b). Accounting for the symmetry with respect to inversion we obtain that six symmetry elements that form the point group $S_6 = (E, C_3, C_3^2) \times (E, I)$, where E is the identical symmetry element, are the common elements of the C_{60} molecule and the point group O_h. In this case, the local symmetry group of a molecule is the group S_6. Since the point group O_h has 48 different symmetry elements the number of different orientational states of C_{60} molecules for such occupation of position in the cubic lattice is 8=48/6. These eight orientational states were used to describe the phase transition in C_{60} fullerites.[13]

If in a cubic crystal the two-fold symmetry axis of a C_{60} molecule is aligned with the [110] crystal direction (see Fig. 3c), the common symmetry elements are $(E, C_2) \times (E, I)$ and consequently, the number of different orientational states of C_{60} molecules is 12=48/4. These 12 orientational states of C_{60} molecules are used to describe the phase transition in AC_{60} fullerides.[20] The local symmetry group of a C_{60} molecule is the group C_{2h}. Note that if in a cubic crystal the two-fold symmetry axis of a C_{60} molecule is oriented along the [001] direction and the three-fold symmetry axes do not coincide

Table 2. The symmetry elements of a free C_{60} molecule and the position in the crystal lattice of the cubic phase *Fm3m* occupied by C_{60} molecule.

Symmetry elements	C_5	C_4	C_3	C_2	I
C_{60} molecule	6	-	10	15	1
Position in the lattice with symmetry O_h	-	3	4	6	1

with ⟨111⟩ directions, a different 12-fold orientational basis arises (see Fig. 3d). Below, we denote these two orientational bases by 12(a) and 12(b), respectively.

If none of the symmetry axes of a C_{60} molecule coincides with the corresponding symmetric direction in the crystal, the only common elements of symmetry are (E, I) and consequently, the number of different orientational states of the C_{60} molecule is 24=48/2 and the local symmetry of the C_{60} molecule is determined by the group I (see Fig. 3e).

The orientational states belonging to one basis have the same continuous degrees of freedom. For example, an orientational state of the 8-fold basis may admit an arbitrary angle of rotation about the three-fold symmetry axis. It is analogous for the 12-fold basis but in this case, the arbitrary rotation can be accomplished about the two-fold symmetry axis. The 24-fold basis admits two independent rotations. Note that for some angle (angles) of rotation, an additional degeneration of the orientational states of C_{60} molecules may take place. This happens due to the fact that for a particular orientation, additional coincidence of the symmetry axes of C_{60} molecules and symmetric directions of crystal may take place. For the 8-fold basis, for example, the two-fold symmetry axes of the C_{60} molecule may coincide with the ⟨100⟩ crystal directions for a particular angle of rotation about the three-fold symmetry axis of the C_{60} molecule. This leads to a degeneration of states of the orientational basis and as a result, instead of eight states only two "standard" orientations remain. For two 12-fold orientational bases, in the situation when the two-fold symmetry axis of C_{60} molecule coincides with the ⟨100⟩ direction for the basis 12(a) or the ⟨110⟩ direction for the basis 12(b), the number of independent orientational states decreases to 6 (see Fig. 3f). In this case, the local symmetry of the orientational state is described by the group D_{2h}. It is apparent that the enumerated bases can be obtained from the 24-fold basis as well. Also, note that the orientational states of 2- and 6-fold bases are the most symmetric states and, in contrast to the orientational states of other bases, do not have rotational degrees of freedom.

The orientational states belonging to one basis transform into one another under the action of symmetry elements of the group *Fm3m* and in this case, form (in the language of the group theory, see, e.g.,Ref.[21]) an orbit.

In Table 3, different types of orientational orbits are summarized with an indication of compounds (fullerite C_{60} or fulleride AC_{60}) for which these orbits are used to describe the phase transition. Let us note that in A_3C_{60} fulleride the orientational phase transition does not take place and in the phase *Fm3m* static disordering of C_{60} molecules over two "standard" orientations is observed[19].

6. THE ORIENTATIONAL PHASE TRANSITION

Diffraction investigations[22] of C_{60} fullerite in the phase *Fm3m* show that rotating C_{60} molecules have an anisotropic orientational distribution in the crystal. Hence it follows that the orientational states of C_{60} are not equally probable. In the previous

Table 3. Different types of the orientational orbits of a C_{60} molecule in the phase *Fm3m*.

Number of orientational states in orbit	Local symmetry of orientational state	Crystal	Comments
2	T_h	A_3C_{60}	static disordering
6	D_{2h}		
8	S_6	C_{60}	dynamic disordering
12(a)	C_{2h}	AC_{60}	dynamic disordering
12(b)	C_{2h}		
24	I		

Table 4. The orientational states of C_{60} molecule in different phases of C_{60} and AC_{60} crystals.

Crystal	High-symmetry phase		Low-symmetry phase
	$T \gg T_c$	$T > T_c$	$T < T_c$
fullerite C_{60}	{2,6,8,...}	{8}	8
fulleride AC_{60}	{2,6,8,...}	{12}	12

Section it is shown that it is possible to combine orientational states into orientational orbits which differ in local symmetry (see Table 3). The orientational states of one orientational orbit are physically equivalent. Therefore, C_{60} molecule occupies them with equal probability. So, to explain the anisotropic distribution of molecular orientations it is necessary to assume that the states of different orbits are occupied by a rotating C_{60} molecule with different probabilities determined by

$$P_n = \frac{\exp(-V_n/k_B T)}{\sum_m \exp(-V_m/k_B T)}, \tag{12}$$

where k_B is the Boltzmann constant, V_n is the potential energy of the state n of the orientational orbit, and the summation runs over all orientational states of all orientational orbits.

Thus, the following model of the orientational phase transition in fullerene crystals can be suggested.[23] At high temperatures ($T \gg T_c$), a rotating C_{60} molecule occupies all orientational states from all orientational orbits. At temperatures close to the phase transition temperature ($T > T_c$), a rotating C_{60} molecule occupies, most of the time, the states of one orientational orbit, i.e., of the orbit that corresponds to a particular orientational potential of the C_{60} molecule in a particular crystal. At phase transition "freezing" of the orbits takes place and the molecule occupies the orientational states of one orbit.

Comparing orientational phase transitions and structural phase transitions of displacement type we can make the following conclusions. The set of states in all orientational orbits is an analogue of possible displacements of all atoms in a crystal cell. The particular orientational orbit can be understood as an analogue of the displacements only of atoms connected with the soft mode of the structural transition. Thus, different orientational orbits are the microscopic realizations of the order parameter of different orientational phase transitions.

Table 4 gives the orientational orbits and states necessary for the description of orientational phase transitions in C_{60} fullerite and AC_{60} fulleride.

Table 4 illustrates the orientational transition. In the C_{60} (AC_{60}) crystal at high temperatures rotating C_{60} molecules occupy all orientational states from all orientational orbits (2-fold, 6-fold orbits, etc.). In the vicinity of the phase transition ($T > T_c$) the molecules occupy, most of the time, the states from {8}- ({12}-) fold orbits. In the low-symmetry phase, orientational ordering of C_{60} molecules in one of the states of 8- (12-) fold orbits takes place.

As we can see, it is shown that in the vicinity of the phase transition temperature rotating C_{60} molecules in the phase $Fm3m$ occupy, most of the time, the orientational states of one orbit and therefore the anisotropic distribution of orientations of a C_{60} molecule in a crystal is determined by the states of one orbit. Let us consider a certain orientational state. For this (first) orientation ψ_1 the atomic density on the surface of a C_{60} molecule at the point determined by the angles $\mathbf{\Omega} \equiv (\theta, \varphi)$ (θ and φ are the polar angles in the coordinate system shown in Fig. 3) is

$$\rho_1(\Omega) = \sum_n \delta_1(\Omega - \Omega_n), \tag{13}$$

where δ - is the Dirac δ-function and the summation runs over 60 carbon atoms in the molecule. Accounting for the remaining orientational states in the selected orbit and modeling the thermal motion of C_{60} molecule according to the normal distribution law with the rotational angle dispersion σ it is possible to calculate the mean anisotropy of the atomic density $\langle\rho(\Omega)\rangle$

$$\begin{aligned} \langle\rho(\Omega)\rangle &= C \int dS' \exp[-(\Omega - \Omega')^2/2\sigma^2] \sum_k \rho_k(\Omega') \\ &= C \int dS' \exp[-(\Omega - \Omega')^2/2\sigma^2] \sum_{k,n} \delta_k(\Omega' - \Omega_n), \end{aligned} \tag{14}$$

where C is a numerical constant, the integration is over the surface of C_{60} molecule, $|\Omega - \Omega'|$ is the distance between the Ω and Ω' points on the surface of C_{60} molecule measured in degrees, and the summation is over the orientational states in the orbit (index k) and 60 carbon atoms in C_{60} molecule (index n).

In Fig. 4, the atomic density distributions of rotating C_{60} molecule in AC_{60} crystal in the phase $Fm3m$ are illustrated. The calculation is performed by Eq. (14) with the orientational states of the 12-fold orbit. We take the orientatonal state shown in Fig. 3c as the initial state ψ_1. Other orientational states of C_{60} are obtained from the initial orientation by means of the following rotations: $\psi_2 = C_4\psi_1$, $\psi_3 = C_4^2\psi_1$, $\psi_4 = C_4^3\psi_1$. The remaining orientational states are obtained with the help of the rotations C_3 and C_3^2 by the following symbolic scheme $(\psi_5, \psi_6, \psi_7, \psi_8) = C_3(\psi_1, \psi_2, \psi_3, \psi_4)$, and $(\psi_9, \psi_{10}, \psi_{11}, \psi_{12}) = C_3^2(\psi_1, \psi_2, \psi_3, \psi_4)$, where C_4 is the rotation by 90^0 about the z-axis and C_3 is the rotation by 120^0 about [111].

The symmetry analysis in the above Section shows that the low-symmetry phase arises as a result of condensation of one of six order parameter components and consequently, the appearance of six domains can be expected in the low-symmetry phase. The atomic density distribution (see Figs. 4a and 4b) is constructed on the basis of all 12 orientational states. However, in a deformed crystal, the 12 orientational states are not physically equivalent. This causes a change of the occupation probability for different states.

The symmetry analysis of the phase transition makes it possible to understand which of 12 orientational states are realized and how they are connected with different types of deformation. For external stresses $\sigma_{xx} = \sigma_{yy} = -2\sigma_{zz} \neq 0$, the atomic density distribution is illustrated in Fig. 4c. It is constructed using the orientational states ψ_1, ψ_2, ψ_3, and ψ_4. Applying the additional shear stress $\sigma_{xy} \neq 0$ to the deformed crystal it

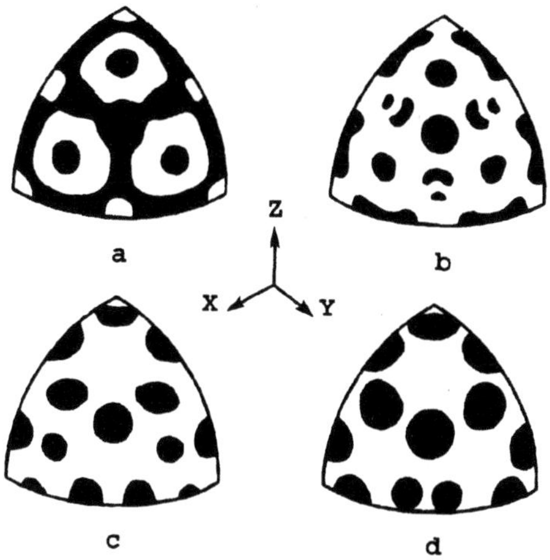

Figure 4. a). The fragment (1/8th of the surface of molecule) of the atomic density distribution of the C_{60} molecule in the phase *Fm3m*. The dark regions correspond to the maximum density (100% to 66%), lighter regions correspond to the atomic density of 66% to 33%, and the light regions - 33% to 0%. The orientational states are distributed according to the normal law with 5^0 dispersion. b), c), and d). The same as for a) but with the dispersion 3^0. In addition, for c) and d) we considered the strain of type $e_{xx} = e_{yy} = -2e_{zz} \neq 0$. In d), the same strain as for (c) plus an additional shear strain $e_{xy} \neq 0$.

is possible to restrict further the orientational degrees of freedom of C_{60} molecule. For this case, the atomic density distribution constructed on the basis of two orientational states ψ_1 and ψ_3 is illustrated in Fig. 4d.

In the next Section the local symmetry and the symmetry of internal vibrations of C_{60} molecule in the crystal are investigated on the basis of the symmetry of orientational states.

7. SYMMETRY OF INTERNAL VIBRATIONS OF C_{60} MOLECULE

It is known that the symmetry of crystal lattice vibrations can be determined in light scattering experiments. In paper[24], Raman light scattering spectra of fullerite C_{60} were reported (see Fig. 5). From these spectra it is seen that the lines in the spectrum obtained at T=259 K in the low-symmetry phase *Pa*3 have a finer structure than the lines obtained at the temperature two degrees higher in the high-symmetry phase *Fm*3*m*. The physical origin of a large width of lines corresponding to the internal vibrations of C_{60} molecule in the phase *Fm*3*m* is not completely clear yet and in principle, can be due to many physical reasons. In pure C_{60} crystal, however, the C_{60} molecules are neutral and this reduces the number of possible reasons. For example, owing to the Jahn–Teller effect the charged C_{60}^{-1} (or C_{60}^{-3}) molecule can go to the strain state and this leads to splitting of the internal vibrations of molecule. In addition, the charged molecules lead to the appearance of a macroscopic electric field in dielectric crystals and this leads to splitting of longitudinal and transverse optic vibrations. So, the large width of lines for C_{60} fullerite can be explained by the anharmonic interaction of internal vibrations of C_{60} molecules with other phonons or the low local symmetry of C_{60} molecules in crystals (splitting in the crystal field) and/or the interaction between randomly disoriented molecules (analogous to Davydov splitting[25]).

In contrast to anharmonic interaction, the last two effects determine the point

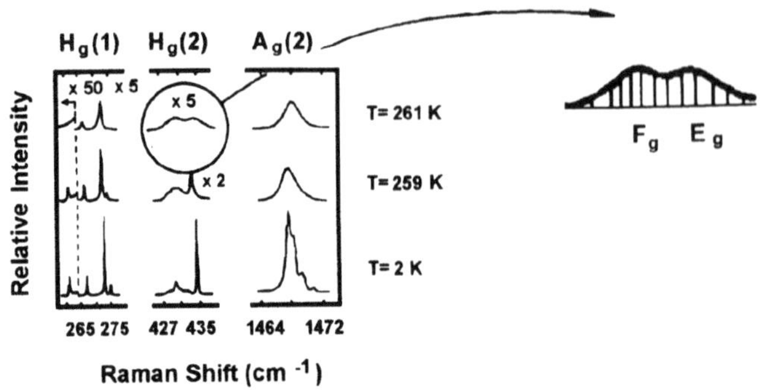

Figure 5. The fragment of Raman scattering spectrum of fullerite C_{60}[24]. In the right hand side a splitting in the crystal field and/or Davydov splitting is schematically shown.

group that can be verified in light scattering experiments by a polarization analysis.

The symmetry analysis[26] shows that the internal vibrations of the C_{60} molecule active in Raman light scattering have the symmetry $2A_g \oplus 8H_g$. As a result, in Raman scattering spectrum ten different lines, two with the symmetry A_g (nondegenerate vibrations) and eight with the symmetry H_g (five-fold degeneration), must be observed. In the crystal field of cubic symmetry degeneration of vibrations with the symmetry H_g is removed: $H_g = E_g \oplus F_g$. In this, the vibration with the symmetry F_g has three-fold degeneracy and E_g – two-fold degeneracy. As is seen from Fig. 5, such splitting is experimentally observed for the vibrations $H_g(1)$ and $H_g(2)$ (T=261 K). Analogous splitting may probably take place for vibrations $H_g(6)$ and $H_g(7)$ as well[24]. According to the polarization analysis[24] the vibration $H_g(1)$ with a lower frequency (266.2 cm^{-1}) has the symmetry E_g and vibration with the frequency 272.4 cm^{-1} has the symmetry F_g. For vibration $H_g(2)$ we have the opposite situation: the vibration with frequency 430.3 cm^{-1} has the symmetry F_g and the vibration with frequency 434.3 cm^{-1} – the symmetry E_g.

Since the vibrations with symmetry A_g are nondegenerate and a primitive cell in the phase $Fm3m$ contains only one C_{60} molecule, the large width of the corresponding lines can be due to the interaction of neighboring molecules with different orientations. In this case, modeling of the strongest interaction between the closest C_{60} neighbors with different orientations in the phase $Fm3m$ (the first coordination sphere contains 12 molecules) is possible using large crystal cells (containing more than one C_{60} molecule). Such crystals with a larger size of cells can be obtained from the phase $Fm3m$ by symmetry analysis of all possible phase transitions leading to an increase of the primitive cell volume. (In the cluster approach, accounting for the interaction of the closest C_{60} neighbors, it can be assumed that the large width of A_g symmetry vibrations is formed by a set of separate lines (1 to 13) with different intensities which, in general, are proportional to time a C_{60} molecule is in a particular orientational state).

Hence it follows that the large width of lines with symmetry A_g is explained by the interaction between the closest C_{60} molecules, i.e., Davydov splitting.[25] In this case, the point group of crystal with an increased size of primitive cell must decrease due to uncorrelated orientations of the closest neighbors.

In what follows, we shall discuss the width of lines with symmetries F_g and E_g. The above symmetry analysis of orientational states of C_{60} molecules in C_{60} and AC_{60} crystals shows that C_{60} molecules mainly occupy the states with symmetries S_6 and C_{2h}, respectively. In this case, further removing of degeneration of F_g symmetry vibrations and the E_g ones takes place due to low symmetry of the crystal field. In addition, in full analogy with the A_g symmetry lines, Davydov splitting due to the interaction of the closest C_{60} neighbors with different orientations also takes place.

Thus, the symmetry of the internal vibrations of C_{60} molecules in C_{60} and AC_{60} crystals cannot be higher than S_6 and C_{2h}, respectively, due to low local symmetry of orientational states and the interaction between misoriented closest neighbors. This means that vibrations with symmetry F_g in the phase $Fm3m$ can be observed not only for the perpendicular orientation of polarization vectors of incident and scattered light but they must be observed for the parallel orientation as well and vice versa the vibrations with symmetry E_g are observed for the perpendicular orientation. Therefore it is interesting to perform an accurate polarization analysis of vibrations with the symmetry F_g and E_g in C_{60} fullerite (phase $Fm3m$) by carrying out Raman light scattering experiments.

8. MOLECULAR STRAIN IN THE PHASE $Pa3$

Below we shall discuss the standard group theory scheme for the symmetry analysis of the lattice dynamics or phase transitions in crystals.[21] Let us discuss a crystal consisting of N crystal cells each containing J atoms. Each atom is characterized by an R–component tensor function. In this case, the state of the whole crystal (with periodic boundary conditions) is determined by the $N \times J \times R$ values. If the state of the crystal is connected with the wave vector $\mathbf{k}$, the basis function can be written as

$$\varphi_{\mathbf{k}}^{jr} = \sum_{n=1}^{N} \varphi^{jr} \exp(i\mathbf{k}\mathbf{t_n}), \tag{15}$$

where $\mathbf{t_n}$ is the translation vector, $\{e^{jr}\}$ is the orthogonal basis, $j \in \{1, ..., J\}$, $r \in \{1, ..., R\}$, and

$$e^{jr} = \begin{cases} 1, & \text{if } j = j_0, \ r = r_0 \\ 0, & \text{otherwise.} \end{cases}$$

Under the action of a $T(g)$ operator, where $g = (h|\tau_h)$ is the symmetry element of the space group, h is the rotation and/or inversion, τ_h is the nontrivial translation, the basis function (15) transforms as

$$T(g)\varphi_{\mathbf{k}}^{jr} = \exp[-i\mathbf{k}\mathbf{a}_{ij}(g)]\delta_{i,hj} \sum_{r'} \{D_R(h)\}^{r'r} \varphi_{\mathbf{k}}^{jr'}, \tag{16}$$

where

$$\delta_{i,j} = \begin{cases} 1, & i = j \\ 0, & i \neq j \end{cases}$$

and $\mathbf{a_{ij}}(g)$ is

$$g\mathbf{r}_j = h\mathbf{r}_j + \tau_h \equiv \mathbf{r}_i + \mathbf{a}_{ij}(g). \tag{17}$$

Eq. (16) can be written as

$$T(g)\varphi_{\mathbf{k}}^{jr} = \sum_{i,r'} \{d_R^{\mathbf{k}}(g)\}^{ir',jr} \varphi_{\mathbf{k}}^{ir'}. \tag{18}$$

The $d_R^{\mathbf{k}}$ matrix is the tensor representation of the wave vector group $G_{\mathbf{k}}$ and in general case, this is a reducible representation:

$$d_R^{\mathbf{k}} = \sum_{\nu}^{\oplus} n^{\nu} d^{\mathbf{k}\nu}. \tag{19}$$

The symbol $\sum^{\oplus}$ denotes a direct sum and n^{ν} is determined by

$$n^{\nu} = |G_{\mathbf{k}}|^{-1} \sum_{g} \chi_R^{\mathbf{k}}(g) \chi^{*\mathbf{k}\nu}(g), \tag{20}$$

where $|G_{\mathbf{k}}|$ is the number of symmetry elements of the $G_{\mathbf{k}}$ group, $\chi^{\mathbf{k}\nu}(g)$ is the character of the νth irreducible representation, $\chi_R^{\mathbf{k}}(g)$ is the character of the tensor representation:

$$\chi_R^{\mathbf{k}}(g) = \sum_{j} \exp[-i\mathbf{k}\mathbf{a}_{ij}(g)] \delta_{i,hj} Sp D_R(h), \tag{21}$$

where Sp denotes the sum of diagonal elements of the D_R matrix. The symmetrized basis function of the νth irreducible representation is determined by the following formula:

$$\psi_{\lambda}^{\mathbf{k}\nu} = \sum_{g} d_{\lambda[\mu]}^{*\mathbf{k}\nu}(g) \sum_{i,\alpha} \exp[-i\mathbf{k}\mathbf{a}_{ij}(g)] \delta_{i,hj} D_R^{\alpha[\beta]}(h) \varphi_{\mathbf{k}}^{j\alpha}, \tag{22}$$

where the indices in brackets can be arbitrary.

To investigate the symmetry of a crystal with orientational degrees of freedom, one can use the above approach. Again, let us investigate a crystal with N crystal cells. This time, however, each cell contains J rigid group atoms. The group atoms occupy L orientational states each characterized by an R-component tensor function. In this case, the basis function (15), formulas (16) and (22) should be written as

$$\varphi_{\mathbf{k}}^{jlr} = \sum_{n=1}^{N} \varphi^{jlr} \exp(i\mathbf{k}\mathbf{t}_{\mathbf{n}}), \tag{23}$$

$$T(g)\varphi_{\mathbf{k}}^{jlr} = \exp[-i\mathbf{k}\mathbf{a}_{ij}(g)] \delta_{i,hj} \delta_{l,hl'} \sum_{r'} \{D_R(h)\}^{r'r} \varphi_{\mathbf{k}}^{jl'r'}, \tag{24}$$

$$\psi_{\lambda}^{\mathbf{k}\nu} = \sum_{g} d_{\lambda[\mu]}^{*\mathbf{k}\nu}(g) \sum_{i,l,r'} \exp[-i\mathbf{k}\mathbf{a}_{ij}(g)] \delta_{i,hj} \delta_{l,hl'} \{D_R(h)\}^{r'r} \varphi_{\mathbf{k}}^{jl'r'}, \tag{25}$$

where

$$\{e^{jlr}\}$$

is the orthogonal basis, $j \in \{1, ..., J\}$, $l \in \{1, ..., L\}$, $r \in \{1, ..., R\}$ and

$$e^{jr} = \begin{cases} 1, \text{if } j = j_0,\ l = l_0,\ r = r_0 \\ 0,\ \text{otherwise.} \end{cases}$$

In papers[17,20] the description of C_{60} molecular strain in the $Pnnm$ phase was done. The phonon basis functions describing the strain of C_{60} molecule have the following form:

$$u_x(i_1) + u_y(i_1) + u_x(i_2) + u_y(i_2) - u_x(i_3) - u_y(i_3) - u_x(i_4) - u_y(i_4), \tag{26}$$

$$u_x(i_1) - u_y(i_1) - u_x(i_2) + u_y(i_2) + u_x(i_3) - u_y(i_3) - u_x(i_4) + u_y(i_4), \tag{27}$$

$$u_z(i_1) - u_z(i_2) + u_z(i_3) - u_z(i_4), \tag{28}$$

where i_1=1,...,16 denote nonequivalent carbon atoms of C_{60} molecule in the $Pnnm$ phase, excluding the carbon atoms with numbers 5 and 6 (see Fig. 6). For these two atoms the basis functions are written as

$$u_x(i_1) - u_y(i_1) - u_x(i_2) + u_y(i_2), \tag{29}$$

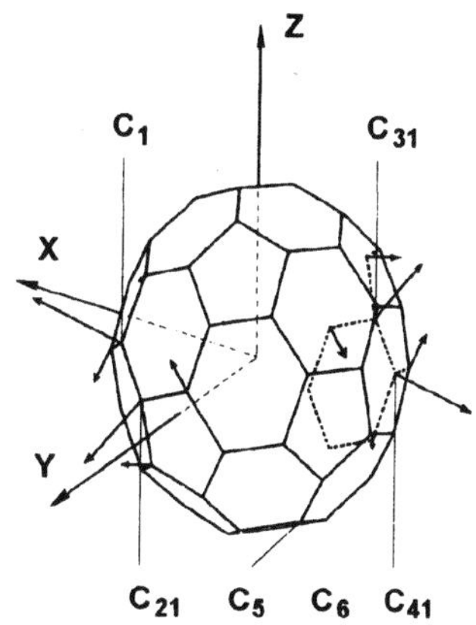

Figure 6. Displacements of carbon atoms in a C_{60} molecule in the *Pnnm* phase. The directions of the atomic displacements are shown by arrows. The arrow lengths are increased five times in comparison with true values[8] normalized to the size of C_{60}.

$$u_z(i_1) - u_z(i_2). \tag{30}$$

In Eqs. (26)–(30) $u_\alpha(i)$ is a unit displacement of the i atom in the α direction. The i_1 atom is connected with the i_2 atom by two-fold symmetry axis and i_3 and i_4 are obtained from i_2 and i_1 by inversion, respectively. For example, for the carbon atoms C_1,C_{21},C_{31} and C_{41} shown in Fig. 6 one obtains

$$u_x(C_1) + u_y(C_1) + u_x(C_{21}) + u_y(C_{21}) - u_x(C_{31}) - u_y(C_{31}) - u_x(C_{41}) - u_y(C_{41}), \tag{31}$$

$$u_z(C_1) - u_z(C_{21}) + u_z(C_{31}) - u_z(C_{41}). \tag{32}$$

In the first approximation, the basis functions (26), (28) describe stretching of the C_{60} molecule in the polymeric link direction and increasing of the covalent bond between two atoms of C_{60} molecule in accordance with experimental observations.[8] Moreover, the C_{60} molecule experiences a small rotation as a whole (basis function (27)).

9. NEUTRON AND X-RAY DIFFRACTION

Neutron and X-ray diffraction in fullerene crystals can be described in the framework of the standard formalism for the static structure factor $S(\mathbf{Q})$ of molecular crystals with orientational degrees of freedom.[27,28] Molecules arranged in an ordered lattice give rise to a coherent elastic intensity in the Bragg positions, orientational molecular motion characterized by fluctuations around the average molecular conformation and orientation and produce diffuse scattering over the entire $\mathbf{Q}$ range. Here $\mathbf{Q} = \mathbf{k}_0 - \mathbf{k}$ is the transfer wave vector and $\mathbf{k}_0$ and $\mathbf{k}$ are the initial and final wave vectors.

The differential scattering cross section of neutrons per unit solid angle $d\Omega$ is

$$\frac{d\sigma}{d\Omega} \sim \langle \sum_{ij} b_i b_j^* \exp[i\mathbf{Q}(\mathbf{r}_i - \mathbf{r}_j)] \rangle, \tag{33}$$

where b_i is the scattering length, the angular brackets denote the space and time average, and the summation is performed over all atoms in the system.

In case of X-ray scattering, in Eq. (33) the atom form-factors $f_i(Q)$ must be used instead of the scattering lengths b_i.

To describe thermal and orientational disorder in molecular crystals the position vector of the ith atom, $\mathbf{r}_i$, is used in the form

$$\mathbf{r}_i = \mathbf{R}_m + \mathbf{r}_{mi}, \tag{34}$$

where $\mathbf{R}_m$ is the position vector of the center of mass of the mth molecule and $\mathbf{r}_{mi}$ is the position of the atom i relative to the center of mass.

The next step is to introduce the molecular form factor F_m

$$F_m = \sum_{i(m)} b_i \exp(i\mathbf{Q}\mathbf{r}_{mi}). \tag{35}$$

Using Eq. (35) and separating intramolecular ($m = m'$) from intermolecular ($m \neq m'$) components we can write Eq. (33) in the form[28]

$$\begin{aligned} \frac{d\sigma}{d\Omega} &\sim \sum_m \langle F_m^2 \rangle \\ &+ \langle \sum_{m \neq m'} \exp[i\mathbf{Q}(\mathbf{R}_m - \mathbf{R}_{m'})] \sum_{j(m)} \sum_{j(m')} b_j b_k^* \exp[i\mathbf{Q}(\mathbf{r}_{mi} - \mathbf{r}_{m'j})] \rangle \\ &= \sum_m \langle F_m^2 \rangle + \sum_{m \neq m'} \langle F_m F_{m'}^* \rangle \exp(-2W) \exp[i\mathbf{Q}\langle \mathbf{R}_m - \mathbf{R}_{m'} \rangle], \end{aligned} \tag{36}$$

where it is assumed that fluctuations of one atom around its equilibrium position split into translational motion of molecular center of mass and intramolecular rotations or vibrations. The Debye–Waller factor $W \equiv 1/2\langle \delta R_m^2 \rangle Q^2$. The term $\langle F_m F_{m'}^* \rangle$ contains information of intermolecular orientational correlations and their range.

We shall investigate a system of identical molecules with similar orientations of nearly spherical molecules and neglect correlation between the orientational fluctuations of neighbor molecules

$$\langle F_m F_{m'}^* \rangle_{m \neq m'} = \langle F \rangle^2 . \tag{37}$$

In this case, Eq. (36) has a simpler form

$$\begin{aligned} \frac{d\sigma}{d\Omega} &\sim \sum_{m,m'} \langle F \rangle^2 e^{-i\mathbf{Q}\langle \mathbf{R}_m - \mathbf{R}_{m'} \rangle} e^{-2W} \\ &+ N[\langle F^2 \rangle - \langle F \rangle^2 e^{-2W}] = S_e(\mathbf{Q}) + S_d(\mathbf{Q}). \end{aligned} \tag{38}$$

After averaging over all possible orientations the structural form factor takes the form

$$\langle F \rangle = 60 b_C \frac{\sin(Q\rho)}{Q\rho} \exp(-W_C) + \frac{1}{N_p} b_A \exp(-W_A) \cos(\mathbf{r}\mathbf{Q}) \sum_{i=1}^{N_p} \cos(\mathbf{r}_i \mathbf{Q}), \tag{39}$$

where ρ is the radius of a C_{60} molecule (ρ = 3.54 Å), $b_c = 0.667 \cdot 10^{-12}$ cm is the coherent length of neutron scattering on carbon. N_p = 1, 6, 8 and 12 is the number of possible locations of a metal atom in the positions **1b**, **6e**, **8f** and **12i**, respectively; $b_K = 0.37 \cdot 10^{-12}$ cm and $b_{Rb} = 0.71 \cdot 10^{-12}$ cm are the coherent lengths of neutron

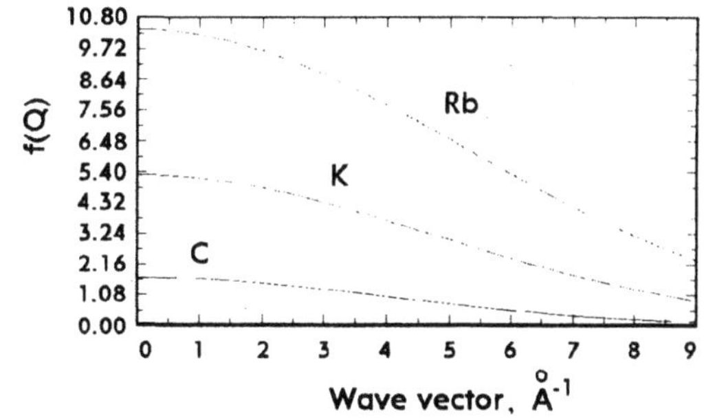

Figure 7. The atomic form-factors $f(Q)$ for X-rays, in 10^{-12} cm.

scattering on K and Rb. The radius-vector $\mathbf{r} = (0, 0, a/2)$ corresponds to central position **1b** and $\mathbf{r}_i = \mathbf{r}_i^A - \mathbf{r}$.

Equation (38) describes both elastic scattering (the first term) and diffuse scattering (the second term). The first term S_e gives rise to sharp Bragg peaks due to molecular ordering over crystal sites and can be written as[29]

$$S_e(Q) = \langle F \rangle^2 \delta(\mathbf{Q} - 2\pi\mathbf{H}) e^{-2W}. \tag{40}$$

The δ-function in Eq. (40), where $\mathbf{H}$ is the inverse lattice vector, determines the diffraction geometry. This makes it possible to determine the lattice symmetry. The peak intensities supply information about atomic positions and are determined by $\langle F \rangle^2$.

It is convenient to write the diffuse scattering term in the form[28]

$$S_d(\mathbf{Q}) = N[(\Delta F)^2 + \langle F \rangle^2 (1 - e^{-2W})], \tag{41}$$

where $(\Delta F)^2 = \langle F^2 \rangle - \langle F \rangle^2$ takes into account the average orientational and conformational disorder of molecules and the second term in Eq. (41) is responsible for thermal fluctuations of the molecular center of mass. For large $\mathbf{Q}$, the leading term in Eq. (41) is the molecular structure factor $N\langle F^2 \rangle$ and the $\langle F^2 \rangle e^{-2W}$ contribution is strongly damped, which results in vanishing Bragg intensity. Estimates[28] show that in general, in molecular crystals there are no more Bragg peaks visible at momentum transfers Q larger than 10 Å^{-1}.

The difference in the scattering of neutrons and X-rays consists of different dependence of scattering length and atomic form-factors on the wave vector momentum transfer. Calculated atomic form-factors for carbon C, potassium K and Rb are shown in Fig. 7. Practically, the neutron scattering lengths do not depend on Q.

Figures 8 and 9 show the calculated dependence of Bragg and diffuse scattering on powder C_{60} fullerite in the $Fm3m$ phase. The intensities of Bragg peaks are calculated by the formula

$$I_{Bragg}(Q) = \sum_{hkl} \frac{I(Q_{hkl})}{(2\pi)^{1/2}\sigma} \exp[-\frac{(Q_{hkl} - Q)^2}{2\sigma^2}],$$
$$I(Q_{hkl}) = [60 b_c \frac{\sin(Q_{hkl}\rho)}{Q_{hkl}\rho}]^2 \exp(-2W). \tag{42}$$

The dependence of the Gaussian resolution function dispersion σ on the momentum transfer for the diffractometer[28] is shown in Fig. 10. We use lattice parameter $a =$ 14.159 Å and the Debye–Waller factor[28] $W = 0.015 \cdot Q^2/2$.

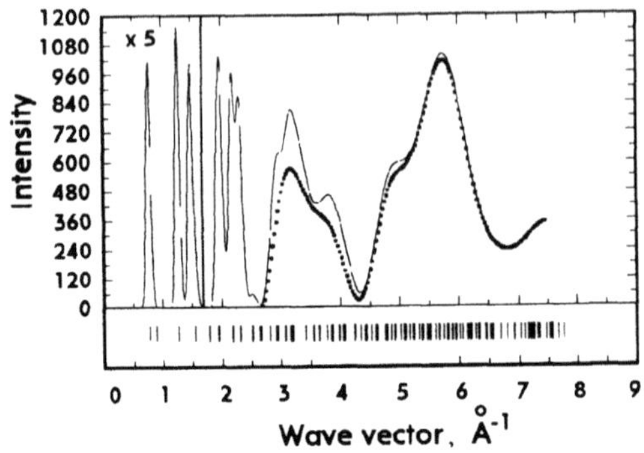

Figure 8. The Bragg and diffuse scattering of neutrons on a powder C_{60} in the $Fm3m$ phase.

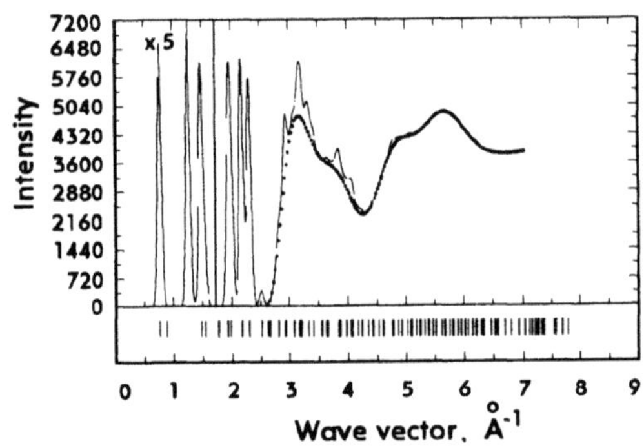

Figure 9. The Bragg and diffuse scatterings of X-rays on a powder C_{60} in the $Fm3m$ phase.

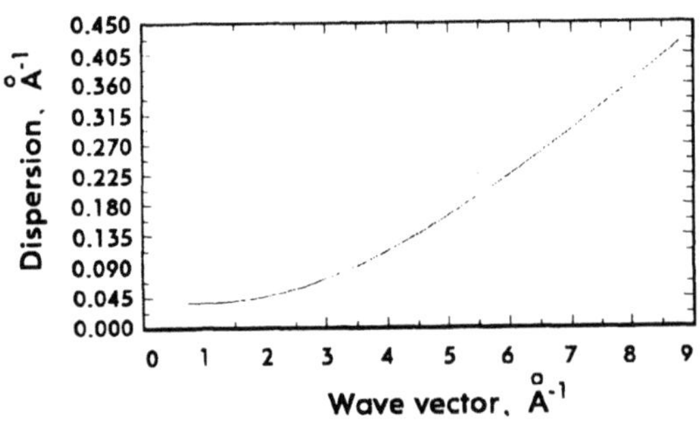

Figure 10. Dispersion of the Gauss resolution function versus the neutron wave vector.

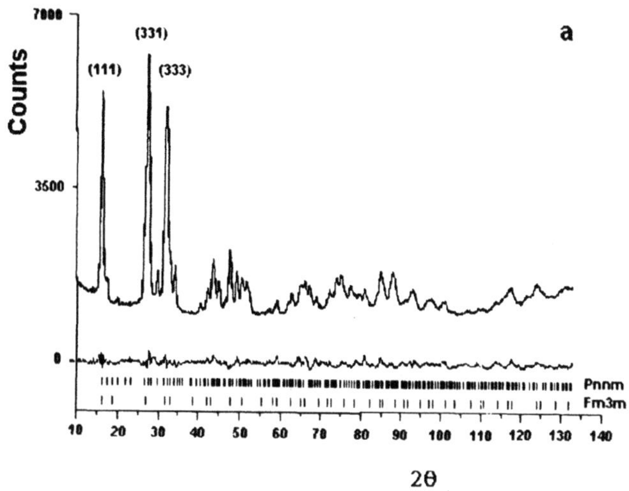

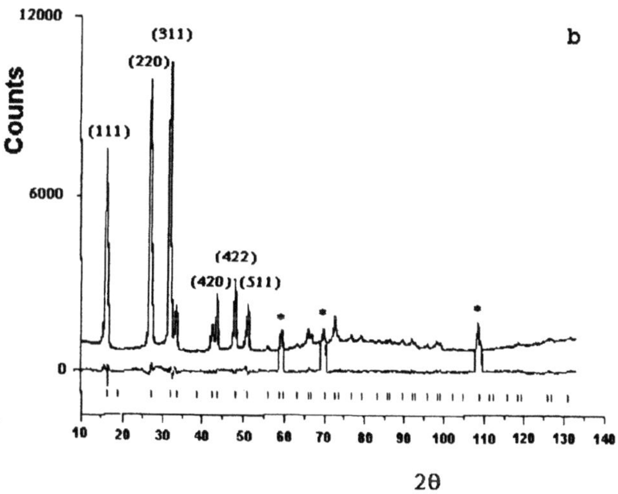

Figure 11. Neutron diffraction spectra of RbC_{60} at 300 K (a) and 450 K (b).

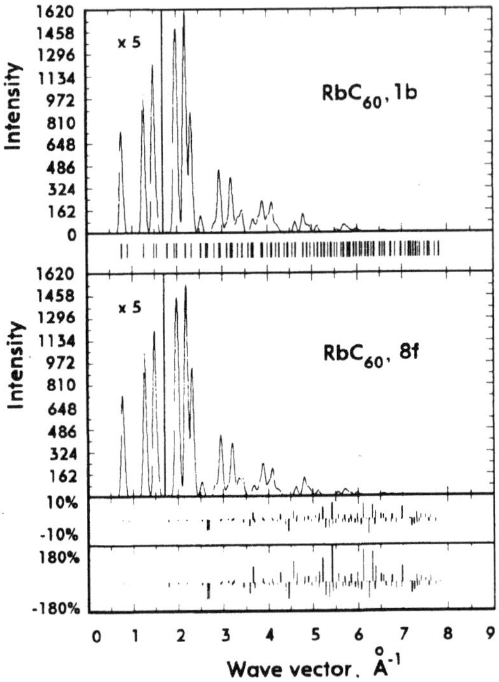

Figure 12. The Bragg neutron scattering on powder RbC_{60} in the $Fm3m$ phase. The rubidium atom occupies the central position **1b** and the position **8f** with displacement 0.5 Å. In the lower part of the figure two dash diagrams are shown for displacements 0.1 Å and 0.5 Å.

Diffuse scattering is calculated in the approximation of large Q when only intramolecular bonds contribute to the diffraction. In this case, one obtains[28]

$$\begin{aligned}\langle F^2\rangle &= \sum_{j(m),k(m)} \overline{b_j b_k^*}\langle \exp[i\mathbf{Q}(\mathbf{r}_{mj}-\mathbf{r}_{mk})]\rangle \\ &= \sum_{j,k(m)} \overline{b_j b_k^*}[\sin(Qr_{jk})/Qr_{jk}]\exp(-W_{jk}) + \sum_{j(m)} \sigma_j/4\pi, \end{aligned} \tag{43}$$

where σ_j is the incoherent scattering cross section of the atom j. The Debye–Waller factor $W_{jk} = 1/2\langle(r_j - r_k)^2\rangle Q^2$ takes into account intramolecular vibrations in the harmonic approximation.

Figure 11 shows the preliminary results of experimental investigations of a RbC_{60} powder sample with the neutron diffractometer G4.2 at the reactor "Orphey" in LLB. The diffractometer G4.2 belongs to Petersburg Nuclear Physics Institute. The sample was prepared by Dr. Forro from EFFL (Switzerland). The diffraction patterns obtained in two phases: *Fm3m* at T=450 K and *Pnnm* at T=300 K, can be used to verify the theoretical predictions discussed in this lecture.

10. CALCULATIONS FOR AC_{60}

First, let us discuss the possibility of experimental verification of results for the orientational states. The calculated atomic density distributions of rotating C_{60} molecules (see Fig. 4) are connected with the diffraction spectra obtained in experiments. It is known[29] that Bragg peak intensities depend on the particular arrangement of atoms in a crystal primitive cell. In our case, the primitive cell contains a rotating C_{60} molecule. The Bragg peak intensities (from C_{60} molecules) are determined by square of the molecular structural factor

$$\begin{aligned} F(Q) = b_c\{c_{con}60\exp[-W_1(Q)]j(Q\rho) \\ + c_{disc}\exp[-W_2(Q)]\int dr'\sum_{k,n}\langle\delta_k(r'-r_n)\rangle\exp(-iQr')\}, \end{aligned} \quad (44)$$

where b_c is the coherent neutron scattering length on a carbon nucleus (or the atomic form factor for X-ray scattering), the continuous c_{con} and discrete c_{disc} weight factors are the fitting parameters that satisfy the condition $c_{con} + c_{disc} = 1$, for C_{60} fullerite $c_{con}=0.31$, $c_{disc}=0.69$[22], $j(Q\rho) = \sin(Q\rho)/(Q\rho)$ is the Bessel function, $W_1(Q)$ and $W_2(Q)$ are the Debye–Waller factors of carbon atom in a C_{60} molecule. Due to different types of averaging the values of $W_1(Q)$ and $W_2(Q)$ can be different from one another. From Eq. (14) it is seen that the averaged atomic density, $\sum_{k,n}\langle\delta_k(r'-r_n)\rangle$, used to describe the atomic density anisotropy, enters into the structural factor.

Note that rotating C_{60} molecules occupy the orientational states from all orientational orbits but only one orientational orbit is connected with the orientational phase transition. The summation in the term $\sum_{k,n}\langle\delta_k(r'-r_n)\rangle$ runs over states of this orbit (index k) and the contribution of other orientational states from remaining orbits is modelled by the Bessel function $j(Q\rho)$.

Thus, the distributions shown in Fig. 4 determine the Bragg peak intensities in the phase *Fm3m* and, therefore, one can verify experimentally by neutron or X-ray diffraction the atomic density distributions of rotating C_{60} molecules.

Second, let us discuss the influence of metal atom displacements on neutron and X-ray diffraction spectra.

Figure 12 shows the calculated Bragg neutron spectra of RbC_{60} in the *Fm3m* phase.[17] In a quantitative analysis we propose to use the modified dash diagrams in the lower part of Fig. 12 which show relative changes in the Bragg peak intensities: $(I_1 - I_2)/I_1$. As is seen, the maximum change in the spectra as the metal atom suffers the displacement of 0.1 Å is 10%. This means that for the experimental determination of this displacement one needs to collect no less than 1000 pulses per peak.

Figure 13 shows the calculated neutron diffraction spectrum of KC_{60} in the *Pnnm* phase. A comparison of modified dash diagrams for the metal atom positions **6e** and **8f** shows that it is possible to determine the position of the metal atom in *Pnnm* phase.

Figure 14 shows the calculated X-ray diffraction spectrum for KC_{60} in the *Pnnm* phase. It is seen that relative changes in intensities are an order of magnitude larger than in neutron diffraction. This is due to the large difference between atomic form-factors of potassium and carbon. At the same time, coherent neutron scattering lengths for these elements have the same order of magnitude. Thus, such experiments take a shorter time for X-ray diffraction.

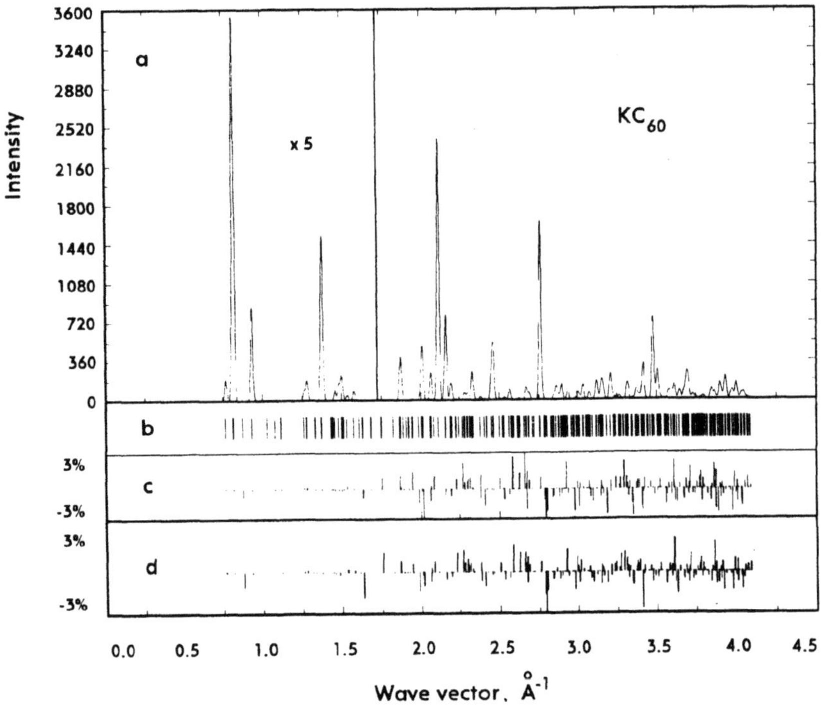

Figure 13. Bragg neutron scattering from powder KC_{60} in the *Pnnm* phase. b) Bragg peak positions. c) and d). Dash diagrams displaying relative changes in Bragg peak intensities for the potassium atom displacement 0.2 Å from the central position to positions **6e** and **8f**, respectively.

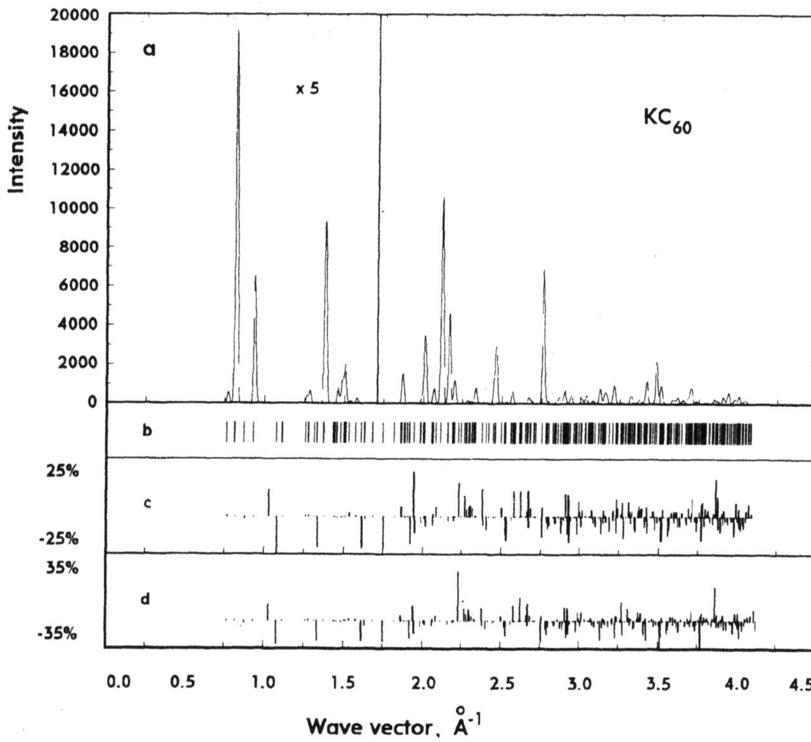

Figure 14. X-ray diffraction spectrum from powder KC_{60} in *Pnnm* phase. b) Bragg peak positions. c) and d). Dash diagrams displaying relative changes in the Bragg peak intensities for the potassium atom displacement 0.2 Å from the central position to positions **6e** and **8f**, respectively.

11. CONCLUSION

Let us briefly summarize the obtained results. In the framework of the discrete method a model of orientational phase transitions in fulleride crystals is proposed. The symmetry classification of all orientational states of rotating C_{60} molecule is done. It is shown that the orientational states of the molecule can be distributed between different orientational orbits with different local symmetries. It is found that particular orientational orbits are connected with corresponding orientational phase transitions and orientational states from one orbit are the analogs of atomic displacements connected with the soft mode of structural phase transition. Namely, the orientational states from one orbit determine the anisotropy of the atomic density distribution of C_{60} molecule in the phase $Fm\bar{3}m$. Atomic density distributions reflect the physical nature of the orientational phase transition, the scheme of which is proposed in this lecture. On the basis of a symmetry analysis of orientational states of C_{60} molecules the effect of increasing both local symmetry and symmetry of internal vibrations of C_{60} in a crystal irrespective of decreasing spatial symmetry at phase transition, is predicted. This prediction can be verified in neutron and X-ray diffraction experiments.

The proposed free energy expansion obtained on the basis of the symmetry analysis shows that the primary order parameter connected with orientational degrees of freedom of C_{60} molecules and secondary order parameters connected with different degrees of freedom of metal atoms interact with each other and lattice strains. The obtained free energy representation describes the experimental observation of spontaneous lattice strains following the phase transition and also makes it possible to predict probable changes in metal atom positions. The fullerides $K_{1.4}C_{60}$, $Rb_{0.9}C_{60}$ and CsC_{60} were investigated[30] with synchrotron X-ray diffraction. It was shown that large Debye–Waller factors for metal atoms in octahedral positions can be explained in the framework of the model of 8-fold position ([111] direction) in the $Fm\bar{3}m$ phase. A more detailed description of the behaviour of metal atoms is discussed in this lecture. The predicted effects can be verified in diffraction experiments as described in the lecture.

Note that for the verification of the proposed theory other experimental techniques can be also used. For example, in Ref.[31] an EXAFS study of the local structure of alkali ions in A_3C_{60} compounds is reported. It points to non-central positions of metal atoms. For A_3C_{60} systems, some problems are connected with the fact that alkali ions occupy two types of sites: tetrahedral and octahedral. From this point of view AC_{60} systems are simpler. Therefore, experiments to conduct investigations in this direction are under preparation at present.

Another experimental technique which can be used for the verification of the symmetry theory is the Raman spectroscopy discussed in Ref.[23]

In conclusion, it should be noted that systematic studies of structural peculiarities of fullerene crystals have been started only recently. We need to verify experimentally the theoretical predictions. Some questions, such as the spatial symmetry of metastable cubic phases and the types of deformation of C_{60} molecules in the dimeric phase in AC_{60} systems are still open.

Acknowledgments

We would like to thank Yu. A. Izyumov, who inspired the group theoretical description of the molecular strain. The present work has been performed within the State Scientific and Technical Program Fullerenes and Atomic Clusters, N 97032, supported by the INTAS–RFFI grant 95–0639 and the INTAS grant 98 MO–45.

REFERENCES

1. H. W. Kroto, J. R. Heath, S. C. O'Brien, R. F. Curl, and R. E. Smalley, *Nature* 318:162 (1985).
2. E. Osawa, Kogaku (Kyoto), *Jap.* 25:854 (1970).
3. D. A. Bochvar, and E. D. Galpern, *Doklady AN SSSR* 209:610 (1973).
4. W. Krätschmer, L. D. Lamb, K. Fostiropoulos, and D. R. Huffman, *Nature* 347:354 (1990).
5. A. F. Hebard, M. J. Rosseinsky, R. C. Haddon, D. W. Murphy, S. H. Glarum, T. T. Palstra, A. P. Ramirez, and A. R. Kortan, *Nature* 350:660 (1991).
6. V. L. Aksenov, and V. V. Kabanov, *Phys. Rev.* B57:608 (1998).
7. J. Winter, and H. Kuzmany, *Solid State Commun.* 84:935 (1992).
8. P. W. Stephens, G. Bortel, G. Faigel, M. Tegze, A. Janossy, S. Pekker, G. Oszlanyi, and L. Forro, *Nature* 370:636 (1994).
9. G. Oszlanyi, G. Bortel, G. Faigel, L. Granasy, G. M. Bendele, P. W. Stephens, and Forro, L., *Phys. Rev.* B54:11849 (1996).
10. M. Kosaka, K. Tanikagi, T. Tanaka, T. Atake, A. Lappas, and K. Prasside, *Phys. Rev.* B51:12018 (1995).
11. O. V. Kovalev, "Irreducible and Induced Representations and Corepresentations of Fedorov Groups," Moscow, Nauka (1986).
12. K. H. Michel, *Z. Phys.* B88:71 (1992).
13. K. Rapcewicz, and J. Przystawa, *Phys. Rev.* B49:13193 (1994).
14. D. Lamoen, and K. H. Michel, *Phys. Rev.* B48:807 (1993).
15. J. Nye, "Physical Properties of Crystals," Moscow, Nauka (1967).
16. V. L. Aksenov, Yu. A. Ossipyan, and V. S. Shakhmatov, *JETP Letters* 62:417 (1995).
17. V. L. Aksenov, Yu. A. Ossipyan, and V. S. Shakhmatov, *J. of Low Temp. Phys.* 107:547 (1997).
18. R. A. Dilanyan, O. G. Rybchenko, and V. S. Shekhtman, *Crystallographia* 40:604 (1995).
19. S. Teslic, T. Egami, and J. E. Fischer, *Phys. Rev.* B51:5973 (1995).
20. V. L. Aksenov, Yu. A. Ossipyan, and V. S. Shakhmatov, *JETP Letters* 64:110 (1996).
21. Yu. A. Izyumov, and V. N. Syromiatnikov. "Phase transitions and symmetry of crystalls," Moscow, Nauka (1984).
22. J. D. Axe, S. C. Moss, and D. A. Neumann, *in*: "Solid State Phys.": Advances in Research and Applications, ed. by H. E. Ehrenreich, and F. Spaepen, 48, 149, New York, Academic Press (1994).
23. V. L. Aksenov, Yu. A. Ossipyan, and V. S. Shakhmatov, *JETP* 113, N3:1081 (1998).
24. P. J. Horoyski, M. L. W. Thewalt, and T. R. Anthony, *Phys. Rev.* B54:920 (1996).
25. A. S. Davydov. "Theory of Solid State," Moscow (1976), p.341.
26. G. Dresselhaus, M. S. Dresselhaus, and P. C. Eklund, *Phys. Rev.* B45:6923 (1992).
27. G. Dolling, B. M. Powell, and V. F. Sears, *Mol. Phys.* 37:1859 (1979).
28. F. Leclercq, P. Damay, M. Foukani, P. Chieux, M. C. Bellissent-Funel, A. Rassat, and C. Fabre, *Phys. Rev.* B48:2748 (1993).
29. V. L. Aksenov, and A. M. Balagurov, *Physics-Uspekhi (Usp. Fiz. Nauk)* 39 (9):897 (1996).
30. Q. Zhu, O. Zhou, J. E. Fisher, A. R. McGhie, W. J. Romanow, R. M. Strongin, M. A. Cichy, and A. B. Smith III, *Phys. Rev.* B47:13948 (1993).
31. G. Nowitzke, G. Wortman, H. Werner, and R. Schlögl, *Phys. Rev.* B54:13230 (1996).

FUNDAMENTAL PROBLEMS OF QUANTUM CRITICAL PHENOMENA

Luigi De Cesare[1] and Dimo I. Uzunov[2]

[1]Dipartimento di Scienze Fisiche "E. R. Caianiello",
Università di Salerno, I–84081 Baronissi, Salerno, Italy,
and Istituto Nazionale per la Fisica della Materia,
Unità di Salerno, Italy.
[2]CPCM Laboratory, G. Nadjakov Institute of Solid State Physics,
Bulgarian Academy of Sciences, BG–1784 Sofia, Bulgaria.

1. INTRODUCTORY NOTES

The present article is an overview of selected topics rather than a thorough review of the classic and recent achievements in the field of quantum critical phenomena (QCP) – critical phenomena produced by quantum statistical effects in condensed matter and gases. The experimental and theoretical research of QCP is a branch of quantum statistical physics and phase transitions physics of a rapidly growing importance for the explanation of essential features of low dimensional fermion and spin systems, dilute Bose fluids, superconductors, quantum Hall systems, ferroelectrics, etc; as references to recent topics, see, e.g., Sonhdi et al. (1996), Schakel (1997a, 1999), and Caramico D'Auria et al. (1997).

The quantum statistics has a substantial influence on the critical behaviour near various (multi)critical points of low temperature (LT) and zero temperature (ZT) continuous phase transitions as well as on thermodynamic and correlation properties near equilibrium points of first order phase transitions; for basic references to problems of quantum statistical physics and phase transitions, see, e.g. Landau and Lifshitz (1980), Huang (1987), Fisher (1967), Pfeuty and Toulouse (1975), Ma (1976), and Uzunov (1993). Numerous real systems exhibit phase transition lines which extend from LT to ZT phase transition point, namely, to temperatures of strong quantum statistical degeneration. In this temperature region the quantum statistical correlations can, in certain cases and under certain experimental conditions, produce observable effects on the phase transition and, in particular, on the critical phenomena near LT critical and multicritical points. Despite the strong statistical degeneration, particular properties of the system or special experimental conditions can quell the quantum statistical correlations and then the LT phase transition properties will remain totally or partially classical up to ZT.

Correlations, Coherence, and Order. Edited by Shopova and Uzunov
Kluwer Academic/Plenum Publishers, 1999

Both of these classical and quantum branches of LT behaviour near phase transitions can be often observed in experiments. The questions concerning the type of systems where the quantum effects will prevail over the usual classical behaviour and the real conditions under which this will happen are a matter of research in this field. Other important problems are related to the difference between the phase transitions at high temperatures (HT), where quantum statistical effects do not exist at all and the LT phase transitions in the regime of a strong quantum statistical degeneration. It seems especially important to establish the real distinction, from one side, between the classical and quantum LT phase transitions, and from the other side, between any of them and the corresponding HT behaviour along the same phase transition line. All these problems present a substantial fundamental and practical interest.

We shall not expand our discussion over the great variety of LT phase transitions. Rather we shall focus our attention on the quantum effects of standard second order phase transitions. We shall consider several topics concerning the change (crossover) of the usual HT critical behaviour when the critical temperature is lowered in the range of the quantum degeneration of system up to ZT. This lowering of the critical temperature is referred to as LT and ZT limiting cases of critical behaviour. It is clear from the numerous experimental and theoretical studies that such a rather nontrivial and general crossover exists, namely, that the HT and LT critical phenomena are quite different from each other

The question of fundamental interest is whether this crossover phenomenon is a result of LT and ZT limits themselves or is produced by quantum effects. This problem can be solved provided both the universal and nonuniversal properties of the LT critical behaviour are thoroughly investigated.

Note, that the critical behaviour universal features are represented by the critical exponents of scaling laws which depend on the system symmetry. The nonuniversal properties, like critical temperature and the size of Ginzburg critical (fluctuation) region, which is a measure of the strength of fluctuation effects, depend on the particular values of material parameters – interaction and particle constants (see, e.g., Landau and Lifshitz, 1980, and Uzunov, 1993). The description of the HT–LT crossover (HLTC) of the critical behaviour near standard phase transitions of second order is undoubtedly important also in order to understand LT multicritical points and properties near LT first order phase transitions.

Although the interest in quantum effects on the critical behaviour in many–body systems dated from the dawn of the quantum statistical physics, the theory of QCP received a substantial development after the late sixties owing to the application of fruitful ideas of universality and scaling by renormalization group (RG) methods; see, e.g., Pfeuty and Toulouse (1975), Ma (1976), Zinn–Justin (1989), and Uzunov (1993). As pointed out by Caramico D'Auria et al. (1997), the contemporary theory of QCP has two distinct periods of development. The first, classical period began with the pioneering papers by Pfeuty and Elliott (1971), Rechester (1971), Young (1975), and the particularly important work of Hertz (1976). This period of relatively quiet research based on quantum field and statistical methods, including RG, continued up to the late eighties when the number of works on QCP has abruptly increased due to the great interest in HT superconductors, low dimensional magnetic systems, and quantum Hall effect. The recent research relies very much on the classical results which will be partly considered here together with related developments and applications. The recent problems in the QCP field are reviewed in this volume by A. M. J. Schakel.

Along with this overview we shall present several new aspects of the QCP theory. In Section 2, we apply the general phenomenological approach in order to outline the size of the classical and quantum critical regions and derive a new criterion for the quantum criticality. These results are valid irrespective of real system specific properties. In Section 5 we suggest a new scheme for the RG rescaling in Bose systems. Other new results are presented also for the nonuniversal critical behaviour of systems described by the transverse Ising model (Section 6). These results confirm the general picture outlined in Section 2. Finally, we discuss the main results for the quantum critical behaviour in disordered systems and justify the method of integration in noninteger spatial dimensionalities in view of applications to QCP and disorder effects (Section 7).

2. CRITERIA FOR QUANTUM CRITICALITY

2.1 General Criterion for Quantum Criticality

The quantum effects due to the overlap of particles wave functions exert influence on the critical phenomena near phase transition critical points in many–body systems provided the de Broglie thermal wavelength λ of quantum statistical correlations (fluctuations) is greater than the correlation length ξ of thermal (classical) fluctuations:

$$\varrho = \frac{\lambda}{\xi} > 1 . \tag{1}$$

Critical phenomena which satisfy the *quantum critical criterion* (1) are called *quantum critical phenomena* (QCP).

The criterion (1) is a direct consequence (Suzuki, 1976a) of the general notion that phenomena at length scales shorter than the length scale ξ of order parameter classical fluctuations are irrelevant to the critical behaviour. Besides, the condition (1) is confirmed by the available studies of quantum statistical models; several of these particular studies are enumerated in the remainder of this paper. The criterion (1) yields both necessary and sufficient conditions for the appearance of QCP but the circumstances under which QCP can actually occur remain hidden in its generality.

In the remainder of this Section we shall present several new aspects of the CQP theory, which are valid irrespectively of particular system specific features. For this reason we shall use a phenomenological approach.

2.2 Quantum Critical Region

The range of temperatures where QCP may occur can be evaluated with the help of the condition (1) and the temperature dependences of lengths ξ and λ. The wavelength λ can be written in the general form

$$\lambda(T) = \left(\frac{2\pi\hbar^2}{mk_BT}\right)^{\theta}, \tag{2}$$

where the familiar exponent $\theta = 1/2$ (see, e.g., Huang, 1987) has been generalized to take values $\theta > 0$ in order to comprise all possible quantum systems (see, e.g., Uzunov, 1993). We shall show in Section 4.2 that for a number of quantum statistical models $\theta = 1/z$, where z is a dynamical critical exponent which describes the intrinsic quantum dynamics of the system. It can be stated as a theorem that the relation $\theta = 1/z$ is valid always when the dynamics of quantum systems is not influenced by other time–dependent phenomena. In Eq. (2), the parameter m denotes either the real particles mass or an effective mass $m_{\text{eff}} \sim \hbar^2/c(g)$ which represents the effect of some interaction constant g; below the suffix "eff" will be omitted.

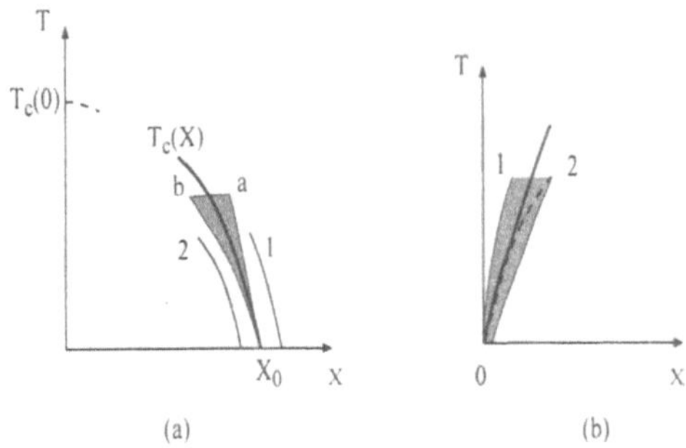

Figure 1. (a) LT part of a critical line with a ZT critical point $T_c(X) = 0$. The shaded part *(a–X_0–b)* of the critical region, marked by the lines 1 and 2, corresponds to a LT classical critical behaviour ($\varrho < 1$). (b) LT critical line with $X_0 = 0$; domains *(a–X_0–b)* and 1–2 coincide.

We shall often use the notation $\lambda = \lambda_0/T^{\theta}$, where $\lambda_0 = (2\pi\hbar^2/mk_B)^{\theta}$ depends on a single material parameter – the effective mass m of particles. Near finite temperature critical points $T_c > 0$, where

$$0 \leq |t(T)| = \frac{|T - T_c|}{T_c} < 1\,, \tag{3}$$

the temperature dependence of correlation length ξ is written by the usual scaling law

$$\xi(T) \;=\; \xi_0(T_c)|t(T)|^{-\nu} \tag{4}$$

with the help of the critical exponent $\nu \geq 1/2$ and the scaling amplitude $\xi_0(T_c) > a$ which is often referred to as the ZT correlation length, $\xi(0) = \xi_0(T_c) > a$; a is the mean interparticle distance (the lattice constant in crystal bodies).

The broad vicinity (3) of critical point T_c, where the phase transition phenomena occur and the Landau expansion of Gibbs free energy in powers of fluctuation order parameter $\phi(\vec{r})$ is valid, includes both mean field (MF) domain of description and Ginzburg critical region of strong fluctuation interactions very close to T_c. This broad critical region* can be denoted also by $\xi(T) > \xi_0(T_c)$; in certain systems, for example, conventional LT superconductors, $\xi_0(T_c) \gg a$. So, as indicated by the condition (1) and Eq. (2), QCP will possibly occur in the regime of quantum degeneration of system ($\lambda > a$), i.e., near LT and ZT critical points, where conditions (1) and (3) are simultaneously satisfied.

QCP may occur only in condensed matter systems and gases, where the critical line $T_c(X)$ extends over the LT region to the ZT critical point $T_c(X_0) = 0$ by variations of additional thermodynamic parameter X. Depending on the particular system, the parameter X may represent some auxiliary intrinsic interaction suppressing the main interaction, which is responsible for the phase transition, or other physical quantities (density, pressure, concentration of impurities, etc.) having the same effect; see Fig. 1, where such critical lines are depicted for $X_0 > 0$ and $X_0 = 0$. The slope of curve $T_c(X)$ and the fact that it is tilted to the left is irrelevant to the present discussion. In certain systems, the critical line $T_c(X)$ is tilted to the right and then the maximal value of T_c will be at some $X > X_0$.

We should keep in mind that Eqs. (3) and (4) describe only the temperature dependence of length ξ. In general, the correlation length ξ depends on the parameter

*Hereforth this region will be referred to as "critical region" and we suppose that the reader will distinguish between it and the Ginzburg fluctuation (critical) region.

X as well: $\xi = \xi(T, X)$. The dependence of thermodynamic and correlation quantities on X must be considered on the same footing as the dependence on T. The reason is that, as seen from Fig. 1, the phase transitions at $T_c(X)$ or, equivalently, at $X_c(T)$, can occur both by variations of T at fixed X (T–driven transitions) and by variations of X at fixed T (X–driven transitions). Phase transitions produced by both T– and X–variations are also possible. We suppose that their main properties can be clarified by the investigation of T– and X–transitions.

These notes are important for the critical region evaluation. The inequality (3) defines the critical region concerning critical phenomena produced by T–transitions and, hence, the critical region width $[T_a(X) - T_b(X)]$ along the T–axis is proportional to the corresponding equality: $|t(T)| \sim 1$; exact calculations of such quantities are senseless. The critical region along the X–axis is produced by X–transitions and this effect will be also considered. For a methodical convenience, we shall consider the (temperature) critical region along the T–axis and the X–critical region along the X-axis separately, although they are a result of one and the same reason – classical and quantum fluctuations.

The classical critical region along the line $T_c(X)$, where the critical behaviour is totally dominated by classical fluctuations, is indicated in Fig. 1 by a shaded domain (a–X_0–b). This region is defined by the condition (3) and that inverse of (1). In Fig. 1 the total critical region, $|t| < 1$, is marked by the lines 1 and 2. The main question is whether the classical domain a–X_0–b is a part of the critical region 1–2, as shown in Fig. 1a, or fills up it completely, as shown in Fig. 1b. We shall consider this problem for T– and X– transitions, separately.

2.3 T–Driven Transitions

According to Eqs. (2)–(4), $\xi \to \infty$ for $T \to T_c > 0$, whereas λ remains finite. This means that the close vicinity of finite temperature critical points $T_c > 0$ will always exhibit a classical behaviour. In particular this is true for the so–called asymptotic critical behaviour corresponding to the infinitesimally narrow distance ($|t| \to 0$) from T_c.

Using the criterion (1) as well as Eqs. (2) and (4), it is easy to show that QCP will occur, i.e., the quantum effects penetrate in the (temperature) critical region $|t(T)| < 1$, if

$$T_c(\xi_0/\lambda)^{1/\nu} \; < \; |(T - T_c)| \; < \; T_c \, . \tag{5}$$

The conditions (5) are well defined for $T_c > 0$, including the ZT limit $T_c \to 0$, provided $T \to 0$ too, so that $|t| < 1$ is satisfied. Obviously, the quantum portion (5) of the critical region will exist if

$$\xi_0(T_c) \; < \; \lambda(T), \tag{6}$$

which corresponds to moderate ($\lambda > a$) or to strong ($\lambda \gg a$) quantum degeneration for ($\xi_0 > a$) and ($\xi_0 \gg a$), respectively.

The inequality (6) may bring more information about the quantum criticality corresponding to T–driven transitions if we remember that, within the present description ($|t| < 1$), so we can freely substitute T with T_c in all formulae except that for $t(T)$. Conversely, the substitution of T_c with T is also allowed in the whole transition domain (3) if the nominator of $t(T)$ is not affected. Besides, the latter variant of the theory seems to be more closely connected with the original parameters of free energy.†

†This is consistent with the Landau expansions (see Section 4).

The further consideration depends on the way in which we shall treat the nonuniversal length $\xi_0(T_c) > a$. Lowering T_c to some extent, depending on the specific properties of a given system, $\xi_0(T_c)$ grows and this is described by the relation

$$\xi_0(T_c) = \xi_{00} T_c^{-\nu_0} , \tag{7}$$

where $a < \xi_{00} < \infty$ is a new (LT) scaling amplitude and $0 < \nu_0 \leq \infty$ is a new LT exponent; ν_0 must be positive because of the criticality at low T_c.

The value of exponent ν_0 and the range of temperatures, where the LT dependence (7) will be valid, can be evaluated from the properties of system. The quantities $\xi_{00} > a$ and $\nu_0 > 0$ appear as a result of a gradual LT crossover in the critical behaviour owing to T_c lowering.

The temperature T is not enough to describe the LT crossover and the entire investigation can be performed in terms of both relevant thermodynamic parameters T and X. The LT critical behaviour limit should be given by the ratio $[T_c(X)/T_c(0)] \ll 1$ or, equivalently, by the ratio $[g(X)/g_0] \ll 1$, where $g_0 = g(0)$ is the interaction responsible for the T–driven transition and $g(X)$ describes the effective decrease of this interaction owing to the auxiliary parameter X; in certain cases, $g(X) \sim T_c(X)$. These remarks, involving the term "interaction" in our phenomenological analysis, are not restricted to systems of interacting particles only; see Section 3, where the phase transition is produced rather by Bose statistics and global thermodynamic constraints than by direct interparticle interactions.

Having in mind these notes we continue our analysis of the condition (6) by substituting $\lambda(T)$ with $\lambda(T_c)$ and $\xi_0(T_c)$ with the scaling form (7):

$$\frac{\xi_{00}}{\lambda_0} < T_c^{\nu_0 - \theta} . \tag{8}$$

When the critical temperature is lowered enough, the condition (8) will be broken unless

$$\theta > \nu_0 . \tag{9}$$

If the inequality (9) is satisfied, QCP will occur at LT in certain part (5) of the critical interval (3) and, moreover, for $T_c \rightarrow 0$, they will prevail in the asymptotic critical behaviour $[t(T) \rightarrow 0]$, too. If the criterion (9) is satisfied, the classical fluctuations will be completely irrelevant to the critical behaviour at $T_c = 0$. Their effect at $T_c \sim 0$ (extremely LT critical points) will be restricted in a negligibly narrow vicinity of T_c, which is practically inaccessible to experiments. In both cases, $T_c \sim 0$ and $T_c = 0$, the experiment should observe QCP only. For $\theta = \nu_0$, the quantum correlations will have an effect on the non–asymptotic LT critical behaviour outside the small distance $T_c(\xi_{00}/\lambda_0)^{1/\nu}$ from T_c, provided $(\xi_{00}/\lambda_0) < 1$, i.e., in case of convenient nonuniversal properties of particular system. QCP do not exist at all if $\theta < \nu_0$.

The same results can be obtained by an alternative consideration which is consistent with our discussion about the possible change of factors T with T_c, and vice versa. The scaling law (4) has the following alternative definition:

$$\xi(T) = \xi_0(T)|t(T)|^{-\nu} , \tag{10}$$

where $t(T) = (T - T_c)/T$. In HT and, to some extent, in LT range of critical temperatures T_c, Eq. (10) will give the same leading scaling dependence as Eq. (4) because the corrections to the scaling are of the order $O(|t|^{1-\nu})$. At lower T_c, the singularity in Eq. (10) may arise only from $\xi_0(T)$. That such a singular behaviour should exist becomes obvious from Fig. 1. For example, when the ZT critical point ($T_c = 0$) is approached by lowering the temperature along the line $X = X_0$, the system will exhibit fluctuation phenomena, typical of the critical behaviour in the paraphase, although it will never reach the ordered state. Let us suppose that $\xi_0(T)$ obeys a scaling law with respect to T having the form (7). Then one can immediately rederive a condition for temperature T which is identical with the condition (8) for T_c and, hence, the criterion (9) is straightforwardly confirmed.

Several notes can be added for the sake of completeness. In the paraphase the criticality condition $|t| < 1$ is redundant because this condition comes as a requirement for the smallness of order parameter with respect to its value at $T = 0$. The singularity (7) is a mere indication for an existence of scaling law for $\xi(T)$ of the type $\xi(T) \sim \xi_0(T) = \xi_{00}/T^{\nu_0}$ corresponding to zero critical temperatures ($T_c = 0$). In the LT limiting case, the critical behaviour singularities are developed as singularities of scaling amplitude, while at HT this scaling amplitude is a slow varying function of temperature.

It is worth noting that the present phenomenological consideration, including the ansatz (7) and the related one for $\xi_0(T)$, is supported by results from RG calculations for particular systems (Vojta et al., 1997; Caramico D'Auria et al., 1997) and experiments (Samara, 1988; Samara et al., 1981); see also Sections 3, 5, and 6.

2.4 X–Driven Transitions

In general, the properties of X–driven transitions are quite different from those of T–transitions. In particular, essential differences may be expected in HT ($X \sim 0$) and LT ($X \sim X_0$) ranges of temperatures. By substituting T with X and, of course, T_c with X_c in Eqs. (3) and (4), one can perform the phenomenological analysis (Section 2.3) in terms of variable X. This analysis straightforwardly yields a form of scaling law for the correlation length $\xi(T, X)$ which is identical to that related with T–transitions in the HT region, where $X \sim 0$. The only difference may come from the values of critical exponents which describe the scaling laws with respect to $t(T)$ and $t(X)$. There is no reason to suppose that the exponents towards $t(T)$ and $t(X)$ should be equal.

In fact, the critical behaviour in the LT limit described in Section 2.3 has a similarity with HT X–transitions because $X_c \sim 0$ at HT. A conformity of type $T \leftrightarrow X$ concerning the form of scaling laws and the way in which the LT limit (for T–transitions) and the HT limit (for X–transitions) are developed from the corresponding scaling amplitudes certainly exists. However, a total correspondence, including the values of respective critical exponents and scaling amplitudes cannot be expected; note, that the length $\lambda(T)$ has no analog in terms of X. In contrast, there is no such similarity between the LT T– and X–transitions.

For X–transitions, the criterion (1) yields

$$|t(X)| > \left[\frac{\xi_0(X_c)}{\lambda(T)}\right]^{1/\nu}, \tag{11}$$

where all quantities are defined by the change of T with X in (3) and (4). Having in mind Eq. (2) and the fact that the critical region (lines 1 and 2 in Fig. 1) along the X–axis is defined by $|t(X)| < 1$, it becomes evident, that in the LT range ($X \sim X_0$), where $[\xi_0(X_c)/\lambda(T) < 1]$, the quantum subdomain gradually enlarges when the temperature is decreased or, equivalently, when $X_c(T)$ approaches X_0. In the ZT limit ($T \to 0$) or, equivalently, for $X_c(T) \to X_0$, this subdomain fills up totally the critical region.

In the range of temperatures, where $[\xi_0(X)/\lambda(T)] > 1$, the critical behaviour is totally classical. We should have in mind that the latter condition for a total classical criticality can be easily satisfied in the temperature range of quantum degeneration in real systems with large ZT correlation length ($\xi_0 \gg a$).

At HT, where $X_0 \sim 0$, the scaling of correlation length ξ can be conveniently investigated by a scaling law with respect to X or $(X - X_c)$, instead of $t(X)$. The problems in the treatment of X-transitions in this HT limit are analogous to those for the LT T-transitions and the analysis can be carried out by following the ideas presented in Section 2.3.

A quite special situation should exist when $X_0 = 0$; see Fig. 1b. In this case both T and X are equal to zero at the ZT critical point, where the correlation length $\xi(T, X)$ may exhibit scaling dependence on both $t(T)$ and $t(X)$. Obviously, such ZT critical points will offer less opportunity for QCP observation or, in some cases, this opportunity may not exist at all. To emphasize that the lack of any QCP may be expected at such ZT critical points, we have depicted in Fig. 1b a classical critical region which totally fills up the critical region 1–2.

However, there exist systems, where the critical temperature T_c depends on X and tends to zero for $X \to X_0 = 0$, but the correlation length ξ does not show any divergence with respect to this auxiliary parameter. In this case the function $\xi(T, X)$, for all possible T and X, has the general form $\xi[t(T), T_c(X))]$, which describes only the scaling law with respect to $t(T)$ as given by Eq. (4) or Eq. (10); see the example considered in Section 3. The critical region width along the T-axis tends to zero for $T_c \to 0$ and a critical region with respect to the parameter X does not exist at all. Thus the lines 1 and 2 in Fig. 1b will definitely terminate at $T = X = 0$; see the dashed line 0–2.

Certainly, the X-transitions at LT offer much more favourable conditions for QCP, although the practical observation of such transitions in certain systems, where the parameter X cannot be gradually varied, is almost impossible.

2.5 Crossovers

We have shown that the LT critical behaviour can be either quantum or classical depending on the specific (nonuniversal) properties of particular system. Besides, there are no general arguments indicating that HT and LT critical properties should be equivalent. Thus we may suppose that, in general, HT classical LT and quantum critical properties present three different types of critical behaviour. So, we can introduce the HT–LT crossover (HLTC) of critical behaviour which corresponds to the change of the critical properties (or of some of them) when λ varies from $\lambda < a$ at HT to $\lambda > a$ (up to $\lambda \gg a$) at LT, and vice versa. A special case of HLTC is the classical ($\varrho < 1$) to quantum ($\varrho > 1$) crossover (CQC) which describes the difference between classical HT and QCP.

Our analysis is not restricted to the MF approximation and the fluctuation effects are not excluded from this quite general scaling consideration. But the fluctuations give rise to a dependence of the critical exponents on the spatial dimensionality d. Therefore, HLTC and CQC will exhibit an effective change of critical fluctuations spatial dimensionality d. Dimensional CQC has been discovered by Pfeuty and Elliott (1971) with the help of series expansions for the transverse Ising model, confirmed by a RG study of the same model (Young, 1975), and comprehensively described by Hertz (1976) on the basis of a number of quantum models, mostly of itinerant magnets. The dimensional crossover in the transverse Ising model has been proven also by the Trotter–Suzuki product formula (Suzuki, 1976b).

The experiment has a finite accuracy and omits very narrow temperature intervals such as the asymptotic classical regions in the close visinity of almost–ZT critical points

($T_c \sim 0$). This should be taken into account in interpretations of experiments. Because of ZT unattainability, QCP produced by ZT X–transitions cannot be observed in real experiment, but certain LT X– transitions might be experimentally investigated. On account of limitations in the accuracy of the equilibrium temperature measurement, the very narrow classical region will remain unattainable for experimental studies and, hence, the experiment will reproduce results which, to some extent, will give an information about asymptotic QCP at ZT.

The present analysis should be considered as a simplified version of a thorough phenomenological theory. The latter should be developed on the basis of scaling analysis of the Gibbs free energy with respect to the variations of scaling parameters $t(T)$ and t(X), and with respect to scaling parameters T and X corresponding to the scaling behaviour in the critical point vicinity ($T_c = 0, X = X_c \geq 0$) and ($X_c = 0, T_c > 0$), provided all these points exist. Such a scaling analysis has been presented by Weichman et al. (1986) with the aim to explain certain properties of dilute Bose fluids. Another attempt based on results for the finite–size crossover in systems of slab geometry (see Section 7.2) has been made by Sachdev (1997) on a quite restricted basis.

In summary, we have shown by general heuristic arguments that:

(a) The finite temperature ($T_c > 0$) continuous transitions always exhibit a classical asymptotic critical behaviour.

(b) The X–driven ZT (continuous) transitions always exhibit QCP, including ones in the asymptotic vicinity of the ZT critical point. While these transitions cannot be reproduced in real experiments, corresponding QCP can be well observed at X–transitions with extremely LT critical points ($T_c \sim 0$).

(c) The T–driven LT (continuous) transitions should exhibit mainly classical critical phenomena. These penomena are quite different from the usual HT critical ones. The reason is in the LT crossover of classical critical behaviour.

Thus pre–asymptotic QCP can be observed in real experiments. The interpretation of these QCP can be given with the help of scaling and RG theory of the asymptotic critical behaviour in the LT limit.

3. IDEAL BOSE GAS

The model of noninteracting bosons – the ideal Bose gas (IBG), describes the phenomenon of Bose–Einstein condensation (BEC). BEC of IBG is important for the undestanding of quantum phenomena in many areas of physics, in particular, in superfluid helium liquids (see, e.g., Lifshitz and Pitaevskii, 1980) and excitonic phases in semiconductors (see, e.g., Halperin and Rice, 1968), although in these and other real systems the interparticle interactions are strong and drastically affect the thermodynamic and correlation properties. While there is a significant difference between the properties of free and strongly interacting bosons, BEC of IBG is quite similar to that in weakly interacting (dilute) bosonic gases (Giorgini et al., 1996; Ensher et al., 1996; Baym and Pethick, 1996; Ketterle et al., 1996; Townsend et al., 1997). Besides, there is a permanent interest to BEC of charged bosons in external magnetic fields (Arias and Joannapoulos, 1989; Bardos et al., 1994) related to the Schafroth–Blatt–Butler concept of superconductivity (Schafroth, 1954; Schafroth et al., 1957; Blatt, 1964). In the context of modern scaling theory of phase transitions IBG has been investigated by Gunton and Buckingham (1968), Cooper and Green (1968), Lacour–Gayet and Toulouse (1974), González et al. (1975), and Busiello et al (1985a); for relativistic and magnetized Bose gases, see Daicic et al. (1994); for other useful sources, see Uzunov (1993).

3.1 Thermodynamic Equations

The thermodynamic properties of d-dimensional IBG, including BEC, are described by equations for the grand canonical potential Ω and number density $\rho = (N/V)$ of bosons:

$$\Omega = k_B T \sum_{\vec{k}} \ln\left[1 - e^{-\beta\varepsilon(k)}\right] , \tag{12}$$

and

$$\rho = \frac{1}{V} \sum_{\vec{k}} \langle a^+_{\vec{k}} a_{\vec{k}} \rangle , \tag{13}$$

respectively, where the brackets $\langle\rangle$ denote a statistical averaging, $\beta = 1/k_B T$, $a^+_{\vec{k}}$ and $a_{\vec{k}}$ are the creation and annihilation Bose operators for a plane-wave state of wave vector $\vec{k} = (k_i;\ i = 1, \ldots, d)$, $V = (L_1 \ldots L_d)$ is the volume of the gas. The self-energy $\varepsilon(k) = \varepsilon_0(k) + r$ is represented by the energy spectrum $\varepsilon_0(k) = \hbar^2 k^2/2m$ of free (noninteracting) real particles or excitations, and the chemical potential $\mu = -r \leq 0$; $k = |\vec{k}|$.

Throughtout this paper we shall use periodic boundary conditions. The wave vector components $k_i = 2\pi l_i/L_i$, $(l_i = 0, \pm 1, \ldots)$, are given by the spatial dimensions L_i. They are supposed to be much larger than any characteristic length of the system, for example, $L_i \gg (\xi, \lambda)$. In this case, it is usually said that the dimensions L_i are "infinite". This condition allows to pass to the continuum limit, i.e., from a summation in Eqs. (12) and (13) to the corresponding integration. The quite uncommon case of composite Bose exitations due to long-range interactions in electron and magnetic systems can be included in consideration by the generalization $\varepsilon_0(k) = \hbar^2 k^\sigma/2m, (0 < \sigma < 2)$ of energy spectrum $\varepsilon_0(k)$; see, e.g., Joyce (1972), Pfeuty and Toulouse (1975), and Uzunov (1993, 1996). A simple dimensional analysis of the exponent $\beta\varepsilon(k)$ in Eq. (12) shows that $\xi = (\hbar^2/2mr)^{1/\sigma}$.

With the help of the Bose distribution

$$n(k) = \langle a^+_{\vec{k}} a_{\vec{k}} \rangle = \frac{1}{e^{\beta\varepsilon(k)} - 1} , \tag{14}$$

Eq. (13) takes the form

$$\rho = \frac{1}{V} \sum_{\vec{k}} \frac{1}{e^{\beta\varepsilon(k)} - 1} . \tag{15}$$

The correct investigation of IBG thermodynamics in continuum limit implies to take the density $\rho_0 = (N_0/V) = n(0)$ of bosons with zero wave numbers out the sum (15). This is important for obtaining of a correct description of properties below finite temperature critical points $T_c > 0$. For all other cases this separation is redundant. As we shall be mainly interested in the latter case, we shall avoid the mentioned separation; see Busiello et al. (1985a).

Substituting the $\vec{k}$-summation in Eqs. (12) and (13) by a d-dimensional integration,

$$\frac{1}{V} \sum_{\vec{k}} \rightarrow \int \frac{d^d k}{(2\pi)^d} \equiv \int_0^\infty dk\, k^{d-1} , \tag{16}$$

$K_d = 2^{1-d}/\pi^{d/2}\Gamma(d/2)$, we obtain

$$P = k_B T A \lambda^{-d} g_{(d/\sigma+1)}\left(\frac{r}{k_B T}\right) \tag{17}$$

and

$$\rho = A\lambda^{-d} g_{(d/\sigma)}\left(\frac{r}{k_B T}\right) , \tag{18}$$

where $P = -(\Omega/V)$ is the pressure, $\lambda = (2\pi\hbar^2/mk_BT)^{1/\sigma}$ is the thermal wavelength with an exponent $\theta = 1/\sigma$, c.f. Eq. (2),

$$g_\nu(\kappa) = \frac{1}{\Gamma(\nu)}\int_0^\infty \frac{x^{\nu-1}dx}{e^{x+\kappa}-1} , \tag{19}$$

and

$$A = \frac{2^{1-d+2d/\sigma}\Gamma(d/\sigma)}{\sigma\pi^{d(1/2-1/\sigma)}\Gamma(d/2)} , \tag{20}$$

(for $\sigma = 2, A = 1$).

The BE condensate is a coherent state of a macroscopic number $N_0 \sim N$ of bosons with momentum $\hbar k = 0$. The critical temperature T_c of the transition to this condensation in momentum ($\vec{k}-$) space is defined by the equation $[r(T_c)/T_c] = 0$. The BEC order parameter $\phi_0 = \langle a_0^+\rangle/\sqrt{V}$ is related to the square root of the number density $\rho_0 = (N_0/V)$ of the condensate bosons: $|\phi_0| = \sqrt{\rho_0}$.

In the present problem, the ratio (1) takes the form

$$\varrho = \left(\frac{4\pi r}{k_B T}\right)^{1/\sigma} . \tag{21}$$

The IBG critical properties can be investigated with the help of Eqs. (17) and (18). They depend on the thermodynamic conditions imposed on IBG, i.e., on the way in which the chemical potential $\mu = -r$ is determined.

We shall briefly consider three cases: a constant pressure P (Lacour–Gayet and Toulouse, 1974), constant density ρ and spatial dimensions $d > \sigma$ (Gunton and Buckingham, 1968; Cooper and Green, 1968), and constant density at dimensions $d < \sigma$ (Busiello et al., 1985a). In all cases, the critical regime is defined by the condition $r \leq k_BT$ and all critical phenomena are almost ($r \sim k_BT$) or completely ($r \ll k_BT$) classical; c.f. Eq. (21).

3.2 Constant Pressure

The parameter r is calculated from Eq. (17) as a function of T and P. The result $\mu(T,P)$ is substituted in Eq. (18) and, hence, one obtains the equation of state $f(T,\rho,P) = 0$. The solution $T_c(P)$ of Eq. (17) with $r = 0$ yields the critical temperature

$$T_c(P) = \left[\frac{\lambda_0^d P}{\zeta(d/\sigma+1)Ak_B}\right]^{\sigma/(d+\sigma)} , \tag{22}$$

where $\lambda_0 = (2\pi\hbar^2/mk_B)^{1/\sigma}$. The shape of the critical line (22) is similar to that depicted in Fig. 1b.

The bosons in the ground state ($k = 0$) do not contribute to the internal pressure and, hence, the phase transition at $T_c(P)$ is of first order (Lacour–Gayet and Toulouse, 1974). At T_c the order parameter ϕ_0 undergoes a jump from $|\phi_0|^2 = 0$ to $|\phi_0|^2 = \rho = (N/V)$, namely, to complete BEC ($N_0 = N$). In contrast to the usual first order phase transitions, IBG under a constant pressure exhibits divergences of correlation length ξ and susceptibility $\chi \sim \xi^{-2} = (2mr/\hbar^2)$. These divergences are given by $r \sim k_Bt$ for $d > \sigma$ and $r \sim k_Bt^{\sigma/d}$, for $d < \sigma$; t is defined by Eqs. (3) and (22). The critical

Table 1. Values of critical exponents (ρ = const).

T_c	d/σ	η	α	α_s	γ	$\tilde{\gamma}$	ν
$T_c > 0$	$d \geq 2\sigma$	$2-\sigma$	0	$(2\sigma - d)/\sigma$	1	-	$1/\sigma$
$T_c > 0$	$\sigma < d < 2\sigma$	$2-\sigma$	0	$(d-2\sigma)/(d-\sigma)$	$\sigma/(d-\sigma)$	-	$1/(d-\sigma)$
$T_c = 0$	1	$2-\sigma$	-1	-	∞	∞	∞
$T_c = 0$	$0 < d < \sigma$	$2-\sigma$	$-d/\sigma$	-	$\sigma/(\sigma-d)$	$d/(\sigma-d)$	$1/(\sigma-d)$

exponents are same as those for the Gaussian model of free (noninteracting) classical fluctuations.

A characteristic feature of the phase transition at constant P is that the critical temperature is finite ($T_c > 0$) for all dimensions $d > 0$ and $P > 0$. The ZT critical point exists only in the limit $P \to 0$. This phase transition does not exhibit a crossover to a LT behaviour when the pressure P tends to zero because the critical exponents remain unchanged. There is a crossover, however, from Gaussian exponents at dimensions $d > \sigma$ to low-dimensional ($d < \sigma$) Gaussian exponents. This crossover cannot be considered as HLTC or CQC.

3.3 Constant Density

At constant density $\rho = (N/V)$, the equilibrium chemical potential $\mu(T.\rho)$ is obtained as a solution of Eq. (18). The equation of state is given by Eq. (17) after the substitution of solution $r(T,\rho)$. Eq. (18) with $r = 0$ yields

$$T_c(\rho) = \left[\frac{\lambda_0^d \rho}{\zeta(d/\sigma)A}\right]^{\sigma/d} \tag{23}$$

and, therefore, $T_c(\rho) > 0$ for $d > \sigma$, provided $\rho > 0$, whereas $T_c = 0$ for $d \leq \sigma$; note, that the zeta function $\zeta(d/\sigma)$ tends to infinity when the ratio d/σ decreases to unity.

The finite temperature critical behaviour ($T_c > 0$) is identical to that of classical spherical model (Gunton and Buckingham, 1968; Cooper and Green, 1968); for the spherical model see, e.g., Joyce (1972), and the brief note in Section 4.2. The reason for this behaviour is in the constant density ρ condition which, as is seen from Eq. (13), is equivalent to the (mean) spherical constraint on the variations of mean densities $\langle a_{\vec{k}}^{+} a_{\vec{k}} \rangle / V$. The critical exponents are shown in Table 1, including the case $d > 2\sigma$ when they have Gaussian values (Ma, 1976). All critical exponents are given by their usual notations and definitions; see, e.g., Fisher (1967) and Ma (1976).

The critical exponent γ describes the off-diagonal susceptibility

$$\chi = -\left(\frac{\partial \Omega}{\partial h \partial h^*}\right)_{h=0} \sim (1/r)\,, \tag{24}$$

where h is a complex number – a fictitions external field conjugate to Bose operator $(\sum_{\vec{k}} a_{\vec{k}}^{+})$. For $T_c = 0$, a new critical exponent $\tilde{\gamma}$ can be introduced to describe the density $n(0) \approx k_B T \chi$. For $d < \sigma$, the values of γ from Table 1 and the relation $\tilde{\gamma} = (\gamma - 1)$ yield: $\tilde{\gamma} = d(\sigma - d)$ for $d < \sigma$, and $\tilde{\gamma} = \infty$ for $d = \sigma$. The exponent $\tilde{\gamma}$ does not exist for $T_c > 0$, where the density $n(0) \sim \chi$ is described by the usual exponent γ.

In Table 1, the exponent α_s is an auxiliary exponent intended to give a more detailed description of IBG specific heat for $d > \sigma$. For $d > \sigma$, the specific heat $C(T)$ of IBG and spherical model can be represented in the form

$$C(T) = C_0 + C'(T - T_c)^{-\alpha_s}\,, \tag{25}$$

where C_0 is a regular constant. This behaviour corresponds to the zero value of usual specific heat exponent α, defined by the scaling law $C(T) = C^{(0)}|T - T_c|^{-\alpha}$, where $C^{(0)}$ is the scaling amplitude. However, several shapes of $C(T)$ are usually ascribed to the value $\alpha = 0$, namely, logarithmic divergence $(\ln|T_c - T|/T_c)$, finite jump as is in MF theory, and cusps of different forms. In order to avoid this arbitrariness, Fisher (1967) has introduced the second term in Eq. (25). The values of α_s in the Table 1 are negative and this next–to–leading term in $C(T)$ is always small but the derivative $dC(T)/dT$ is divergent at T_c and this is an information about the cusp shape at T_c.

3.4 Zero Temperature Condensation

The ZT ($T_c = 0$) BEC at constant density has been investigated by Busiello et al. (1985a). This condensation is possible at a finite constant density ($\rho > 0$) and a low spatial dimensionality $d \leq \sigma$. The results for the critical exponents at $d = \sigma$ and $0 < d < \sigma$ are shown in Table 1. These critical exponents describe the scaling laws with respect to the variations of temperatute, for example, $C(T) = C_0/T^{\alpha}$.

It is seen from Table 1 that the ZT condensation at constant density exhibits quite unusual critical properties. The critical exponents γ and ν for $d < \sigma$ can be obtained from the familiar values at $\sigma < d < 2\sigma$ by the change of sign of $(d - \sigma)$. The Fisher exponent η has its known value but the exponent α is negative rather than zero. as is for $d > \sigma$. For ZT T–driven transitions (Section 2.3), as is in our case of $d \leq \sigma$ spatial dimensions, the negative values of critical exponent α are consistent with the Nernst theorem. In the case of dimensionalities $d > \sigma$, $\alpha = 0$ and this theorem is satisfied in the limit $T_c(\rho) \to 0$ of an extreme dilution ($\rho \sim 0$) because the scaling amplitude $C^{(0)}$ tends to zero.

The infinite values of ν and γ at $d = \sigma$ indicate the exponential divergence of correlation length $\xi = (\hbar^2/2mr)^{1/\sigma}$ and susceptibility $\chi = (1/r)$ at $T_c = 0$. These exponential divergences are known from the mechanism of critical fluctuations in classical systems at their lower borderline (critical) dimensionality d_L. Here the exponential divergence of ξ and χ comes from the same low dimensional effect. The exponents given in the last two lines of Table 1 can be referred to as *low dimensional spherical exponents.*

The next question is why this ZT critical behaviour exists. It is intuitively clear that the low dimensional ($d \leq \sigma$) IBG should have a ground state of total $\vec{k}$–space condensation ($N_0 = N$) at $T = 0$. This means that the solution $T_c = 0$ of equation $[r(T_c)/T_c] = 0$ should exist. For $T \sim 0$, $(r/k_B T) \ll 1$ and, hence, the contributions to the sum (15) are given by terms with low energies $\epsilon(k) \sim r$, i.e., with small wave numbers k.

There is an onset of BEC at $T \sim 0$, where the bigger part of bosons are at states $\varepsilon(k) \sim 0$ but the macroscopic condensation occurs only at $T = 0$ (Busiello et al., 1985a). At $T = 0$ the order parameter ϕ_0 jumps from zero to $|\phi_0| = \rho^{1/2}$ and all N bosons enter in the ground state ($k = 0$). This peculiar critical behaviour results from the constant density ρ condition (the spherical constraint).

The same behaviour can be obtained from the classical spherical model. Although the $\vec{k}$–space condensation is a pure quantum effect due to Bose statistics, the critical behaviour around critical points ($T_c \geq 0$) is totally ruled by classical fluctuation effects described by the classical spherical model.

The IBG critical behaviour can be exactly derived from thermodynamic Eqs. (12) and (15) because this model is exactly solvable. The RG analysis of IBG has been carried out in an investigation (Weichman et al., 1986) of crossover of the IBG critical behaviour to that of nonideal Bose gas. The RG investigation of spherical model has been performed by González et al. (1975).

3.5 Classical and Quantum Critical Regimes

The critical behaviour of IBG in a close vicinity ($r \ll k_B T$) of phase transition points ($T_c \geq 0$) to BEC of IBG is definitely classical. All our calculations have been performed by keeping the leading power in $(r/k_B T)$ in the corresponding series for Bose integrals (19) in powers of $(r/k_B T) < 1$. This approximation has been used to determine the leading scaling behaviour, namely, the scaling laws. This essentially classical expansion yields the critical regime ($\mu \sim 0$).

The critical region (Section 2.2), where the phase transition phenomena occur, can be defined in broad interval of temperatures around $T_c \geq 0$, i.e., by $r < k_B T$. In this region but not very near ($r \ll k_B T$) to T_c we must take into account some next–to–leading terms in the same series expansion of Bose integral (19) in powers of $r/k_B T$. These secondary terms have a quantum origin and, therefore, taking several of them as corrections to the leading scaling powers, we shall obtain the quantum corrections to the main scaling behaviour.

Thus the classical region *(a–0–b)* discussed in Section 2 can be defined by the inequality $r \ll k_B T$. For both $T_c(P)$ and $T_c(\rho)$ this region will look like the shaded wing $1-0-2$ in Fig. 1b; see the dashed line $0-2$. The quantum region will be outside this vicinity of critical line, in the rest part of critical domain ($r < k_B T$) discussed in Section 2.

The parameters P and ρ rule the critical temperature but they have no participation in the scaling laws. Thus they have not a direct effect on the critical region size. This is a particular IBG property. The size of classical critical region is reduced with the decrease of P or ρ, i.e., of T_c. This is valid both for widths of classical ($r \ll k_B T$) and total ($r < k_B T$) critical regions along the $X = (P, \rho)$–axis. In fact, at fixed X we should consider the inequalities $r < k_B T_c(X)$ and $r \ll k_B T_c(X)$. When $T_c(X)$ is lowered the critical regions width along the X–axis will decrease to zero for $X \to 0$.

If the ZT critical point is approached by variations of T and X in the ordered phase (below the critical linc), no critical effects will be observed. This is consistent with the intuitive notion that, in this system, the ZT critical point should belong to the ordered phase. If this ZT critical point, containing the complete BE condensate of the gas is approached by T–variations in the disordered phase, critical phenomena, mainly classical, will be observed, because of the real phase transition to BEC at $T_c = 0$.

The phase diagram of ZT BEC in low dimensional systems at constant density (Section 3.4) is very simple because the critical line coincides with the ρ–axis. The critical line $T_c(0 \leq \rho \leq \infty)$ can be approached only by T–variations, and the critical phenomena are purely classical.

Finally we shall mention that the criterion (9) is not satisfied for the above mentioned transitions. There is no real HLTC in the IBG because of the lack of fluctuation interactions, but we can consider the crossover from HT to ZT at constant density $\rho > 0$ produced by the change of the dimensionality from $d > \sigma$ to $d < \sigma$. This formal dependence describes the difference between the high–dimensional ($d > \sigma$) critical phenomena at finite critical temperatures and those of ZT BEC at constant density $\rho > 0$ and low dimensionality ($d < \sigma$).

4. QUANTUM FLUCTUATION FUNCTIONALS

The interacting bosons can be investigated by propagator expansions within the Green functions method; see, e.g., Kadanoff and Baym (1962), and Abrikosov et al. (1963). These infinite perturbation series are truncated on the basis of several approximations. The perturbative approach breaks down in a close vicinity of the critical point because of the strong fluctuation interactions. In this situation, after accomplishing the MF analysis, the strong fluctuation interactions in the close vicinity (Ginzburg fluctuation region) of critical point and their effects on the critical behaviour are investigated by RG. The latter is specially designed for studies of fluctuation critical phenomena. This outline of general investigation methods of critical phenomena is common for both classical and quantum statistical models.

4.1 Classical Systems

In the classical picture we must know the quasimacroscopic Gibbs–Landau free energy. This is a generalized grand canonical potential Ω, which includes all equilibrium thermodynamic properties as well as the relevant quasimacroscopic (largelength–scale) fluctuations of order parameter $\phi(\vec{r})$. The generalized free energy is a functional of field $\phi(\vec{r})$, $\Omega[(T...); \tilde{a}; \phi(\vec{r})]$, in which all microlength– and mesolength–scale phenomena are ignored. This functional depends also on the relevant thermodynamic parameters $(T...)$ such as temperature T, pressure P, and magnetic field $\vec{H}$, as well as on several material constants – the so–called Landau parameters $\tilde{a}$; hereforth the symbols $(T...)$ and $\tilde{a}$ will be omitted.

There are two general approaches to the derivation of the functional $\Omega[\phi(\vec{r})]$:

(a) The Landau phenomenological approach within which the grand canonical Gibbs free energy $\Omega[\phi(\vec{r})]$ is first expanded in symmetry allowed powers (Landau invariants) of order prameter field $\phi(\vec{r})$ and then truncated to several first relevant terms. This approach is general and defines quite loosely the Landau parameters so that the theory has the flexibility to be fitted to nonuniversal properties of real systems.

(b) The second systematic and widely applicable method is based on the Hubbard–Stratonovich transformation of original (microscopic) statistical degrees of freedom to continuum $[\phi(\vec{r})]$ or lattice $[\phi(\vec{r}_i)]$ fields; see, e.g., Brout (1974), Hertz (1976, 1985), Uzunov (1993, 1996), and Section 2 of the lecture by A. M. J. Schakel in this volume. Applying this transformation to a particular microscopic model together with the long–wavelength approximation ($ka \ll 1$) one obtains the two–point correlation function in the Ornstein–Zernicke form and then the quasimacroscopic fluctuation functional $\Omega[\phi(\vec{r})]$ appears in transcedental series like Langevin or Brillouin functions in magnetism. The redundant terms are neglected on the same thermodynamic grounds as this is done within the Landau approach. This neglecting is without any consequences for the results in further investigations.

The truncation of the shortlength–scale phenomena (particle or spin short–range correlations) is an apparent approximation which yields results for the Landau parameters within the MF approximation. Therefore, the advantage of this approach, namely, the possibility for a derivation of Landau parameters, comes together with a MF approximation for the corresponding ralationship between microscopic and macroscopic parameters of system. This approximation is not a disadvantage in the broad MF region of description as well as for investigations of critical phenomena universality features in the Ginzburg region. In fact, the mentioned approximation yields wrong predictions mainly for the critical temperature T_c, which is obtained too higher than the real one (it is a characteristic property of the MF theory to yield the upper bound for the critical temperature). Thus the fluctuation free energy functionals obtained by Hubbard–Stratonovich transformations include the longlength–scale fluctuations into consideration and exclude the effects of (micro– and mesoscopic) shortlength–scale fluctuation correlations; the latters are normally accounted for by cumulant expansions, where the small parameter is the inverse interaction radius (see, Stinchcombe et al., 1963; Brout, 1974; Uzunov, 1996).

The approaches (a) and (b) are consistent with each other. Up to now there are no results from the Hubbard–Stratonovich approach which contradict to Landau phenomenological scheme.

Next, one performs the MF analysis in order to describe the stable and metastable phases (ground states). This means complete ignoring of spatial fluctuations of field $\phi(\vec{r})$ by setting $\phi(\vec{r}) \approx \phi_0$. The minimization of function $\Omega(\phi_0)$ should yield the (stable, metastable and unstable) thermodynamic equilibria $\bar{\phi}(T..., \tilde{a})$, and this procedure cuts the uniform (macroscopic) fluctuation $\delta\phi = (\phi_0 - \bar{\phi})$, too. The field $\phi(\vec{r})$ and, in particular, its uniform configuration ϕ_0, is often referred to as "nonequilibrium order parameter" with the aim to distinguish between it and the equilibrium order parameter $\bar{\phi}$ which is a function of thermodynamic parameters and material constants.

Finally, one should consider the fluctuations $\delta\phi(\vec{r}) = \phi(\vec{r}) - \bar{\phi}$, and this is a problem only for fluctuation critical phenomena in the Ginzburg region; for $T > T_c$, in the paraphase ($\bar{\phi} = 0$), $\delta\phi(\vec{r}) = \phi(\vec{r})$. Then, the fluctuation $[\delta\phi(\vec{r})]$ dependent part of $\Omega[\delta\phi(\vec{r}) + \bar{\phi}]$ is considered as an effective fluctuation Hamiltonian in the grand canonical partition function $\mathcal{Z} = \mathrm{Tr}[\exp(-\mathcal{H})]$, which yields, according to the standard statistical rules, the fluctuation contributions to the critical thermodynamics. The resulting fluctuation field variant of the theory is investigated by RG (see Section 4.3).

The RG results depend on the spatial dimensions d and tensor structure of field $\phi(\vec{r})$. Here we shall assume that the order parameter field is a n–component vector with real components: $\phi(\vec{r}) = [\phi_\alpha(\vec{r}); \alpha = 1, ..., n]$. This choice covers the prevailing part of known cases. The symmetry of ordering is given by n– the symmetry index (this term for the number n of the order parameter components is often used, too). The dependence of the thermodynamic quantities on d and n is an important fluctuation effect. At a given mathematical structure of the functional $\Omega[\phi(\vec{r})]$. the critical behaviour depends on n and d. Then the couple (d, n) defines the possible classes of critical behaviour described by a given generalized free energy $\Omega[\phi(\vec{r})]$. So, the possible types of critical behaviour are classified in universality classes, denoted by (d, n), but one should know to which effective model $\Omega[\phi(\vec{r})]$ they are associated.

Note that $\Omega[\phi(\vec{r})]$ is a free energy part of total free energy $\Omega_{\mathrm{tot}}[\phi(\vec{r})] = \Omega_d + \Omega[\phi(\vec{r})]$, which contains the effects of the ordering field $\phi(\vec{r})$ and nothing else; Ω_d is the equilibrium free energy of the disordered phase ($\bar{\phi} = 0$). While we are interested only in the fluctuations $\delta\phi(\vec{r}) = \phi(\vec{r})$ in the disordered phase, all $\phi(\vec{r})$– independent terms of Ω_{tot} are assigned to Ω_d rather than to $\Omega = \Omega_{\mathrm{tot}} - \Omega_d$. Therefore, $\phi(\vec{r})$– independent ("vacuum") terms which often appear as a result of transformations of field $\phi(\vec{r})$ or other treatments of $\Omega[\phi(\vec{r})]$ are omitted with the argument that they belong to Ω_d or, in the field theoretical language, to the "vacuum".

As the generalized free energy functional $\Omega[\phi(\vec{r})]$ is a subject of statistical investigations, it is often called "the effective Hamiltonian", "(quasimacroscopic) fluctuation Hamiltonian", or "Landau–Ginzburg– Wilson Hamiltonian" (for V. L. Ginzburg contribution to the investigation of fluctuation effects, and for K. G. Wilson RG approach); see, e.g., Pfeuty and Toulouse (1975), Ma (1976), Zinn-Justin (1989), and Uzunov (1993). Usually, the notation for $\Omega[\phi(\vec{r})]$ is $H_{\mathrm{eff}}[\phi(\vec{r})]$ or, simply, $H[\phi(\vec{r})]$.

The RG studies of field models are performed in $\vec{k}$-space. Several field theoretical schemes can be used, including high-order loop expansions (J. Zinn-Justin, 1989). For our purposes it is convenient to apply the Wilson–Fisher method of RG recursion relations and ϵ-expansion (Ma, 1976; Uzunov, 1993). This method is most frequently used in the QCP field. Let us mention that the main features of fluctuation critical behaviour are reliably described within the one-loop expansion (the first order in ϵ). In particular this is true for systems with competing effects like the quantum systems, where we must investigate the outcome of the simultaneous effect of quantum and thermal fluctuations. In relatively rare cases, when the one-loop approximation is not sufficient for definite predictions, the higher-loop investigations also do not suffice; see, e.g., Lawrie et al. (1987).

Note, that the present scheme does not include the critical dynamics (Ma, 1976). Within the classical approach static and dynamic critical phenomena are treated separately. One must first know the critical statics and then construct the corresponding dynamic model $H[t, \phi(t,\vec{r})]$ in order to describe the time-dependent fluctuations $\delta\phi(t,\vec{r}) = \phi(t,\vec{r}) - \bar{\phi}$; see, also, Hohenberg and Halperin (1977), and Lifshitz and Pitaevskii (1981).

4.2 Quantum Systems

For interacting bosons we must use the Hamiltonian $H[\hat{\psi}(\vec{r})]$ in terms of second-quantized Bose field operators $\hat{\psi}(\vec{r})$ which, for practical calculations are expanded in terms of annihilation and creation operators mentioned in Section 3.1. Within this formulation, which includes the opportunity to use the powerful Green functions method, the quantum effects are "hidden" in the commutation relations for the field operators $\hat{\psi}(\vec{r})$ and, hence, in the time ordering procedure under the trace for partition function and statistical averages. For finite temperatures, the "time" ordering procedure, i.e. the account of quantum effects, is performed with the help of auxiliary (inverse temperature) "time" τ, $(0 \leq \tau \leq \infty)$. This quantity is referred to as "imaginary time" because of the relation $\tau = it$ with the real time t (these notations correspond to $\hbar = 1$).

Alternatively, the theory can be formulated by the Feynman path integrals (see Casher et al, 1968; Wiegel, 1975; Popov, 1983; Negele and Orland, 1988). Using the coherent state representation, the grand canonical partition function can be written as a functional integral over a (c-complex) field $\phi(x)$ depending on a $(d+1)$-dimensional vector $x = (\tau, \vec{r})$ in restricted "time"-space: $[(0 \leq \tau \leq \infty), \vec{r} \in V]$. So, we may introduce a $(d+1)$-dimensional "volume" $V_{d+1} = (\beta L_1 ... L_d)$, where $\beta (= L_0)$ is the finite size ("thickness") of time-space "hyperslab". Thus the quantum effects bring an extra dimension β along the new (τ-) axis.

The τ-variations of field $\phi(x)$ are created by quantum correlations. These quantum variations, often called quantum fluctuations are given in a net form by $\delta_q\phi(x) = \phi(x) - \phi(0,\vec{r})$. The way in which the quantum fluctuations take part in this picture is very similar to that in which the classical fluctuations appear as spatial variations of field $\phi(x)$. Because of the analytical relation $\tau = it$, these purely quantum fluctuations reveal the intrinsic quantum dynamics of system.

The dependence of physical quantities on real time t is obtained by analytical continuation ($\tau \to it$) of the results from calculations to the real time axis t. The resulting intrinsic quantum dynamics can be substantially changed under the action of external time dependent potential(s). The quantum critical dynamics has been investigated by a number of authors; see, e.g., Abe (1974), Abe and Hikami (1974), Kondor and Szépfalusy (1974), Suzuki and Igarashi (1974), Suzuki (1975), and Hertz

(1976).

Adopting periodic bondary conditions along the τ–axis, we can expand the field components $\phi_\alpha(x)$ in Fourier series

$$\phi_\alpha(x) = \frac{1}{\sqrt{\beta V}} \sum_q e^{iqx} \phi_\alpha(q) , \tag{26}$$

where the $(d+1)$–dimensional vector $q = (\omega_l, \vec{k})$ is given by the Matsubara frequency $\omega_l = 2\pi l k_B T$; $l = (0, \pm 1, ...)$. Setting $\omega_l = 0$ in Eq. (26) we neglect the quantum fluctuations and the amplitudes $[\phi(0,\vec{k})/\sqrt{\beta} \rightarrow \phi(\vec{k})]$ will describe classical fluctuations only. In the x–representation, this corresponds to the limit $\beta \rightarrow 0$, in which the system size ("thickness") along the τ–direction tends to zero. This mechanism yields the analogy with the critical phenomena in classical systems of slab geometry; see. e.g., Lawrie (1978a,b).

The complex Bose field $\phi(x)$ is often referred to as "classical field". This term points to the difference from the field operators, which obey commutation rules. In fact, the classical fluctuation field is $\phi(0, \vec{r}) \equiv \phi(\vec{r})$. A formal generalization of the complex scalar $\phi(x)$ to a complex $(n/2)$–component vector $\phi(x) = \{\phi_\alpha(x)\}$, equivalent to a n–component real field, can always be made in order to compare the results with those for classical systems.

The functional formulation of Bose systems in terms of (c–number) complex functions $\phi(x)$ is performed directly for the original microscopic second–quantized Hamiltonian because the commutation relations of operators $\hat{\psi}(\vec{r})$ have a direct classical limit ($\hbar \rightarrow 0$). The functional formulation of fermionic systems is performed with the help of Grassmann integrals (for this topic and applications of RG to fermionic systems, see Shankar, 1994; Belitz and Kirkpatrick, 1994).

The general approach to the treatment of critical phenomena in fermionic and spin systems is based again on the Hubbard–Stratonovich transformations and Feynman path integration. In contrast to Bose systems, here we should derive the quasi-macroscopic effective Hamiltonian which has the same meaning and role in the theory of quantum critical phenomena as the effective Hamiltonian for classical systems. According to this type of transformations the Fermi and spin operators are transformed to Bose fields of the type considered in this Section (see, e.g., Hertz, 1976; Lawrie, 1978a; Micnas, 1978; Leibler and Orland, 1981; Izyumov and Scryabin, 1988; Kolokolov and Podivilov, 1989; Manousakis, 1991). The final result is a formulation in terms of the Bose field $\phi(x)$ and an effective Bose Hamiltonian which contains additional information about the original symmetries of microscopic system and effective (Landau) parameters related to the microscopic ones. For spin systems one may first perform a direct transformation to Bose fields and then apply the path integral method (see, e.g., Goldhirsch et al., 1979; Lewis, 1979; Goldhirsch, 1980; Chubukov, 1984a,b; Kaganov and Chubukov, 1988a,b; Manousakis, 1991).

As our interest is focussed on standard second order phase transitions, we shall consider ϕ^4–effective Bose Hamiltonians of fermionic and spin systems, as well as same type microscopic Hamiltonians of genuine Bose systems. A quite general quantum effective Hamiltonian ($\mathcal{H} = \beta H$) comprising a number of systems can be written in the form

$$\mathcal{H}[\phi] = \sum_{\alpha,q} G_0^{-1}(q)|\phi_\alpha(q)|^2 + \frac{v}{2\beta V} \sum_{\alpha,\beta;q_1,q_2,q_3} \phi^*_\alpha(q_1)\phi^*_\beta(q_2)\phi_\alpha(q_3)\phi_\beta(q_1+q_2-q_3) . \tag{27}$$

where $v > 0$ is the interaction constant. The (bare) correlation (Green) function

$G_0(q) = \langle|\phi_\alpha(q)|^2\rangle_0$ is given by

$$G_0^{-1}(q) = i\omega_l + \varepsilon(k) \tag{28}$$

for interacting real bosons and transverse XY model (Gerber and Beck, 1977; Goldhirch, 1979). For other effective Bose models of quantum systems, the bare correlation function $G_0(q)$ can be written in the form (Hertz, 1976)

$$G_0^{-1}(q) = \frac{|\omega_l|^m}{k^{m'}} + ck^\sigma + r \tag{29}$$

with positive exponents m, m', $0 < \sigma \leq 2$, and Landau parameters $c > 0$ and $r \sim t(T)$ or $r \sim t(X)$. The parameter c can be always presented as $c = \hbar^2/2m(g)$, where the effective mass $m(g)$ of the composite bosons (Bose excitations) depends on some interaction constant g.

The discussion presented in Section 4.1 for classical systems with respect to the meaning of the effective classical Hamiltonians is valid also for the effective Bose Hamiltonians given by Eqs. (27)–(29); note, that this scheme may include also original microscopic Bose Hamiltonians. The microscopically formulated Bose Hamiltonians contain an upper cutoff $\Lambda = (\pi/a)$ corresponding to the Brillouin zone ends $(-\pi/a) < k_i \leq (\pi/a)$. An example of such Hamiltonians is considered in Section 5.

The effective quasimacroscopic Hamiltonians derived from microscopic Hamiltonians contain large–scale spatial fluctuations of field $\phi(q)$ and, hence, the corresponding cutoff Λ is relatively small: $0 < \Lambda \ll (\pi/a)$. As far as we are discussing critical phenomena ($\xi \gg a, \lambda \gg a$) the precise value of the cutoff Λ is not important. It is however important to have the conditions $\xi > (1/\Lambda)$, and $\lambda > (1/\Lambda)$ satisfied and, then, the relevant phenomena included into consideration. The original or effective Hamiltonians (27) are related with the action $\mathcal{S}[\phi]$ of system by $\mathcal{S}[\phi] = -\mathcal{H}[\phi]$; see, e.g., Popov (1983), Negele and Orland (1988). In the usual terms of statistical physics $\mathcal{S}[\phi]$ is the entropy.

The statistical treatment implies a calculation of the *generalized* grand canonical partition function

$$\mathcal{Z}(T..., \tilde{a}) = \int \mathcal{D}\phi\, e^{-\mathcal{H}[\phi(x)]}\,, \tag{30}$$

which yields the Gibbs free energy

$$\Omega = -\beta \ln \mathcal{Z}(T..., \tilde{a})\,. \tag{31}$$

In Eq. (30), $\int \mathcal{D}\phi$ denotes the functional (infinite dimensional) integration over all allowed field $\phi(x)$ configurations

$$\int \prod_{\alpha=1}^{n/2} \prod_{x \in V_{(d+1)}} d\phi_\alpha^*(x) d\phi_\alpha(x)\,, \tag{32}$$

or, in the $\vec{k}$–space,

$$\int \prod_{\alpha;q} d\phi_\alpha^*(q) d\phi_\alpha(q) \tag{33}$$

over all allowed Fourier amplitudes $\phi_\alpha(q)$. The Jacobians of field transformations change the measure of functional integration but this is without consequences for the treatment of critical phenomena, where we neglect any contributions to the free energy of paraphase (see Section 4.1).

Constraints on the field configurations lead to a critical behaviour change. Within the present functional formulation the constraint of constant density (13) is given by the mean spherical condition

$$\frac{nN}{2} = \sum_{\alpha,q} \langle |\phi_\alpha(q)|^2 \rangle , \tag{34}$$

which means that the mean square of the $Nn/2$ dimensional complex vector $\tilde{\phi} = [\phi_\alpha(q)]$ (with components given by all possible α and q) describes a sphere of a radius $Nn/2$. In the ϕ^4–Hamiltonian of the nonideal Bose gas (NBG) given by Eqs. (27) and (28) with $r = -\mu$, this constraint can be taken into account in the so-called large–n limit ($n \to \infty$); see Busiello and De Cesare (1980a), and Busiello et al. (1983). The alternative approach (Busiello and Uzunov, 1987) is the mapping of original Bose system at constant density to classical models describing compressible magnets (Rudnick et al., 1974; Achiam and Imry, 1975). In both cases one should consider the classical spherical model generalizations (see, e.g., Joyce, 1972) which belong to classes of universality defined by the Fisher phenomenological renormalization scheme for systems with hidden variables (Fisher, 1968).

Let us note, that the effect of constraints like (34) can be taken into account in Eq. (27) by adding an auxiliary ϕ^4–interaction term: $\sim \tilde{u}\phi^4$. In contrast to the u–interaction terms $\phi^*_\alpha(q_1)\phi^*_\beta(q_2)\phi_\alpha(q_3)\phi_\beta(q_1+q_2-q_3)$ in Eq. (34), the auxiliary interaction is given by two summation q–vectors: $\phi^*_\alpha(q_1)\phi^*_\beta(q_2)\phi_\alpha(q_1)\phi_\beta(q_2)$ (Busiello and Uzunov, 1987).

Within an exact statistical treatment, which is not the case for realistic models, the free energy (31) should be the exact thermodynamic potential of system. The exact treatment implies both an exact calculation of functional integral (30) and an exact representation of effective Hamiltonian $\mathcal{H}$. This remark points the important fact that in practical calculations with realistic models, which do not allow an exact treatment, we should distinguish between the generalized (nonequilibrium) Gibbs–Landau free energy $\Omega[\phi] \equiv \beta^{-1}\mathcal{H}[\phi]$, its equilibrium conterpart $\Omega[\bar{\phi}] \equiv \beta^{-1}\mathcal{H}[\bar{\phi}]$, and Ω from Eq. (31). The free energies $\Omega[\bar{\phi}]$ of stable equilibria are lower than more realistic ones, given by Eq. (31), where the fluctiation effects are taken into account.

As an exact treatment of nontrivial microscopic model is impossible we remain with a doubt about the possibility to reach or approach the exact free energy by a method based on effective Hamiltonians which are treated by a minimization (self–consistency condition) and, as a second step, by an account of fluctuations about the incorrect MF equilibria. We are not aware of papers presenting a profound investigation of the important question whether this approach is originally correct, but we are certain that it has received thousands of excellent applications to various problems of physics.

The practical calculations are carried out by a substitution of summation over the wave vector components k_i with an integration as shown by Eq. (16), provided the corresponding dimensions L_i satisfy the criterion $L_i \gg \xi$ for a quasi–infiniteness. The summation over the Matsubara frequencies ω_l can be substituted with an integration

$$\frac{1}{\beta}\sum_{\omega_l} \to \int_{-\infty}^{\infty} \frac{d\omega}{(2\pi)} , \tag{35}$$

provided the temperature T is low enough or, in exact mathematical sense, if $T \to 0$.

In order to determine the condition under which the integration (35) can be used without substantial errors in the final results, let us take the model (27) corresponding to the correlation function $G_0(q)$ given by Eq. (29) with $m' = 0$ and $m = 1$. Then the

ω_l-dependent modes $\phi(\omega_l \neq 0, \vec{k})$ in Eq. (27) will yield relevant contributions to the partition function if they have a relatively high statistical weight. This may happen provided $\omega_1 = 2\pi k_B T < r$; near the critical point $r \sim 0$. In terms of characteristic lengths $\lambda = (4\pi c/k_B T)^{1/\sigma}$, and $\xi = (c/r)^{1/\sigma}$, we have $\lambda > (8\pi^2)^{1/\sigma}\xi$, which is consistent with the criterion (1); here we have used Eq. (2) and $c = \hbar^2/2m$.

If the criterion (1) is satisfied we can substitute the summation over ω_l with an integration as shown by the rule (35) without introducing a substantial error in the calculations. If the criterion (1) is not satisfied the frequencies ω_l can be neglected and the quantum fluctuations ignored.

We have already mentioned that the d–dimentional quantum systems resemble $D = (d+1)$–dimensional classical ones. This is valid for the model (29) with $m' = 0$ and $m = \sigma$. In this case, by the formal substitution $k_B T = c^{1/\sigma}/L_0$ in ω_l, where $L_0 \equiv \lambda = (c^{1/\sigma}/k_B T)$ is the thermal wavelength corresponding to this model, we obtain a $D = (d+1)$–dimensional momentum $q = [q_i; i = 0, 1, ..., d]$ with components $q_i = (2\pi l_i/L_i)$. This representation yields a total formal correspondence between the present time–space geometry and a hyperslab with $(d+1)$ spatial dimensions. Together with this analogy, we have shown, that for this effective model, the notation (2) yields the exponent $\theta = 1$. The temperature dependence $\lambda = \lambda_0/T^{1/\sigma}$, as in IBG, is valid for other effective models, for example, for the model (28) and that given by (29) with $m' = 0$ and $m = 1$.

Thus the form of the scaling law $\lambda = \lambda_0/T^\theta$ depends on the values of exponents m, m' and σ: $\theta = \theta(m, m', \sigma)$. A simple dimensional analysis of Eqs. (28) and (29) shows that $\theta = 1/z$. The same result ($\theta = 1/z$) can be obtained by the RG rescaling transformation which is discussed in Section 4.3. Therefore, the criterion (9) becomes $z\nu_0 < 1$. Usually, these (bare) values of exponents θ and $\nu = 1/\sigma$ of λ and ξ, respectively, receive perturbation corrections, which are calculated with the help of RG.

It has been shown (Hertz, 1976) in the above mentioned way as well by a thorough RG analysis that

$$z = \frac{\sigma + m'}{m} \tag{36}$$

and, hence, that the real CQ dimensional crossover is given by

$$D = d + z . \tag{37}$$

The CQC is always associated with the dimensional crossover (37). That is why CQC is called "dimensional CQC" (Hertz, 1976). Thus we should expect that CQP in d–dimensional (quantum) systems are described by the universality classes (D, n) known for $D = (d+z)$–dimensional classical systems. If the shift $d \to D$ of spatial dimensionality d is the only result of quantum effects, we can say that QCP in the particular d–dimensional quantum system (or a class of systems) obey a form of universality, namely, that they are identical with the critical phenomena in the corresponding D–dimensional classical system (or class of systems). Under the term "corresponding classical system" of quantum model (27) we should understand the system described by the classical variant ($\omega_l = 0$) of Eq. (27).

This definition of the universality of QCP allows a comparison of critical properties of classical and quantum systems. There are other points of view on the definition of the QCP universality. For example, sometimes it is said that CQP in a system are universal if a classical critical behaviour can be found at the same or another spatial dimensionality, which describes the same critical phenomena; see, e.g., Goldhirsch (1979), Busiello and De Cesare (1980b), Kopeć and Kozlowski (1983). This point is important, for example, in the interpretation of special critical behaviour discussed in Section 5.2.

The Bose models (27)–(29) can be used for investigations of intrinsic quantum dynamics at critical temperatures $T_c \geq 0$, and the LT phase transition phenomena, in particular, CQC. Certainly, these studies can be performed within the RG approach. For example, interesting results, which we shall not be able to discuss in the next Sections of this Lecture have been obtained for itinerant magnets (Hertz, 1976; Busiello and De Cesare, 1980a; Millis, 1993; Sachdev et al., 1994; Vojta et al., 1997), structural phase transitions (Oppermann and Thomas, 1975; Holz and Medeiros, 1975; Beck et al., 1975; Morf et al., 1977; Millev and Uzunov, 1982), excitonic phases in semiconductors (Baba et al., 1979), Bose systems (Zannetti, 1980; Creswick and Wiegel, 1982, 1983; Walasek, 1984; Uzunov and Walasek, 1985; Kolomeisky and Straley, 1992), and magnetized bosonic systems (Caramico D'Auria et al., 1996), quantum sine–Gordon model (Busiello et al., 1985b).

4.3 Renormalization Group

From a purely technical point of view the RG method does not present a special interest because its technical tools are standard. It is a method which can be easily used by everyone who has some experience with perturbation expansions. The significance and the success in applications which RG certainly holds are a result of the profound ideas implemented in this approach. One of them is the idea of homogeneity of critical state or, in other words, the invariance of critical properties in a close (asymptotic) vicinity of the critical point with respect to length–scale transformations (scale invariance). The particular way in which these ideas are applied in the constructive RG method for practical studies of given system, can be demonstrated without any calculations.

Let us suppose that the relatively shortlength–range modes $\phi(q)$ with wave numbers in the interval $(\Lambda/b) < k < \Lambda$, defined by $b > 1$, are irrelevant to the critical behaviour and we can neglect them. This supposition is consistent with the concept that the shortlength–scale ($k\xi > 1$) phenomena do not affect the critical properties. Then the rest part of modes $\phi_\alpha(q)$ with $0 \leq k < \Lambda/b$ will be described by the same Hamiltonian but the cutoff will be Λ/b. However, our aim is to show that the truncated Hamiltonian is similar, if not equivalent to the initial one in the same range of length scales from infinity to $1/\Lambda$. This is the only way to verify the initial expectation that the short–range phenomena are irrelevant. In order to restore the cutoff Λ we have no choice but to perform the lengh scale transformation $L_i = bL_i'$, which directly leads to the desired result for the cutoff, and to scaling $\vec{k}' = b\vec{k}$ for the wave vector. Now we should work with the transformed wave vectors $\vec{k}'$. In terms of $\vec{k}'$, the field components take the form $\phi_\alpha(\omega_l, \vec{k}'/b)$.

Within the framework of a quite general scheme, namely, the phenomenological scaling theory, we may assume that the field amplitudes $\phi_\alpha(\omega_l, \vec{k}'/b)$ are homogeneous functions. In fact, this is the content of the whole scaling theory of critical phenomena, which has a firm experimental verification. According to the standard mathematical definition, the homogeneity of field with respect to the wave vector is denoted by $\phi(\omega_l \vec{k}'/b) = b^y \phi(\omega_l, \vec{k}')$, which yields the scale transformation of the Fourier amplitudes. The exponent y is usually substituted by another exponent η: $y = 1 - \eta/2$. In the quantum electrodynamics η is known as "anomalous dimensionality of the field". In the statistical physics the exponent η has been introduced by Fisher (1967) to describe the anomalous scaling properties of two point correlation function in a close vicinity of the critical point.

Further, we transform the volume $V = b^d V'$ and substitute all these transformations in the Hamiltonian. In result we obtain a Hamiltonian in terms of the transformed wave vector $\vec{k}'$, and a new frequency $\omega_l' = b^z\omega_l$ as well as new Landau parameters (c', r', v'). The Landau parameters are related to the original ones by powers of the rescaling factor b, for example, $c' = b^{y_c}c$. However, one of these parameters should be kept invariant with respect to the length–scale transformation. The reason is that the difinition of Landau parameters depends on the definition of field $\phi(q)$, and vice versa. Because of the mathematical form of the Hamiltonian (27), transformations of the type $\phi(q) \rightarrow \Upsilon\phi(q)$, where Υ is some parameter, change also the definition of the Landau parameters. The definition of the field $\phi(q)$ can always be changed so that to fix or eliminate one of the Landau parameters during a particular treatment. Another way to do the same is to put this parameter in the scale of units. For example, the transformation $\sqrt{c}\phi = \tilde{\phi}$ leads to $c = 1$ and a slight redefinition of parameters r and v, as well as frequency ω_l. It is convenient to take into account this gauge in our length–scale transformation by the invariance of parameter c. In the quantum electrodynamics this choice is called "gauge of invariant charges".

This simple transformation outlined above can be depicted in the parameter space (ω_l, c, r, v) as a movement of the Hamiltonian from the initial point (ω_l, c, r, v) to the point (ω_l, c, r', v'). As this always happens on the surface $c =$ const we may exclude the c–axis from our consideration and reduce the parameter space to (ω_l, r, v), or, to the subspace (r, v) for the classical case $\omega_l \equiv 0$; in the remainder of this paragraph we shall discuss the classical case. Varying the rescaling factor b or applying successive transformations with rescaling factors $b, b', b'', ...$, the Hamiltonian moves on a line (RG flow line). The exponent y_r corresponding to the parameter r is always positive, which means that our primitive RG trajectory will run away to infinity. The same behaviour is exhibited by the parameter v' for spatial dimensionalities $d < d_U = 2\sigma$, where d_U denotes the upper critical (borderline) dimensionality. For $d > d_U$, the interaction parameter v' tends to zero. In both cases, the only finite fixed point (FP) of the transformation, i. e. the point in the parameter space (r, v), where the transformation terminates, $(r', v') = (r, v)$, is the point (0,0).

In more complex cases, established by the proper RG transformation, the system exhibits several FPs. The main task in RG theory, after the derivation of RG equations is to obtain their FPs and investigate their stability properties.

We have considered a primitive variant of the semi–group transformation called RG. In this simplest variant the transformation has all group properties but in the more realistic realizations it is a semi–group (there is no inverse element in RG).

For quantum models the initial rescaling procedure has an alternative variant. One may extend the initial rescaling by including the variable τ through $\tau = b^{y_\tau}\tau'$ and, hence, the Matsubara frequency $\omega_l' = b^{-y_\tau}\omega_l$, where the exponent $y_\tau > 0$ should be determined from the requirement for the scale invariance of ω_l–term in the Hamiltonian; the first term in $G_0(q)$ given by Eqs. (28) and (29). This requirement has the same justification as that for the invariance of the parameter c. In result, one obtains $y_\tau = -z$, which means that the scale transformations with and without initial rescaling of the frequency ω_l yield equivalent results (Hertz, 1976). In Section 5.1 we shall use the variant without an initial rescaling of frequency ω_l.

The primitive (zero–order) variant of RG does not imply any integration of modes $\phi(q)$ in the partition function. It is only a scale transformation. For this reason the scale invariance has been analysed but we are unable to establish any new properties or calculate useful quantities.

The actual RG schemes includes the procedure of partition function integration over the short-range modes, which we have neglected in the simple length–scale transformation. The integration cannot be done exactly but one may use the loop expansion to first or higher order approximation. This expansion yields an effective Hamiltonian for the relatively long–range modes $\phi(q)$ with wave numbers in the interval $0 \leq k < \Lambda/b$. Then we should perform the length–scale transformation explained above, in order to restore the initial value Λ of the cutoff.

In this RG variant, which is used in numerous investigations, the new element is that the Landau parameters receive perturbation corrections. The corrections depend on the order of loop expansion to which the calculation has been performed. Within this general scheme the length–scale transformation, discussed above in which the short–range modes are neglected, is equivalent to the so–called "tree approximation" – a term from the "diagrammatic language", denoting that the diagrams within it are "branchy like trees" (see, e.g. the diagram in Fig. 2a) and perturbative insertions are not present. In the tree approximation the short–scale modes are neglected, too. The real application of the RG scheme starts with the one–loop approximation, the lowest order approximation of short–range fields integration. In this nontrivial order of theory, we receive a definite notion about the divergences of the perturbation terms and, hence, about the necessity of a nonstandard approach.

Here we shall use the Wilson–Fisher $\epsilon = (d_U - d)$ expansion, where d_U is the so called upper critical (or borderline) dimensionality below which the perturbation series develop powerwise divergences at $r = 0$. This method can be used up to the two–loop approximation. As main results of the RG transformation of Hamiltonian, one obtains the recursion relations for the Hamiltonian parameters. The differentiation of these relations with respect to the rescaling factor yields a system of first–order differential equations (differential RG equations), which describes the "evolution" of the Hamiltonian parameters, i.e., of the Hamiltonian itself, with respect to successive RG transformations. This evolution has been illustrated by the simple length–scale transformation. In the present case one has to obtain the FPs and to investigate the RG flow lines in the parameter space.

The RG equations are nonlinear and cannot be solved analytically except when they can be linearized. The linearization is made by considering small variations of the parameters around their FP values. This approach is well justified and yields the opportunity for a reliable calculation of critical exponents. The latters are obtained as perturbative corrections to the MF critical behaviour below the upper critical dimensionality d_U. Performing the analysis of the RG equations one obtains all types of critical behaviour allowed by the symmetry of the particular effective Hamiltonian togehter with the associated critical properties.

The main part of this programme can be relaibly accomplished within the one–loop approximation, namely, by the RG analysis to first order in $\epsilon = (d_U - d)$. The great experience from RG numerous applications, in particular, to complex models with competing effects, shows that in the rare cases where the one–loop approximation does not give definite answers about the critical behaviour, the higher orders in the loop expansion are also helpless (Lawrie et al., 1987); for references to RG, see, e.g., Pfeuty and Toulouse (1975), Ma (1976), Uzunov (1993), for other RG schemes, intended to calculations in high orders in the loop expansion, see Zinn–Justin (1989) and the lecture of R. Folk and Yu. Holovatch in this volume.

5. BOSE SYSTEMS

Here we shall consider CQC in the model of nonideal Bose gas (NBG) (Huang, 1987; Lifshitz and Pitaevskii, 1980), which is used in studies of real interacting bosons such as the ^{4}He atoms in liquid helium. The first RG study of IBG by RG was performed by Singh (1975, 1976) within the framework of the operator formalism. Other works within the same formalism are, for example, those of Olinto (1985, 1986), and Weichman et al (1986), intended to explain experiments on ^{4}He in Vycor glass (Crooker et al., 1983) and spin–polarized hydrogen (Reppy, 1984). Weichman et al. (1986) have also developed a phenomenological scaling theory of dilute and dense NBG, for a purpose to describe the crossover from the limiting case of an extreme dilution (IBG) to the dense limit of strongly interacting bosons. Other, relatively early papers on critical properties of real bosons systems are mentioned by Uzunov (1993), and Weichman et al. (1986).

The direct RG applications to the second–quantized NBG Hamiltonian lead to a scaling law for the commutation relations of the field operators $\hat{\psi}(\vec{r})$. In the pioneering papers by Singh (1975, 1976) this was interpreted as a rescaling of boson mass m which grows with the successive RG transformations. The fact that the quantity to which such a rescaling should be ascribed is the thermal length λ or, equivalently, the factor $(1/mT)$ associated with it, has become clear later. However, this circumstance was undoubtedly irrelevant to the main topic considered and correctly solved by Singh. So, he demonstrated for the first time, that the finite temperature ($T_c > 0$) BEC should exhibit a universal critical behaviour corresponding to universality class $(d, 2)$ of classical XY model (see Section (4.2).

The Singh results have been confirmed by Stella and Toigo (1976) and De Cesare (1978), and extensively discussed and applied by Busiello and De Cesare (1980b,c), Uzunov (1981), Olinto (1985, 1986) and, in particular, by Weichman et al. (1986) for both $T_c > 0$ and $T_c = 0$; an error in the Singh calculations of importance for the ZT critical behaviour was pointed out by Uzunov (1981) and noticed also by Weichman et al. (1986).

Here we shall apply RG to the functional formulation of IBG (Section 4.2); see Casher et al. (1968), and Popov (1983). This approach has been used in the RG study of NBG for the first time by Baldo et al. (1976), and then by Wiegel (1978a,b), De Cesare (1978, 1980), Busiello and De Cesare (1980b,c), and Uzunov (1981); other papers with a relevant contribution have been enumerated in the end of Section 4.2. In the present Section we shall briefly discuss the LT critical behaviour of NBG. The relation to other model systems with similar properties and recent research will be also mentioned.

5.1 Renormalization Group Equations

The functional NBG formulation is given by Eqs. (27) and (28), where the parameter $r = |\mu|$ is related to the (bare) unrenormalized chemical potential $\mu \leq 0$. In order to avoid lengthy mathematical formulae we shall set $\sigma = 2$. The generalization of results to $0 < \sigma \leq 2$ is not difficult.

The critical exponents corresponding to the LT critical behaviour of NBG have been obtained for the first time by Busiello and De Cesare (1980c) within the one–loop approximation. Moreover, the critical exponents η and z have been calculated (Busiello and De Cesare, 1980a) within the two–loop approximation, i. e., up to the second order in $\epsilon = (2 - d)$–expansion.

The Wilson–Fisher recursion relations appropriate for a correct and thorough treatment of the critical properties of IBG at LT and ZT critical points have been derived by Uzunov (1981) and in Sections 5.1–5.3 we shall follow his paper. The perturbation series and the diagrammatic representation of perturbation terms are quite standard and we shall not dwell on technical details; for a more instructive explanation of corresponding calculations, see Uzunov (1982, 1993).

Because of the model particular properties, the LT limit yields an extraordinary opportunity for an exact summation in all orders of the loop expansion and, therefore, for obtaining of the LT critical behaviour without approximations (Uzunov, 1981). This LT critical behaviour was later established in the one–loop approximation and applied to dilute Bose systems (Weichman et al. 1986; Fisher and Hohenberg, 1988). Weichman et al. (1986) presented the precise RG recursion relations (Uzunov, 1981) in form of differential equations and obtained solutions by the Rudnick–Nelson (1976) method of intergation. More recently, the simple structure of the perturbation series discussed below, was revealed also in the nonrelativistic scalar field theory and appropriately called "scaling anomaly" (Bergman, 1992). Sachdev et al. (1994) introduced another term – "zero scale–factor universality" – for the same exact solution (Uzunov, 1981). In spite of the wide application of the results presented in Section 5.2, these authors have overlooked the opportunity to take advantage of perturbation series exact summability; see also comments by Caramico D'Auria et al. (1997). The same disadvantage characterizes also the works of Weichman et al. (1986), and Fisher and Hohenberg (1988).

The RG recursion relations within the one–loop approximation (to first order in $\epsilon = d_U - d$) are:

$$k_i' = bk, \qquad \omega_l' = b^{2-\eta}\omega_l, \qquad m' = b^{\eta} m\,, \tag{38}$$

$$r' = b^{2-\eta}\left[r + \frac{1}{2}(n+2)(v/\beta)I_1(r)\right]\,, \tag{39}$$

and

$$(v/\beta)' = b^{4-d-2\eta}\left\{(v/\beta) - \frac{1}{2}(v/\beta)^2\left[(n+6)I_2(r) + 2\tilde{I}_2(r)\right]\right\}\,, \tag{40}$$

with

$$I_1(r) = \beta K_d \int_{\Lambda/b}^{\Lambda} dk\, k^{d-1} n(k)\,, \tag{41}$$

$$I_2(r) = \left(\frac{\beta}{2}\right)^2 K_d \int_{\Lambda/b}^{\Lambda} dk \frac{k^{d-1}}{\mathrm{sh}^2[\beta\varepsilon(k)/2]}\,, \tag{42}$$

and

$$\tilde{I}_2(r) = \frac{\beta}{2} K_d \int_{\Lambda/b}^{\Lambda} dk\, k^{d-1} \frac{\mathrm{cth}[\beta\varepsilon(k)/2]}{\varepsilon(k)}\,, \tag{43}$$

where the integration is performed up to an upper cutoff Λ.

In contrast to other RG studies, the problem for the cutoff Λ has been found to be nontrivial (Uzunov, 1981; see also Weichman et al., 1986, who directly follow the same ideas, and the discussion of the cutoff problem in Section 4.2). The problem arises in the treatment of the dilute ($d > 2$)–dimensional Bose gas in the LT limit ($T_c \to 0$) which corresponds to an extreme dilution $\rho \to 0$; see Eq. (23), which indicates that $\Lambda \sim (\pi/a) \to 0$ for $\rho = a^{-d} \to 0$. It seems at first sight, that in this situation the interparticle distance a cannot be used in the definition of finite cutoff Λ or as a length unit of dimensionless characteristic lengths ξ/a and λ/a. For this reason, the fourth (and last) characteristic length in the problem – the finite scattering length $v \sim (\hbar^2/m)a_{sc}$ (Huang, 1987, Lifshitz and Pitaevskii, 1980) has been discussed as a possible inverse cutoff, $(1/\Lambda) \sim a_{sc}$, by Uzunov (1981) and later applied in RG by Weichman et al. (1986); for $d \neq 3$, $v \sim (\hbar^2/m)a_{sc}^{(d-2)}$.

Let us note however that the RG investigation can be reliably performed by the standard cutoff $\Lambda = \pi/a$, although a tends to infinity for ZT critical behaviour. In fact, according to Eq. (23), $\lambda > a$ for all temperatures in the interval $0 \leq T < A\zeta(d/2)T_c(\rho)$ including the close vicinity of T_c, where $\xi > a$, too; note that $A\zeta(d/2) > 1$. In the temperature range of interest the characteristic lengths ξ and λ are always greater than a and, therefore, there is no danger of shortcomings in the description; see also the brief discussion after Eq. (50) in Section 5.2.

The summation over the frequencies ω_l of the internal lines of perturbation diagrams formed by the legs of one and the same Hamiltonian is performed with the help of the rule

$$\sum_{\omega_l} e^{(+0)\omega_l} G_0(q) , \tag{44}$$

where $(+0)$ denotes a positive number which is set equal to zero after the calculation. The correlation function $G_0(q)$ is given by Eq. (28); c.f. Uzunov (1981)–(1983), where the Green function $G_0(q)$ has an opposite sign.

The summation (44) leads to a difference between the integrand $n(k)$ in $I_1(r)$ and the corresponding integrand in the RG relations presented by Busiello and De Cesare (1980b,c), but coincides with the corresponding result in the papers of Singh (1975), and Stella and Toigo (1976). Moreover, the integrand in $\tilde{I}_2(r)$ differs from the corresponding integrand in the RG equations of Singh (1975), but coincides with that of Busiello and De Cesare (1980c) as mentioned by Uzunov (1981). The mentioned differences disappear in the classical limit $(\varrho < 1, T_c \neq 0)$.

However, in the LT limiting case $(T_c \to 0)$, i.e., the QCP limit, these differences are essential: the pure calculational error in the Singh paper leads to a completely wrong result, whereas the error in neglecting of rule (44) in the papers by Busiello and De Cesare (1980b,c) affects the FP coordinates but the predictions for ZT critical behaviour are correct; see also the discussion by De Cesare (1982) and Caramico D'Auria et al. (1997).

Owing to this formulation of the RG treatment, the relation (38) for ω_l should be referred to the temperature:

$$T' = b^{2-\eta} T . \tag{45}$$

For the same reason, the mass m should be kept invariant. This point has been explained in Section 4.3. We shall proceed with this choice up to the end of this subsection, where we shall propose a more elegant scheme of scaling. The mass $(m-)$ invariance implies $\eta = 0$. This is the usual result within the one-loop approximation. Apart from quite special cases, the exponent value $\eta = 0$ receives ϵ–corrections of order $O(\epsilon^2)$, i. e., in the two– and higher–loop approximations.

The task is to solve RG Eqs. (38)–(40). This means to obtain FPs and investigate their stability. Eq. (45) shows two temperature FP values: $T_C^* = \infty$ (classical) and $T_Q^* = 0$ (quantum). The infinite FP value T_C^* of temperature has nothing common with infinite temperatures. Rather, it describes the classical critical behaviour near finite temperature critical points $T_c > 0$.

Remember that the classical limit of quantum Hamiltonians (27) is taken. in a strict mathematical sense, by decreasing the interval $[0, \beta]$ of variations of Matsubara time τ to zero, which corresponds to the limit $T \to \infty$. Obviously, the FP coordinate $T_Q^* = 0$ should correspond to the LT regime. As FPs lay on the critical surface $[\xi(T_c) = \infty]$ in the Hamiltonian parameter space, it is clear that the zero FP value corresponds to a ZT critical point $(T_c = 0)$. However, it can be shown that, for HT critical points, the parameter T is absolutely redundant together with the relation (45) and the corresponding infinite FP value of temperature (see Section 5.4). The infinite

FP value $T_C^* = \infty$ merely indicates that the ZT critical behaviour is unstable towards temperature fluctuations.

Eq. (38)–(40) can be solved analytically in two limiting cases:

(i) $\beta\varepsilon(k) \ll 1$ (HT behaviour),

(ii) $\beta\varepsilon(k) \gg 1$ (LT behaviour).

These conditions will be discussed in Section 5.3. Here we shall take the leading terms of the integrals (41)–(43) for each of these limiting cases and perform the formal analysis of RG equations.

The limit (i) yields the classical universality as shown by Singh (1975); see also Busiello and De Cesare (1980c), De Cesare (1982), Creswick and Wiegel (1982, 1983), and Weichman et al. (1986). In this case ($\epsilon = 4-d$) the FP value of temperature is $T_C^* = \infty$. This allows small values of factor β in the condition (i). The RG relations have two stable FPs. For dimensionalities $d > 4$, the Gaussian FP with (r,u)–coordinates $r_G = u_G = 0$ is stable and describes the usual MF behaviour; $u = v/\beta$. For dimensionalities $2 < d < 4$, the Heisenberg FP is stable. The coordinates of HFP are given by

$$r_H = -\frac{(n+2)}{2(n+8)}\frac{\hbar^2\Lambda^2}{2m}\epsilon, \qquad u_H = \frac{16\pi^2}{(n+8)}\left(\frac{\hbar^2}{2m}\right)^2\epsilon\,. \tag{46}$$

As the temperature FP coordinate is infinity, the FP value v_H of parameter $v = \beta u$ is equal to zero, because the respective FP value u_H of u is finite, as given by Eq. (46). This behaviour of interaction parameter (v or u) is not strange because it precisely describes the behaviour of the system at HT, where the modes $\phi(q)$ with $\omega_l \neq 0$ can be neglected. Neglecting these quantum fluctuations we obtain a classical Hamiltonian with an interaction constant of the form $u = v/\beta$, which plays the role of interaction constant in the usual classical Hamiltonians ($\mathcal{H} = \beta H$).

However this is not the final answer. As shown in Section 4.2, we must substitute the modes $\phi(0,\vec{k})$ with the true classical modes $\sqrt{\beta}\phi(\vec{k})$, and this yields a factor $\beta \approx (1/k_BT_c)$ in front of parameters in the ϕ^2–part as well as the square of same factor in front of interaction constant. So, in the classical variant of the theory the actual interaction constant is βv. Therefore, the parameter $u = (v/\beta)$ is the interaction in the classical Hamiltonian in terms of field $\phi(0,\vec{k})$, and this parameter transforms to $u = \beta v$ when the classical Hamiltonian is written by the field $\phi(\vec{k}) = \phi(0,\vec{k})/\sqrt{\beta}$.

This discussion can be used to explain the infinite value of FP coordinates of temperature: $T_G^* = T_H^* \equiv T_C^* = \infty$. The field $\phi(0,\vec{k})$ has a rescaling factor b (for $\eta = 0$). The field crossover $[\phi(0,\vec{k}) = \sqrt{\beta}\phi(\vec{k})]$ implies that this rescaling factor should be associated with the proper classical field $\phi(\vec{k})$ rather than with temperature. This is just what happens in the classical limit (i) where the relation (45) does not exist at all. Therefore, within the limiting case (i), the temperature is a redundant parameter and the relation (45) can be suspended from the HT analysis, although it remains in the general RG scheme. Up to the stage for which the investigation is performed in terms of field $\phi(0,\vec{k})$, i.e., in the LT (classical or quantum) regime, we must consider the relation (45) and the FP temperature value $T_Q^* = 0$.

5.2 Low Temperature Behaviour

In the LT limiting case (ii), Eqs. (38)–(40) are solved with the help of $\epsilon = (2-d)$–expansion. This expansion reflects the dimensional CQC which is given by $D = (d+2)$. The upper critical dimensionality d_U is changed from $d_U = 4$ for the classical case (i) to $d_U = 2$ for the quantum case (ii) (Hertz, 1976). Up to now there is no evidence that the case (ii) describes quantum critical phenomena but it is clear that the LT limit can be taken within the RG scheme and that it will bring to a new critical behaviour. Within RG this limit always exists because of the lower cutoff $(\Lambda/b) > 0$, which permits the condition (ii) irrespective of the value of ratio (r/k_BT); see also Eq. (21). In this limit, the integral $I_2(r)$ tends exponentially to zero but the integral $\tilde{I}_2(r)$ is finite and exhibits a power–law infrared divergence for $d < 2$. This yields the borderline value $d_U = 2$. Performing standard calculations, we obtain Eqs. (39) and (40) in a simple form

$$r' = b^2 r\,, \qquad v' = b^{2-d}v(1 - a_0 v)\,, \tag{47}$$

where

$$a_0 \equiv \left[\tilde{I}_2(r)\right]_{\beta\varepsilon(k)\to\infty} = \frac{1}{4\pi}\int_{\Lambda/b}^{\Lambda} \frac{k\,dk}{\varepsilon(k)}\,. \tag{48}$$

The straightforward calculation yields

$$a_0 = \frac{m}{2\pi\hbar^2}\ln b\,. \tag{49}$$

The relevant parameters are T and r. The variations of these parameters near the FP values $T^* = r^* = 0$ drive the system away from the ZT critical state. The inclusion of temperature T as a second relevant parameter corresponds to a real physical situation, namely, that the ZT critical state is approached when both T and r tend to zero. This ZT critical behaviour is described by two stable FPs: the Gaussian FP ($T_G = r_G = v_G = 0$), which is stable for $d > 2$, and the *Gaussian–like* FP (Uzunov, 1981),

$$T_{\rm Gl} = r_{\rm Gl} = 0\,, \qquad v_{\rm Gl} = \frac{2\pi\hbar^2}{m}\epsilon\,. \tag{50}$$

The equation for $v_{\rm Gl}$ shows that the renormalized value of s–wave scattering length $a_{sc} \sim (1/\epsilon)^{1/\epsilon}$; $\epsilon = (2-d)$. The scattering length $a_{sc} \to \infty$ for the important case $\epsilon \to 0$ of $2d$ Bose fluids and, therefore, the cutoff $\Lambda \sim (1/a_{sc})$ applied by Weichman et al. (1986) seems to be inconvenient.

In the remainder of this Section we shall discuss the nontrivial Gaussian–like FP, which will be referred to as GlFP. This FP gives Gaussian values for the main critical exponents: $\eta = 0$, $\nu = 1/2$, $z = 2$. Busiello and De Cesare (1980c) have obtained the Gaussian values of the critical exponents $(\eta, \nu, \gamma, \alpha, z)$ to first order in $\epsilon = (2-d)$ from RG recursion relations, in which the rule (44) has not been taken into account; see also a discussion by De Cesare (1982). Moreover, Busiello and De Cesare (1980b) have shown by a Green function calculation to the two–loop approximation (second order in $\epsilon = 2 - d$), that the critical exponents η and z do not receive ϵ–corrections. On the basis of these results, the crossover of Bose system critical behaviour from the usual HT universality class (d, n) to the ZT universality class $(d+2, n = -2)$ was predicted (Busiello and De Cesare, 1980b). Note, that the Gaussian universality class is sometimes denoted by $n = -2$ (see Pfeuty and Toulouse, 1975; Goldhirsch, 1979, and the discussion in Section 4.2).

In the present case the ϵ–analysis can be extended to any order in ϵ. The problem has an exact solution (Uzunov, 1981) because of the great simplification of the perturbation series in the limit (ii). In this limit, all self–energy perturbation contributions (see Fig. 2b) exponentially tend to zero. The bigger part of perturbation contributions to the interaction vertex v' tend to zero too. The only exception is the ladder series shown in Fig. 2d. This is the so–called superconductivity channel for the interaction vertex, where the Green function lines are oriented in one and the same direction; for

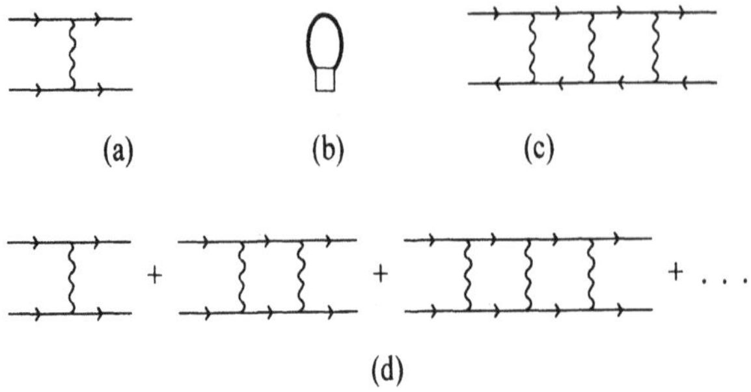

Figure 2. (a) A tree diagram denoting the interaction part of the Bose Hamiltonian, (b) The compact self-energy diagramme which is equal to zero in the LT limit [the thick loop denotes the full (renormalized) correlation function $G(q)$], (c) An example of a diagram from the perturbation series which gives a zero contribution in the LT limit, (d) The infinite ladder series of diagrams which yields the geometric progression (51).

more details, see (Uzunov, 1981–1993). The final result of the summation of the ladder in Fig. 2d is an infinite geometric progression

$$v' = b^{2-d}\frac{v}{1+a_0 v} . \tag{51}$$

This consideration proves that within the grand canonical formalism, the dynamical and static critical exponents of main scaling behaviour described by GIFP have Gaussian values. This is true for the scaling laws written in terms of chemical potential $|\mu| = r$. As the parameter r does not receive a renormalization, it remains equal to that of IBG. The ZT critical point is defined by $[r(T_c)/T_c] = 0$ as is for IBG (Section 3).

The FP value $v_{\rm GI}$ can be obtained from Eq. (51) to any order in $\epsilon = (2-d)$. To do this one should expand the integral (48) and Eq. (51) to the corresponding order in ϵ by having in mind that the FP value $v_{\rm GI}(\epsilon)$ is expanded too. For the sake of convenience in calculations to second order in ϵ, the FP value $v_{\rm GI}(\epsilon)$ is usually written in the form

$$v_{\rm GI}(\epsilon) = \frac{\hbar^2 \epsilon}{mK_d}[1 + \epsilon \ln\Lambda] \tag{52}$$

Expanding K_d in ϵ we have

$$v_{\rm GI}(\epsilon) = \frac{2\pi^2}{m}\epsilon\left\{1 + \epsilon\left[\frac{c_E}{2} + \ln\frac{\Lambda}{\sqrt{4\pi}}\right]\right\} , \tag{53}$$

(c_E is the Euler constant). The FP coordinate v_{Gl} receives corrections in all orders of ϵ expansion. Within the one–loop approximation, the stability exponent y_v associated with the interaction parameter v has a negative value $y_v = -\epsilon = (d-2)$ for $0 < d < 2$. This exponent does not receive higher order ϵ corrections and, therefore, the value $(d-2)$ is exact.

This completes the proof (Uzunov, 1981). The above results has been confirmed by Monte Carlo calculations for one–dimensional interacting bosons (Batrouni et al., 1990).

We should emphasize that GIFP is a non–Gaussian FP. It is a product of the thermal and quantum fluctuation interactions represented by the interaction parameter v. In fact, GIFP is conjugate to actual GFP ($v_G = 0$) within the same RG analysis. Moreover, the total critical behaviour includes corrections to the main scaling laws which are described by GIFP. The interpretation of results within the classical universality class $(d+2, n = -2)$ has no heuristic significance. GIFP constitutes a new universality class of ZT QCP which has numerous applications (see Sections 5.3, 5.5, and 7.1).

The theoretical significance of results presented in this Section is in establishing an *exact* solution for a *nontrivial* model of real systems. The exact solution is possible as both the classical and the quantum bosons fluctuations vanish at $T = 0$. Then the lowering of temperature to zero merely produces a gradual decrease of fluctuation effects.

5.3 Discussion of Limiting Cases

The limiting cases (i) and (ii) are analysed under the general condition $(\xi/\Lambda) \gg 1$. For a microscopic model, $\Lambda \sim (\pi/a)$ and, hence, we have $(\xi/a) \gg 1$. Then the inequality (i) yields $\lambda \ll a$, i.e., the HT condition.[‡]

The HT condition is certainly satisfied near the infinite FP value $T_C^* = \infty$ and well below it. The condition (ii) is valid either in the classical LT critical region

$$a \ll \lambda < \xi , \tag{54}$$

or in quantum one, given by

$$a \ll \xi < \lambda ; \tag{55}$$

c.f. the general criterion (1). The inequalities (54) and (55) can be written in terms of parameters T and r which enter in the RG relations. The condition (54) will be valid near and at LT FPs given by the zero values of T and r provided in the ZT limit $T \to 0$ we have $(r/k_BT) \to 0$. In the same limit, the quantum condition (55) will be valid if $(r/k_BT) \to \infty$. In particular, to answer the question which of these two types of LT critical behaviour is described by GIFP (50), we must know the value of ratio $(r_{\rm GI}/T_{\rm GI})$. If the latter is greater than unity, GIFP will describe quantum critical phenomena but if it is less than unity, the low dimensional $(d \leq 2)$ LT critical behaviour will be classical.

A standard example is given by interacting bosons at constant density (the case of noninteracting particles has been discussed in Section 3.3). As the perturbation series does not yield self-energy contributions at all, the parameter r for ideal and interacting Bose systems is the same. The value of correlation length critical exponent ν, cited in Table 1, is greater than the exponent $1/\sigma$ of thermal length λ. So in the ZT limit the ratio $(r/k_BT) \sim (\lambda/\xi)^2$ will tend to zero; c.f. Eq. (21). Therefore, the LT critical behaviour described by GIFP will be classical. This is true for T–driven transitions (Section 2.3). The ρ–driven transitions will exhibit QCP. These results are consistent with the general criterion (9).

In the same way one can show that the critical effects near the phase transition points of interacting bosons under the condition of a very low constant pressure will be influenced by quantum effects (see also Section 3.2). The application of the results from Section 5.2 to dilute Bose systems in the low density limit $\rho \to 0$ will be briefly discussed in Section 5.5.

The NBG properties at finite LT are very similar to those of IBG at constant density (Section 3.3). It is then convenient to investigate the corrections to the (spherical) Hartree limit. For this aim RG studies in the large–n limit have been performed (Busiello and De Cesare, 1980a; De Cesare, 1982). The alternative approach (Busiello and Uzunov, 1987) has been mentioned in Section 4.2.

[‡]The values $\lambda \ll a$ have a statistical meaning in the continuum limit $(V/a^d) \to \infty$.

5.4 Formulation by Thermal Wavelength

An alternative RG treatment can be performed by the transformation

$$\sqrt{\hbar^2/2m}\phi(q) \to \phi(q) \tag{56}$$

in the Bose Hamiltonian (27)–(28). In the new notations for theory parameters the mass m is absent and the Matsubara frequency ω_l is substituted by $8\pi^2 l/\lambda^2$, i.e., by $(1/\lambda^2)$. The parameters r and v are multiplied by factors $(2m/\hbar^2)$ and $(2m/\hbar^2)^2$, respectively. The important point is that the recursion relation (38) for mass m and the recursion relation (45) for temperature are now substituted by

$$1 = b^\eta \,, \tag{57}$$

and

$$\lambda' = b^{-1}\lambda \,, \tag{58}$$

respectively. Eq. (58) can be written in the form

$$m'T' = b^2 mT \,. \tag{59}$$

The comparison of Eq. (38) for m and Eq. (45) for T with Eqs. (57) and (59) shows that this second formulation is not completely equivalent to the usual one although the formal results for the critical behaviour are equivalent. Within the present formulation Eqs. (57) and (59) do not imply mass invariance. Rather the mass enters as a factor in the parameters λ, r and v and, hence, participates in the renormalization scheme but in a way different from that suggested by Singh (1975). From Eq. (59) one can conclude nothing about the individual behaviour of T and m towards the RG transformation. Rather, the correct conclusion is that the thermal wavelength λ is renormalized, as given by Eq. (58), and that RG transformation drives this quantity to zero unless it is at the HT FP value $\lambda_C^* = 0$; another FP value of λ is $\lambda_Q^* = \infty$.

Note that the renormalized thermal wavelength $\lambda = \lambda(b)$ does not belong to the parametric space (r, v) of the Hamiltonian. Hence, the temperature T is also excluded from this space. It is convenient to consider these parameters as describing HLTC. The renormalized wavelength $\lambda(b)$ drives the system from the HT ($\lambda \ll a$) to ZT ($\lambda \gg a$) Gaussian-like critical behaviour; see Eq. (58). The rescaling factor cannot be taken as such crossover parameter because of the RG restriction $\epsilon \ln b < 1$. So, the RG flows from usual Heisenberg FP, where $\lambda_C^* = 0$, to GlFP, where $\lambda_Q^* = \infty$ are produced by successive RG transformations. If the temperature is equal to zero the system will be exactly at GlFP and the FP Hamiltonian is invariant towards RG. The system remains in this ground state of total BEC ($\rho_0 = \rho$). For any $T > 0$, the RG flow will drive the system to the usual classical behaviour. Now it is not difficult to rederive all RG results in the present formulation by avoiding the unnecessary mass-invariance condition and the unnatural HT FP value $T_C^* = \infty$.

5.5 Related Topics and Applications

The above results for interacting real bosons are straightforwardly extended to systems described by the quantum XY model in a transverse field. Using RG calculations to the one-loop order, Gerber and Beck (1977) and Goldhirsch (1978) have drawn the attention to the possibility for a Gaussian ZT critical behaviour of this model. In contrast to spin systems with different symmetry, for example, the transverse Ising model (Section 6), in XY systems above the critical transverse field the spins at $T = 0$

have maximal projections on the field axis and there are no quantum spin fluctiations at all. Below the critical field, in the ferromagnetic phase, such quantum fluctuations exist; for the ground state properties of spin systems see, e. g. Kaganov and Chubukov (1988). Gerber ans Beck (1977) have ignored the real symmetry of XY systems and obtained incorrect RG equations. The correct RG equations are given by (38)–(43).

These symmetry properties are related to the particular form of bare correlation function $G_0(q)$ which has been obtained by Goldhirsch (1979), and Goldhirsch et al. (1979) exactly of the form (28). The main interaction term in the effective Hamiltonian of the XY model is of the form given by Eq. (27), but there are also additional interaction terms of type $|\phi|^{2m}$; $m > 2$. It has been shown (Chubukov, 1984b), that the additional interaction terms do not change the critical behaviour predicted with the help of the ϕ^4–Hamiltonian (27)–(28) and, therefore, the RG results from Sections 5.1–5.4, can be straightforwardly applied to XY systems; see, e.g., Kopeć and Kozlowski (1983).

We should emphasize, that the transverse field in the XY model plays the role of the auxiliary parameter X, introduced in Section 2. The X-transitions considered in Sections 2.3 are easily performed in XY systems by variations of transverse field around its critical value at fixed LT. Similar field variations in another class of systems are discussed in Section 6. The measurements should indicate the Gaussain-like critical behaviour. Therefore, according to our considerations in Section 2 as well as the concrete studies of nonuniversal characteristics of XY systems (Chubukov, 1984a), QCP should be observed in LT critical experiments. The arguments presented so far, make us think that the LT phase transitions in XY systems is a promising area of experimental studies of Gaussian-like critical behaviour.

The theoretical predictions about the LT critical behaviour presented in this Section may also be helpful in the interpretation of experiments on three dimensional ($3d$) and two dimensional ($2d$) systems of ^{4}He in porous media, such as Vycor and xerogel glasses (Chan et al., 1988; Finotello et al., 1987), superfluid ^{3}He aerogels (Matsumoto et al., 1997), and in spin-polarized hydrogen; see also experimental papers cited in the beginning of Section 5. In fact, the same results have been a subject of further developments, rederivations, justifications, and applications to experimental and theoretical studies of real superfluids; see, e. g., Weichman et al. (1986), Fisher and Hohenberg (1988), Fisher and Fisher (1988), and Fisher et al. (1989). As mentioned in Section 5.1, Sachdev et al. (1994) also applied the results from Section 5.2 to low dimensional quantum antiferromagnets.

Another interesting application of the results from Section 5.2 is quantum Hall liquids. For example, in the limiting case of zero statistical gauge (Chern–Simons) field, the generalized effective Hamiltonian describing the fractional quantized Hall effect is similar to the model (27)–(28) and yields the critical behaviour presented in Section 5.2; see Bergman (1992), and Schakel (1995).

6. TRANSVERSE ISING MODEL

In this Section we shall discuss the LT critical properties of the quantum transverse Ising model (TIM) with the help of MF, Ginzburg criterion for the validity of MF, and RG. The results from MF and Ginzburg criterion confirm the general picture of HLTC outlined in Section 2. RG is discussed in order to demonstrate the difference between the classical and the quantum fluctuations (see also De Cesare et al., 1998a).

The TIM describes essential features of phase transitions in ferromagnets with a strong uniaxial anisotropy and displacive phase transitions in certain types of quantum ferroelectrics (Rechester, 1971; Beck et al., 1975; Schneider et al., 1976; Morf et al., 1977; A. Bruce, 1980); for experiments, see, e.g., Rytz et al. (1980), and Samara et al. (1981); for other applications, see Stinchcombe (1973), and Chakrabatri et al. (1996).

Here we shall mention the MF results of De Gennes (1963), Brout et al. (1966), and Tokunaga and Matsubara (1966), the cumulant expansion of Stinchcombe (1973), the RG investigations of Young (1975) and Hertz (1976), the description of CQC in Hartree limit (Lawrie, 1978a), and the formal interrelationship between the finite–size crossover in systems of slab geometry and CQC in TIM (Lawrie, 1978b); see also Section 7.2.

TIM is given by the Hamiltonian (De Gennes, 1963)

$$H = -\frac{1}{2}\sum_{ij} J_{ij} S_i^z S_j^z - \Gamma \sum_i S_i^x \,, \tag{60}$$

where S_γ, $\gamma = (x, y, z)$, are the components of spins taken to have a magnitude $S = 1$, J_{ij} is the exchange interaction and Γ is the transverse field magnitude. We shall assume that TIM is defined on a d–dimensional regular lattice with a lattice spacing $a = 1$. The assumption for nearest–neighbour (nn) interactions, which we shall use to define the parameters of the effective Hamiltonian, does not restrict the generality of results in Section 6.

The effective field TIM Hamiltonian has been derived by Young (1975); see also Lawrie (1978a). The quadratic (ϕ^2–) part of this Hamiltonian ($\mathcal{H} = H/T$; $k_B = 1$) is given in the form

$$\mathcal{H}_2[\phi] = \frac{1}{2}\sum_{\alpha,q} G_0^{-1}(q)|\phi_\alpha(q)|^2 \,, \tag{61}$$

where $\phi(q)$ is a real scalar field with a (bare) correlation function of the form

$$G_0^{-1}(q) = |\omega_l|^2 + k^2 + t_0 \tag{62}$$

with

$$t_0 = \left[1 - \frac{J}{\Gamma}\text{th}\left(\frac{\Gamma}{T}\right)\right] \,. \tag{63}$$

In Eq. (63), $J \equiv J(0) = 2dJ_0$ is a product of the number $z = 2d$ of nearest neighbour nn spins and the constant J_0 of the single $(i - j)$ nn interaction ($J_0 = J_{ij}$ for nn sites i and j). The ϕ^4–interaction is given by the corresponding term in Eq. (27), provided the parameter $(v/2\beta)$ is substituted with the interaction parameter

$$u_0 = \frac{J^2T}{8\Gamma^3}\left[\text{th}\left(\frac{\Gamma}{T}\right) - \frac{\Gamma}{T} + \frac{\Gamma}{T}\text{th}^2\left(\frac{\Gamma}{T}\right)\right] \,. \tag{64}$$

It is convenient to use units in which all quantities in the effective Hamiltonian $\mathcal{H}$ are dimensionless. Dimensionless wave components k_i are given by $k_i = 2\pi\tau l_i/L_i$, where

$$\tau = \sqrt{\frac{J\text{th}(\Gamma/T)}{2d\Gamma}} \,. \tag{65}$$

In the critical region $t_0 < 1$, $\tau \sim 1$. The upper cutoff for the wave numbers k is $\Lambda = \gamma\pi\tau$, where γ is a small number ($\gamma \ll 1$); see Uzunov (1996).

The characteristic lengths ξ and λ are given by $\xi = 1\sqrt{|t_0|}$ and $\lambda = (\Gamma/T)$. We shall consider only the LT range of temperatures defined by $\lambda \gg 1$. In this case, $u_0 = (J^2T/8\Gamma^3)$.

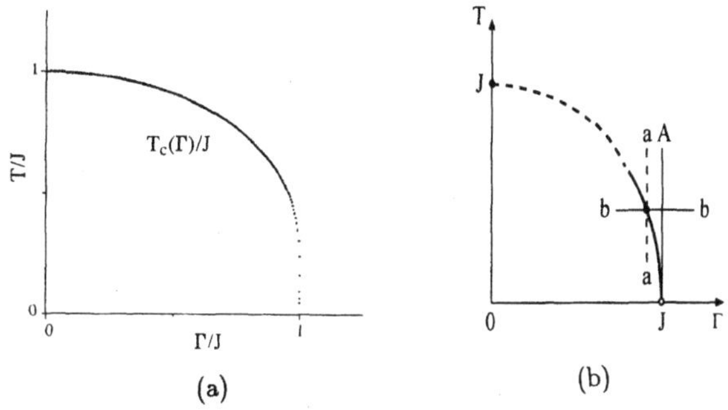

Figure 3. (a) The graphical representation of Eq. (67) for $J = 1$. (b) The LT part (the notations are explained in the text).

6.1 Mean Field

Within the standard MF approximation we shall consider only the uniform mode $\phi(0)$ of the filed ϕ. It will be more convenient to use the MF free energy $\Omega = T\mathcal{H}[\phi(0)]$ instead of the dimensionless MF free energy $\mathcal{H}[\phi(0)]$. This choice, after a change of the nonequilibrium order parameter from $\phi(0)$ to $\phi_0 = (T/V)^{1/2}\phi(0)$, will produce an extra factor $1/T$ in front of ϕ_0^4-term of free energy Ω, which makes possible to avoid difficulties in our further analysis connected with the definition of the order parameter ϕ_0 at $T = 0$. The problem is that the parameter u_0 is proportional to T, see Eq. (64), and the investigation with the original parameter $\phi(0)$ will give for the equilibrium order parameter $\phi(0) \sim (1/T)^{1/2}$ which is divergent for $T \to 0$. This is an example of order parameter HLTC (Section 2.5).

The Gibbs thermodynamic potential in the form

$$\Omega = V\left[\frac{t_0}{2}\phi_0^2 + \frac{u_0}{T}\phi_0^4\right] \tag{66}$$

allows a correct MF analysis at low temperatures. The analysis of the free energy (66) can be straightforwardly performed and we shall not enter in details. The critical line (Fig. 3a) is defined by $t_0(T_c, \Gamma) = 0$, i.e.,

$$T_c(\Gamma) = \frac{\Gamma}{\mathrm{Arth}(\Gamma/J)} \tag{67}$$

or, equivalently, by $t_0(T, \Gamma_c) = 0$, which yields

$$\frac{\Gamma_c}{J} = \mathrm{th}\left(\frac{\Gamma_c}{T}\right) . \tag{68}$$

Despite of the simple form of Eq. (66) the MF critical properties of TIM cannot be investigated analytically for the whole curve $T_c(\Gamma)$ because of the quite complex dependence of parameters t_0 and u_0 on T and Γ. We shall consider the LT critical behaviour in the vicinity ($|\Gamma - J| < J$) of ZT critical point $[T_c(J) = 0, \Gamma_c(0) = J]$ and the neighbour critical points with coordinates $T_c \ll J$ and $\Gamma_c \sim J$. In this LT region we can distinguish four types of phase transitions along the lines: *0J*, *AJ*, *aa*, and *bb*; see Fig. 3b. We shall suppose that the couples of parallel lines are very near to each other. Following the terms introduced in Section 2, the transitions along the lines AJ and aa are T-transitions whereas the transitions along the lines $0J$ and bb can be thought of as Γ-transitions.

In MF, we shall consider the critical exponents β, γ, and ν, of the order parameter $|\phi_0| = (T|t_0|/4u_0)^{1/2}$, the susceptibility $\chi = 1/|\bar{t}_0|$, and the correlation length $\xi = 1/|\bar{t}_0|^{1/2}$, respectively, where $\bar{t}_0 = t_0$ for $t_0 > 0$, and $\bar{t}_0 = 2t_0$ for $t_0 < 0$. Obviously, the behaviour of these quantities with respect to variations of T and Γ depends on that of t_0. Note, that parameter t_0 as given by Eq. (63) does not change in the Gaussian approximation for the quantum and classical fluctuations. This makes possible to extend the consideration of the parameter t_0 in the paraphase ($\phi_0 = 0$), where the results should be interpreted as pure fluctuation effects.

The summary of results reads:

Line 0J: We obtain $t_0 \approx (\Gamma - J)/\Gamma$, $\phi_0 \sim |t_0|^{1/2}$, $\chi \sim \Gamma/|\Gamma - J|$ and, therefore, $\gamma = 1$ and $\beta = 1/2$. There exists a full coincidence between the standard MF behaviour with respect to variations of T around J at $\Gamma = 0$ and the present critical behaviour due to variations of Γ around J at $T = 0$. This correspondence can be written as $T \leftrightarrow \Gamma$.

Line bb: The parameter t_0 is

$$t_0 \approx \frac{J(\Gamma - \Gamma_c)}{\Gamma_c^2}(1 - \frac{4\Gamma_c}{T}e^{-2\Gamma_c/T} + ...) . \tag{69}$$

Here t_0 tends to the value $(\Gamma - J)/\Gamma$ when $\Gamma_c(T) \to J$ for $T \to 0$. The critical exponents are the same as those along the line *0J*. The exponential correction in Eq. (69) can be neglected.

Line AJ: There is no equilibrium ordering along this line but there is a criticality in the paraphase. The susceptibility above ZT critical ($\Gamma_c = J$) is

$$t_0 = 2e^{-2J/T} . \tag{70}$$

This exponential behaviour corresponds to the critical exponents $\gamma = \nu = \infty$.

Line aa: Along this line

$$t_0 \approx \frac{4J|T - T_c|}{T_c^2}e^{-2\Gamma/T_c} \tag{71}$$

and, therefore, for both $T < T_c$ and $T > T_c$, the susceptibility $\chi \sim 1/|t_0|$ obeys the scaling law $\chi = \chi_0/|T - T_c|$ with an exponentially increasing scaling amplitude χ_0. There is a smooth crossover between the pure exponential behaviour (70) and the power law (71). As $(u_0/T) \approx J^2/8\Gamma^3$ for $\Gamma \sim J \gg T$, we have $|\phi_0| \sim |t_0|^{1/2}$, where t_0 is given by Eq. (71). The order parameter exponent β has the classical value $\beta = 1/2$ but the scaling amplitude $[\sim T_c^{-1}\exp(-\Gamma/T_c)]$ of the order parameter ϕ_0 exponentially decreases for $T_c \to 0$; for a comparison of these LT MF properties with HT ones, see De Cesare et al. (1998a).

6.2 Lowest Order Perturbation Theory

The (T, Γ) domains of validity of the MF results in Section 6.1 can be investigated by the Ginzburg criterion; see, e.g., Uzunov (1993). Here we shall use a standard derivation of this criterion from the first-order perturbation contribution to the "self-energy" t_0:

$$\tilde{t}_0 = t_0 + 12u_0A_1(t_0) \tag{72}$$

with

$$A_1(t_0) = \int \frac{d^dk}{(2\pi)^d}S_1(t_0, k) , \tag{73}$$

where

$$S_1(t_0, k) = \frac{\Gamma\ \text{cth}[(\Gamma/T)\sqrt{k^2 + t_0}]}{T\ \sqrt{k^2 + t_0}} \ . \tag{74}$$

Neglecting a term of order u_0^2 we can substitute t_0 in the intergal $A_1(t_0)$ with $\tilde{t_0}$. The Ginzburg criterion for the validity of the MF results is given in the general form

$$|\tilde{t_0}| > 12u_0[A_1(0) - A_1(\tilde{t_0})] \ . \tag{75}$$

The same criterion defines the validity of Gaussian approximation of noninteracting fluctuations above and below the critical point. The general criterion (75) is valid for both the paramagnetic ($\tilde{t_0} > 0$) and the ferromagnetic ($\tilde{t_0} < 0$) phases; in the ferromagnetic phase a factor 1/2 should be added to the r.h.s. of (75).

Obviously the results from this first-order perturbation approximation depend on the properties of the integral (73). We shall write this integral in the form

$$A_1(\lambda, \xi) = K_d \lambda^{2-d} \int_0^{\lambda\Lambda} dz z^{d-1} \frac{\text{cth}\sqrt{z^2 + \varrho^2})}{\sqrt{z^2 + \varrho^2}} \ , \tag{76}$$

where $z = \lambda k$, and ϱ is given by Eq. (1).

For HT ($\lambda < 1$, $\varrho \ll 1$), the main contributions in the integral are given by the small wave numbers $z = \lambda k \ll 1$, and the approximation $\text{cth}y \sim 1/y$ yields the standard perturbation integral A_1 known from the classical theory (the Curie–Weiss limit). One can rederive the classical theory ($\omega_l = 0$) by using the HT value of the interaction parameter: $u_0 = (J^2/12T^2)$; see Eq. (64) for $\Gamma \ll T$.

In the LT range of temperatures ($\lambda \gg 1, \lambda\Lambda \sim 1$), we can distinguish between the quantum limit $\varrho \gg 1$ and the classical limit $\varrho \ll 1$. In the quantum limit ($\varrho \gg 1$) the *coth* can be approximated with unity by neglecting exponentially small correction terms. They are very similar to those given by Eq. (69)–(71). These corrections enter in the renormalized parameter $\tilde{t_0}$ and give small ($\sim u_0$) corrections to the coefficients of corresponding exponential terms in t_0; see Eqs. (70) and (71).

Let us denote the integral which is obtained from A_1 for $\text{cth}y \sim 1$ by A_{01}. The difference $u_0[A_1 - A_{01}]$ has been estimated (Rechester, 1971) to be of order $\lambda^{-2} \sim T^2$ for $3d$–systems; note that in the LT limit, $u_0 \sim T$. This type of temperature corrections to the pure quantum limit has been widely used in interpretations of experimental results for quantum ferroelectrics (Rytz et al, 1980; Samara, 1981). The same corrections were also derived by Schneider et al (1976), and Morf et al (1977). Oppermann and Thomas (1975) have obtained the temperature corrections from a model of structural phase transitions corresponding to the Hartree limit.

Taking into account the temperature corrections to the ZT limiting case is equivalent to an estimation of difference arising from the substitution of summation over the Matsubara frequencies with an integration according to the rule (35). It is then clear that such corrections will come from the counterterm series in powers of ϱ^{-2} in the Euler–Maclaurin summation formula. Therefore these corrections could not be given by d–dependent powers in T as is in the Hartree limit considered by Oppermann and Thomas (1975).

The temperature corrections are important for the phase transition properties along lines like AJ and aa, where $\Gamma \sim J$. Within the present lowest order perturbation theory the corrections in powers of T should be considered small. It has been however proven by the "parquet" summation (Rechester, 1971) that they are quite big and essentially influence the critical behaviour in the classical LT region ($\varrho < 1, \lambda \gg 1$).

Now we shall consider the Ginzburg criterion (75). In the quantum limit we must substitute the integral A_1 with A_{01}. As a result of CQC, the integral A_{01} yields the upper and lower borderline dimensionalities: $d_U = 3$ and $d_L = 1$. In order to simplify the calculations we shall consider the case $d = 2$ which shows the main features of the quantum critical behaviour for all dimensionalities $1 < d < 3$.

The straightforward calculation yields the criterion (75) in a simple form:

$$|t_0| > \left(\frac{3}{4\pi}\right)^2 \left(\frac{J}{\Gamma}\right)^4 . \qquad (77)$$

This criterion cannot be applied to the T-transitions along the lines AJ and aa, where the behaviour is classical ($\lambda < \xi$). The reason is that the strong quantum condition ($\varrho \gg 1$) has been used in the derivation of inequality (77).

Along the line bb we obtain the criterion

$$|\Gamma - \Gamma_c| > 0.06\Gamma_c \left(\frac{J}{\Gamma_c}\right)^3 \sim 10^{-2}\Gamma_c . \qquad (78)$$

If we set in (78) $\Gamma_c = J$ we shall find a criterion along the line $0J$. Because of the very sharp slope of transition curve $T_c(\Gamma)$ near $\Gamma = J$, the quantum criterions along the lines bb and $0J$ are practically the same. The Ginzburg critical region ($\sim 10^{-2}\Gamma_c$) given by (78) is well established. It enlarges at $T > 0$ by T-correction terms.

The quantum condition (1) along the line bb can be written in the simple form $J(|\Gamma - \Gamma_c|) \gg T^2$ provided $(\Gamma/\Gamma_c)^2 \approx 1$. As $J \gg T$, this condition becomes

$$|\Gamma - \Gamma_c| > T , \qquad (79)$$

which is obviously consistent with (78).

The condition (92) shows that the quantum region approaches the critical point Γ_c when T decreases to zero. The whole surrounding of ZT critical point ($\Gamma_c = J$) is influenced by quantum effects which produce the quantum Ginzburg region. A more detailed analysis including the onset of thermal fluctuations can be obtained by the exact calculation of the integral A_1 at $d = 2$:

$$A_1 = \frac{1}{2\pi} \ln \frac{\mathrm{sh}\sqrt{\lambda\Lambda + \varrho^2}}{\mathrm{sh}\varrho} . \qquad (80)$$

The approximation $\mathrm{sh}y \sim \exp(y)/2$ corresponds to $A_1 \approx A_{01}$. The thermal corrections are obtained in powers of $(\varrho^2/\lambda\Lambda)$ which, together with the factor $u_0 \sim T$ yields the lowest order correction term to $u_0 A_{01}$ of the type $T^2|t_0|$.

For any fixed $|t_0| > 0$, the T-transitions will exhibit critical exponents $\gamma = 2\nu = 2$ instead of classical exponents $\gamma = 2\nu = 1$. As we see these exponents do not depend on the dimensionality d as is in the Hartree limit (Oppermann and Thomas, 1975). For $t_0 = 0$, i.e., on the line AJ, the integral A_1 has a logarithmic divergence which is an evidence of strong thermal fluctuations.

HLTC of the T-transitions from classical critical exponents at HT to LT critical exponents $\gamma = 2\nu_0 = 2$ shows that $z\nu_0 = 1$. These bare values of z and ν_0 have ϵ corrections from RG and we should have in mind that the corrections to z are of order $O(\epsilon^2)$ whereas that of ν are of first order in ϵ and, hence, the criterion (9) is not satisfied.

The asymptotic critical behaviour corresponding to the $T-$transitions will not exhibit CQP but rather a LT classical behaviour which is different from HT one. The only quantum effect is related with CQC at ZT $\Gamma-$transitions (Section 6.3).

6.3 Renormalization Group Arguments

The RG recursion relation for the interaction parameter u_0 in the one–loop approximation will be

$$u_0' = b^{4-d} u_0 [1 - 36 u_0 A_2(0, b)] , \tag{81}$$

where

$$A_2(t_0, b) = -\frac{\partial A_1(t_0, b)}{\partial t_0} . \tag{82}$$

In Eq. (82) $A_1(t_0, b)$ is the integral (73) with lower ($0 < b^{-1} < 1$) and upper ($\Lambda = 1$) cutoffs of the wave number k.

The other recursion relations are

$$t_0' = b^{4-d} [t_0 + 12 u_0 A_1(t_0, b)] , \tag{83}$$

and

$$\lambda' = b^{-z} \lambda , \tag{84}$$

where the dynamical critical exponent z is equal to unity in this order of the theory.

In the HT region ($\lambda < a$) the integrals A_1 and A_2 can be substituted with the classical integrals by setting $\text{cth}(y) \sim 1/y$. In this case the integral $A_2(0, b)$ has a logarithmic infrared divergence at the upper critical dimensionality $d_U = 4$. For $d = 4$, $A_2(0, b) = K_d \ln b$.

Further, using the standard RG analysis (Hertz, 1976) one reveals the usual universality class of critical behaviour of classical Ising model ($\Gamma = 0$). Besides, there is a possibility to perform a calculation of dynamical critical exponent z and, hence, to reveal the quantum dynamics of TIM (Millev and Uzunov, 1983). The ϵ corrections to z are calculated from the q–dependent self–energy diagrams in two– and higher–loop approximations. These corrections are small compared with the ϵ–corrections to the static exponents, for example, ν and γ.

The LT limit is treated as shown in Section 6.2. In this case we should make the approximation $\text{cth}y \sim 1$ in Eqs. (81)–(83). Thus we recover the dimensional CQC $d \to (d+1) = D$. The analysis in $\epsilon = (3 - d)$ yields results for the quantum critical behaviour corresponding to the nontrivial Ising universality class $(D, 1)$; Hertz (1976). For all dimensionalities $d \geq 3$, the QC behaviour will be described by the classical MF universality class, while for dimensionalities $1 < d < 3$ the QC behaviour will be nontrivial. This result is straightforwardly generalized for a n–component real field with a bare correlation function of the form (62).

Thus we obtain that in TIM CQC satisfies the universality property discussed in Section 4.2. An important point in the mechanism of this crossover is that the interaction parapeter u_0 in the recursion relation (81) is changed to the parameter $v_0 = (u_0/\lambda)$:

$$v_0' = b^{3-d} v_0 [1 - 36 v_0 K_3 \ln b] . \tag{85}$$

The quantum effects play the crucial role for CQC in TIM. If the RG equations are treated in the classical scheme, as shown by Beck et al (1975) and Morf et al. (1977), the transformation of Eq. (81) for u_0 to that for v_0 cannot be performed in the way shown by Eq. (85). An extra–factor λ^{-1} that remains in the second term of Eq. (85) leads to the prediction for a Gaussian critical behaviour at LT for dimensionalities $d > 2$.

Another important feature of the QC behaviour is that it is unstable with respect to any perturbation of the temperature from zero. This feature is common to QCP in all systems. Of course, QCP will belong to the nontrivial (d–dependent) class of critical behaviour in the quantum critical region shown by Eq. (78).

The ZT transitions outside this narrow region will exhibit the usual MF behaviour. It becomes clear from this picture that like in the XY model discussed in Section 5.3, the systems described by TIM, mainly quantum ferroelectrics, are a good area for experimental observation of QCP near Γ–transitions at low and extremely low temperatures.

7. DISORDER EFFECTS AND δ–INTEGRATION

7.1 Instability of Quantum Critical Behaviour in Disordered Systems

Here we shall briefly consider the effects of quenched disorder of "random critical temperature" type which is produced by randomly distributed impurities and inhomogeneities; for references, see, e.g., Ma (1976), Hertz (1985), Grinstein (1985), De Cesare (1989), and Uzunov (1993). The disorder effects change the local interaction responsible for the phase transition and, hence, the critical temperature. The effective Hamiltonians of such quenched systems referred to as systems with quenched impurities are derived in the way outlined in Section 4. In the continuum limit the local (nonequilibrium) critical temperature depends on the spatial vector $\vec{r}$. This effect is taken into account in the effective Hamiltonian by an additional ϕ^2–term containing a random function $\varphi(\vec{r})$ which obeys the Gaussian distribution

$$[\varphi_\alpha(\vec{r})\varphi'_\alpha(\vec{r}')]_R = \Delta\delta_{\alpha,\alpha'}\delta(\vec{r}-\vec{r}') \ , \tag{86}$$

or, in the $\vec{k}$–space,

$$[\varphi_\alpha(\vec{k})\varphi'_\alpha(\vec{k}')]_R = \Delta\delta_{\alpha;\alpha'}\delta(\vec{k}+\vec{k}',0) \ . \tag{87}$$

The distribution (86) describes quenched impurities with the so–called short–range random correlations. For the case of long–range random correlations, the δ–function should be substituted by a function of type $f(R) \sim 1/R^a$, where $R = \vec{R} = |\vec{r}-\vec{r}'|$, $0 < a < d$.

The new term which should be added to the quantum Hamiltonian (27) is written in the form

$$\mathcal{H}_R[\phi] = \frac{1}{\sqrt{V}} \sum_{\alpha,\omega_l;\vec{k}_1,\vec{k}_2} \varphi(\vec{k}_1-\vec{k}_2)\phi_\alpha(\omega_l,\vec{k}_1)\phi_\alpha(\omega_l,\vec{k}_2) \ . \tag{88}$$

This term can be derived by the Hubbard–Stratonovich transformation (Section 4.2).

The thermodynamic behaviour strongly depends on the properties of random potential represented by the random function $\varphi(\vec{r})$. The random potential $\varphi(\vec{r})$ is not a thermodynamic variable and for this reason, the theoretical treatment includes two steps: firstly, one should try to calculate the thermal averages as functionals of $\varphi(\vec{r})$ and then perform the averaging over the random function. Finally, one obtains the thermodynamic quantities as functions of disorder parameter Δ; for details and references to original papers, see Ma (1976), Grinstein (1985), and Hertz (1985).

The RG investigation of quantum effective Hamiltonians (27)–(29) with the additional term (88) for all possible values of m, m' and σ was done by Korutcheva and Uzunov (1983, 1984). The case (28) of Bose systems and the case (29) with $m' = 0, m = 2$, corresponding to TIM and structural phase transitions in the displacive limit has been investigated by Busiello et al. (1984).

The finite temperature critical behaviour of systems with symmetry indices $n > 4$ is described by the usual Heisenberg FP of corresponding pure systems ($\Delta = 0$). The disorder effects in these systems are irrelevant to the critical behaviour. For systems with $n < 4$ the disorder effects essentially influence the critical behaviour (see e.g., Ma, 1976; Grinstein, 1982). In this case the critical behaviour is described by a "random" FP (Lubensky, 1975).

The new information which can be obtained for the finite temperature critical behaviour of disordered quantum systems is the value of dynamical critical exponent z. The results (Korutcheva and Uzunov, 1983, 1984) for the dynamical critical exponent z of finite temperature critical points can be written in the form:

$$z = \sigma + \frac{(4-n)}{8(n-1)}\epsilon \,, \qquad 1 < n < 4 \,, \tag{89}$$

where $\epsilon = (4-d)$. Eq. (89) has been obtained for the XY model and Bose systems which are described by the bare correlation function (28). In the case of models described by Hamiltonians (27) with the correlation function (29), the dynamical exponent has been obtained (Korutcheva and Uzunov, 1983, 1984) in the form

$$z = \frac{\sigma}{m} + \frac{(4-n)}{8(n-1)}\epsilon \,, \quad m' = 0, \qquad 1 < n < 4 \,. \tag{90}$$

For $m' > 0$ the exponent z is given by Eq. (36). In the latter case the critical exponent z has no correction to first order in ϵ (Korutcheva and Uzunov, 1984).

Eq. (89) coincides with that obtained by Grinstein et al. (1977) for disordered classical systems within an approach based on a time dependent Landau–Ginzburg equation (Ma, 1976; Hohenberg and Halperin, 1977). The results (89) and (90) demonstrate that the dynamical critical exponent z in systems with random impurities receives corrections in the one–loop (first order in $\epsilon = 4-d$) approximation whereas dynamical exponent of the corresponding pure system ($\Delta = 0$) has ϵ–corrections of order $O(\epsilon^2)$. This is a direct consequence of the fact that the random function $\varphi(\vec{k})$ does not depend on the Matsubara frequency ω_l.

The ZT critical behaviour exhibits an instability with respect to the quenched disorder (Korutcheva and Uzunov, 1983, 1984). This instability has been demonstrated for long–range quenched imputities as well (Busiello et al., 1984). The mechanism of the instability is related to CQC. We shall show this by the simple rescaling transformation of Hamiltonians (27), i.e. within the tree approximation (Section 4.3).

In the limit $T_c \to 0$ the parameters rescaling yields the following transformations of interaction parameters v and Δ:

$$v' = b^{2\sigma-d-z_0}v \,, \tag{91}$$

and

$$\Delta' = b^{2\sigma-d}\Delta \,, \tag{92}$$

where z_0 denotes the bare value $z(0)$ of the dynamical exponent $z(\epsilon)$. Because of CQC, the interaction parameter v is relevant to the ZT critical behaviour in spatial dimensionalities $d < (2\sigma - z_0)$, whereas the disorder parameter Δ is relevant to $d < 2\sigma$. This difference is produced by CQC and the lack of ω_l–dependence of random function $\varphi(\vec{k})$. On one side, at the classical borderline dimensionality $d_U = 2\sigma$, the parameter v is irrelevant and, hence, the RG equations in $d = (4-\epsilon)$ dimensions describe a simple Gaussian instability towards the disorder parameter Δ. On the other side, the ZT ϵ–expansion in terms of $\epsilon_0 = (2\sigma - d - z_0)$ yields pure FPs of the type $\Delta^* = 0$ which are unstable with respect to the disorder parameter Δ. In these both variants of theory the FPs of corresponding pure system are unstable with respect to disorder effects for dimensionalities less than the upper borderline dimensionality d_U. Moreover, the disorder itself does not generate new stable FPs. This is a clear indication for the lack of standard (pure or random) ZT critical behaviour and, therefore, one should expect either some unconventional (multi)critical behaviour in these systems at ZT or a first–order phase transition.

Long–Range Random Correlations and Extended Impurities. The instability of the quantum critical behaviour in pure systems with respect to disorder effects is a result of dimensional CQC which suppresses the thermal fluctuation interactions but does not affect the disorder. The reason is simply in the fact that the quenched disorder is considered as time ($\tau-$) independent.

Owing of the same mechanism the quantum critical behaviour is unstable towards disorder effects produced by long–range random correlations (Busiello et al., 1984). In this case the exponent $(2\sigma - d)$ in Eq. (92) for Δ should be changed to $[2\sigma - d + (d-a)]$; $d > a$. Then the upper critical dimensionality of corresponding classical system becomes $d_U = (2\sigma + a)$.

As the difference $(d - a)$ is considered to be of order of unity, an instability of the critical behaviour will occur also in classical system, where CQC is not present and the rescaling factor of the interaction parameter v is given by Eq. (92) with $z_0 = 0$. The mechanism of this instability of pure classical behaviour with respect to impurities with long–range random correlations is again a dimensional crossover $d \to D = (d+a)$. The critical behaviour in such classical systems will be then unstable for all dimensionalities $d < d_U = (2\sigma + a)$.

The same problem exists in systems with the so–called extended impurities. The disorder of type extended impurities is described by infinitely–ranged random correlations along one or more spatial directions, say, $\bar{d} < d$, and short–range random correlations along the other $(d - \bar{d})$ dimensions. For example, in systems having a film geometry the appropriate choice of extended impurities is the one–dimensional (linear) impurities oriented along the direction of small size L_0 and randomly distributed along the other spatial directions (De Cesare et al., 1998b).

The extended quenched impurities are described by a modification of the model for short–range correlated point impurities, in which the short–range correlations along the "small size" L_0 are substituted with infinite–range correlations. The length scale of latters is much larger than the correlation length ξ and the thickness L_0 of the slab. So, the strongly correlated along the small size point impurities behave like continuous uniform strings. In regard to the critical behaviour this disorder acts like point impurities with a short–range random distribution along the large (infinite) dimensions L_i and an uniform distribution along the small size L_0.

In infinite systems the extended impurities can be assumed also to be $2d$ planes or, generally, of any dimensionality $\bar{d}$ less than the number d of system dimensions. In all cases the extended impurities will generate a critical behaviour instability for dimensionalities less than $d < D = (2\sigma + \bar{d})$.

The critical behaviour instability of pure classical systems with respect to extended impurities has been investigated by a special theoretical construction, which seems consistent with the ideas of the ϵ–expansion. According to the $\epsilon-$ expansion basic idea we calculate the relevant quantities for small values of ϵ and then the results are extrapolated for real dimensions corresponding to $\epsilon \sim 1$. This approach is fruitful and gives excellent results except in case of singularities in the ϵ–series; see, e.g., Lawrie et al. (1987). Following this idea Dorogovtsev (1980) has performed a RG treatment of extended impurities by supposing that $\bar{d} \sim \epsilon$. Thus denoting $\bar{d} = \epsilon_d$ as a second small parameter, he has developed a double (ϵ, ϵ_d)–expansion. The final results are interpolated to the real values of ϵ and ϵ_d. This treatment yields a prediction for a new type of random critical behaviour which has been established and thoroughly described (Dorogovtsev, 1980; Boyanovsky and Cardy, 1982). The same method of a double ϵ–expansion, $(d - a \sim \epsilon_d)$, has been applied by Weinrib and Halperin (1983) to the quenched impurities with long–range random correlations.

The formal similarities between dimensional CQC and the dimensional crossovers established in disordered classical systems can be used in the development of a unified approach to the description of dimensional crossover phenomena. It has been pointed out for the first time by Boyanovsky and Cardy (1982) that the related with the time ($\tau-$) extra dimensionality of the quantum Ising model can be supposed throughout a double (ϵ, ϵ_d)–expansion equal to ϵ_d in a complete analogy with the case of linear random impurities in classical models. We shall consider purely methodical problems of the double ϵ–expansions in disordered, quantum and classical finite–size systems in Section 7.2.

Zero Temperature Disorder. The instability of quantum critical behaviour is related to interesting disorder phenomena in real systems as shown, for example, by experiments (Chan et al., 1988) on luquid ^{4}He in porous media: Vycor, xerogel and aerogel glasses; see also the scaling theory of Fisher and Fisher (1988). Besides, the LT instability with respect to disorder is related to the possibility for a destruction of the ZT superfluid state in ^{4}He and existence of a "Bose glass" (insulating) phase (Giamarchi and Schulz, 1988); see the comprehensive representation of the problem by Fisher et al. (1989). A similar problem about the destruction of ZT superconductivity in electronic systems at ZT by random potentials was considered by Ma et al. (1986). Schakel (1997b) addressed the same problem to the disorder effects in the fractional quantum Hall efect. The latter is described by an effective Bose Hamiltonian, where the Bose field $\phi(\vec{r})$ is coupled to a statistical gauge field governed by a Chern–Simons term; see also the lecture by A. M. J. Schakel in this volume. Scalettar et al. (1991) have investigated an effective Bose Hamiltonian of the type (27) for $d = 1$ by the quantum Monte Carlo techniques with the aim to clarify the competition between strong interaction and disorder effects.

The main theoretical efforts were pointed to the explanation of ZT critical behaviour of disordered quantum systems by means of phenomenological scaling theory. An attempt for solving the problem for the superfluid liquids instability with respect to random impurities has been made by Weichman and Kim (1989). These authors have applied the idea of Boyanovsky and Cardy (1982) mentioned above for a formal similarity between the crossovers in classical systems with linear quenched impurities and CQC in TIM and related systems. An artificial upper cutoff Λ_ω of Matsubara frequences has been introduced in the Bose Hamiltonian (27) with a bare Green function given by Eq. (28); this cutoff has been avoided in a paper of Schakel (1997b) devoted to the same problem.

Weichman and Kim considered linear impurities along the time ($\tau-$) axis instead one of the spatial ones. Thus the impure strings along the time direction are of size $\sim \beta = 1/T$. For $\lambda \sim 1/\sqrt{T} \to 0$ (in the continuum approximation), i.e., when the temperature time ($\tau-$) dimension reduces to zero these lines shrink to randomly distributed points in the d–dimensional volume of the system. For $T \to 0$, their length β tends to infinity.

In fact, in the model considered by Weichman and Kim (1989) and later by Schakel (1997b), the dynamical critical exponent z is equal to two and, hence, the term "lines" should be used if we consider a (d + 1)–dimensional space–time. However, if we consider this space–time representation equivalent to $(d+2)$ dimensional space, as is in CQC, we should stretch the "lines" (or strings) of "length" $\beta \sim \lambda^2$ in planes of area $\lambda \times \lambda$. In general, we shall have hyperplanes of dimensionality z and "area" $\sim \lambda^z$. For $\lambda \to 0$ these randomly distributed objects will shrink to randomly distributed point impurities. These are two different geometrical representations of same phenomenon. The formal analysis can be performed with the additional small parameter ϵ'_d which runs from zero to unity and thus describes the classical ($\epsilon'_d = 0$) and the quantum ($\epsilon'_d = z$) limits of Bose system with point impurities. Alternatively one may consider the small parameter $\epsilon_d = \epsilon'_d/z$ (Weichman and Kim, 1989; Schakel, 1997b).

In result of their investigation, Weichman and Kim have obtained the following interesting picture of Bose systems critical behaviour. In the Curie–Weiss temperatures range, where $\lambda < a$, the critical behaviour is described by HT Lubensky (1975) random FP corresponding to short–range (point–like) impurities. In the ZT limit, the Lubensky critical behaviour undergoes a crossover to a critical behaviour governed by extended (plane) impurities. This ZT critical behaviour is described by the corresponding ZT variant of the Dorogovtsev (1980) random FP with complex stability exponents. In the Hamiltonian parameter space (β, v, Δ), HT Lubensky random FP is conjugate to pure ($\Delta^* = 0$) Heisenberg FP and both of them are far from the ZT plane $\beta = \infty$. ZT Dorogovtsev FP is conjugate to GIFP ($\Delta_{GI} = T_{GI} = 0$) considered in details in Section 5.2. Like Heisenberg FP, GIFP is stable only for pure Bose fluids (for real bosons, $n = 2$). For any $\Delta > 0$ and $T > 0$, the RG flows tend to HT Lubensky random FP, whereas ZT Dorogovtsev random FP is attainable only if the system is at $T = 0$. These four conjugate FPs describe the critical phenomena in disordered superfluids and related systems, as well as the superfluid–insulator transition at LT. This picture is quite special because of the GIFP presence which introduces an extraordinary crossover from the usual HT Heisenberg critical behaviour of pure systems to that of Gaussian–like critical behaviour at ZT (Section 5). It is valid for all systems described by microscopic or effective Bose Hamiltonians of the type (27)–(28).

In other systems such as disordered quantum antiferromagnets, the ZT pure critical behaviour obeys the exact universality rule (Section 4), namely, that the d–dimensional ZT critical behaviour is described exactly by the HT universality class $(d{+}2, n)$ to which the system belongs. This is the case of disordered itinerant quantum antiferromagnets investigated by Kirkpatrick and Belitz (1996a, 1997); within the notations given by Eqs. (27)–(28) these systems correspond to Eq. (29) with $m' = 0, m = 1$. These authors have also generalized the result (90) for the dynamical critical exponent z with the help of (ϵ, ϵ_d)–expansion.

The effect of nonmagnetic quenched impurities on the electron spin interactions in itinerant ferromagnets at LT has been considered by Belitz and Kirkpatrick (1996) and Kirkpatrick and Belitz (1996b). It has been established that quenched disorder produces a diffusive electron dynamics which induces an effective long–range electron spin interaction of the form $1/R^{(2-2d)}$. This leads to a change in the energy spectrum of the effective Bose field $\phi(q)$. The bare Green function G_0 is described with the help of the terms given by Eq. (29) for $m' = 2$, $m = 1$, and $\sigma = \min(2, d-2)$. The d–dependence of exponent σ leads to interesting results for $d < 4$.

7.2 Parity with Classical Systems of Slab Geomerty

Now we shall consider the parity between the dimensional CQC and dimensional crossover in $D = (d+1)$–dimensional hyperslabs with a finite size ("thickness") L_0 and other d infinite dimensions $L_i \geq \xi$ for any value of the correlation length ξ. The strong similarity between the description of the finite–size crossover in slabs and CQC was poined out for the first time by Suzuki (1976b) and Pfeuty (1976).

In particular a more precise correspondence between these two different crossovers is valid for the effective quantum Hamiltonians (27) given by Eq. (29) with $m' = 0$, and $m = 2$, i. e. effective quantum models of TIM type (Section 6). This case has been thoroughly investigated by Lawrie (1978b) and applied in a number of studies (see, e.g., O'Connor and Stephens, 1991; Sachdev, 1997); for a review of the theory of finite–size critical phenomena, see Barber (1983).

Here we shall continue our discussion in terms of short range interactions ($\sigma = 2$) but the results can be straightforwardly generalized for $0 < \sigma \leq 2$. For a small hyperslab thickness $L_0 < \xi$, the critical behaviour corresponds to that of a quasi–d dimensional hyperslab, whereas the critical behaviour of thick slabs $L_0 > \xi$ coincides with that of infinite D–dimensional system. As $\xi \to \infty$ for $T \to T_c$, the finite–size crossover ($d \to D$) in the asymptotic critical behaviour will always occur, provided the number d of "infinite" dimensions $L_i \gg \xi$, ($i = 1, ..., d$) is larger than the lower borderline (critical) dimensionality D_L.

If the coefficient c in Eq. (29) is chosen equal to unity, the Matsubara frequency ω_l enters in the correlation function $G_0(q)$ on equal footing with the components k_i of wave vector $\vec{k}$. Then the formal change of temperature T with $1/L_0$ will yield the effective fluctuation functional of D–dimensional slab. Conversely, the substitution of one of dimensions L_i with $1/T$ will yield TIM; see also Section 4.2. Here we shall use this correspondence. In particular, we shall discuss the applicability of the double ϵ–expansions to the RG description of dimensional crossovers; see also De Cesare (1998b).

The RG investigations of films reveal the two limiting cases of finite–size crossover: (i) thick (quasi–D–dimensional) films, where the ratio $y = (L_0/\xi)$ tends to infinity ($y \gg 1$), and (ii) thin (quasi–d–dimensional) films, where the ratio y tends to zero ($y \ll 1$). In the former case the upper borderline dimensionality is $D_U = 4$, whereas in the latter case this dimensionality is $D_U = 5$. The ϵ–expansions for cases (i) and (ii) are performed for $\epsilon = (4 - D) = (3 - d)$ and $\epsilon = (5 - D) = (4 - d)$, respectively.

An obvious disadvantage of all existing descriptions of finite–size and other dimensional crossovers is that the limiting cases can be easily proven and described but the intermediate case ($0 < y < 1$) presents a difficult and unresolved task. The systematic way of investigation of intermediate cases ($y \sim 1$) is to use the Euler–Maclaurin summation formula and take into account the corrections in inverse powers of $y(k)$ to the continuum limit. Such a treatment requires a numerical analysis.

Alternatively, one may perform RG studies by an integration in noninteger dimensionalities (Section 7.1). In this Section we shall discuss advantages and disadvantages of this method. We shall show that it can be applied as an interpolation between the limiting cases $y \ll 1$ and $y \gg 1$ only to specific theoretical schemes such as RG. We shall work with TIM as example, but our consideration can be easily generalized to any quantum model from the scheme (27)–(29).

δ–integration. Let us rewrite Eqs. (73) and (74) in the present notations:

$$A_1[y(k)] = \int \frac{d^d k}{(2\pi)^d} S_1[y(k)] \,, \tag{93}$$

$$S_1[y(k)] = \frac{1}{L_0} \sum_{k_0} \frac{1}{(k_0^2 + k^2 + r)} = \frac{L_0}{2y(k)} \mathrm{cth} \left[\frac{y(k)}{2} \right] , \tag{94}$$

where $r \equiv t_0$, $y(k) = \sqrt{k^2 + r}$, $y(0) = y$, and Γ has been set equal to unity.

The integrand $S_1[y(k)]$ in Eq. (93) exhibits a single-power behaviour [$\sim y^{\sigma}(k)$] only for $y(k) \ll 1$ and $y(k) \gg 1$. The existence of a leading power dependence of the integrals $A_{m+1} = -(\partial A_m/\partial r)$ on $y = (L_0\sqrt{r})$ and irrelevance of the correction terms lead to a simple structure of RG equations and, hence, to their scale invariant solutions. The problem is to obtain such solutions for the intermediate cases of $y \sim 1$, too.

It is impossible to construct an exact integral counterpart of Eq. (93) with a power law behaviour with respect to y; that is why here we shall consider an approximate solution of the problem. We shall substitute the sum S_1 with a δ–dimensional integral

$$S_1' = L_0^{\delta-1} \int \frac{d^\delta x}{(2\pi)^\delta} \frac{1}{(x^2 + k^2 + r)} , \qquad 0 \leq \delta \leq 1 . \tag{95}$$

Accordingly, the integral (93) will be substituted by a double (δ, d)–dimensional integral

$$A_1'(r, b) = L_0^{\delta-1} \int \frac{d^d k}{(2\pi)^d} \int \frac{d^\delta x}{(2\pi)^\delta} \frac{1}{(x^2 + k^2 + r)} . \tag{96}$$

The alternative is to substitute the integral (93) with the $(d + \delta)$–dimensional integral

$$A_1''(r, b) = L_0^{\delta-1} \int \frac{d^{d+\delta} q}{(2\pi)^{d+\delta}} \frac{1}{(q^2 + r)} . \tag{97}$$

Certainly, these substitutions are not exact counterparts of the original quantities and their utility in our attempts to present a reliable interpolation between the limiting cases should be justified.

The integral (97) can be derived after the conjecture that the wave vector component k_0 is a $\delta(< 1)$–dimensional (sub)vector, $\vec{k}_0 = (k_{\mu 0};\ \mu = 1, ..., \delta)$ and, accordingly, that the total wave vector $\vec{q} = \{k_{\mu 0}; k_i\}$ and the volume $V_D = L^\delta V_d$ correspond to a $(d + \delta)$–dimensional system. The integrals A_1'' and $A_{(m+1)}'' = -(\partial A_m''/\partial r)$ have been used in studies of quantum systems and extended impurities for $\delta > 1$ varying in the broad interval from zero to $\epsilon_d < d$ (Section 7.1).

The integrals S_1' and $A_{(m+1)}' = -(\partial A_m'/\partial r)$ given by Eqs. (95) and (96) are defined with the help of another conjecture, namely, that one may perform a smooth interpolation between the integral values in the continuum limits ($\delta = 0$) and $\delta = 1$ for d- and $D = (d + 1)$–dimensional cases, respectively, with the help of the formal rule

$$\frac{1}{L_0^\delta} \sum_{k_0} \rightarrow \int \frac{d^\delta x}{(2\pi)^\delta} \equiv K_\delta \int_0^\infty dx\, x^{\delta-1} , \tag{98}$$

where K_δ has been defined by Eqs. (16). The limit $\delta \rightarrow 0$ in the last integral in Eq. (98) should be taken with a special attention because of the gamma function divergence; the area $S_\delta = (2\pi)^\delta K_\delta$ of the zero dimensional ($\delta = 0$) unit sphere is ill-definite. At first one should perform the integration over x of the integrand, say, $x^{+0} f(x)/x$, and then take the limit $\delta \rightarrow 0$. Usually, the integrands [$\sim f(x)$] which appear by the perturbation series are such that no divergences arise in the final results for $\delta \sim 0$. This is confirmed by a direct calculation of integrals (95) and (96).

For $\delta \rightarrow 0$ and $\delta \rightarrow 1$, integrals (96) and (97) exactly reproduce the results from (93) and (94) for $y \rightarrow 0$ and $y \rightarrow \infty$, respectively. The same is valid for the integral (95) with respect to Eq. (94). So, there are some grounds for the supposition that the intermediate states ($y \sim 1$) could be interpolated by the values $0 < \delta < 1$.

Effective Dimensionality of the Fluctuation Modes. The coincidence of results for the original integrals $A_m(r)$ in the limits $y \to 0$ and $y \to \infty$ with the results from the integrals A'_m and A''_m could be used to suppose that there exists a continuous increasing function $\delta(y)$ with the properties $\delta(y \to 0) \to 0$ and $\delta(y \to \infty) \to 1$. This supposition presents the opportunity to introduce a new dimensionality – *the effective spatial dimensionality* of fluctuation modes $\phi(\vec{q})$ given by

$$D_{\text{eff}}(y) = d + \delta(y) . \tag{99}$$

Eq. (99) is a straightforward generalization of the known from previous finite–size studies fact that a finite–size system abruptly changes its D–dimensional behaviour to the corresponding d–dimensional behaviour when the thickness L_0 is lowered to values less than ξ.

The idea of fluctuation modes effective dimensionality D_{eff} could not be considered as a totally new concept but in this Section we shall consider it more explicitly within the generalized form given by Eq. (99). Moreover, the effective dimensionality D_{eff} can be used in new variants of calculations in all cases when a secondary relevant length like L_0 or λ competes with the correlation length ξ.

Note, that the interpolation between the limiting cases $y \to 0$ and $y \to \infty$ can be performed with the help of a new $\tilde{\epsilon} = (D_{\text{eff}} - D_{\text{eff}}^{(U)})$–expansion in the terms of effective dimensionality D_{eff} around the upper effective borderline dimensionality $D_{\text{eff}}^{(U)}$. This expansion is identical to the $\epsilon = (4-d)$ and $\epsilon = (3-d)$ expansions only in the limiting cases of quasi–d and $D = (d+1)$–dimensional systems. In general, this is a way to describe the RG scaling in terms of fundamental ratio $y = (L_0/\xi)$. Unfortunately, the present investigation does not give an opportunity to obtain the function $\delta(y)$ and, hence, $D_{\text{eff}}(y)$. The reason is in the fact that, as we shall see below, the method of δ–integration is an approximation for all $0 < \delta < 1$. The $\tilde{\epsilon} = (D_{\text{eff}} - D_{\text{eff}}^{(U)})$–expansion has been applied to the description of phase transitions in thin films with quenched disorder, including randomly distributed extended impurities (De Cesare et al, 1998b). It yields the previously known results in a unified way. Such studies make possible to describe the critical behaviour in the whole crossover region from $y \ll 1$ to $y \gg 1$.

Validity. Despite the formal difference in their definitions, the integrals A'_m and A''_m lead to the same results in many practical calculations. The problem is that the values of the original sum (94) and integral (93) do not coincide with the values of corresponding integrals (95) and (96)–(97) in noninteger dimensionalities ($\delta \neq 0$). Therefore, the application of integration in noninteger dimensionalities to particular problems requires a special attention. The reliability of such applications should be justified for any particular case.

Despite the numerous RG studies which have been already performed with the help of integrals A''_m, the question about the limitations of corresponding results is not considered. In order to justify the RG investigations, we shall consider this problem. We shall demonstrate the degree of approximation by comparing the original sum S_1 and integral A_1 with corresponding quantities S'_1, A'_1, and A''_1.

By calculating the integral (95) for S'_1 and comparing the result with S_1 from (94) we obtain

$$\text{th}[y(k)/2] \leftrightarrow g_\delta[y(k)] \tag{100}$$

with

$$g_\delta[y(k)] = A(\delta)\left[\frac{y(k)}{2}\right]^{1-\delta} , \tag{101}$$

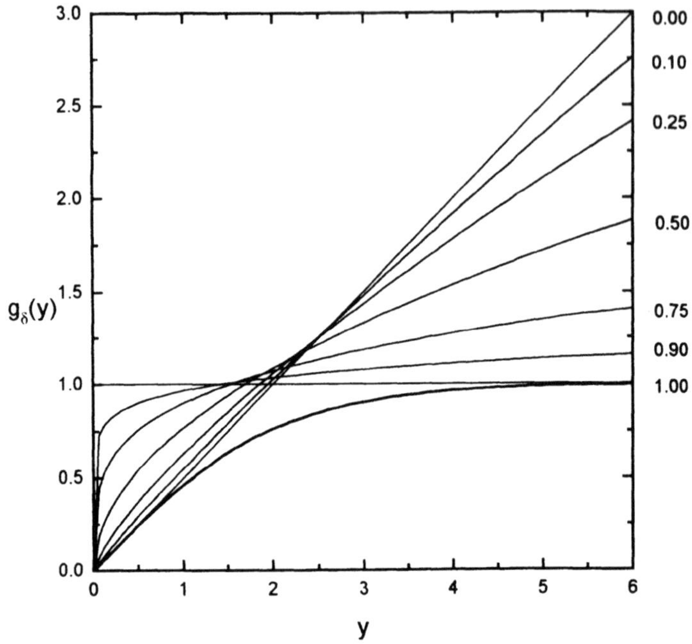

Figure 4. A graphical representation of the comparison (102): the th$(y/2)$ is shown by the line $0-1$, the function $g_\delta(y)$ is drawn for $\delta = 0, 0.1, 0.25, 0.5, 0.75, 0.9, 1$.

where $A(\delta) = \pi^{\delta/2}/\Gamma(1-\delta/2)$; hereafter the symbol "$\leftrightarrow$" will denote a "comparison" and nothing else. The comparison (100) can be done for any $k < \Lambda$. For $k = 0$, we have $y(0) = y$, for $k\xi = 1$, $y(1/\xi) = \sqrt{2}$, and for the third wave number $k\xi = \sqrt{3}$, which also has an essential contribution in the integrals over $\vec{k}$, we have $y(\sqrt{3}/\xi) = 1$. As we shall see our approximations are not precise enough in order to distinguish between the fits of S_1 and S_1' for different wave numbers. Besides, there is a strong argument that the most important value of k, for which we should make the comparison (100), is $k = 0$. This value corresponds to the uniform mode $\phi(\vec{k} = 0)$ describing the spontaneous symmetry breaking in the d–dimensional subsystem.

By setting $k = 0$ in (100) and (101) we have

$$\mathrm{th}(y/2) \leftrightarrow A(\delta)\left(\frac{y}{2}\right)^{1-\delta}. \tag{102}$$

The l.h.s and the r.h.s. of (102) are depicted in Fig. 4. On one side it is obvious that $S_1'(y)$ is a good approximation to $S_1(y)$ in quite broad intervals of y values, for example: $0 < y < 1$ and $y > 4$. On the other side, it becomes evident that δ values which give the best fit of $S'(y)$ to $S(y)$ are: $\delta = 0$ – for $y < 1$, and $\delta = 1$ – for $y \gg 1$. So, within the present consideration, the function $\delta(y)$ can be approximated with zero for all $y < 1$ corresponding to a relatively good fit of the curves th$(y/2)$ and $g_\delta(y)$ and that, on behalf of same arguments, $\delta(y) \approx 1$ for $y \gg 1$.

In a broad region of y values around $y = 2$ the approximation of $S(y) \sim S'(y)$ breaks down and we could not deduce any reliable conclusion about the exact critical value y_c of y at which the finite–size crossover occurs. According to the present picture one may speculate that this value is probably $y_c = 2$ whereas the intuitively appealing value is $y = 1$. The value $y_c = 2$ comes from the factor $(1/2)$ in front of $y(k)$ in Eqs. (100)-(102).

The picture outlined from the comparison (102) is valid also for the more detailed comparison denoted by (100). In this case one should change y by $y(k)$. For $k = (\sqrt{3}/\xi)$, $y_c = 1$, and the points of intersection of g_δ–lines will be located around the coordinate $y_c = 1$.

The calculations within the present approach yield that the locations of minimal values of the difference between the r.h.s. and the l.h.s. of (102) are $\delta = 0$ for all $y < 2$ and $\delta = 1$ for all $y > 2$, namely, that $\delta(y)$ coincides with the Θ–function $\Theta(y-2)$. Certainly this qualitatively wrong result is due to the obvious fact, see Fig. 4, that in a broad region around the point $y_c = 2$, the approximation of $S_1(y)$ to $S_1'(y)$ is not valid. We are not aware of other treatments of this "$\Theta(y-2)$ ansatz". Despite the uncertainty about the location of y_c, it can be definitely concluded from our rough considerations that the real function $\delta(y)$ will have a steep increase from very small values $\delta \sim 0$ up to $\delta \sim 1$ in a relatively close vicinity ($|y - y_c| \sim y_c$) of real critical point y_c.

The comparison of integral $A_1(r)$ with $A_1'(r)$ and $A_1''(r)$ is a more difficult task. In order to avoid the cutoff (Λ–) dependence of the results and simultaneously to avoid the irrelevant to our problem ultraviolet divergences, we shall set $\Lambda = \infty$ and consider the differences of type $\Delta A_1(y) = [A_1(0) - A_1(y)]$ rather than the integrals themselves. Using Eq. (93) and (94) we obtain

$$\Delta A_1(y) = \frac{1}{2} K_d L_0^{1-d} I_d(y) \tag{103}$$

with

$$I_d(y) = \int_0^\infty dz\, z^{d-1} \left[\frac{\mathrm{cth}(z/2)}{z} - \frac{\mathrm{cth}(\sqrt{z^2+y^2}/2)}{\sqrt{z^2+y^2}}\right], \tag{104}$$

where $z = L_0 k$.

The difference $\Delta A_1'(y) = [A_1'(0) - A_1'(y)]$ is obtained from Eq. (96):

$$\Delta A_1'(y) = \frac{K_d J_d'(\delta, y)}{2^\delta A(\delta) L_0^{d-1}}, \tag{105}$$

where $A(\delta)$ is given by Eq. (101), and

$$J_d'(\delta, y) = \int_0^\infty dz\, z^{d-1} \left[\frac{1}{z^{2-\delta}} - \frac{1}{(z^2+y^2)^{1-\delta/2}}\right]. \tag{106}$$

Finally, from Eq. (97) we have that the difference $[A_1''(0) - A_1''(y)]$ is given by

$$\Delta A_1''(y) = K_{d+\delta} L_0^{1-d} y^2 \int_0^\infty dz\, \frac{z^{d+\delta-3}}{z^2+y^2}. \tag{107}$$

The integrals (104), (106) and (107) have obvious ultraviolet and infrared divergences at the corresponding lower and upper borderline dimensionalities. In order to avoid unnecessary complications in our calculation we shall consider the case of real films ($d = 2$). For ($d = 2$) Eq. (107) becomes

$$\Delta A_1''(y) = \frac{y^\delta}{\pi \delta 2^{\delta+1} A(\delta) L_0}. \tag{108}$$

Because of the obvious infrared divergence at $\delta = 0$, we shall consider the derivative $(\partial\Delta A_1''/\partial y)$ instead of the difference $\Delta A_1''$ itself:

$$\frac{\partial\Delta A_1''(y)}{\partial y} = \frac{y^{\delta-1}}{\pi 2^{\delta+1}A(\delta)L_0} \, . \tag{109}$$

The ultraviolet divergences in the two parts of the integral $J_2'(\delta, y)$ exactly compensate each other and this integral takes the simple form $J_2'(\delta, y) = y^\delta/\delta$. Thus we obtain that $\Delta A_1'(y)$ coincides with $\Delta A''(y)$.

The integral $\partial I_2(y)/\partial y$ can be represented in the form

$$\frac{\partial I_2(y)}{\partial y} = y\int_y^\infty dt \left(\frac{\mathrm{cth}(t/2)}{t^2} + \frac{1}{2t\mathrm{sh}^2(t/2)}\right), \tag{110}$$

which directly yields the result $\mathrm{cth}(y/2)$. For $d = 2$, the derivative of the difference $\Delta A_1(y)$ becomes

$$\frac{\partial\Delta A_1(y)}{\partial y} = \frac{1}{4\pi L_0}\mathrm{cth}\left(\frac{y}{2}\right) \, . \tag{111}$$

The correspondence (102) straightforwardly follows from the comparison of Eqs. (109) and (111).

Thus we have shown that the derivatives of integrals A_m' and A_m'' are quite different from the derivatives of original integrals A_m, in particular for $y \sim y_c$. Having in mind that all integrals coincide at the limiting points $\delta(y = 0) = 0$ and $\delta(y = \infty) = 1$, the same conclusion can be made for the integrals themselves.

The demonstrated deviation of integrals in noninteger dimensionalities from the initial integrals A_m does not mean that the RG analysis based on them is unreliable for $0 < \delta < 1$. The argument here is that the RG predictions about the critical behaviour follow from the RG transformation which reflects the structure and the symmetry of Hamiltonian rather than from the values of perturbation integrals. The latters determine the location of FPs of RG equations and, therefore, might be relevant only to problems of special interest. Hopefully, RG investigations discussed in Section 7.1 do not come upon such problems.

Acknowledgements

D.I.U. thanks the hospitality of Dipartimento di Scienze Fisiche "E. R. Caianiello", Università di Salerno, where this article has been written. A grant Ph.560 from NFSR (Sofia) is also acknowledged.

REFERENCES

Abe, R., 1974, *Progr. Theor. Phys.* 52:1135.
Abe, R. and Nikami, S., 1974, *Phys. Lett.* 47A:341.
Abrikosov, A. A., Gor'kov, L. P., and Dzyaloshinskii, I. E., 1963, "Methods of Quantum Field Theory in Statistical Physics," Prentice–Hall, Englewood Cliffs, New Jersey.
Achiam, Y., and Imry, Y., 1975, *Phys. Rev.* B12:2768.
Arias, T. A., and Joannopoulos, J. D., 1989, *Phys. Rev.* B39:4071.
Baba, Y., Nagai, T., and Kawasaki, K., 1979, *J. Low Temp. Phys.* 36:1.
Baldo, M., Catara, E., and Lombardo, U., 1976, *Lett. Nouvo Cimento* 15:214.

Barber, M. N., 1983, *in*: "Phase Transitions and Critical Phenomena," vol. 8, ed. by Domb, C., and Lebowitz J. L., Academic Press, London, p. 145.
Bardos, D. C., Hines, D. F., and Frankel, N. E., 1994, *Phys. Rev.* B49:4082.
Batrouni, G. G., Scaletter, R. T., and Zimanyi, G. T., 1990, *Phys. Rev. Lett.* 65:1765.
Baym G., and Pethick, C. J., 1996, *Phys. Rev. Lett.* 76:6.
Beck, H., Schneider, T., and Stoll, E., 1975, *Phys. Rev.* B12:5198.
Belitz, D., and Kirkpatrick, T. R., 1994, *Rev. Mod. Phys.* 66:261; 1996, *Europhys. Letts.* 35:201.
Bergman, 0., 1992, *Phys. Rev.* D46:5474.
Blatt, J. M., 1964, "Theory of Superconductivity", Academic Press, New York.
Boyanivsky, D., and Cardy, J. I., 1982, *Phys. Rev.* B26:154.
Brout, R., 1974, *Phys. Reps.* 10:1.
Brout, R., Muller, K. A., and Thomas, H., 1966, *Phys. Reps.* 4:507.
Bruce, A., 1980, *Adv. Phys.* 29:111.
Busiello, G., and De Cesare, L., 1980a, *J. Phys. A: Math. Gen.* 13:3779; 1980b, *Phys. Lett.* 77A:177; 1980c, *Nuovo Cimento* 59B:327.
Busiello, G., De Cesare, L., Rabuffo, I., 1983, *Physica A* 117:445; 1984, *Phys. Lett.* 102A:41; 1985b, *Phys. Rev.* B32:5918.
Busiello, G., De Cesare, L., and Uzunov, D. I., 1985a, *Physica A* 132:199.
Busiello, G., and Uzunov, D. I., 1987, *in*: "Advances on Phase Transitions and Disorder phenomena," ed. by Busiello, G., De Cesare, D., Mancini, F., and Marinaro, M., World Scientific, Singapore, p. 130.
Caramico D'Auria, A., De Cesare, L., and Rabuffo, I., 1996, *Physica A* 225:36; (E) 1997, 236:550.
Caramico D'Auria, A., De Cesare, L., Esposito, U., and Rabuffo, I., 1997, *Physica A* 243:152.
Casher, A., Lurié, D., and Revzen, M., 1968, *J. Math. Phys.* 9:1312.
Chakrabarti, B. K., Dutta, A., and Sen, R., 1996, "Quantum Ising Phases and Transitions in Transverse Ising Models," Springer, Berlin.
Chan, M. H. W., Blum, R. I., Murthy, S. Q., Wong, G. K. S., and Reppy, J. D., 1988, *Phys. Rev. Lett.* 17:1950.
Chubukov, A. V., 1984a, *Fizika Nizkih Temp.* 10:381; 1984b, *Teor. Mat. Fiz.* 60:145.
Cooper, M. J., and Green, M. S., 1968, *Phys. Rev.* 176:302.
Creswick, R. J., and Wiegel, F. W., 1982, *Phys. Lett.* 92A:189; 1983, *Phys. Rev.* A28:1579.
Crooker, B. C., Hebral, B., Smith, E. N., Takano, Y., and Reppy, J. D., 1983, *Phys. Rev. Lett.* 51:666.
Daicic, J., Frankel, N. E., Gailis, R. M., and Kowalenko, V., 1994, *Phys. Reps.* 237:64.
De Cesare, L., 1978, *Lett. Nuovo Cimento* 22:325; 22:632; 1982, *Nuovo Cimento* 1D:289; 1989, *Reviews of Solid State Sciences* 3:71.
De Cesare, L., Craco, L., Rabuffo, I., and Uzunov, D. I., 1998a, *J. Phys. Studies* (Lviv), vol. 2. No 2.
De Cesare, L., Craco, L., Rabuffo, I., Takov, I. P., and Uzunov, D. I., 1998b, preprint, Salerno University (Italy).
De Gennes, P. G., 1963, *Solid State Commun.* 1:132.
Dorogovtsev, S. N., 1980, *Phys. Lett.* 76A:169; 1980, *Fiz. Tv. Tela* 22: 321, 3658 [*Sov. Phys.-Solid State* 22:168,2141].
Ensher, J. R., Jin, D. S., Mathews, M. R., Wieman, C. E., and Cornell, E. A., 1996, *Phys. Rev. Lett.* 77:4984.
Finotello, D., Wong, G. K. S., Gillis, K. A., Awschalom, D. O., and Chan, M. H. W., 1987, *Jap. J. of Appl. Phys.* Suppl. 26–3, 26:283.
Fisher, M. E., 1967, *Rep. Progr. Phys.* 30:615; 1968, *Phys. Rev.* 176:257.
Fisher, D. S., and Fisher, M. P. A., 1988, *Phys. Rev. Lett.* 61:1847.
Fisher, D. S., and Hohenberg, P. C., 1988, *Phys. Rev.* B37:4936.
Fisher, M. P. A., Weichman, P. B., Grinstein, G., and Fisher, D. S., 1989, *Phys. Rev.* B40:546.
Gerber, P. R., and Beck, H., 1977, *J. Phys. C: Solid State Phys.* 10:4013.
Giamarchi, T., and Schulz P., 1988, *Phys. Rev.* B37:325.
Giorgini, S., Pitaevskii, L. P., and Stringari, S., 1996, *Phys. Rev.* A54:R4633.
Goldhirsch, I., 1979, *J. Phys. C: Solid State Phys.* 12:5345; 1980, *J. Phys. A: Math. Gen.* 13:453.
Goldhirsch I., Levich, E., and Yaknot, V., 1979, *Phys. Rev.* B19:4780.
Grinstein, G., 1985, *in*: "Fundamental Problems in Statistical Mechanics," vol VII, ed. by E. G. D. Cohen, North–Holland, Amsterdam, p. 147.
Grinstein, G., Ma, S. K., and Mazenko, G., 1977, *Phys. Rev.* B15:258.
González, J. J., Hauge, E. H., and Hemmer, P. C., 1975, *Phys. Rev.* B11:1952; B12:198.
Gunton, J. D., and Buckingham, M. J., 1968, *Phys. Rev.* 166:152.

Halperin, B. I., and Rice, T. M., 1968, *in*: "Solid State Physics," ed. by Ehrenreich, H., Seitz, F., and Turnbull, D., Academic, New York, 21:115.
Hertz, J. A., 1976, *Phys. Rev.* B14:1165; 1985, *Phys. Scripta* T10:1.
Hohenberg, P. C., and Halperin, B. I., 1977, *Rev. Mod. Phys.* 49:435.
Holz, A., Medeiros, J. T. N., 1975, *J. Phys. A: Math. Gen.* 8:1115.
Huang, K., 1987, "Statistical Mechanics," Wiley, New York.
Izyumov, Yu. A. and Skryabin, Yu. N., 1988, "Statistical Mechanics of Magnetically Ordered Systems," Consultant Bureau, New York.
Joyce, G. S., 1972, *in*: "Phase Transitions and Critical Phenomena," vol. 2, ed. by Domb, C. and Green, M. S., Academic Press; p. 375.
Kadanoff, L. P., Baym, G., 1962, "Quantum Statistical Mechanics," Benjamin, Menlo Park, California.
Kaganov, M. I., and Chubukov, A. V., 1988, *in*: "Spin Waves and Magnetic Excitations," ed. by Borovik-Romanov A. S., and Shinha, S. K., North-Holland, Amsterdam, p. 1.
Kalokolov, I. V., and Podivilov, E. V., 1989, *Zh. Eksp. Teor. Fiz.* 95:211 [*Sov. Phys.-JETP* 68:119].
Ketterle, W., Andrews, M. R., Davis, K. B., Durfee, D. S., Kurn, D. M., Mewes, M.-O., and Druten, N. J., 1996, I., *Phys. Scripta* T66:31.
Kirkpatrick, T. R., and Belitz, D., 1996a, *Phys. Rev. Lett.* 76:2571; 1996b, *Phys. Rev.* B53:14364; 1997, *Phys. Rev. Lett.* 78:1197 (E).
Kolomeisky, E. B., and Straley, J. P., 1992, *Phys. Rev.* B46:13942.
Kondor, I., and Szépfalusy, P., 1974, *Phys. Lett.* 47A:393.
Kopec, T. K., and Kozlowski, G., 1983, *Phys. Lett.* 95A:104.
Korutcheva, E. R., and Uzunov, D. I., 1983, preprint (BAS, Sofia); 1984, *Phys. Lett.* 106A:175.
Lacour-Gayet, P., and Toulouse, G., 1974, *J. Physique* 35:426.
Landau, L. D., and Lifshitz, E. M., 1980, "Statistical Physics," part I, revised edition by Pitaevskii, L. P., Pergamon, London.
Lawrie I. D., 1978a, *J. Phys. C: Solid State Phys.* 11:1123.
Lawrie I. D., 1978b, *J. Phys. C: Solid State Phys.* 11:3857.
Lawrie I. D., Millev, Y. M., and Uzunov, D. I., 1987, *J. Phys. A: Math. Gen.* 20: 1599; (E) 20:6159.
Leibler, S., and Orland, H., 1981, *Ann. Phys.* 132:277.
Lewis, A. L., 1979, *Phys. Rev. Lett.* 42:907.
Lifshitz E. M., and Pitaevskii, L. P., 1980, "Statistical Physics," part II, Pergamon, London; 1981, "Physical Kinetics," Pergamon, Oxford.
Lubensky, T. C., 1975, *Phys. Rev.* B11:3573.
Ma, S. K., 1976, "Modern Theory of Critical Phenomena," Benjamin, London.
Ma M., Halperin, B. I., and Lee, P. A.,1986, *Phys. Rev.* B34:3136.
Manousakis, E., 1991, *Rev. Mod. Phys.* 63:1.
Matsumoto, K., Porto, J. V., Pollack, L., Smith, E. N., Ho, T. L., and Parpia, J. M., 1997, *Phys. Rev. Lett.* 79:253.
Micnas, R., 1979, *Physica* 98A:403.
Millev, Y. M., and Uzunov, D. I., 1983, *J. Phys. C: Solid State Phys.* 16:4107.
Millis, A. J., 1993, *Phys. Rev.* B48:7183.
Morf, R., Schneider, T., and Stoll, E., 1977, *Phys. Rev.* B16:462.
Negele, J. W., and Orland, H., 1988, "Quantum Many-Particle Systems," Addison-Weley, New York.
O'Connor, D., and Stephens, C. R., 1991, *Nucl. Phys.* B360:297.
Olinto, A. C., 1985, *Phys. Rev.* B31:4279; 1986, *Phys. Rev.* B33:1849.
Oppermann, R., and Thomas, H., 1975, *Z. Physik* B22:387.
Pfeuty, P., 1976, *J. Phys. C: Solid State Phys.* 9:3993.
Pfeuty, P., and Elliott, R. J., 1971, *J. Phys. C: Solid State Phys.* 4:2370.
Pfeuty, P., and Toulouse, G., 1975, "Introduction to the Renormalization Group and Critical Phenomena," Wiley, Chichester.
Popov, V. N., 1983, "Functional Integrals in Quantum Field Theory and Statistical Physics," Reidel, Dordrecht.
Rechester, A. B., 1971, *Zh. Eksp. Teor. Fiz.* 60:782 [*Sov. Phys.-JETP* 33:423].
Reppy, J. D., 1984, *Physica* 126B:336.
Rytz, D., Hochli, U. T., and Bilz, H., 1980, *Physica* B22:359.
Rudnick, J., Bergman, D. J., and Imry, Y., 1974, *Phys. Lett.* 146A:449.
Rudnick, J., and Nelson D. R., 1976, *Phys. Rev.* B13:2208.
Sachdev, S., 1997, *Phys. Rev.* B55:142.

Sachdev, S., Senthil, T., and Shankar, R., 1994, *Phys. Rev.* B50:258.
Samara G. A., 1988, *Physica B* 150:179.
Samara G. A., Massa, N. E., and Ullman, F. G., 1981, *Ferroelectrics* 36:335.
Scalettar, R., Batrouni, G. G., and Zimanyi G. T., 1991, *Phys. Rev. Lett.* 66:3144.
Schafroth, M. R., 1954, *Phys. Rev.* 96:1149.
Schafroth, M. R., Blatt, J. M., and Butler, S. T., 1957, *Helv. Phys. Acta* 30:93.
Schakel, A. M. J., 1995, *Nucl. Phys.* B 453[FS]:705; 1997a, "Boulevard of broken symmetries," preprint, Freie University (Berlin); 1997b, *Phys. Lett.* A224:287; 1999, "Quantum phase transitions in 2d quantum liquids," see this volume.
Schneider, T., Beck, H., and Stoll, E., 1976, *Phys. Rev.* B13:1123.
Shankar, R., 1994, *Rev. Mod. Phys.* 66:129.
Singh, K. K., 1975, *Phys. Lett.* 51A:27; *Phys. Rev.* B12:2819; 1976, *Phys. Rev.* B13:3192; *Phys. Lett.* 57A:309.
Sondhi, S. L., Girvin, S. M., Carini, J. P., and Shahar, D., 1997, *Rev. Mod. Phys.* 69:315.
Stella, A. L., and Toigo, F., 1976, *Nouvo Cimento* 34B:207.
Stinchcombe, R. B., 1973, *J. Phys. C: Solid State Phys.* 6:2459.
Stinchcombe, R. B., Horwitz, G., Englert, F., and Brout, R., 1963, *Phys. Rev.* 130:155.
Suzuki, M., 1975, *Progr. Theor. Phys.* 53:97; 1976a, *Progr. Theor. Phys.* 56:1007; 1976b, *Progr. Theor. Phys.* 56:1454.
Suzuki, M. and Igarashi, 1974, *Phys. Lett.* 47A:361.
Tokunaga, M., and Matsubara, T., 1966, *Progr. Theor. Phys.* 35:581.
Tounsend, C., Ketterle, W., and Stringari, 1997, *Phys. World*, March, No3:29.
Uzunov, D. I., 1981, *Phys. Lett.* 87A:11; 1982, *Communication*, No: E17–82–21 (JINR, Dubna); 1993, "Introduction to the Theory of Critical Phenomena," World Scientific, Singapore: 1996, *in*: "Lectures on Cooperative Phenomena in Condensed Matter," ed. by Uzunov, D. I., Heron Press, Sofia, p. 46.
Uzunov, D. I. and Walasek, K., 1985, *Phys. Lett.* 107A:207; (E) 110A:482.
Vojta, T., Belitz, D., Narayanan, R., and Kirkpatrick, T. R., 1997, *Z. Phys.* 103:451.
Walasek, K., 1984, *Phys. Lett.* 101A:343.
Weichman, P. B., Rasolt, M., Fisher, M. E., and Stephen, M. B., 1986, *Phys. Rev.* B33:4632.
Weichman, P. B., and Kim, K., 1989, *Phys. Rev.* B40:813.
Weinrib, A., and Halperin, B. I., 1983, *Phys. Rev.* B27:413.
Wiegel, F. W., 1975, *Phys. Rep.* 16:57; 1978a, *Physica A* 91:139; 1978b, *in*: "Path Integrals in Quantum Field Theory and Statistical Physics," ed. by Papadopoulos, G. J., and Devreese, J. T., (NATO ASI Series), Plenum, New York, B34:419.
Young, A. P., 1975, *J. Phys. C: Solid State Phys.* 8:L309.
Zannetti, M., 1980, *Phys. Rev.* B22:5267.
Zinn–Justin, J., 1989, "Quantum Field Theory and Critical Phenomena," Clarendon, Oxford.

CRITICAL FLUCTUATIONS IN NORMAL–TO–SUPERCONDUCTING TRANSITION

R. Folk[1] and Yu. Holovatch[2]

[1]Institut für Theoretische Physik, Johannes Kepler Universität Linz, A–4040 Linz, Austria
[2]Institute for Condensed Matter Physics, Ukrainian Academy of Sciences, UA–290011 Lviv, Ukraine

1. INTRODUCTION

Recent advances in our understanding of critical phenomena due to the application[1] of the renormalization group (RG) approach[2] are now well known (see, e.g., the textbooks[3-5]). The scale invariance at the critical point and the universality of certain features of critical phenomena can be explained by RG transformation, and lead to theory which provides a quantitative description of the critical behaviour of various thermodynamic quantities of interest.

Specifically, the critical point in the RG "language" (in the context of our lecture we specify it as an equilibrium second order phase transition point) corresponds to the stable fixed point of RG transformation, where the system is scale invariant. The system asymptotic properties are governed solely by the stable fixed point coordinate whereas non-asymptotic ones are defined in the region of approach to fixed point. In addition, the RG approach may permit the order of phase transition occurring in a certain model to be determined. This is explored by studying the stability of fixed points of RG transformation: an absence of stable fixed point is interpreted as an evidence of a fluctuation-induced first-order phase transition in the system under consideration.* For models with no exact solutions or rigorous proofs of existence of second order phase transition, that is to say, for the majority of realistic models in statistical physics, RG provides a tool to check the order of transition.

Now let us define the kind of problem we are going to discuss in this lecture. It can be formulated as: *What is the order of the normal-to-superconducting phase transition?* According to the Bardeen–Cooper–Schrieffer theory of superconductivity, the normal-to-superconducting (NS) phase transition is a classical second order phase transition described by the Landau–Ginzburg Hamiltonian with complex order parameter corresponding to the wave function of Cooper pairs. Taking into account order parameter fluctuations one can find values of corresponding critical exponents which in this case will coincide with critical exponents of $O(n)$ symmetric field-theoretical model for the

**Editor's note*: This interpretation should be supported by non–RG arguments, too.

Correlations, Coherence, and Order. Edited by Shopova and Uzunov
Kluwer Academic/Plenum Publishers, 1999

case $n = 2$ (the XY model). Consequently, this leads to assertion that the NS phase transition is described by the same set of critical exponents as the phase transition in normal-to-superfluid liquid. The latter set of exponents has been measured to a high accuracy[6] and also calculated by various methods.[7-14]

Taking into account that the corresponding "superfluid liquid" is charged in case of NS transition, however, complicates the problem. This was considered first of all by B. I. Halperin, T. C. Lubensky, and S. Ma[15] and since then different ways of tackling it have been suggested. We shall discuss some of them briefly in the subsequent section.

From an experimental viewpoint, when the question of order of the superconducting phase transition was first discussed it was considered more or less academic because due to large correlation length $\xi_0 \sim 10^3$Å, the first order characteristics can be seen only very near the phase transition; otherwise mean field behaviour is to be expected. This situation was changed after the discovery of high-T_c superconductors with correlation lengths of order of lattice distances ($\xi_0 \sim 1$ Å).[16] Since then, critical effects have been observed in several experiments.[17-24]

Our main result presented in this lecture is that within the framework of RG method applied to original superconductor model, minimally coupled to the gauge field[15], one still can demonstrate the existence of a *second order phase transition with critical exponents distinct from those of superfluid liquid.* To prove this we shall consider two-loop renormalization group functions for the model, paying particular attention to the fact that the loop expansion is asymptotic. [27-29] In this way we find several fixed points with new scaling exponents and a rich crossover behavior. Some of our results have been previously published in Ref.[30-32]

The lecture is, therefore, organized as follows. In the following section we shall give a brief review of methods used to study the problem we are interested in. Then we describe the model of a superconductor, provide results of its study by the mean field approximation, obtain expressions for renormalization group functions in a two loop approximation and describe the results obtained subsequently without applying any resummation procedure. Then in the next section we shall discuss several well-established examples in the modern theory of critical phenomena where the resummation of asymptotic series is applied. After that we shall present the pivotal section of our investigation: it is devoted to the study of RG functions and corresponding flows on the basis of resummation technique. We obtain as a result that a stable fixed point is present, which is an evidence of second order phase transition in the model. In the remainder of the paper asymptotic and effective values for critical exponents are calculated and we give expressions for amplitude ratios. To conclude, these results are discussed in the closing section.

2. NORMAL–TO–SUPERCONDUCTING TRANSITION: 1st OR 2nd ORDER?

The order of a phase transition may have severe effects on physical quantities of a material. This is illustrated by the first-order liquid-gas transition phenomena like overheating and undercooling connected with the metastability at transition. For second order phase transitions, divergences in physical quantities occur (in the thermodynamic limit, of course) leading to a dramatic increase of specific heat or scattering of light (the critical opalescence) near the liquid-gas critical point. Similar dramatic changes are associated with phase transitions which have occurred in early stages of the universe where the questions discussed here for superconductors also find relevance (for a recent review, see Ref.[33]).

As mentioned in the Introduction, the question of NS phase transition order becomes complicated when one accounts for fluctuations of the order parameter being coupled to the "gauge field" (the vector potential of fluctuating magnetic field created by Cooper pairs), the fluctuations of which also diverge at long distances. Posed for the first time more than 20 years ago[15], this problem has remained a challenging one in the physics of superconductivity until now.[30-59]

The theoretical model of Halperin, Lubensky, and Ma[15] describing relevant critical behaviour is the usual $O(n)$ symmetric ϕ^4 model with $n/2$-component complex field ϕ coupled to a gauge field which describes the fluctuating magnetic field created by Cooper pairs. Due to coupling to the gauge field, in the mean field approximation a third order term appears in the free energy of superconductor and the NS phase transition is of first order.[15] This mean field analysis is appropriate for type–I superconductors[34] where fluctuations in the order parameter have no significant effect on the thermodynamics of transition.

The case of type–II superconductors is more complicated, however, because fluctuations cannot be neglected. Studying the problem within Wilson–Fisher recursion relations[35] in first order of ε it has been found[15] that a stable fixed point (necessary, but not sufficient for a second order phase transition) exists only for the order parameter components number $n > 365.9$, far exceeding superconductor case $n = 2$. The crossover near the first order phase transition has been examined[26] and the expression for the crossover function of specific heat is given by one-loop perturbation theory.

The kinetics of fluctuations arising from vortex pairs in a superconductor has been studied by means of numerical simulations.[36] The result leads to the conclusion that there exist nucleation processes typical of the first order phase transition, confirming mean field and RG results.[15] Note, however, that the mean field analysis applied to Ginzburg–Landau free energy of superconductor[37] including a Chern–Simons term leads to quantitatively different behaviour: for different values of the topological mass, there occurs in system either a fluctuation-induced first order phase transition or a second-order transition. This result is also confirmed in the framework of one-loop RG calculations.[37]

The occurrence of a first order phase transition has been also found in the massless scalar electrodynamics[38, 39] and confirmed to a linear order in ε for n-component Abelian Higgs models by an explicit construction of coexistence curve and equation of state.[40] In addition, the CP^{N-1} non-linear sigma model which is related to the superconductor model in the limit of infinite charge by means of $(2+\varepsilon)$ expansion, has been shown[41] to posses a similar behaviour.

When the study is performed by means of strict expansion results for type–II superconductors obtained by ε -expansion methods[15, 26, 40] appear to be stable towards the influence of different physical factors such as the possibility of another (non-magnetic) ordering, the presence of disorder and crystal anisotropy. The scaling behaviour of a superconducting system with additional non-magnetic ordering investigated by ε-expansion methods provides one more example of system where a weak first-order phase transition occurs.[45] The analysis of the quenched impurities influence on the critical behaviour of superconductors when taking account the magnetic field fluctuations demonstrates[46, 47] the appearance of a new stable fixed point for $1 < n < 366$. It has been shown[47], however, that this describes the critical behaviour in the range of space dimensionalities $d_c(n) < d < 4$ with $d_c(2) = 3.8$ and results in a first order phase transition.

The RG flow for the superconductor model with quenched impurities has been found[48] to exhibit a stable focus surrounded by an unstable limit cycle. The second order phase transition behaviour appears inside the limit cycle. Introducing quenched random fields with short and long range correlations does not lead to a second order behaviour in the region of (d, n) near $(3, 2)$.[49] Note however, that studies of the influence of quenched and annealed gauge fields on the spontaneous symmetry breaking, performed in terms of Helmholtz free energy[42] lead to the conclusion that in the first nontrivial or one-loop approximation in the annealed model the spontaneous symmetry breaking occurs through a first order transition for $d = 2, 3$ whereas the quenched model displays a continuous phase transition. A more complicated account of fluctuations in the annealed model changes the nature of transition to continuous one, but the spontaneous symmetry breaking is absent in the model with quenched disorder.[42] The combined influence of crystal anisotropy, magnetic fluctuations, and quenched randomness on the critical behaviour of unconventional superconductors[50] studied by means of RG analysis within the ε-expansion[51] gives that only fluctuation-induced first order transitions should occur in unconventional superconductors in the vicinity of critical point.

The mean field results however, have been questioned by Lovesey[42], mentioned above, who has shown that taking into account the gauge field fluctuations in the free energy calculations leads back to a second order phase transition. A further indication of a second order phase transition came several years later when this question has been studied as a lattice problem by means of Monte–Carlo calculations and duality arguments.[43] These results have confirmed that there are scenarios of NS transition that differ from those obtained in Ref.[15] Namely, the NS transition has been found to be of second order, asymptotically equivalent to that of superfluid with a reversed temperature axis. Subsequent Monte–Carlo simulations[44] performed in different regions of couplings lead to the result that NS transition is strongly first order deep in the type–I region and becomes more weakly first order moving in the direction of type–II region. Beyond a certain point the data reported in Ref.[44] suggest a second-order transition. The corresponding $O(n)$ nonlinear σ-model coupled to Abelian gauge field studied near two dimensions by $2 + \varepsilon$ expansion[53] does not show a first order phase transition either.

The existence of a tricritical point, where the order of phase transition changes from second to first, has been predicted[58] by representing the $3d$ superconductor model by a disordered field theory. The position of tricritical point is located slightly in the type-I region for values of Ginzburg parameter[34] $k < 0.8/\sqrt{2}$. Starting from the dual formulation of Landau-Ginzburg theory by means of RG arguments it has been shown that the critical exponents of NS transition coincide with those of superfluid transition with a reversed temperature axis[59]. But while the correlation length critical exponent of normal-to-superconducting transition is predicted to coincide with the ordinary 3D XY model, the divergence of the renormalized penetration depth is characterized by mean field value[59] $\nu = 1/2$.

The influence of critical fluctuations on the order of NS transition has been reconsidered on the basis of field theoretical RG ideas in Ref.[30] Here two-loop flow equations[30] for static parameters and ζ-functions[52] are obtained. It is shown that a stable fixed point possibly exists and, as a consequence, a second order phase transition may occur. An attractive feature of the RG flow found in Ref.[30] is that it discriminates between type–I and type–II superconductors, depending on initial (background) values of couplings. For small values of ratio (coupling to gauge field)/(fourth order coupling), appropriate for type-II superconductors, the flow comes very near to the fixed point of uncharged model but ends in a new superconducting fixed point. For large values of same ratio corresponding to type–I superconductors the flow runs away. For intermediate values of ratio, the critical behaviour may be influenced by a second (unstable) superconducting fixed point with scaling exponents quite different from those for the uncharged model.

A flow picture qualitatively similar to that in Ref.[30] has been obtained[54] by solving the model of charged superconductor approximately with the help of nonperturbative flow equations – a method which appears to give very encouraging results for critical scalar field theories.[55, 56] Depending on the relative strength of the ratio (coupling to gauge field)/(fourth order coupling) first or second order phase transition has been found. An approximate description of tricritical behaviour has been given as well as an estimate of correlation length critical exponent ν and pair correlation function critical exponent η, which give us a second order phase transition, has been reported. Depending on three different assumptions for the stable fixed point value of coupling to gauge field, the following values are obtained in two successive truncations of the potential: $(\eta, \nu) = [(-0.13, 0.50); (-0.20, 0.47)], [(-0.13, 0.53); (-0.17, 0.58)], [(-0.13, 0.59); (-0.15, 0.62)]$, indicating that critical exponents belong to the physical region $\eta > 2 - d$ and $\nu > 0$, independent of the truncation, clearly pointing towards a second order phase transition.

In the context of baryogenesis the question of NS phase transition order was considered within the two-loop approximation and the effective potential was calculated.[57] The ε-expansion has been applied to the electroweak phase transition in order to estimate various parameters in leading and next-to-leading orders in ε, including scalar correlation length, latent heat, surface tension, free energy difference, bubble nucleation rate, and baryon nonconservation rate. Of course, the transition is found to be first order since only run away flows occur in the strict ε-expansion perturbation theory. Note, that in the electroweak scenario of baryogenesis there exists a so-called Sakharov requirement which is met when the transition is strongly first order rather than second order.

The NS transition problem has been also studied by an analytical method which is not based on ε or $1/n$ expansions. Using a non-perturbative method of solving the approximate Dyson equation for arbitrary d and n[60] it has been found[61] that NS phase transition is governed by a "charged" fixed point. The value of pair correlation function critical exponent η at $d = 3, n = 2$ is $\eta(3,2) = -0.38$. It is interesting to note that although the result for η appears to be a well-behaved function of d and n, the same result for η breaks down at the critical value $n_c \simeq 18$ when expanded in ε. Hence, the conclusion is drawn that results of ε-expansion obtained in Ref.[15] and, in particular, the absence of stable fixed point solution for $n < n_c \simeq 365.9$ are to be interpreted as a breakdown of ε-expansion rather than a fluctuation-induced first-order phase transition. On the other hand, near $d = 4$ the results in Ref.[61] are in a good agreement with ε-expansion data[15] for high n ($n > 366$).

Recently the same problem has been studied with the RG technique in a fixed dimension $d = 3$ within the one-loop approximation.This study[62] gives an evidence of attractive charged fixed point distinct from that of neutral superfluid and corresponding to correlation length critical exponents values $\nu \simeq 0.53$ and $\eta \simeq -0.70$. However, the magnetic penetration depth has been shown to diverge with the XY exponent, contradicting to results mentioned above[30, 54, 61, 62] considered in the form of continuum dual theory.[63] To investigate this discrepancy, Monte–Carlo simulations of 3D isotropic lattice superconductor in a zero external magnetic field have been performed. This results in the conclusion that there is a single diverging length scale consistent with the universality of ordinary 3D XY model.[65] Further applications of the model containing coupling to gauge field have been suggested in the context of quantum Hall effect.[66]

Now let us give a brief resume of experimental data relevant to our study. As mentioned in the Introduction, the effects of thermodynamic fluctuations are generally small in conventional low-T_c superconductors because of their low transition temperatures and large coherence length. In contrast, high transition temperatures and small coherence lengths mean that critical fluctuations are relevant in high-T_c superconductors. Though critical fluctuations in high-T_c superconductors have been observed in a series of experiments[17-24] their interpretation is changed somewhat. Deviations from the mean field (i.e. first order) behaviour have been accounted for either by $3d$ Gaussian fluctuations (giving, in particular, values for specific heat critical exponent α and correlation length critical exponent ν: $\alpha = \nu = 1/2$)[17, 19] or by a non-trivial XY behaviour characteristic for uncharged superfluid (with $\nu \simeq 2/3$ and logarithmic divergences in α).[18, 20, 22, 24] Measurements of heat capacity,[22, 24] magnetization and electric conductivity[22] of single-crystal samples of $YBa_2Cu_3O_{7-x}$ in magnetic fields near T_c support the existence of a critical regime governed by XY-like critical exponents;[18, 20, 22, 24, 25] a similar conclusion follows from the crossover analysis of the zero-field heat capacity on a comparable sample.[21] The maximum applied magnetic field for which 3D XY scaling is valid, however, differs for various materials.[67]

To conclude this brief review it is worth mentioning one more physical interpretation of charged field coupled to gauge vector potential. Namely, this is the nematic-smectic–A transition in liquid crystals.[68-73] The nematic phase is an orientationally ordered but translationally disordered phase, rodlike molecules are aligned with their long axes parallel to the director and the smectic–A phase contains layers of molecules with their long axes perpendicular to the layer. It has been proposed[68, 69] that this transition can be described by a model similar to that describing NS transition in the charged case.[15] Now the smectic order parameter (being a complex field $\Psi(\vec{r})$ that specifies amplitude and phase of density modulation induced by layering) is coupled to director fluctuations. Contrary to NS transition, the nematic-smectic–A transition is characterized by a critical region in experimentally accessible range. For certain materials it has been indeed shown[72] that both the latent heat data obtained through an adiabatic scanning calorimetry as well as independent interface velocity measurements near the Landau tricritical point can be fitted by a crossover function consistent with mean field free energy density that has a cubic term[15], implying that the nematic-smectic–A transition is a weakly first order. Many liquid crystals, though, appear to exhibit a continuous nematic-smectic–A transition (see Ref.[73] and references therein). High-resolution heat-capacity and X-ray studies of nematic-smectic–A transition performed during the past twenty years[73] show complex systematic trends to crossover from three-dimensional XY to tricritical behaviour and anisotropic behaviour due to coupling between the smectic order parameter and director fluctuations.

3. THE MODEL AND ITS "NAIVE" ANALYSIS

Now it is well–known that the influence of order parameter fluctuations on the NS transition can be described by Landau-Ginsburg free energy functional:

$$F[\phi] = \int d^3x \{ \frac{t_0}{2}|\phi_0|^2 + \frac{1}{2}|(\nabla\phi_0|^2 + \frac{u_0}{4!}|\phi_0|^4 \}, \tag{1}$$

t_0 is temperature-dependent, u_0 is a coupling constant and the complex order parameter ϕ_0 is connected with the wave function of Cooper pairs. The Cooper pairs are charged and therefore create a fluctuating magnetic field which leads to the appearance of additional terms in the free energy functional. Note, that this is not the case of normal-to-superfluid transition in neutral (uncharged) fluid, which is well described by (1) without any modification. Describing the fluctuating magnetic field $\mathbf{B}$ by the vector potential $\mathbf{A}$ ($\mathbf{B} = \mathrm{rot}\mathbf{A}$) and adding to (1) minimal coupling between fluctuating vector potential and order parameter one obtains the free energy functional $F[\Psi, \mathbf{A}]$, originally considered in Ref.[15] for generalized superconductor in d dimensions with a d-dimensional vector potential $\mathbf{A}$ and order parameter Ψ consisting of $n/2$ complex components.

One can now describe fluctuation effects by an Abelian Higgs model with the gauge invariant Hamiltonian[15]:

$$H = \int \mathrm{d}^d x \{ \frac{t_0}{2} |\Psi_0|^2 + \frac{1}{2} |(\nabla - i e_0 \mathbf{A}_0)\Psi_0|^2 + \frac{u_0}{4!} |\Psi_0|^4 + \frac{1}{2} (\nabla \times \mathbf{A}_0)^2 \}, \tag{2}$$

which depends on bare parameters t_0, e_0, u_0. The parameter t_0 changes its sign at some temperature, the rest of parameters are considered temperature-independent. When the coupling constant $e_0 = 0$ no magnetic fluctuations are induced and the model reduces to the usual field theory (1) describing a second-order phase transition and corresponding to particular case $n = 2$ of superfluid transition in ^{4}He.

The mean field results for the critical behaviour of model with free energy functional $F[\Psi, \mathbf{A}]$ corresponding to the Hamiltonian (2) have been already reported in the original paper of Halperin, Lubensky and Ma.[15] In the framework of mean field theory one can determine that systems characterized by free energy functionals $F[\phi]$ (1) and $F[\Psi, \mathbf{A}]$ ($n = 2$) possess a qualitatively different critical behaviour.

Neglecting the order parameter fluctuations (in accordance with the Ginzburg criterion this may be done for a good type–I superconductor) shows that depending on the sign of t_0 the free energy (1) is minimized by the order parameter $\phi = 0$ value for $t_0 > 0$ or by a non-zero value, when $t_0 < 0$, and that the appearance of a non-zero order parameter is continuous; in the system under consideration a second order phase transition occurs.

When applied to the free energy functional $F[\Psi, \mathbf{A}]$, however, the mean field theory predicts a qualitatively different behaviour. Defining the effective free energy $F_{\mathrm{eff}}[\Psi]$ as a function of single variable Ψ by taking the trace over configurations of vector potential $\mathbf{A}$ one finds[15] that the expression for $F[\Psi]$ will contain a term which has a negative sign and is proportional to $|\Psi|^3$. Such term inevitably leads to a first order transition; $F_{\mathrm{eff}}[\Psi]$ develops a minimum at a finite value of Ψ when the quadratic term coefficient is still slightly positive.

As has been already mentioned, the above reasoning is appropriate for a type–I superconductor. The case of type–II superconductors is considerably more complicated. Here, fluctuations in Ψ cannot be neglected and one must choose an appropriate technique to study the problem. Originally, the critical behaviour of the model (2) in the presence of order parameter fluctuations has been studied[15] with the help of Wilson-Fisher recursion relations[35] in first order of $\varepsilon = 4 - d$ and the result shows, in particular, that the second order phase transition is absent for $n = 2$ in the region of couplings appropriate for the type–II superconductor. We will reproduce these ε-expansion results below and then continue to analyze the problem further.

In order to describe long-distance properties of model (2) arising in the vicinity of the phase transition point we shall use a field-theoretical RG approach. Two-loop results[30] for RG functions corresponding to (2) are obtained on the basis of dimensional regularization and minimal subtraction scheme[74], defining the renormalized quantities so as to subtract all poles at $\varepsilon = 4 - d = 0$ from the renormalized vertex functions. Renormalized fields, mass and couplings are introduced by

$$\Psi_0 = Z_\Psi^{1/2}\Psi, \qquad \mathbf{A}_0 = Z_A^{1/2}\mathbf{A}, \qquad t_0 - t_{0c} = Z_t Z_\Psi^{-1} t,$$
$$e_0^2 = Z_e^1 Z_A^{-1} Z_\Psi^{-1} e^2 \mu^\varepsilon S_d^{-1}, \qquad u_0 = Z_u Z_\Psi^{-2} u \mu^\varepsilon S_d^{-1}, \tag{3}$$

where μ is an external momentum scale, t_{0c} is a shift which for the results considered here can be set to zero, and $S_d = 2^{1-d}\pi^{-d/2}/\Gamma(d/2)$ is related to the surface $S_d' = (2\pi)^d S_d$ of the unit d-dimensional hypersphere. The Z–factors are determined by the condition that all poles at $\varepsilon = 0$ are removed from renormalized vertex functions.

The RG equations are written bearing in mind the fact that the bare vertex functions $\Gamma_0^{N,M}$ are calculated with the help of bare Hamiltonian (2) as a sum of one-particle irreducible (1PI) diagrams[75]:

$$\Gamma_0^{N,M}(\{r\}, \{R\}) = \langle \Psi_0(r_1)\ldots\Psi_0(r_N)A_0(R_1)\ldots A_0(R_M)\rangle_{1PI}, \tag{4}$$

which do not depend on the scale μ and their derivatives with respect to μ at fixed bare parameters are equal to zero. So one gets

$$\mu\frac{\partial}{\partial\mu}\Gamma_0^{N,M}|_0 = \mu\frac{\partial}{\partial\mu}Z_\Psi^{N/2}Z_A^{M/2}\Gamma_R^{N,M}|_0 = 0, \tag{5}$$

where the index 0 means a differentiation at fixed bare parameters. The RG equations for the renormalized vertex function $\Gamma_R^{N,M}$ will be

$$\Big(\mu\frac{\partial}{\partial\mu} + \beta_u\frac{\partial}{\partial u} + \beta_f\frac{\partial}{\partial f} + \zeta_\nu t\frac{\partial}{\partial t} - \frac{N}{2}\zeta_\Psi - \frac{M}{2}\zeta_A\Big)\Gamma_R^{N,M}(t,u,f,\mu) = 0, \tag{6}$$

where $f = e^2$, $\zeta_\nu = \zeta_\Psi - \zeta_t$ and RG functions read

$$\beta_u(u,f) = \mu\frac{\partial u}{\partial\mu}|_0, \qquad \beta_f(u,f) = \mu\frac{\partial f}{\partial\mu}|_0,$$
$$\zeta_\Psi = \mu\frac{\partial \ln Z_\Psi}{\partial\mu}|_0, \qquad \zeta_A = \mu\frac{\partial \ln Z_A}{\partial\mu}|_0, \qquad \zeta_t = \mu\frac{\partial \ln Z_t}{\partial\mu}|_0. \tag{7}$$

Using the method of characteristics the solution of RG equation may be written formally as

$$\Gamma_R^{N,M}(t,u,f,\mu) = X(l)^{N/2}(X'(l))^{M/2}\Gamma_R^{N,M}(Y(l)t, u(l), f(l), \mu l), \tag{8}$$

where the characteristics are solutions of following ordinary differential equations:

$$l\frac{d}{dl}\ln X(l) = \zeta_\Psi(u(l), f(l)), \qquad l\frac{d}{dl}\ln X'(l) = \zeta_A(u(l), f(l)),$$
$$l\frac{d}{dl}\ln Y(l) = \zeta_\nu(u(l), f(l)), \tag{9}$$
$$l\frac{d}{dl}u(l) = \beta_u(u(l), f(l)), \qquad l\frac{d}{dl}f(l) = \beta_f(u(l), f(l)) \tag{10}$$

with

$$X(1) = X'(1) = Y(1) = 1, \qquad u(1) = u, \qquad f(1) = f. \tag{11}$$

For small values of l, Eq. (8) is mapping the large length scales (the critical region) to the noncritical point $l = 1$. In this limit the scale-dependent values of couplings $u(l)$, $f(l)$ will approach the stable fixed point, if it exists.

The fixed points u^*, f^* of differential Eqs. (10) are given by the solutions of system of equations:

$$\begin{aligned} \beta_f(u^*, f^*) &= 0, \\ \beta_u(u^*, f^*) &= 0. \end{aligned} \tag{12}$$

The stable fixed point is defined as a fixed point where the stability matrix

$$B_{ij} = \frac{\partial \beta_{u_i}}{\partial u_j}, \qquad u_i = \{u, f\} \tag{13}$$

possesses positive eigenvalues (or if complex, the eigenvalues with positive real parts). The stable fixed point corresponds to the critical point of the system. As we have mentioned above, in the limit $l \to 0$ (corresponding to the limit of infinite correlation length) renormalized couplings reach the values they have in the stable fixed point.

Now we can write the results for RG functions obtained in a two-loop approximation[30] following the above described procedure in frames of dimensional regularization and minimal subtractions schemes. From a Ward identity one has $Z_\Psi = Z_e$, and the remaining Z–factors are to be found from corresponding vertex functions $\Gamma^{2,0}$, $\Gamma^{0,2}$, and $\Gamma^{4,0}$. Since the gauge field is massless, the renormalization has been performed at a finite wave vector. The results in two-loop order read

$$Z_\Psi = 1 + \frac{1}{\varepsilon}\{3e^2 - u^2(n+2)/144 + e^4[(n+18)/4\varepsilon - (11n+18)/48]\}, \tag{14}$$

$$Z_{\mathrm{A}} = 1 + \frac{1}{\varepsilon}\{-ne^2/6 - ne^4/2\}, \tag{15}$$

$$\begin{aligned} Z_t = 1 &+ \frac{1}{\varepsilon}\{(n+2)u/6 + u^2[(n+2)(n+5)/36\varepsilon - (n+2)/24] \\ &+ ue^2[-(n+2)(1/2\varepsilon - 1/3)] + e^4[(3n+6)/2\varepsilon + (5n+1)/4]\}, \end{aligned} \tag{16}$$

$$\begin{aligned} Z_u = 1 &+ \frac{1}{\varepsilon}\{(n+8)u/6 + 18e^4/u + u^2[(n+8)^2/36\varepsilon - (5n+22)/36] \\ &+ ue^2[-(n+8)/2\varepsilon + (n+5)/3] + e^4[(3n+24)/\varepsilon + (5n+13)/2] \\ &+ e^6/u[3(n+18)/\varepsilon - 7n/2 - 45]\}. \end{aligned} \tag{17}$$

Following the standard procedure one then shows that the expressions for β-functions in two-loop approximation will be

$$\beta_f = -\varepsilon f + \frac{n}{6}f^2 + nf^3, \tag{18}$$

$$\begin{aligned} \beta_u &= -\varepsilon u + \frac{n+8}{6}u^2 - \frac{3n+14}{12}u^3 - 6uf + 18f^2 \\ &+ \frac{2n+10}{3}u^2 f + \frac{71n+174}{12}uf^2 - (7n+90)f^3. \end{aligned} \tag{19}$$

The previous analysis of equations like (18), (19) either on one-loop[15] or two-loop level[30] has been based on direct solutions of the fixed point equation. In the present study we want to point out that the series have a zero radius of convergence and they are known to be asymptotic at best. Therefore, some additional mathematical methods have to be applied in order to obtain a reliable information on their basis.

We shall start by recalling the results of an ε^2-expansion for β-functions.[15, 30] For second order in ε one obtains three fixed points: Gaussian ($u^{*G} = f^{*G} = 0$), "Uncharged" ($u^{*U} \neq 0, f^{*U} = 0$) and "Charged" ($u^{*C} \neq 0, f^{*C} \neq 0$), to be denoted as G, U, C, respectively. The expressions for them read

$$\mathrm{G}: \quad u^{*G} = 0, \qquad f^{*G} = 0, \tag{20}$$

$$\mathrm{U}: u^{*U} = u_1^U \varepsilon + u_2^U \varepsilon^2, \; f^{*U} = 0, \tag{21}$$

$$\mathrm{C}: u^{*C} = u_1^C \varepsilon + u_2^C \varepsilon^2, \; f^{*C} = f_1^C \varepsilon + f_2^C \varepsilon^2, \tag{22}$$

where

$$u_1^U = \frac{6}{n+8}, \qquad u_2^U = \frac{18(3n+14)}{(n+8)^3},$$

$$u_1^C = \frac{3(n+36) + (n^2 - 360n - 2160)^{1/2}}{3n(n+8)},$$

$$u_2^C = \frac{a_2}{a_1}, \qquad f_1^C = \frac{6}{n}, \qquad f_2^C = -\left(\frac{6}{n}\right)^3 n$$

with

$$a_1 = 1 + \frac{n+8}{3} u_1^C - \frac{36}{n},$$

$$a_2 = \frac{3n+14}{12}\left(u_1^C\right)^3 - 6nu_1^C\left(\frac{6}{n}\right)^3 + 36n\left(\frac{6}{n}\right)^4 - \frac{(n+5)4}{n}\left(u_1^C\right)^3 - \frac{3(71n+174)}{n^2}u_1^C + \left(\frac{6}{n}\right)^3(7n+90).$$

Almost all physical results concerning the phase transition described by the field theory (2) are to some extent based on the information given by (20)–(22). The main ones are:

(i) The fixed point U is unstable with respect to the presence of f-symmetry at $d < 4$ with a stability exponent

$$\lambda_f(u = u^{*U}, f = f^{*U} = 0) = \frac{\partial \beta_f}{\partial f}|_U = -\varepsilon.$$

(ii) The fixed point C appears to be complex already for $n < n_c = 365.9$ 15 on a one-loop level. The stability exponent is given by

$$\lambda_u(u = u^{*C}, f = f^{*C}) = \frac{\partial \beta_u}{\partial u}|_C$$

and on the two-loop level it reads

$$\lambda_u = -\varepsilon s, \qquad s = \left[\left(1 + \frac{36}{n}\right)^2 - \frac{432(n+8)}{n^2}\right]^{1/2},$$

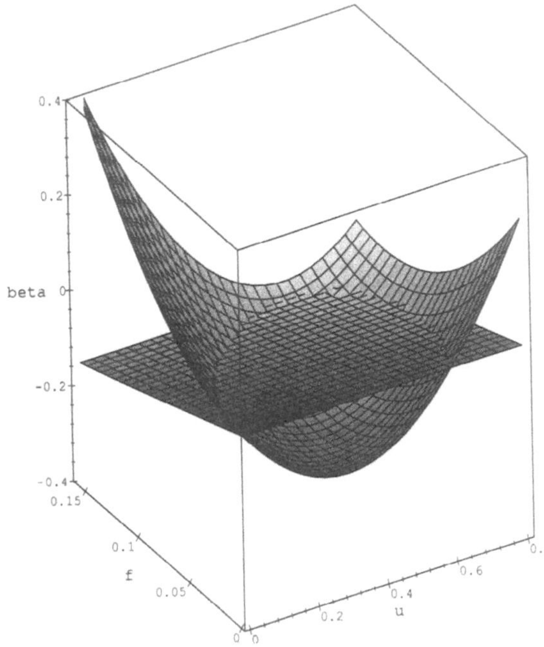

Figure 1. β-functions of the model of superconductor $\beta_u(u, f)$, $\beta_f(u, f)$ in one-loop approximation for $d = 3$, $n = 2$. The fixed points have coordinates $(u^* = f^* = 0)$, $(u^* = 0.6, f^* = 0)$.

leading to an oscillatory flow of u in one-loop order below n_c described by the solution[26, 30]

$$f(l) = \frac{6fl^{-\varepsilon}}{6 + n\varepsilon f(l^{-\varepsilon} - 1)}, \tag{23}$$

$$u(l) = f(l)\frac{n}{2(n+8)}\left\{s \tan\left[\frac{s}{2}\ln\left(f(l)f^{-1}l^{\varepsilon}\right) + \arctan\left(\frac{2(n+8)}{sn}\frac{u}{f} + \frac{n+36}{ns}\right)\right] - \frac{n+36}{n}\right\}; \tag{24}$$

here f and u are the initial parameters for $l = 1$.

(iii) From the condition of positiveness of the fixed point coordinate f^* $(f = e^2)$ it follows that at $\varepsilon = 1$, n has to be larger that 36. This questions the applicability of ε-expansion for $n = 2$ to $d = 3$.

So, it follows that for the "superconductor" case $n = 2$ which is most interesting from a physical point of view a stable fixed point does not exist and, therefore, the observed phase transition is of first order.

Now we shall study the RG equations in the minimal subtraction scheme in the framework of three dimensional ($3d$-) superconductors ("$3d$-theory"[11-13]) by putting $\varepsilon = 1$ in expressions for RG functions and studying the perturbation theory in powers of coupling constants. These powers correspond to the number of loops in Feynman diagrams and hence one develops a perturbation theory in successive number of loops. Direct calculations based on Eqs. (18) and (19) at fixed $d = 3$ do not bring qualitatively new features to this analysis. In the one-loop approximation, leaving square terms in

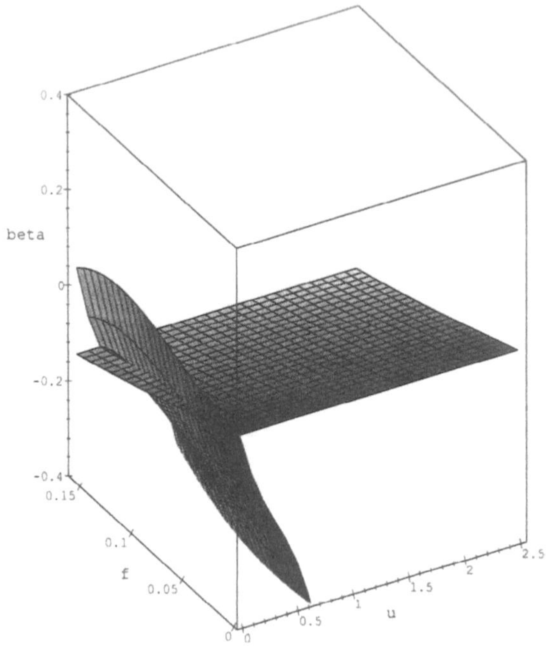

Figure 2. β-functions of the superconductor model $\beta_u(u, f)$, $\beta_f(u, f)$ in the two-loop approximation for $d = 3$, $n = 2$. Only the Gaussian fixed point $u^* = f^* = 0$ survives.

Eqs. (18) and (19) one finds that only one nontrivial fixed point $u^* = 6/(n+8)$, $f^* = 0$ exists. The β-functions $\beta_u(u, f)$, $\beta_f(u, f)$ in one-loop approximation at $d = 3$, $n = 2$ are shown in Fig. 1. The simultaneous intersection of surfaces corresponding to both functions with the plane $\beta = 0$ results in fixed points; for $n = 2$ they have coordinates $u^* = f^* = 0$ and $u^* = 0.6$, $f^* = 0$ as seen in the picture. In two-loop approximation only the Gaussian fixed point survives, as one may see from Fig. 2.

Nevertheless one should note that such straightforward interpretation of above expansion data has been questioned and the way of analyzing the series (18) and (19) for β-functions has been proposed[30] in which the strict ε-expansion is avoided and the information about the accurate solution for pure model at $d = 3$ is exploited. Also, from the comparison of ε-expansion data for f^* (giving positive values of f^* only for $n > 36$) with the value of f^* obtained without ε-expansion (remaining positive for all n) it has been conjectured that the lower boundary for n resulting in negative f^* may be an artifact of the expansion procedure. Let us now consider expressions for RG functions more carefully, paying attention to their possible asymptotic nature and treating them with a resummation procedure.

4. RESUMMATION

The resummation technique appropriate for critical phenomena and applied to the asymptotic series for RG functions enables one to obtain extremely accurate values of critical exponents.[76] In fact, the asymptotic nature of series for RG functions has been only proven in case of the ϕ^4 model containing one coupling of $O(n)$-symmetry (the n-vector model). The high-order asymptotes for these series are known[27-29] in analytical form as well. These results give the possibility of obtaining precise values for the critical exponents of the n-vector model by a resummation of corresponding series for the renormalization group functions (see, e.g., Refs.[7, 8, 11]). As far as we know, no information similar to that obtained[27-29] for the "uncharged" case ($f = 0$) is available for the "charged" model. For models containing several couplings of different symmetries the asymptotic nature of corresponding series for the RG functions is a generally accepted belief rather than a proven fact.

As it will be important in the course of our future analysis we shall mention here the weakly diluted n-vector model, describing ferromagnetic ordering in a system of N_1 classical n-component "spins" located in N sites of a lattice ($N_1/N < 1$) and quenched in a certain configuration. Using the replica trick in order to perform quenched averaging one concludes[82] that an effective Hamiltonian of such model contains two fourth order terms of different symmetry and reads:

$$H = \int \mathrm{d}^d x \Big\{ \frac{1}{2} \sum_{\alpha=1}^{m} \left[|\nabla \vec{\phi}^\alpha|^2 + m_0^2 |\vec{\phi}^\alpha|^2 \right] - \frac{v_0}{8} \left(\sum_{\alpha=1}^{m} |\vec{\phi}^\alpha|^2 \right)^2 + \frac{u_0}{4!} \sum_{\alpha=1}^{m} \left(|\vec{\phi}^\alpha|^2 \right)^2 \Big\}, \tag{25}$$

where $\vec{\phi}^\alpha$ is an n-component vector $\vec{\phi}^\alpha = (\phi^{\alpha,1}, \phi^{\alpha,2}, \ldots, \phi^{\alpha,n})$; $u_0 > 0, v_0 > 0$ are bare coupling constants; m_0 is the bare mass and in the final results a replica limit $m \to 0$ is to be taken. The RG functions for this model are obtained in the form of double series in renormalized couplings u, v and the asymptotic nature of series has not been proven for this model till now.[84] Nevertheless, the appropriate resummation technique (applied *as if* these series are asymptotic) enables one to obtain accurate values for critical exponents in three dimensions[77-81] and to describe (in the $n = 1$ case) the experimentally observed crossover to a new type of critical behavior caused by weak dilution.[85, 86] These results have been confirmed by Monte-Carlo[87, 88] and Monte-Carlo RG[89] calculations.

The two main ways of resummation commonly used for the asymptotic series arising in the RG approach are: (i) resummation based on the conformal mapping technique and (ii) the Padé-Borel resummation. Case (i) is based on the conformal transformation, which maps part of the analytical domain containing the real positive axis onto a circle centered at the origin and the asymptotic expansion for a certain function is thus re-written in the form of new series.[8] This resummation, however, is based on the knowledge of subtle details of asymptotes (the location of pole, high-order behaviour) which are not available in our case.

In the absence of knowledge about the singularities of series, the most appropriate method which can be used to perform the analytical continuation is the Padé approximation resulting in Padé-Borel resummation technique (ii) (see, e.g., Ref.[7]). We are going to apply it to the special case, $f = 0$, so let us disucss it in detail.

Starting from the Taylor series for the function $f(u)$:

$$f(u) = \sum_{j \geq 0} c_j \, u^j, \tag{26}$$

one constructs the Borel–Leroy transform

$$F(ut) = \sum_{j \geq 0} \frac{c_j}{\Gamma(j+p+1)} \, (ut)^j, \tag{27}$$

where $\Gamma(x)$ is the Euler gamma-function and p is some positive number.[90] Then one represents (27) in the form of Padé approximant $F^{\text{Padé}}_{[L/M]}(ut)$:

$$F^{\text{Padé}}_{[L/M]}(x) = \frac{\sum\limits_{i=0}^{L} a_i x^i}{\sum\limits_{j=0}^{M} b_j x^j} \tag{28}$$

(in the subsequent analysis, proceeding in two-loop approximation we shall use the [1/1] Padé approximant) and the resummed function will be given by

$$f^{\text{Res}}(u) = \int\limits_0^\infty dt\ e^{-t}\ t^p\ F^{\text{Padé}}(ut). \tag{29}$$

The resummation scheme (27) – (29) of (asymptotic) series in one variable (26) is easily generalized for two-variable case when the series is given in the form

$$f(u,v) = \sum_{j,j\geq 0} c_{i,j}\ u^i\ v^j, \tag{30}$$

with the Borel-Leroy transform

$$F(u,v,t) = \sum_{i,j\geq 0} \frac{c_{i,j}}{\Gamma(i+j+p+1)}\ (ut)^i (vt)^j. \tag{31}$$

The procedure is aimed to help how to choose an appropriate form of series analytic continuation (31). Two most common methods to proceed are the Borel resummation combined with Chisholm approximants and the Borel resummation of resolvent series, presented in the form of Padé approximant. For the first method, in order to write an analytic continuation of series (31) one uses the rational approximants of two variables: so-called Canterbury approximants or generalized Chisholm approximants[91, 93] which are a generalization of Padé approximants in the case of several variables, representing (31) in the form

$$F^{\text{Chisholm}}(u,v,t) = \frac{\sum\limits_{i,j} a_{i,j} u^i v^j t^{i+j}}{\sum\limits_{i,j} b_{i,j} u^i v^j t^{i+j}}, \tag{32}$$

(sums in numerator and denominator are limited by the condition of correspondence between known numbers of terms in the initial series and those in the approximant). Again, the resummed function is given by the integral (29):

$$f^{\text{Res}}(u,v) = \int\limits_0^\infty dt\ e^{-t}\ t^p\ F^{\text{Chisholm}}(ut). \tag{33}$$

Proceeding with the second method, one writes for series of two variables (30) the so-called resolvent series $\mathcal{F}(u,v,\tau)$[92, 93] introducing an auxiliary variable τ, which allows the separation of contributions from different orders of perturbation theory in variables u, v:

$$\begin{aligned} \mathcal{F}(u,v,\tau) &= \sum_{i,j\geq 0} c_{i,j}\, (u\tau)^i\, (v\tau)^j, \\ f(u,v) &= \mathcal{F}(u,v,\tau=1). \end{aligned} \tag{34}$$

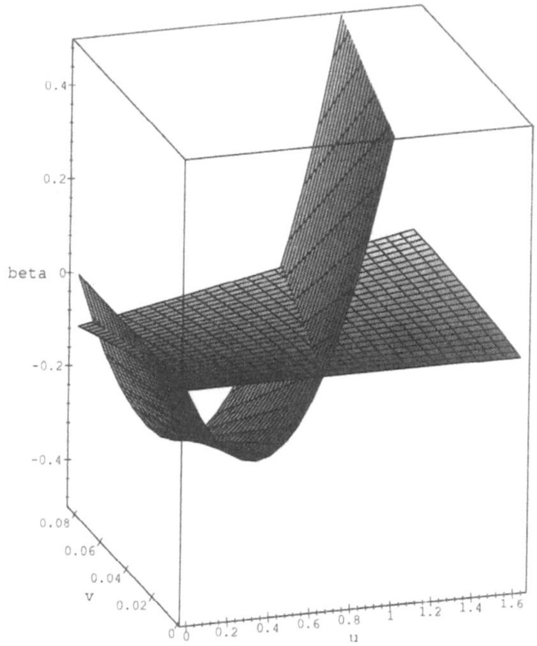

Figure 3. β-functions of diluted Ising model $\beta_u(u,v)$, $\beta_v(u,v)$ in the one-loop approximation for $d=3$. The fixed points have coordinates $(u^*=v^*=0)$, $(u^*=0.667,\ v^*=0)$.

Now the resummation of series $\mathcal{F}(u,v,\tau)$ is performed with respect to the variable τ as for series in a single variable, applying the above described scheme (27)–(29).

Let us illustrate how the resummation procedure works in case of the effective Hamiltonian (25). In order to make a direct comparison with the superconductor case, let us take β-functions obtained for the model (25) in the minimal subtraction scheme in two-loop approximation, though high-order results are available for this model[81, 95] as well as results[77-80] obtained in the $d=3$ massive field theoretical approach.[94] The expressions for β-functions, corresponding to renormalized couplings u, v in the replica limit $m \to 0$ for the Ising model ($n=1$) read

$$\beta_u = -\varepsilon u + \frac{3}{4}u^2 - 6uv - \frac{17}{12}u^3 + \frac{23}{2}u^2 v - \frac{41}{2}uv^2, \tag{35}$$

$$\beta_v = -\varepsilon v + uv + -4v^2 - \frac{5}{12}u^2 v + \frac{11}{2}uv^2 - \frac{21}{2}v^3. \tag{36}$$

We shall not present the expressions for the other RG functions here, as we are going to study only the fixed point equations.

Looking for the solutions of the fixed point equations for functions (35), (36) one can show that in the one-loop approximation, in addition to the Gaussian fixed point $u^*=v^*=0$, there exist two more solutions $u^*=2/3$, $v^*=0$ and $u^*=0$, $v^*=-1/5$ and the solution $u^*\neq 0$, $v^*\neq 0$ is absent.[96] The fixed point with $v^*<0$ is beyond the region of parameters describing the diluted magnet[97] and the pure model fixed point $u^*\neq 0, v^*=0$ appears to be unstable with respect to v-coupling (we propose to the reader to check this by looking at the stability matrix $B_{ij}(u,v)$ (13) eigenvalues at the fixed points). The corresponding plot of functions β_u, β_v in one-loop approximation is shown in Fig. 3. Passing to the two-loop approximation makes the result even worse: only the Gaussian fixed point is present (see Fig. 4). Returning back to the initial problem statement one should conclude that the obtained picture corresponds to an absence of second order phase transition in $d = 3$ Ising model with a weak dilution as well as without a dilution (the absence of the fixed point $u^* \neq 0$, $v^* = 0$). Of course, this is contrary to the real situation. Let us further note that the behaviour obtained for β-functions of the model (25) in one- and two-loop approximations (Figs. 3, 4) resembles those for the superconductor case in corresponding approximations (Figs. 1, 2).

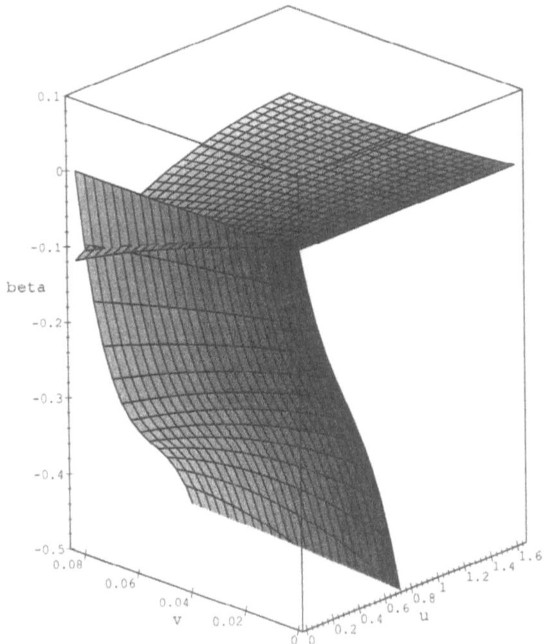

Figure 4. β-functions of diluted Ising model $\beta_u(u,v)$, $\beta_v(u,v)$ in the two-loop approximation for $d = 3$. Only the Gaussian fixed point $u^* = v^* = 0$ survives.

Applying, however, the resummation procedure to series (35) and (36) in two loop approximation one reconstitutes fixed points ($u^* \neq 0$, $v^* = 0$), ($u^* = 0$, $v^* \neq 0$) and obtains a new stable fixed point $u^* \neq 0$, $v^* \neq 0$ which governs second order phase transition in a weakly diluted Ising model. The picture obtained appears to be stable with respect to successive accounts of higher order terms in the perturbation theory, when the appropriate resummation technique is applied. As we have already claimed above, these RG results are confirmed by other different theoretical approaches and correspond to the experimentally observed second order phase transition in a weakly diluted Ising magnet with critical exponents differing from those of pure case. In Fig. 5 we show the crossing of $\beta_u(u,v)$ and $\beta_v(u,v)$ surfaces for the resummed function. The calculations have been performed by means of Padé-Borel resummation technique for resolvent series (34) of two-loop functions (35), (36) as described above.[98] Gaussian ($u^* = v^* = 0$) and pure ($u^* = 1.3146$, $v^* = 0$) fixed points can be seen at the cube rear. The cross-section of u and v planes in the picture are chosen to pass through the stable fixed point ($u^* = 1.6330$, $v^* = 0.0835$) corresponding to a new critical behaviour.

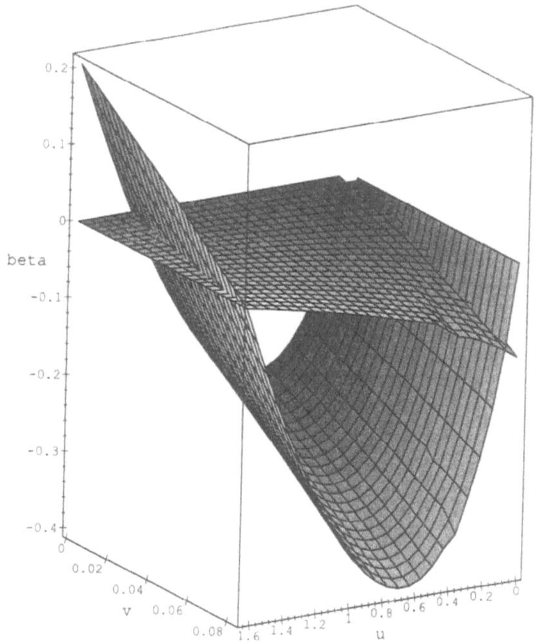

Figure 5. β-functions of diluted Ising model $\beta_u(u, v)$, $\beta_v(u, v)$ in the two-loop approximation for $d = 3$ obtained by applying the Padé-Borel resummation technique. The resummation restores the presence of fixed point $u^* \neq 0$, $v^* = 0$ and results in the appearance of new stable fixed point $u^* \neq 0$, $v^* \neq 0$.

The example we have considered above is a typical situation corresponding to the RG $3d$-theory: when considered without an appropriate resummation technique, the RG analysis might not only give imprecise values for critical exponents but also a qualitatively wrong answer about the absence of stable fixed point for a certain model, resulting in the absence of second order phase transition. Now with this information in hand let us pass to the analysis of superconductor model described in two-loop approximation by RG functions (35) and (36).

5. FIXED POINTS AND FLOWS IN THREE DIMENSIONS

We shall continue by considering the flow equations (10) directly for $d = 3$. We shall look for the solutions of fixed point equations at $d = 3$ paying attention to the possible asymptotic nature of corresponding series (18), (19). Consider first the equation for the uncharged fixed point U. Substituting $f^* = 0$ in (19) one obtains the following expression for the function $\beta_u^U \equiv \beta_u\,(u, f^* = 0)$:

$$\beta_u^U = -u + \frac{n+8}{6}u^2 - \frac{3n+14}{12}u^3. \tag{37}$$

Solving this polynomial for the fixed point one obtains for non-trivial $u^* > 0$:

$$u^{*U} = \frac{n+8}{3n+14} + \frac{\sqrt{n^2 - 20n - 104}}{3n+14} \tag{38}$$

and immediately the "condition of existence of a non-trivial solution u^{*U}" follows qualitatively, quite similar to those, appearing in the ε-expansion technique (see Refs.[15, 30]

and formula (21) of the present article as well); the solution exists only for certain values of $n > n_c = 24.3$!

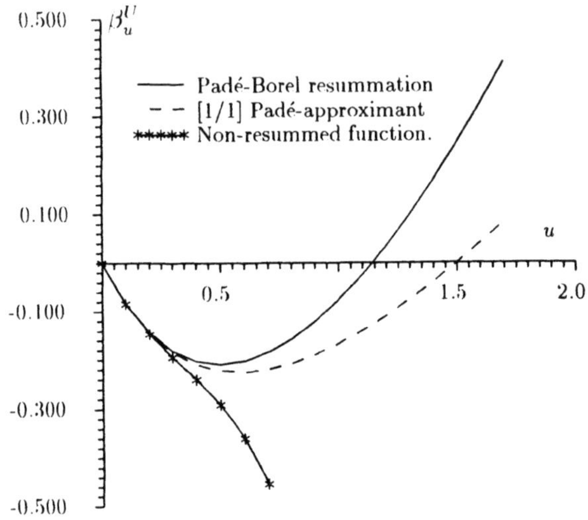

Figure 6. β_u-function of the uncharged model β_u^U at $d = 3$, $n = 2$.

From Fig. 6 one can see that the function β_u^U given by Eq. (37) does not intersect the u-axis for any non-zero value of u for $n = 2$. In the $O(n)$-symmetric ϕ^4-theory at $d = 3$ this situation is well known (see, e.g., Refs.[94, 99]). The β-function calculated directly at $d = 3$ does not possess a stable zero for realistic values of n, nevertheless within the three-loop order the presence of stable fixed point is restored. To avoid this artifact appearing in the two-loop calculation one can either resum the series for β-function or construct appropriate Padé approximant[100] in order to perform the analytical continuation of Eq. (37) out of the domain of convergence (which is equal to zero for series in the right-hand side of (37)). Let us try both methods. Representing Eq. (37) in the form of [1/1] Padé approximant,

$$\beta_u^{U,Padé} = u\frac{-1 + A_u u}{1 + B_u u}\,, \tag{39}$$

one obtains

$$A_u = \frac{n^2 + 7n + 22}{6(n+8)}, \qquad B_u = \frac{3n + 14}{2(n+8)}, \tag{40}$$

and solving the equation for the fixed point

$$\beta_u^{U,Padé}(u^{*P,Padé}) = 0\,, \tag{41}$$

one obtains

$$u^{*U,Padé} = \frac{6(n+8)}{n^2 + 7n + 22}. \tag{42}$$

So we have obtained a qualitatively different situation. The behaviour of the function $\beta_u^{U,Padé}(u)$ for $n = 2$ is shown in Fig. 6 by the dashed curve. If one is interested in more accurate values of u^* some resummation has to be applied. Choosing the Padé–Borel

Table 1. The fixed point U coordinate u^{*U} as a function of n. $u^{*U,Padé}$ is obtained on the basis of [1/1] Padé approximant; $u^{*U,Res}$ is obtained by the Padé-Borel resummation.

n	1	2	3	4	5	6	7	8
$u^{*U,Padé}$	1.800	1.500	1.269	1.091	0.951	0.840	0.750	0.676
$u^{*U,Res}$	1.315	1.142	1.002	0.888	0.794	0.717	0.652	0.597

resummation technique[101] and following the scheme (26) – (29) one obtains for the resummed function $\beta_u^{U,Res}$ (see Ref.[98])

$$\beta_u^{U,Res} = u\left[2\left(1 - A_u/B_u\right)\left(1 - E\left(\frac{2}{uB_u}\right)\right) - 1\right]. \tag{43}$$

In Eq. (43) the coefficients A_u, B_u are given by Eqs. (40), $E(x) = xe^x E_1(x)$, where the function

$$E_1(x) = e^{-x}\int_0^\infty dte^{-t}(x+t)^{-1}$$

is connected with the exponential integral by the relation[102]

$$E_1(x \pm i0) = -Ei(-x) \mp i\pi.$$

The behaviour of the function $\beta_u^{U,Res}(u)$ is shown in Fig. 6 by the solid curve and the fixed point coordinate $u^{*U,Res}$ is obtained by solving the non-linear equation

$$\beta_u^{U,Res}(u^{*U,Res}) = 0. \tag{44}$$

The coordinates of the fixed point u^{*U} obtained on the basis of Padé approximation and Padé-Borel resummation ($u^{*U,Padé}$, $u^{*U,Res}$) for different n are given in Table 1.

From this analysis we conclude that in the $d = 3$ theory Padé approximants (as an analytical continuation of β-functions) may change the picture qualitatively and lead to values of fixed points comparable to those obtained by the Padé-Borel resummation technique.

Consider now the equation for the charged fixed point C applying the above considerations to β_f, from Eq. (18), for which the expression at $d = 3$ reads

$$\beta_f = -f + \frac{n}{6}f^2 + nf^3. \tag{45}$$

The behaviour of β_f as a function of f is shown in Fig. 7 by asterisks. Note, however, that in this case the function β_f even without any resummation possesses non-trivial zero f^{*M} (its value $f^{*C,Dir}$ is given in the 2nd row of Table 2). Representing (45) in the form of [1/1] Padé approximant,

$$\beta_f^{Padé} = f\frac{-1 + A_f f}{1 + B_f f}, \tag{46}$$

one has for A_f, B_f:

$$A_f = \frac{n+36}{6}, \quad B_f = -6 \tag{47}$$

Table 2. The fixed point C coordinate f^{*C} as a function of n. $f^{*C,Dir}$ is obtained by a direct solution of equation for the fixed point; $f^{*C,Padé}$ is obtained on the basis of [1/1] - Padé approximant; $f^{*C,\varepsilon}$ is the ε-expansion result with a linear accuracy in ε; f^{*C,ε^2} is the ε-expansion result with a square accuracy in ε.

n	1	2	3	4	5	6	7	8
$f^{*C,Dir}$	0.920	0.629	0.500	0.424	0.372	0.333	0.304	0.280
$f^{*C,Padé}$	0.162	0.158	0.154	0.150	0.146	0.143	0.140	0.136
$f^{*C,\varepsilon}$	6.000	3.000	2.000	1.500	1.200	1.000	0.857	0.750
f^{*C,ε^2}	−210.000	−51.000	−22.000	−12.000	−7.440	−5.000	−3.551	−2.625

and, solving the equation for the fixed point coordinate $f^{*C,Padé}$:

$$\beta_f^{Padé}(f^{*C,Padé}) = 0 \tag{48}$$

one obtains

$$f^{*C,Padé} = \frac{6}{n+36} \,. \tag{49}$$

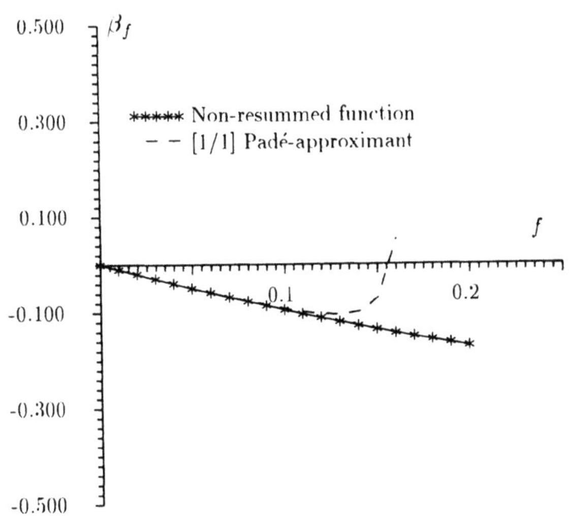

Figure 7. β_f-function at $d = 3, n = 2$.

The function $\beta_f^{Padé}(f)$ is shown in Fig. 7 by the dashed line, the coordinate $f^{*C,Padé}$ is given in the 3rd row of Table 2. But now the series (45) is not alternating and this results in the presence of a pole (at $f = 1/6$) in the approximant (46). Therefore,

Eq. (46) correctly represents the function $\beta_f(f)$ only for $f < 1/6$. Let us note however that for all positive n a fixed point exists and its coordinate $f^{*M,Padé}$ lies within limits $0 < f^{*C,Padé} < 1/6$, where there are no poles in (46). Comparing this result with those obtained for the uncharged fixed point one can conclude that the β_f representation in the Padé approximant form does not qualitatively change the picture (a solution for $\beta_f(f) = 0$ exists at $d = 3$ even without an analytical continuation) but results in a decrease of fixed point coordinate. In contrast to the ε-expansion values (22) there are no borderline values of n for the positivity of f^{*C}. Unfortunately, we cannot check this result by means of Padé-Borel resummation technique. The above mentioned presence of a pole in the Padé approximant denominator makes the corresponding integral representation problematic.[103]

In order to find the u-coordinate of fixed point C, u^{*C}, we have to deal with a function of two variables, $\beta_u(u, f)$, represented by rather short series (19). Another problem arises due to the fact that the function $\beta_u(u, f)$ contains generating terms (i.e. $\beta_u(u = 0, f) \neq 0$). In order to perform some kind of analytic continuation of function of two variables one can use Chisholm approximants[91, 93] as shown by Eq. (32). But the presence of generating terms makes this choice rather ambiguous. The most reliable way in such case is the representation of $\beta_u(u, f)$ in the form of "resolvent" series $B(u, f, \tau)$, see Eq. (34) and Ref.[92, 93], introducing an auxiliary variable τ, which allows the separation of contributions from different orders of the perturbation theory in the coupling constant. The series for $B(u, f, \tau)$ then reads

$$B(u, f, \tau) \equiv \beta_u(u\tau, f\tau) = \sum_{j\geq 0} b_j\, \tau^j, \tag{50}$$

with an obvious notation for the coefficients b_j. Now one considers (50) as a series in a *single* variable τ. This series can be represented in a form of Padé approximant $B^{Padé}(u, f, \tau)$ as an analytical continuation of the function $B(u, f, \tau)$ for general value of τ. In particular at $\tau = 1$ the equality holds $B(u, f, \tau = 1) = \beta_u(u, f)$ and the approximant

$$B^{Padé}(u, f, \tau = 1) \equiv \beta_u^{Padé}(u, f)$$

represents the initial function $\beta_u(u, f)$. In our case the expression for $B(u, f, \tau)$ reads

$$B(u, f, \tau) = \tau(b_1 + b_2\tau + b_3\tau^2), \tag{51}$$

where

$$b_1 = -u, \qquad b_2 = \frac{n+8}{6}u^2 - 6uf + 18f^2,$$

$$b_3 = -\frac{3n+14}{12}u^3 + \frac{2n+10}{3}u^2 f + \frac{71n+174}{12}uf^2 - (7n+90)f^3. \tag{52}$$

Representing the expression in brackets in the right-hand side of (51) in the [1/1] Padé approximant form we have

$$B^{Padé}(u, f, \tau) = \tau\, b_1 \frac{1 + A_{u,f}\tau}{1 + B_{u,f}\tau}, \tag{53}$$

where

$$A_{u,f} = \frac{b_2}{b_1} - \frac{b_3}{b_2}, \qquad B_{u,f} = \frac{-b_3}{b_2}. \tag{54}$$

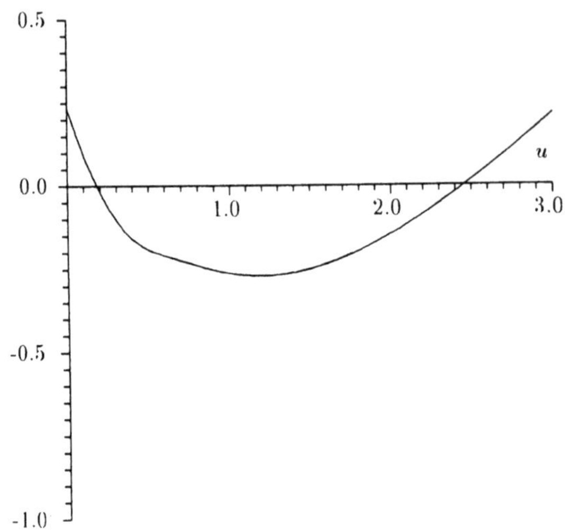

Figure 8. The intersection of function $\beta_u^{Padé}(u,f)$ at $d=3$, $n=2$ with the plain $f=f^{*C,Padé}$ in two-loop approximation.

Table 3. The fixed point C coordinates $u^{*C,Padé}$ obtained on the basis of [1/1] Padé approximant for "resolvent" series as a function of n. C1 : the unstable fixed point; C2 : the stable fixed point.

n	1	2	3	4	5	6	7	8
C1	0.184	0.181	0.179	0.177	0.175	0.175	0.176	0.179
C2	3.309	2.457	1.781	1.150	0.473	0.369	0.305	0.256

Let us note here that the function $B(u,f,t)$ obtained in this way as an approximant for the function of two variables $\beta_u(u,f)$ obeys certain projection properties in the single-variable case: substituting $f=0$ or $u=0$ into (53) one obtains the [1/1] Padé approximant for $\beta_u^U(u)$ or the [0/1] Padé approximant for $\beta_u(u=0,f)$. Finally the expression for $\beta_u(u,f)$ approximated in such way reads

$$\beta_u^{Padé}(u,f) = b_1 \frac{1+A_{u,f}}{1+B_{u,f}}. \tag{55}$$

Substituting into the equation for the fixed point $\beta_u(u^{*C},f^{*C})=0$ the value of coordinate $f^{*C}=f^{*C,Padé}$, see Eq. (49), one obtains the non-linear equation for $u^{*C,Padé}$:

$$\beta_u^{Padé}(u, f=f^{*C,Padé}) = 0. \tag{56}$$

Solving Eq. (56) with respect to u one obtains the values $u^{*C,Padé}$ given in Table 3. The intersection of the function $\beta_u^{Padé}(u,f)$ given by Eq. (55) with the plane $f=f^{*C,Padé}$ is shown for $n=2$ in Fig. 8. The first fixed point (C1) given in the 2nd row of Table 3 turns out to be unstable, while the fixed point C2 is stable in the case $n=2$ we are predominantly interested in.

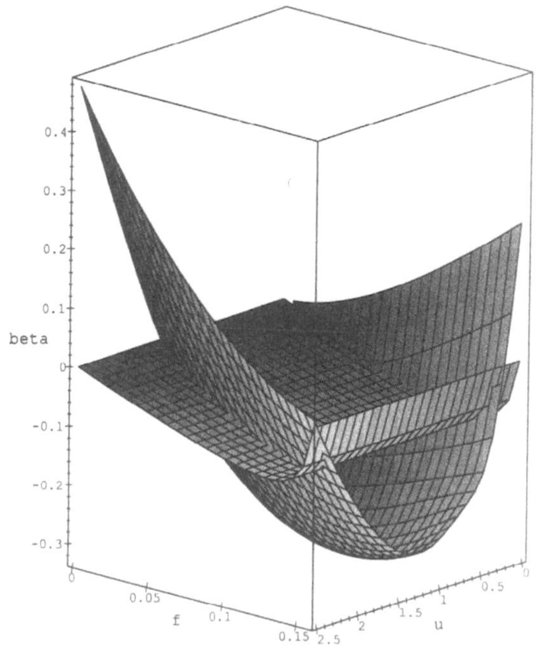

Figure 9. β-functions of the model of superconductor $\beta_u(u,f)$, $\beta_f(u,f)$ in the two-loop approximation for $d=3$, $n=2$ obtained by a Padé analysis for the resolvent series. The stable "charged" fixed point C2 with coordinates $u^* = 2.457$, $f^* = 0.158$ as well as the unstable fixed point C1 $u^* = 0.181$, $f^* = 0.158$ are seen at the cube front side. Gaussian ($u^* = f^* = 0$) and "uncharged" ($u^* = 1.500$, $f^* = 0$) fixed points are located at the cube rear.

The resulting picture of β-functions surfaces is shown in Fig. 9. The Gaussian and uncharged fixed points may be seen at the picture rear, whereas the intersection of u- and f-planes has been chosen in the picture to cross the stable fixed point C2. The unstable fixed point C1 is seen as well.

The crossover to asymptotic critical behavior is described by the solutions of flow equations (10) with initial values of $u(\ell_0)$ and $f(\ell_0)$ at $\ell = \ell_0$.[104] Substituting the analytical continuation of β-functions in the right-hand side of Eq. (10) with Padé approximants (46) and (55), we get the following system of differential equations:

$$l\frac{df}{dl} = f\frac{-1 + A_f f}{1 + B_f f}, \tag{57}$$

and (58)

$$l\frac{du}{dl} = -u\frac{1 + A_{u,f}}{1 + B_{u,f}},$$

where A_f, B_f and $A_{u,f}$, $B_{u,f}$ are given by Eqs. (47) and (54), respectively.

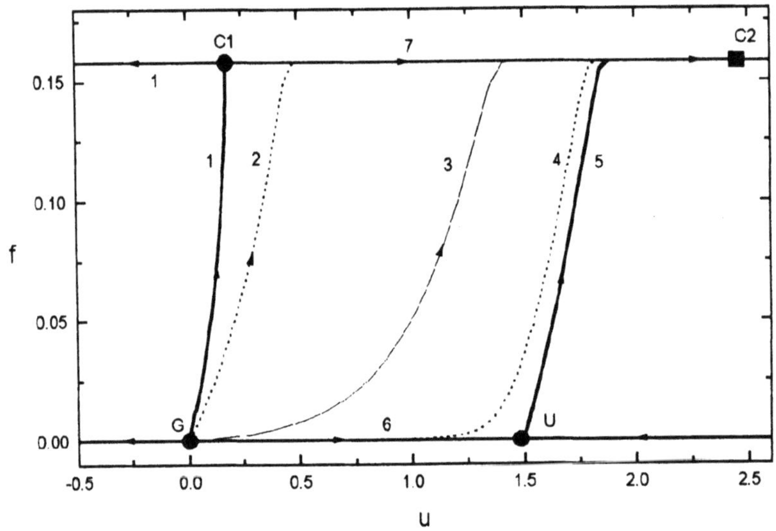

Figure 10. Flow lines for the case $n = 2$, $d = 3$ given by Eqs. (57) and (59). Fixed points G, U, C1 are unstable, fixed point C2 (shown by a box) is the stable one (for a further description see the text).

Solving Eqs. (57) and (59) numerically one obtains the flow diagram shown in Fig. 10 for the case $n = 2$. The space of couplings is divided into several parts by separatrices (thick lines in Fig. 10) connecting the fixed points. Besides the Gaussian (G) there exist three fixed points, one corresponding to the uncharged (U) and two others corresponding to the charged (C1, C2) cases. The fixed points G, C1 and U are unstable (solid circles in Fig. 10) and the fixed point C2 is stable (shown as a solid box in Fig. 10). Several different flow lines are shown in Fig. 10. They can be compared with the corresponding flow picture obtained by a direct solution of flow equations for two-loop β-functions expressed by the third-order polynomials in couplings u, f; see Eqs. (18) and (19). It is shown in Fig. 2a of Ref.[30] that no stable fixed point exists and furthermore that the fixed point U is absent. Comparing Fig. 10 and Fig. 2b from

Ref.[30] one can see how an analytical continuation of β-functions (10), (18), done only partly[30] and performed here in the form of Padé approximants restores the presence of the fixed point U (unstable) and leads to the appearance of a new stable fixed point C2 for the charged model. The coordinates of the fixed points U, C1 and C2 given in the corresponding rows of Tables 1, 2, 3 and for $n = 2$, are equal to:

$$\begin{array}{lll} \mathrm{U}: & u^* = 1.500, & f^* = 0, \\ \mathrm{C1}: & u^* = 0.181, & f^* = 0.158, \\ \mathrm{C2}: & u^* = 2.457, & f^* = 0.158. \end{array}$$

6. CRITICAL EXPONENTS

The critical exponent values can be determined with the help of fixed point values of ζ-functions defined on the basis of renormalizing Z-factors (14)–(17) by

$$\zeta_i = \mu \partial \ln Z_i / \partial \mu, \tag{59}$$

where the derivative is taken at fixed unrenormalized couplings. The expressions for the ζ-functions related to order parameter and temperature field renormalizations in two-loop approximation read[30]

$$\zeta_\psi = -3f + \frac{(n+2)}{72}u^2 + \frac{(11n+18)}{24}f^2, \tag{60}$$

$$\zeta_t = \frac{-(n+2)}{6}u + \frac{(n+2)}{12}u^2 - \frac{2(n+2)}{3}uf - \frac{(5n+1)}{2}f^2, \tag{61}$$

$$\zeta_A = \frac{n}{6}f + nf^2. \tag{62}$$

If there exists a stable fixed point, the critical exponent of correlation length ν, order parameter susceptibility γ and specific heat α will be given by

$$\nu = (2 - \zeta_\nu^*)^{-1}, \tag{63}$$

$$\gamma = (2 - \zeta_\nu^*)^{-1}(2 - \zeta_\psi^*), \tag{64}$$

$$\alpha = (2 - \zeta_\nu^*)^{-1}(\varepsilon - 2\zeta_\nu^*), \tag{65}$$

$$\eta = \zeta_\psi^*, \tag{66}$$

where $\zeta_\nu = \zeta_\psi - \zeta_t$. The exponents (63)–(66) are related by the familiar scaling laws. From the analysis above it follows that the charged fixed point $C2$ is stable and this results in values for exponents (63)–(65) that are different from values of uncharged fixed point U, i.e., they are not given by ^{4}He values as is sometimes stated.[16,25,59]

Recently, an interesting consequence of the existence of a stable charged fixed point (C_2) has been observed.[62] According to the charge renormalization (3) the β_f-function reads

$$\beta_f = f\ (\varepsilon - \zeta_A(f, u)). \tag{67}$$

Thus at a fixed point with nonzero f^* the value of gauge field the ζ-function is given by $\zeta_A^* = \varepsilon$ **exactly**. This means that the penetration depth λ and the correlation length ξ are proportional and the temperature dependence follows a power law with an exponent ν.[62] This is not the case, though, at the fixed point with $f^* = 0$. There we have $\zeta_A^* = 0$ (each loop contribution to the ζ_A-function contains at least one f-factor) and the penetration depth behaves as $\lambda \sim \xi^{(2-\epsilon)/2}$. Thus one would have two different critical length scales.

Trying to obtain numerical values of critical exponents on the basis of values of fixed point $C2$ coordinates $f^{*C,Padé}$, $u^{*C2,Padé}$ given in Tables 1, 2, in order to be self-consistent let us perform the same type of analytical continuation for ζ-functions series as those which we have applied to β-functions (18), (19). So, introducing the auxiliary variable τ let us represent functions (63)–(65) in form of resolvent series in τ and then chose the [1/1] Padé approximants for these series, which at $\tau = 1$ will give us the requested analytical continuation of series. Thus, the expression for the critical exponent $\phi(\phi \equiv \{\nu, \gamma, \alpha)\}$ reads

$$\phi = a_\phi^{(0)} \frac{1 + A_\phi}{1 + B_\phi}. \tag{68}$$

The expressions for coefficients A_ϕ, B_ϕ in (68) read

$$A_\phi = a_\phi^{(1)} + B_\phi; \quad B_\phi = -a_\phi^{(2)}/a_\phi^{(1)}, \tag{69}$$

and $a_\phi^{(i)}$ are to be determined from resolvent series in τ:

$$\phi = \sum_{i \geq 0} a_\phi^{(i)} \tau^i|_{\tau=1}. \tag{70}$$

Substituting (60) and (61) into (63)–(65) and representing (63)–(65) in the form of (70) one finds

$$\begin{aligned}
a_\nu^{(0)} &= 1/2, \\
a_\nu^{(1)} &= (n+2)/12\ u - 3/2\ f, \\
a_\nu^{(2)} &= (n^2 - n - 6)/144\ u^2 + (71n + 138)/48\ f^2 + (n+2)/12\ uf, && (71) \\
a_\gamma^{(0)} &= 1, \\
a_\gamma^{(1)} &= (n+2)/12\ u, \\
a_\gamma^{(2)} &= (n^2 - 2n - 8)/144\ u^2 + (5n+1)/4\ f^2 + 5(n+2)/24\ uf, && (72) \\
a_\alpha^{(0)} &= 1, \\
a_\alpha^{(1)} &= -3(n+2)/12\ u + 9/2\ f, \\
a_\alpha^{(2)} &= (-3n^2 + 3n + 18)/144\ u^2 - (71n + 138)/16\ f^2 \\
&\quad -(n+2)/4\ uf. && (73)
\end{aligned}$$

Considering the case $n = 2$ and substituting the coordinates of fixed point $C2$ ($f^{*C,Padé} = 0.158$), $u^{*C2,Padé} = 2.457$ (see Tables 1, 2) into (71)–(73) one obtains for the critical exponents[105] (63)–(66):

$$\begin{aligned}
\nu &= 0.86, \quad \gamma = 1.88, && (74) \\
\alpha &= -1.14, \quad \eta = -0.19.
\end{aligned}$$

The application of Padé approximants to the analytical continuation of functions may result in the appearance of poles in these functions. If the pole is located in a region of expansion parameters which has no physical meaning, e.g. negative coupling u or f, the analysis is not complicated. This is the case for β-functions in the region of couplings less than the fixed point values. For the ζ-functions, however, considering the non-asymptotic behaviour (and thus being far from the stable fixed point) one passes through a region of couplings where the Padé approximation for the ζ-functions becomes ambiguous which results in the appearance of pole. Therefore, studying the crossover behaviour in the next subsection we will still keep the polynomial representation for ζ-functions instead of Padé approximants. Then for asymptotic values of critical exponents one gets

$$\begin{aligned} \nu = 0.77, \quad \gamma = 1.62, \\ \alpha = -0.31, \quad \eta = -0.10. \end{aligned} \tag{75}$$

The comparison of values (75) and (76) shows a numerical difference of 15% in ν and γ and a considerable increase of α. However, there is no qualitative change (e.g., the sign of the specific heat exponent remains the same). These results should be compared with critical exponent values given by Ref.[62] ($\nu = 0.53$ and $\eta = -0.70$) and Ref.[61] ($\eta = -0.38$).

In conventional superconductors the experimentally accessible regime lies in the precritical region away from T_c. For this reason we shall discuss some non-asymptotic quantities such as effective exponents and amplitude ratios.

7. AMPLITUDE RATIO FOR THE SPECIFIC HEAT

One of the most interesting measurable quantities is the specific heat. Near a second order phase transition, asymptotically it follows a power law

$$C_0^{\pm} = \frac{A^{\pm}}{\alpha}|t|^{-\alpha} + \text{const}, \tag{76}$$

where $\pm$ indicates the specific heat C and its non universal amplitude A above (+) and below (−) T_c. The amplitude ratio A^+/A^- found from the ratio $C^+(t^+)_0/C^-(t^-)_0$ after subtracting the non-singular background value constitutes a universal quantity at T_c depending only on the dimension and the number of order parameter components.

The calculation of this ratio can be extended to the non-asymptotic region[106, 107] resulting in a temperature dependent measurable quantity. This also tests the description of nonasymptotic behaviour by a certain flow in the interaction space of Hamiltonians, discussed for effective exponents. The starting point in the calculation is the renormalization group equation for the specific heat $C^{\pm}$,

$$\left[\mu\frac{\partial}{\partial\mu} + \beta_u\frac{\partial}{\partial u} + \beta_f\frac{\partial}{\partial f} + \zeta_\nu\left(2 + t\frac{\partial}{\partial t}\right)\right] C^{\pm}(t,u,f,\mu) = \mu^{-\varepsilon}B(u,f), \tag{77}$$

where inhomogeneity B comes from the additive renormalization. The formal solution reads

$$C^{\pm}(t,u,f,\mu) = \mu^{-\varepsilon}\exp\left[-\int_1^l(\varepsilon - 2\zeta_n u(x)\frac{dx}{x}\right]$$

Table 4. Asymptotic values of the specific heat amplitude ratio at various fixed points. Exponents in first two lines correspond to the procedure leading to (75). Third and fourth lines correspond to the procedure leading to (76).

F.P.	A^+/A^-	α	ν	u^*
U	1.81	−0.33	0.72	1.500
C2	1.07	−1.14	0.86	2.457
U	0.78	0.15	0.62	1.500
C2	1.06	−0.31	0.77	2.457

$$\times \left\{ F^{\pm}(l) - \int_1^l \frac{dy}{y} B(y) \exp\left[- \int_l^y (\varepsilon - 2\zeta_\nu(x)) \frac{dx}{x} \right] \right\}. \tag{78}$$

The amplitude ratio is most easily calculated by choosing same value for the flow parameter both above and below T_c, which means[108] $t^+ = -2t^-$. We then recover the asymptotic expression found in Ref.[107]

$$\frac{A^+}{A^-} = 2^\alpha \frac{B\nu + F^+\alpha}{B\nu + F^-\alpha}, \tag{79}$$

where functions B and $F^{\pm}$ are taken at the fixed point. For the ratio (79) we use the lowest order result known from Ψ^4 theory neglecting coupling to the gauge field; $B = 2n$, $F^+ = -n$ and $F^- = 12/u^* - 4$. Then we have for $n = 2$

$$\frac{A^+}{A^-} = 2^\alpha \frac{2\nu - \alpha}{2\nu - 2\alpha + 6\alpha/u^*}. \tag{80}$$

In Table 4 we have collected the values obtained for different fixed points. It is interesting to note the reasonable estimate for this ratio[109] for the superfluid phase transition. The value[109] $A^+/A^- = 1.067$ is surprisingly near to that and corresponds to the stable charged fixed point as shown by both calculation schemes, although the exponents are very different; note, that we have not taken into account changes in scaling functions due to coupling f.

In comparison with experimental results[18], the amplitude ratio $A^+/A^- = n/2^{3/2}$ corresponding to the Gaussian n-vector model without coupling to the gauge field has been used since the order parameter dimension is unclear. Besides, the amplitude ratio has been calculated for cases other than the isotropic symmetry. This leads to a dependence of ratio on higher order couplings.[110]

8. EFFECTIVE EXPONENTS

Effective exponents are usually defined by logarithmic temperature derivatives of corresponding correlation functions.[26, 111] They can be found from the solutions of renormalization group equation for renormalized vertex functions. The effective exponents contain two contributions: one from the corresponding ζ-functions now taken at values of $u(\ell)$, $f(\ell)$ of the particular flow curve considered ("the exponent part"), and one from the change of corresponding scaling function ("the amplitude part"). For the analysis below, we neglect these contributions as they are expected to be smaller than the differences for the fixed point values of exponents coming from the different treatments discussed above. Thus we have

$$\nu = (2 - \zeta_\nu(\ell))^{-1}, \tag{81}$$
$$\gamma = (2 - \zeta_\nu(\ell))^{-1}(2 - \zeta_\psi(\ell)), \tag{82}$$
$$\alpha = (2 - \zeta_\nu(\ell))^{-1}(\varepsilon - 2\zeta_\nu(\ell)). \tag{83}$$

The flow parameter ℓ can be related to the relative temperature distance T_c by the matching condition $t(\ell) = (\xi_0^{-1}\ell)^2$, where ξ_0 is the correlation length amplitude.

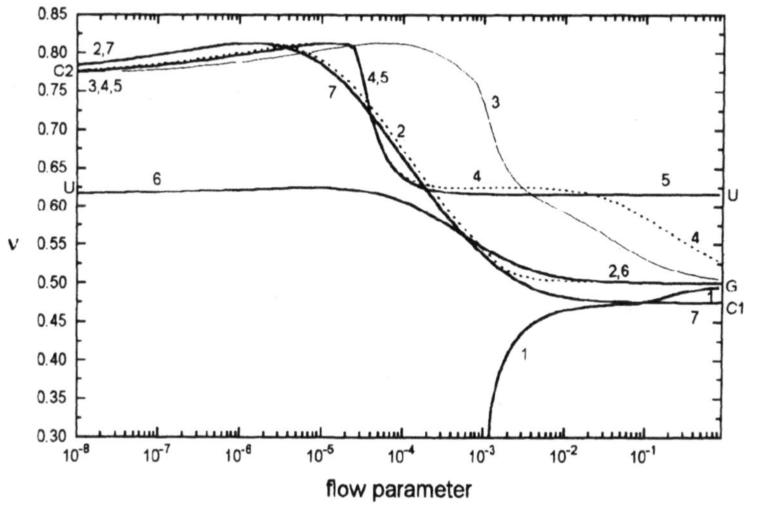

Figure 11. The effective exponent ν for flows shown in Fig.10.

We have computed these effective exponents, see Figs. 11–13, along the flow lines shown in Fig. 10 by inserting[112] the couplings $u(\ell)$ and $f(\ell)$ in Eqs. (81)–(83). For the separatrix 1 we have started with initial conditions leading to a flow, which does not coincide with the fixed point C1 but slightly misses it although the flow curve does not differ from the separatrix within the thickness of lines shown in Fig. 10. For the curve number 4, we start somewhat further away from the Gaussian fixed point G leading to initial values of the effective exponents between their Gaussian values and their values for the uncharged fixed point U. Note that values of effective exponent γ for the uncharged fixed point U and the charged fixed point C1 are the same within the accuracy given by the figure scale.

Note that when coupling f to gauge field fluctuations is small (i.e. for extreme type-II superconductors) the RG flow passes very closely to the uncharged fixed point U and the effective exponents, and within some region of temperatures, they coincide with those of uncharged superfluid liquid. In this region the effective Hamiltonian (2) may be considered as that of a superconductor in a constant magnetic field neglecting

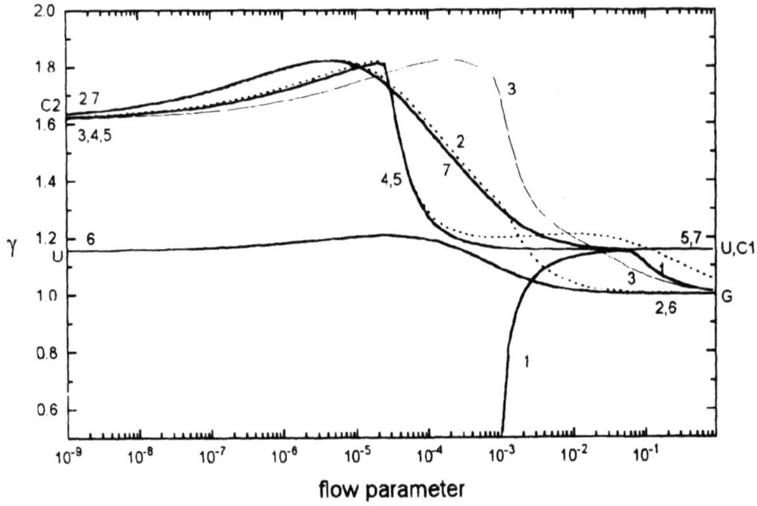

Figure 12. Effective exponent γ for the flows shown in Fig.10.

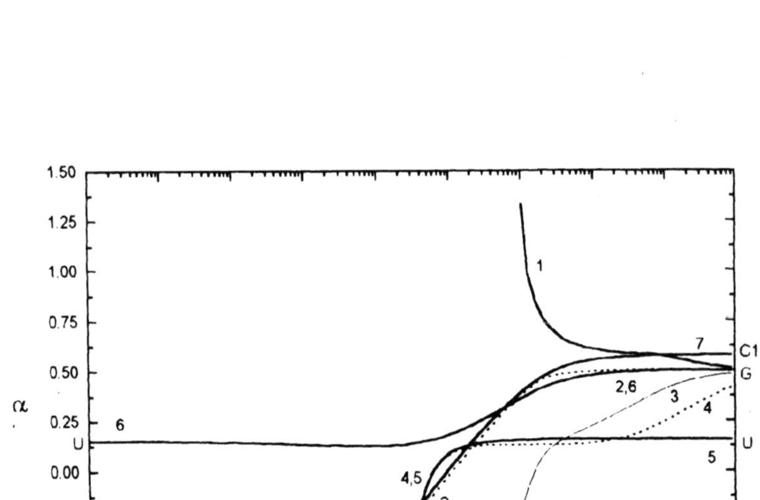

Figure 13. Effective exponent α for the flows shown in Fig.10.

magnetic field fluctuations. For such model it has been recently shown[113] that near the zero-field critical point the singular part of free energy scales as $F_{sing} \simeq |t|^{2-\alpha}\mathcal{F}(B|t|^{-2\nu})$ with ν being the coherence length exponent.

9. CONCLUSIONS

Does the above account give a definite conclusion about the phase transition order occurring in a model of superconductor minimally coupled to the gauge field? First of all one should keep in mind that such answer may be given in an analytical theory framework only by obtaining an exact result or a rigorous proof. Here, the problem has been treated by a perturbation theory approach and the account of the influence of fluctuations on the phase transition order is studied within the field theoretical RG technique. We have shown that remaining within this approach one may get an answer about the second order phase transition occurring in the above mentioned model.

The main point discussed in this context is whether the equations for β-functions possess a stable fixed point or not. The absence of stable fixed point is often interpreted as a change of phase transition order (caused by the presence of magnetic field fluctuations) and the evidence of a fluctuation-induced first-order phase transition. This change of phase transition order (being of second-order in the absence of coupling to gauge), however, is confirmed only by perturbation theory calculations in low orders[15], (see Ref.[30] and references therein.)

Applying a simple Padé analysis to the series under discussion[114] we have shown how one can recover a stable fixed point in the RG equations. In the case of one coupling, such approach gives a qualitatively correct picture of phase transition and restores the presence of a stable fixed point (see Ref.[94] as well). The same occurs in case of two couplings: for $n = 2$ the "uncharged" fixed point U ($f^{*U,Padé} = .158$, $u^{*U,Padé} = 2.457$) appears to be stable, which leads to a new set of critical exponents. We should note, however, that the pair correlation function critical exponent η, calculated by familiar scaling relations on the basis of sets of values (75) or (76), remains negative which agrees with the result in Refs.[61, 62, 54] Being calculated only in a two-loop approximation with the application of Padé analysis, these values for critical exponents are to be considered as preliminary ones.

The main point we wish to emphasize is that within the framework of renormalization group analysis for the superconductor model there exists a possibility for a second-order phase transition characterized by a set of critical exponents differing from those of ^{4}He. Another important task could be to calculate the nonasymptotic specific heat in order to compare with experiments within the region of crossover to the background.

Acknowledgements

We acknowledge the useful discussions with A. M. J. Schakel, D. I. Uzunov, and A. E. Filippov. We are grateful to O. J. Poole who read the manuscript and made suggestions for improving the text.

REFERENCES

1. K. G. Wilson, *Phys. Rev.* B 4:3174 (1971); ibid 4:3184 (1971).

2. N. N. Bogoliubov, and D. V. Shirkov. "Introduction to the Theory of Quantized Fields," Wiley & Sons, New York (1959).
3. D. J. Amit. "Field Theory, the Renormalization Group, and Critical Phenomena," World Scientific, Singapore (1984).
4. M. Le Bellac. "Quantum and Statistical Field Theory," Claredon Press, Oxford (1991).
5. J. Zinn–Justin. "Quantum Field Theory and Critical Phenomena," Oxford University Press, Oxford (1996).
6. J. A. Lipa, D. R. Swanson, J. A. Nissen, T. C. P. Chui, and U. E. Israelsson, *Phys. Rev. Lett.* 76:944 (1996).
7. G. A. Baker Jr., B. G. Nickel, and D. I. Meiron, *Phys. Rev.* B 17:1365 (1978).
8. J. C. Le Guillou, and J. Zinn-Justin, *Phys. Rev.* B 21:3976 (1980).
9. C. Bagnuls, and C. Bervillier, *Phys. Rev.* B 32:7209 (1985).
10. C. Bagnuls, C. Bervillier, D. I. Meiron, and B. G. Nickel, *Phys. Rev.* B 35:3585 (1987).
11. R. Schloms, and V. Dohm, *Europhys. Lett.* 3:413 (1987).
12. R. Schloms, and V. Dohm, *Nucl. Phys.* B 328:639 (1989).
13. R. Schloms, and V. Dohm, *Phys. Rev.* B 42:6142 (1990).
14. I. A. Vakarchuk, *Theor. Math. Phys.* (Moscow) 36:122 (1978).
15. B. I. Halperin, T. C. Lubensky, and S. Ma, *Phys. Rev. Lett.* 32:292 (1974).
16. J. Lobb, *Phys. Rev.* B 36:3930 (1987).
17. S. E. Inderhees, M. B. Salamon, N. Goldenfeld, J. P. Rice, B. G. Pazol, D. M. Ginsberg, J. Z. Liu, and G. W. Crabtree, *Phys. Rev. Lett.* 60:1178 (1988).
18. M. B. Salamon, S. E. Inderhees, J. P. Rice, B. G. Pazol, D. M. Ginsberg, and N. Goldenfeld, *Phys. Rev.* B 38:885 (1988).
19. S. E. Inderhees, M. B. Salamon, J. P. Rice, and D. M. Ginsberg, *Phys. Rev. Lett.* 66:232 (1991).
20. S. Regan, A. J. Lowe, and M. A. Howson, *J. Phys.: Condens. Matter* 3:9245 (1991).
21. G. Mozurkewich, and M. B. Salamon, *Phys. Rev.* B 46:11914 (1992).
22. M. B. Salamon, J. Shi, N. Overend, and M. A. Howson, *Phys. Rev.* B 47: 5520 (1993).
23. A. Junod, E. Bonjour, R. Calemczuk, J. Y. Henry, J. Muller, G. Triscone, and J. C. Vallier, *Physica* C 211:304 (1993).
24. N. Overend, M. A. Howson, and I. D. Lawrie, *Phys. Rev. Lett.* 72:3238 (1994).
25. I. D. Lawrie, *Phys. Rev.* B 50:9456 (1994).
26. J. H. Chen, T. C. Lubensky, and D. R. Nelson, *Phys. Rev.* B 17:4274 (1978).
27. J. C. Le Guillou, E. Brézin, and J. Zinn-Justin, *Phys. Rev.* D 15:1544 (1977).
28. L. N. Lipatov, *Sov. Phys. JETP.* 45:216 (1977).
29. E. Brezin, and G. Parisi, *J. Stat. Phys.* 19:269 (1978).
30. S. Kolnberger, and R. Folk, *Phys. Rev.* B 41:4083 (1990)
31. R. Folk, and Yu. Holovatch, *J. Phys. A: Math. Gen.* 29:3409 (1996).
32. R. Folk, and Yu. Holovatch, *J. Phys. Stud.* (Lviv) 1:343 (1997).
33. H. Meyer-Ortmanns, *Rev. Mod. Phys.* 68:473 (1996)
34. Recall that for a given Landau-Ginsburg parameter (ratio of penetration depth to coherence length) k a superconductor with $k < 1/\sqrt{2}$ is called type–I and one with $k > 1/\sqrt{2}$ is called type–II.
35. K. G. Wilson, and M. E. Fisher, *Phys. Rev. Lett.* 28:240 (1972).
36. A. E. Filippov, A. V. Radievsky, and A. S. Zeltser, *Phys. Lett.* A 192:131 (1994); A. S. Zeltser, A. E. Filippov, *J. Exp. Theor. Phys.* 106:1117 (1994) (in Russian).
37. A. P. C. Malbouisson, F. S. Nogueira, and N. F. Svaiter, *Europhys. Lett.* 41:547 (1998).
38. S. Coleman, and E. Weinberg, *Phys. Rev.* D 7:1988 (1973).
39. J. S. Kang, *Phys. Rev.* D 10:3455 (1974).
40. I. D. Lawrie, *Nucl. Phys.* B 200:1 (1982).
41. S. Hikami, *Progr. Theor. Phys.* 26:226 (1979).
42. S. W. Lovesey, *Z. Physik B Condensed Matter* 40:117 (1980)
43. C. Dasgupta, B. I. Halperin, *Phys. Rev. Lett.* 47:1556 (1981).
44. J. Bartholomew, *Phys. Rev.* B 28:5378 (1983).
45. N. C. Tonchev, and D. I. Uzunov, *J. Phys.* A 14:521 (1981).
46. D. Boyanovsky, and J. L. Cardy, *Phys. Rev.* B 25:7058 (1982).
47. D. I. Uzunov, E. R. Korutcheva, and Y. T. Millev, *J. Phys.* A 16:247 (1983).
48. C. Athorne, and I. D. Lawrie, *Nucl. Phys.* B 265:551 (1986)
49. G. Busiello, L. De Cesare, and D. I. Uzunov, *Phys. Rev.* B 34:4932 (1986).
50. E. J. Blagoeva, G. Busiello, L. De Cesare, Y. T. Millev, I. Rabuffo, and D. I. Uzunov, *Phys. Rev.* B 42:6124 (1990).

51. G. Busiello, L. De Cesare, Y. T. Millev, I. Rabuffo, and D. I. Uzunov, *Phys. Rev.* B 43:1150 (1991).
52. The expression of ζ-function for the mass aggrees with C. Ford, I. Jack, and D. Jones, *Nucl. Phys.* B387:373 (1992).
53. I. D. Lawrie, and C. Athorne, *J. Phys. A: Math. Gen.* 16:L587 (1983).
54. B. Bergerhoff, F. Freire, D. F. Litim, S. Lola, and C. Wetterich, *Phys. Rev.* B 53:5734 (1996).
55. N. Tetradis, and C. Wetterich, *Nucl. Phys.* B 422:541 (1994).
56. M. Gräter, and C. Wetterich, *Phys. Rev. Lett.* 75:378 (1995).
57. P. Arnold, and L. G. Yaffe, *Phys. Rev.* D 49:3003; P. Arnold, and L. G. Yaffe, *Phys. Rev.* D 55:1114 (1997) (erratum).
58. H. Kleinert, *Lett. Nouvo Cimento* 35:405 (1982) (we are thankful to Prof. H. Kleinert for attracting our attention to this reference).
59. M. Kiometzis, H. Kleinert, and A. M. J. Schakel, *Phys. Rev. Lett.* 73:1975 (1994), and *Fortschr. Phys.* 43:697 (1995).
60. A. J. Bray, *Phys. Rev. Lett.* 32:1413 (1974).
61. L. Radzihovsky, *Europhys. Lett.* 29:227 (1995).
62. I. F. Herbut, and Z. Tešanović, *Phys. Rev. Lett.* 76:4588 (1996).
63. I. F. Herbut, *J. Phys.* A 30:423 (1997).
64. In fact an extensive two loop calculation have been already performed earlier by M. Machacek and M. Vaughn, *Nucl. Phys.* B222:83 (1983); B236:221 (1984); B249:70 (1985).
65. P. Olsson, and S. Teitel, *Phys. Rev. Lett.* 80:1964 (1998).
66. see e.g. X. Wen, and Y. Wu, *Phys. Rev. Lett.* 70:1501 (1993) and L. Pryadko, and S. Zhang, *Phys. Rev. Lett.* 73:3282 (1994).
67. K. K. Nanda, B. Kalta, *Phys. Rev.* B 57:123 (1998).
68. P. G. de Gennes, *Solid State Commun.* 10:753 (1972).
69. B. I. Halperin, and T. C. Lubensky, *Solid State Commun.* 14:997 (1974).
70. T. C. Lubensky, and J.-H. Chen, *Phys. Rev.* B 17:366 (1978).
71. G. B. Kasting, K. J. Lushington, and C. W. Garland, *Phys. Rev.* B 22:321 (1980).
72. M. A. Anisimov, P. E. Cladis, E. E. Gorodetskii, D. A. Huse, V. E. Podneks, V. G. Taratuta, W. van Saarloos, and V. P. Voronov, *Phys. Rev.* A 41:6749 (1990).
73. For the most updated comprehensive review of the experimental data on effective critical exponents governing nematic-smectic-A phase transitions see: C. W. Garland, and G. Nounesis, *Phys. Rev.* E 49:2964 (1994).
74. G. t'Hooft, and M. Veltman, *Nucl. Phys.* B 44:189 (1972).
75. The first index is the number of Ψ fields, the second index is the number of A fields.
76. As an example for determination of critical exponents values for models with complicated symmetry by applying the resummation technique in different RG schemes see Refs.[77-81]; N. A. Shpot, *Phys.Lett.* A 142:474 (1989); S. A. Antonenko, and A. I. Sokolov, *Phys. Rev.* B 49:15901 (1994); C. von Ferber, and Yu. Holovatch, *Europhys. Lett.* 39:31 (1997), *Phys. Rev.* E 56:6370 (1997); H. Kleinert, S. Thoms, and V. Schulte-Frohlinde, preprint (1997).
77. G. Jug, *Phys. Rev.* B 27:609 (1983).
78. I. O. Mayer, A. I. Sokolov, and B. N. Shalaev, *Ferroelectrics* 95:93 (1989).
79. I. O. Mayer, *J. Phys.* A 22:2815 (1989).
80. Yu. Holovatch, and M. Shpot, *J. Stat. Phys.* 66:867 (1989); Yu. Holovatch, and T. Yavors'kii, (1998) submitted to *J. Stat. Phys.*
81. H. K. Janssen, K. Oerding, and E. Sengespeick, *J. Phys. A: Math. Gen.* 28:6073 (1995).
82. G. Grinstein, and A. Luther, *Phys. Rev.* B 13:1329 (1976)
83. G.H. Hardy. "Divergent Series," Oxford University, Oxford (1948).
84. Only when one of couplings is equal to zero one does obtain a series which is proven to be asymptotic.
85. P. W. Mitchell, R. A. Cowley, H. Yoshizawa, P. Böni, Y. J. Uemura, and R. J. Birgeneau. *Phys. Rev.* B 34:4719 (1986).
86. T. R. Thurston, C. J. Peters, R. J. Birgeneau, and P. M. Horn, *Phys. Rev.* B 37:9559 (1988).
87. J.-S. Wang, M. Wöhlert, H. Mühlenbein, and D. Chowdhury, *Physica* A 166:173 (1990).
88. J.-S. Wang, W. Selke, Vl. S. Dotsenko, and V. B. Andreichenko, *Europhys. Lett.* 11:301 (1994); A. L. Talapov, and L. N. Shchur, *Europhys. Lett.* 27:193 (1994).
89. T. Holey, and M. Fähnle, *Phys. Rev.* B 41:11709 (1990).
90. The results of resummation appear to be quite insensitive to the choice of p.
91. J. S. R. Chisholm, *Math. Comp.* 27:841 (1973).
92. P. J. S. Watson, *J. Phys.* A 7:L167 (1974).

93. G. A. Baker Jr., and P. Graves-Morris. "Padé approximants," Addison-Wesley Publ. Co., Reading, Mass. (1981).
94. G. Parisi, *in*: "Proceedings of the 1973 Cargrése Summer School," unpublished. G. Parisi, *J. Stat. Phys.* 23:49 (1980).
95. H. Kleinert, V. Schulte-Frohlinde, *Phys. Lett.* B 342:284 (1995).
96. For this model the system of fixed point equations is degenerate at one-loop level, resulting in particular in the $\sqrt{\varepsilon}$-expansion for critical exponents: D. E. Khmelnitskii, *Zh. Eksp. Theor. Fiz.* 68:1960 (1975); T. C. Lubensky, *Phys. Rev.* B 11:3573 (1975); Ref.[82]
97. It corresponds to $n = 0$ fixed point of $O(n)$ symmetrical model, described by the de Gennes limit of self avoiding walk problem.
98. The results are given for the value of parameter $p = 0$ in the Borel–Leroy image.
99. Yu. Holovatch, Preprint, and C.E.Saclay, *Service de Physique Theorique*; S Ph T / 92 – 123); Yu. Holovatch, *Int. J. Mod. Phys.* A 8:5329 (1993).
100. The last possibility has been chosen by G. Parisi[94] in order to restore the presence of stable solution for the fixed point in the two-loop approximation.
101. The series in (37) appears to be alternating and this scheme can be applied without any difficulties.
102. M. Abramowitz, and A. I. Stegun, (editors) "Handbook of Mathematical Functions with Formulas, Graphs and Mathematical Tables," National Bureau of Standards (1964).
103. In this case the principal value of integral (29) could be taken, but generally speaking it is preferable to avoid such situations (see Ref.[7] as well).
104. We take $\ell_0 = 1$.
105. The value of η has been found by the scaling law: $\eta = 2 - \gamma/\nu$.
106. V. Dohm, *Phys. Rev. Lett.* 53:1379 (1984).
107. V. Dohm, *in*: "Application of field theory to statistical mechanics," L. Garrido Ed. p 263, Berlin, Heidelberg, New York, Tokyo, Springer (1985).
108. V. Dohm, *Z. Physik Condensed Matter* 60:61 (1985).
109. A. Singsaas and G. Ahlers, *Phys. Rev.* B 30:5103 (1984).
110. J. F. Annett, S. R. Renn, *Phys. Rev.* B 38:4660 (1988).
111. E. K. Riedel, and F. J. Wegner, *Phys. Rev.* B 9:294 (1974).
112. In fact we have solved flow equations (57) and (59) starting near unstable fixed points, for the initial value of flow parameter we have taken $\ell = 1$. The use of different initial values ($u(1)$ and $f(1)$ on the seperatrix) would amount to rescale the flow parameter.
113. I. D. Lawrie, *Phys. Rev. Lett.* 79:131 (1997); see also K. K. Nanda, and B. Kalta, *Phys. Rev.* B 57:123 (1998).
114. Because they are double series in two coupling constants we have represented them in a form of resolvent series which has enabled us then to pass to the Padé analysis.

SPIN DYNAMICS OF SPIN POLARIZED FERMI LIQUIDS

Igor A. Fomin

P. L. Kapitza Institute for Physical Problems,
ul. Kosygina 2, 117334 Moscow, Russia

1. INTRODUCTION

The properties of spin-polarized Fermi liquids have been discussed theoretically and investigated experimentally for several decades already. The main objects are the liquid ^{3}He and ^{3}He -^{4}He mixtures. Many original papers and several review articles have been written on this subject[1], there is no need to repeat these earlier results here. A new rise of interest in the spin-polarized liquids in the last few years is greatly stimulated by the progress in methods of producing a high polarization at low temperatures. The simplest method is the direct polarization by a magnetic field. For a degenerate liquid the orientational effect of magnetic field on magnetic moments of ^{3}He nuclei competes with the increase of the kinetic energy of quasiparticles.

The polarization is defined as $P = (N_\uparrow - N_\downarrow)/(N_\uparrow + N_\downarrow)$, where $N_\uparrow$ and $N_\downarrow$ are respectively numbers of spins oriented along and opposite to the magnetic field **H**. For not too strong fields the polarization is proportional to the strength of field and to the magnetic susceptibility χ:

$$P = \frac{2\chi}{\gamma\hbar N}H = \frac{3\gamma\hbar}{2(1+F_0^a)}\frac{1}{p_F v_F}H, \tag{1}$$

where γ is the gyromagnetic ratio for ^{3}He nuclei; F_0^a is an expansion amplitude which will be defined below and p_F and v_F are the Fermi momentum and Fermi velocity, respectively.

For a degenerate liquid, the coefficient of proportionality between P and H is roughly inversely proportional to its Fermi energy ϵ_F, which makes this method favorable for dilute solutions of ^{3}He in ^{4}He. A polarization of 65% was achieved in the solution with a concentration $x_3 = 3.5 \cdot 10^{-4}$ for 9.2 T magnetic field.[2] For pure liquid ^{3}He at a zero pressure the corresponding coefficient is $2.6 \cdot 10^{-3}$1/T and even fields up to 20 T produce a quite modest polarization of $\approx 5\%$. For a stronger polarization of pure liquid ^{3}He indirect methods are used.

The method of rapid melting of solid ^{3}He suggested by Castaing and Nozieres[3] is a good example of an indirect polarization of liquid ^{3}He. It exploits the high magnetic susceptibility of ^{3}He solid phase at low temperatures. This susceptibility obeys the Curie law down to temperature of antiferromagnetic ordering $\simeq$ 1mK, unlike the

Correlations, Coherence, and Order. Edited by Shopova and Uzunov
Kluwer Academic/Plenum Publishers, 1999

susceptibility of liquid phase which saturates at the Fermi temperature $\sim 0.1 \div 1$K. A field of 11 T polarizes solid ^{3}He up to 80% at 5 mK. When the pressure is reduced (from 34 to 29 bar in a real experiment[4]) the solid melts preserving most of its polarization. In the cited experiments the liquid polarization just after melting was as high as 70%. In the process of melting the heat is released because of inverse Pomeranchuk effect and the polarized liquid in these experiments has temperature of 75 mK. The nonequilibrium magnetization relaxes in several minutes. The highest polarization in the liquid is achieved by this method but it has also certain drawbacks. For getting rid of heat released at melting, the experiments are conducted in the silver sinter and the effect of restricted geometry is always present. The relaxation of magnetization on a scale of several minutes imposes a restriction on possible measurements. For a recent review of this technique and of the obtained results cf. Ref.[5]

A high stationary polarization of liquid ^{3}He at low temperatures can be achieved with the aid of another indirect method. This method exploits the Leiden dilution refrigerator situated in a magnetic field. The polarizer-refrigerator with the Leiden dilution machine was constructed and successfully applied by G. Vermeulen. The underlying physics and description of method and refrigerator are given in Refs.[6, 7] At the moment a 15% stationary polarization of bulk liquid ^{3}He at temperature of 10 mK is achieved with the use of magnetic field of 7 T. Even higher polarizations can be reached when stronger fields are applied.

High stationary polarizations but also at higher temperatures are obtained by the method of optical pumping of Helium vapor and liquefying of polarized gas.[8] A more detailed description of all these methods has been given recently by D. M. Lee.[9] It is essential for the present discussion, that the methods of producing a high polarization both in pure liquid ^{3}He and in liquid solutions of ^{3}He in ^{4}He in different temperature ranges are already available and a further progress is possible.

The effort to reach high polarizations approaching 100% is justified by a hope to find new phenomena in the extreme regions of parameters. A change of some physical properties of Fermi liquid is expected at much lower polarizations, when the separation of two Fermi surfaces, corresponding to two spin orientations exceeds the temperature T, i.e., when

$$\mu B_{eff} \geq T, \tag{2}$$

where B_{eff} is the field, which would produce a given polarization and $\mu = \gamma\hbar/2$ is the magnetic moment of ^{3}He nucleus. For pure liquid ^{3}He at a temperature of few millikelvin a field of 10 T is sufficiently high in that sense. There is already some interesting experimental data in that region of parameters.[10, 11]

Nontrivial magnetic phenomena are observed in Fermi liquids even at quite low polarizations, provided a collisionless region is reached, i.e. the characteristic frequency of spin motions $\omega_{eff} = \gamma B_{eff}$ and the time between collisions of quasiparticles τ_D meet a strong inequality $\omega_{eff}\tau_D \gg 1$. This condition is much less restrictive than Eq. (2), since $\tau_D \sim \epsilon_F/T^2$, i.e. the product $\omega_{eff}\tau_D \sim \omega_{eff}\epsilon_F/T^2 \sim (\mu B_{eff}/T)(T_F/T)$ contains extra large factor (T_F/T), which for ^{3}He at $T \sim 1$mK is $10^{-2} \div 10^{-3}$ and the collisionless region is reached already at a field of $H \sim 0.1$T. The phenomena, observed in that region of parameters are well described by the Landau Fermi liquid theory. This region can serve as a good starting point for the discussion of properties of spin polarized Fermi liquid.

2. EQUATIONS OF SPIN DYNAMICS

The polarization of Fermi liquid lowers its symmetry. This is the most important qualitative change in the state of liquid introduced by the polarization. The longitudinal relaxation time both in liquid ^{3}He and in mixtures is very long at low temperatures. The bulk relaxation is due to the magnetic dipole interaction of ^{3}He nuclei which is very weak. In the millikelvin temperature range the bulk relaxation time can be as long as several days and it can be safely neglected. In real conditions the relaxation takes place on the container walls, it depends on many factors but usually the relaxation time can be kept of order of minutes, which is sufficiently long for performing NMR experiments. At polarization in the Leiden dilution refrigerator the difference between chemical potentials for two opposite orientations of spins is kept stationary and the polarization can be kept constant indefinitely long for a realistic rate of relaxation. It means that when interpreting NMR data one can treat the magnetization as "frozen" and the symmetry of spin polarized state is that of a ferromagnet. A polarized Fermi liquid like the ferromagnet is degenerate with respect to the direction of frozen magnetization. This degeneracy gives rise to a "soft" dynamics which in that case is the spin dynamics. The macroscopic equations of motion of magnetization $\mathbf{M}$ in an isotropic ferromagnet in the limit of long wavelength and low frequencies are the equations of Landau and Lifshitz[12]:

$$\frac{\partial \mathbf{M}}{\partial t} = \frac{\gamma e}{2mc}\mathbf{M} \times (C\Delta\mathbf{M} + \mathbf{H}). \tag{3}$$

These equations apply for an arbitrary polarization but the standard argument assumes that the system is close to its local equilibrium at given $\mathbf{M}$. This assumption generally does not hold for a liquid with a "frozen" polarization.

In the limit of weak polarization the equations of the spin hydrodynamics can be derived from the kinetic equation of Landau Fermi liquid theory[13]. For this derivation there is no need in the apriori assumption for a local equilibrium. The coefficients, entering in the resulting equations are expressed in terms of parameters, which determine the properties of a given Fermi liquid, i.e., the density of particles N, the Fermi velocity v_F, the average time between collisions of quasiparticles τ_D and the coefficients in the expansion of function, which describes in the Landau theory the interaction between quasiparticles $F(\theta)$, in spherical harmonics $P_l(\cos\theta)$

$$F^{s,a}(\theta) = \sum_l F_l^{s,a} P_l(\cos\theta).$$

The indices s, a denote here the spin-symmetric and spin-antisymmetric parts of this function. Only the spin-antisymmetric part of this function F^a enters in equations of spin motion. The above parameters can be found from independent experiments. The true hydrodynamic motion is realized when all characteristic frequencies ω, entering in the problem, satisfy the strong inequality $\omega\tau_D \ll 1$. Nontrivial Fermi-liquid effects manifest themselves in the opposite limit $\omega\tau_D \gg 1$. Characteristic frequencies in that limit are the frequencies of oscillations of spatially uniform deviations of spin part of distribution function from the equilibrium. A deviation having the symmetry of l-th spherical harmonics oscillates with the frequency[13]:

$$\omega_l = \pm\frac{\gamma^2 S}{\chi(1 + F_0^a)}\left(F_0^a - \frac{F_l^a}{2l+1}\right). \tag{4}$$

The evolution of a general perturbation of spin distribution function in the collisionless limit is described by an infinite set of coupled equations for different spherical harmonics of perturbation. Leggett has shown[14] that the description of spin motion can be greatly simplified in a very important limit when the motion of averaged spin density is smooth and slow, i.e. if the characteristic frequency of this motion Ω meets the inequality $\Omega \ll \omega_l$ and the characteristic scale of its spatial variation λ meets the inequality $\lambda \ll v_f/\omega_l$. In that case the spin motion does not generate higher spherical harmonics of the spin distribution function and a closed system of equations can be written for only two first harmonics, which determine the spin density $\mathbf{S}$ and the spin current density $\mathbf{J}_i$. The system of spin hydrodynamics equations has the following form:

$$\frac{\partial \mathbf{S}}{\partial t} + \frac{\partial \mathbf{J}_i}{\partial x_i} = \mathbf{S} \times \boldsymbol{\Omega}_L, \tag{5}$$

$$\frac{\partial \mathbf{J}_i}{\partial t} + \frac{1}{3} v_F^2 (1 + F_0^a)(1 + \frac{F_1^a}{3}) \frac{\partial \mathbf{S}}{\partial x_i} = \mathbf{J}_i \times \boldsymbol{\Omega}_L - \frac{\gamma^2 S}{\chi(1 + F_0^a)} \left(F_0^a - \frac{F_1^a}{3} \right) \mathbf{J}_i \times \mathbf{S}$$
$$- (1 + F_1^a) \frac{\mathbf{J}_i}{\tau_D}. \tag{6}$$

D. M. Lee has suggested to call this system Leggett and Rice equations to distinguish it from the system of Leggett equations for superfluid ^{3}He. Eq. (5) is a spin conservation law, it follows from the above assumption for an infinite longitudinal relaxation time. The second term in the r.h.s. of the second equation gives rise to all nontrivial magnetic Fermi liquid effects, including Silin waves[13] and the Leggett and Rice effect.[15] Only two first harmonics of function $F^a(\theta)$: F_0^a and F_1^a enter the spin hydrodynamics equations. To simplify the further analysis of these equations, it is convenient to introduce here the shorthand notations $w^2 = v_F^2(1+F_0^a)(1+F_1^a/3)$, $\kappa = -(F_0^a - F_1^a/3)/(1+F_0^a)$, where v_F is the Fermi velocity and $\tau_1 = \tau_D/(1+F_1^a)$. We shall use also the variables $\vec{\sigma} = \gamma^2 \mathbf{S}/\chi$, $\mathbf{j}_i = \gamma^2 \mathbf{J}_i/\chi$, where χ is the magnetic susceptibility and γ - the gyromagnetic ratio for ^{3}He nuclei. The index $i = 1, 2, 3$ specifies spatial component of spin current tensor and the boldface type or an arrow denote a vector in the spin space. With these notations the equations of motion have the following form:

$$\frac{\partial \vec{\sigma}}{\partial t} + \frac{\partial \mathbf{j}_i}{\partial x_i} = \vec{\sigma} \times \vec{\omega}_L, \tag{7}$$

$$\frac{\partial \mathbf{j}_i}{\partial t} + \frac{w^2}{3} \frac{\partial}{\partial x_i} (\vec{\sigma} - \vec{\omega}_L) = \mathbf{j}_i \times \vec{\omega}_L + \kappa \mathbf{j}_i \times \vec{\sigma} - \frac{\mathbf{j}_i}{\tau_1}; \tag{8}$$

here $\omega_L = \gamma H$ is the Larmour frequency of the magnetic field H. Two conservation laws follow from these equations :

$$\frac{\partial \vec{\sigma}_z}{\partial t} + \frac{\partial j_i^z}{\partial x_i} = 0, \tag{9}$$

$$\frac{\partial}{\partial t} \left[\frac{(\vec{\sigma} - \vec{\omega}_L)^2}{2} + \frac{3\mathbf{j}^2}{2w^2} \right] + \frac{\partial}{\partial x_i} (\vec{\sigma} \cdot \mathbf{j}_i) = -\frac{3\mathbf{j}^2}{w^2 \tau_1}. \tag{10}$$

An integral relation which is useful for the evaluation of the relaxation rate of a given pattern in a closed volume follows from Eq. (10):

$$\frac{\partial}{\partial t}\int\left[\frac{(\vec{\sigma}-\vec{\omega}_L)^2}{2}+\frac{3\mathbf{j}_i^2}{2w^2}\right]dV=-\frac{3}{w^2\tau_1}\int\mathbf{j}_i^2 dV\le 0. \tag{11}$$

If $\tau^{-1}\neq 0$, only states without a spin current can be steady states. According to the inequality in formula (11), the energy has its minimum at equilibrium. The true equilibrium is reached for $\vec{\sigma}=\vec{\omega}_L$. If we take into account the conservation of spin longitudinal projection by introducing a constant vector $\vec{\omega}_p$, which is parallel or antiparallel to z-axis as a Lagrange multiplier, we shall find the set of steady-states $\vec{\sigma}=\vec{\omega}_L-\vec{\omega}_p$, $\mathbf{j}_i=0$.

The equations of the spin hydrodynamics have been successfully applied for the description of the spin motion in spin echo experiments.[14, 16] Recently a new phenomenon – the formation of coherently precessing two-domain spin pattern has been observed in spin-polarized solutions of ^{3}He in ^{4}He and in pure liquid ^{3}He. The interpretation of this phenomenon will be discussed in the next section.

3. COHERENTLY PRECESSING SPIN PATTERN

In the extensive experimental study of 350 ppm solutions polarized by magnetic field of 9.2 T up to a 65 % polarization with the aid of pulsed NMR technique, Nunes et al.[2] have observed at temperature $\sim$ 10 mK an anomalously long lived induction signal. The signal duration exceeds 10 seconds when the magnetic-field gradient is in the range 0.02 - 0.10 Oe/cm. With such a gradient for independently precessing spins, a dephasing will occur in time $\sim 10^{-2}$ s. The anomaly is similar to that previously observed in the superfluid B-phase of ^{3}He.[17-19] It has been shown that in ^{3}He-B even in a nonuniform magnetic field, spins precess coherently. The coherence of the precession is maintained by the long-range correlations in the Cooper pairs condensate. A deviation of condensate from the uniformity drives the spin current which redistributes spin over the experimental cell volume and eventually leads to formation of two-domain pattern. The same argument cannot be directly applied to the normal phase where a condensate is absent. Nevertheless the motion of individual spins in a spin polarized liquid is correlated because of exchange part of Fermi liquid interaction. This correlation manifests itself, in particular, in the existence of spin waves. The numerical simulation made by Nunes et al.[2] has shown, that in pulsed NMR experiments, when the d.c. magnetic field is slightly nonuniform a coherently precessing spin pattern consisting of two domains can be formed. In one of domains the magnetization is parallel to the d.c. field while in the other it is antiparallel. Nunes et al.[2] have suggested that the long lived signal in their experiments originates from the precessing domain wall which separates two domains. The existence of a two domain precessing pattern has been also suggested by Akimoto et al.[20] on the basis of numerical simulations of their pulsed NMR experiments with 6.4% ^{3}He -^{4}He solutions at temperature $\sim$ 1 mK in a field of 56 mT. In both cases the simulations have been made with the aid of the spin hydrodynamics equations. A quite different evidence of the formation of coherently precessing pattern in 6.4% solutions has been obtained by Dmitriev et al.[21] at temperature $\sim$ 1 mK in a field of 28 mT.

It has been shown later, that the existence of coherently precessing pattern follows directly from the equations of the hydrodynamics.[22, 23] In the limit $\tau_1\to\infty$ the pattern is described by the steady-state solution of these equations. The long lived induction signal indicates the existence of a solution of Eqs. (7) and (8) corresponding to the precession of $\vec{\sigma}$ and $\mathbf{j}_i$ in a given volume with a frequency $\vec{\omega}_p$ which is constant throughout the volume. This means that at each point the motion of $\vec{\sigma}$ and $\mathbf{j}_i$ is described by the formulae

$$\frac{\partial \vec{\sigma}}{\partial t} = \vec{\sigma} \times \vec{\omega}_p \quad \text{and} \quad \frac{\partial \mathbf{j}_i}{\partial t} = \mathbf{j}_i \times \vec{\omega}_p. \tag{12}$$

To avoid unnecessary complications, let us assume, that both the container holding the liquid and the d.c. magnetic field have an axial symmetry. The symmetry axis will be taken as the z-axis direction. Let us assume also, that the walls of container do not transmit the spin current. In other words the boundary condition is imposed:

$$\mathbf{j}_i n_i = 0, \tag{13}$$

where n_i is the normal to the wall.

Under these conditions, the solution of interest depends only on one coordinate – z. For such solution, there is no spin transport in directions perpendicular to $\vec{\omega}_L$. The longitudinal component of spin current $\mathbf{j}$ and spin $\vec{\sigma}$ satisfy the equations following from Eqs. (7), (8) and (12)

$$\frac{\partial \mathbf{j}_3}{\partial z} = \vec{\sigma} \times (\vec{\omega}_L - \vec{\omega}_p), \tag{14}$$

$$\frac{w^2}{3}\frac{\partial}{\partial z}(\vec{\sigma} - \vec{\omega}_L) = \mathbf{j}_3 \times (\vec{\omega}_L - \vec{\omega}_p + \kappa\vec{\sigma}) - \frac{\mathbf{j}_3}{\tau_1}. \tag{15}$$

For passage to the limit $\tau_1 \to \infty$ let us resolve Eq. (15) with respect to $\mathbf{j}_3$:

$$\mathbf{j}_3 = \frac{w^2}{3}\frac{\tau_1^2}{1 + u^2\tau_1^2} \times \left[\mathbf{u} \times \frac{\partial(\vec{\sigma} - \vec{\omega}_L)}{\partial z} - \tau_1 \mathbf{u}\left(\mathbf{u} \cdot \frac{\partial(\vec{\sigma} - \vec{\omega}_L)}{\partial z}\right) - \frac{1}{\tau_1}\frac{\partial(\vec{\sigma} - \vec{\omega}_L)}{\partial z}\right], \tag{16}$$

where $\mathbf{u} = \vec{\omega}_L - \vec{\omega}_p + \kappa\vec{\sigma}$. In most applications $|\ \kappa\sigma\ | \gg |\ \omega_L - \omega_p\ |$ then the obtained formula coincides with the formula (24) of Ref.[14].

Three terms in the square brackets in the r.h.s. of Eq. (16) have different dependence on τ_1. The last term is a contribution to the spin current from the usual diffusion, it vanishes in the limit $\tau_1 \to \infty$. The second term is also dissipative, it is growing with τ_1 and transports the component of spin, which is parallel to $\mathbf{u}$:

$$\mathbf{j}_{\|} = -\frac{w^2}{3u^2}\mathbf{u}\left(\mathbf{u} \cdot \frac{\partial(\vec{\sigma} - \vec{\omega}_L)}{\partial z}\right)\tau_1. \tag{17}$$

When $\vec{\omega}_L = \vec{\omega}_p = \text{const}$, $\mathbf{j}_{\|}$ is proportional to $\partial\vec{\sigma}^2/\partial z$. This part of spin current tends to keep $\vec{\sigma}^2$, or the difference between chemical potentials of spin-up and spin-down quasiparticles, constant over the volume occupied by liquid. As a result in the limit $\tau_1 \to \infty$, $\vec{\sigma}^2$ is nearly constant. The first term in the square brackets in Eq. (16) is reactive, it does not depend on τ_1 and for $\vec{\omega}_L = \vec{\omega}_p$ it is perpendicular to $\vec{\sigma}$:

$$\mathbf{j}_\perp = \frac{w^2}{3u^2}\mathbf{u} \times \frac{\partial(\vec{\sigma} - \vec{\omega}_L)}{\partial z}. \tag{18}$$

Now Eqs. (14) and (15) can be solved in principal order in τ_1. The r.h.s. of Eq. (14) does not contain terms, growing with τ_1, so we have

$$\frac{d\mathbf{j}_\parallel}{dz} = 0, \tag{19}$$

or $\mathbf{j}_\parallel = \text{const}$. The container is assumed to be closed either at the top or at the bottom (or at both sides). Combining Eq. (18) with the boundary condition (13) one arrives at

$$\mathbf{j}_\parallel = 0. \tag{20}$$

In a zeroth order in τ_1, Eq. (14) reads as

$$\frac{\partial \mathbf{j}_\perp}{\partial z} = \vec{\sigma} \times (\vec{\omega}_L - \vec{\omega}_p). \tag{21}$$

Taking the scalar product of this equation with $\vec{\omega}_p$ and combining the result with the boundary condition (13) we conclude that $(\mathbf{j}_\perp \cdot \vec{\omega}_p) = 0$. Multiplying both sides of Eq. (18) by $\vec{\omega}_p$, one can see that $\partial\vec{\sigma}/\partial z$ lies in the $(\vec{\omega}_p, \vec{\sigma})$ plane. This means that in the coordinate system rotating with the frequency $\vec{\omega}_p$, the variation of spin $\vec{\sigma}$ as a function of z is restricted to a single plane. Let us assume this single plane to be the (y, z) plane and denote by θ the angle between $\vec{\sigma}$ and the z axis. Then $\vec{\sigma}$ has the following components: $(0, \sigma\sin\theta, \sigma\cos\theta)$. The current will have only a single nonvanishing component:

$$j_3^x = -\frac{w^2}{3u^2}\left[(\vec{\sigma}\cdot\mathbf{u})\frac{d\theta}{dz} + \kappa\sigma\sin\theta\frac{d\omega_L}{dz} + (\omega_L - \omega_p)\sin\theta\frac{d\sigma}{dz}\right]. \tag{22}$$

The current and the spin density are defined completely by two functions $\sigma(z)$ and $\theta(z)$, which must satisfy two equations. One is found by substituting Eq. (22) in Eq. (14), and the other is Eq. (20). The most interesting case corresponds to $|\kappa\sigma| \gg |\omega_L - \omega_p|$, when the Fermi-liquid interaction is dominant. In this limit the expression for the spin current simplifies greatly

$$j_3^x = -\frac{w^2}{3\kappa}\frac{d\theta}{dz}. \tag{23}$$

Substituting this expression in Eq. (14) and using the expansion $\omega_L - \omega_p = (d\omega_L/dz)(z - z_0)$, where z_0 is defined by the condition $\omega_L(z_0) = \omega_p$, one arrives at an equation for θ:

$$\frac{d^2\theta}{dz^2} = -\frac{1}{\lambda^3}(z - z_0)\sin\theta. \tag{24}$$

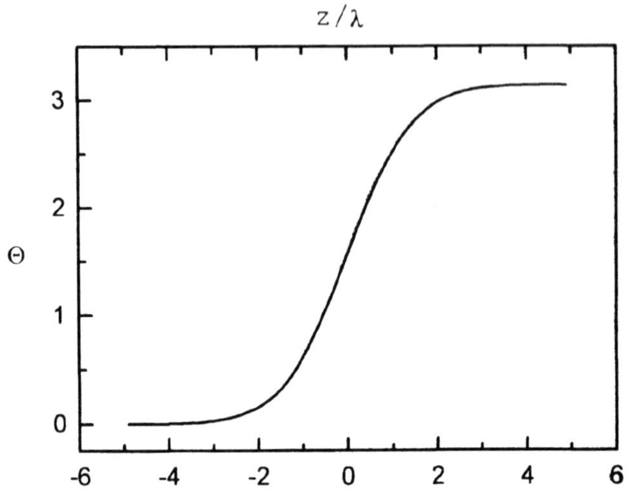

Figure 1. The variation of θ within the domain wall.

The combination of parameters $\lambda = [w^2/3\kappa\sigma(d\omega_L/dz)]^{1/3}$ plays the role of a length scale here. The equation (20) in the same approximation will be

$$\frac{d\sigma}{dz} = \frac{d\omega_L}{dz}\left[\cos\theta - \frac{1}{\kappa}(z - z_0)\frac{d\cos\theta}{dz}\right]. \tag{25}$$

The boundary condition to Eqs. (24) and (25) is Eq. (13) applied at the upper and lower walls of container. If the distance between z_0 and both upper and lower walls of the container is much greater than λ the boundary condition at the walls of container can be replaced by the condition at infinity; i.e., $d\theta/dz \to 0$ as $z \to \pm\infty$. The asymptotic behavior of σ and θ far from the domain wall can be easily found with the help of Eqs. (24) and (25): $\sigma \cong \sigma_0 \pm [\omega_p - \omega_L(z)]$ as $(z - z_0) \to \pm\infty$. The constant σ_0 is determined by the initial conditions. On the strength of assumption $|\kappa\sigma| \gg |\omega_L - \omega_p|$ the overall change of $|\kappa\sigma|$ is small. The absolute value of spin density is kept nearly constant by the current $\mathbf{j}_{\|}$, which is largest in the limit $\tau_1 \to \infty$. For θ one obtains

$$\theta \simeq C\exp -\frac{2}{3}\left|\frac{z - z_0}{\lambda}\right|^{3/2} \qquad \text{as} \frac{z - z_0}{\lambda} \to -\infty,$$

$$\theta \simeq \pi - C\exp -\frac{2}{3}\left|\frac{z - z_0}{\lambda}\right|^{3/2} \quad \text{as} \frac{z - z_0}{\lambda} \to +\infty.$$

The constant C is determined in the process of solution. For λ one can use its value at $\sigma = \sigma_0$. The numerical solution of Eq. (24) which satisfies the boundary conditions is presented in Fig. 1.

This solution has a form of two domains which are separated by a domain wall with a thickness of order of λ. The wall is situated at $z = z_0$. The position of domains in space is determined by the sign of λ, i.e. by the sign of product $\kappa(d\omega_L/dz)$. If $\kappa > 0$, then the domain with $\vec{\sigma}\|\vec{\omega}_L$ (the parallel domain) is situated in the region of relatively weaker magnetic field. The absolute value of $\vec{\sigma}$ in that case is slowly decreasing towards both infinities.

4. RELATION BETWEEN SPIN WAVES AND PRECESSING PATTERN

The formation of coherently precessing pattern which has been discussed in the previous section is an effect of the Fermi liquid interaction. Formally when the interaction vanishes ($\kappa \to 0$), the characteristic length $\lambda \to \infty$ and the wall disappears. For a formation of the wall one needs also the Larmor frequency gradient; the wall is localized by this gradient. The previously investigated standing spin waves localized by a magnetic field gradient and a wall of container[24] have the same physical origin. It is instructive to follow the relation between these two types of stationary modes.

Consider liquid with $\kappa > 0$. Let us place the origin $z = 0$ at the boundary of container, let us assume that the liquid occupies the region $z < 0$ and ω_L increases in the direction of positive z. Let us also introduce a dimensionless coordinate $\zeta = z/\lambda$ and a dimensionless frequency shift $\nu_p = (\omega_p - \omega_L(0))/\lambda\nabla\omega_L$. Then Eq. (24) reads

$$\frac{d^2\theta}{d\zeta^2} + (\zeta - \nu_p)\sin\theta = 0. \tag{26}$$

The solutions of interest satisfy the following boundary conditions $d\theta/d\zeta = 0$ at $\zeta = 0$ and $d\theta/d\zeta \to 0$ at $\zeta \to -\infty$. By changing the independent variable $\zeta = \nu_p + s$, we exclude ν_p from Eq. (26):

$$\frac{d^2\theta}{ds^2} + s\sin\theta = 0. \tag{27}$$

The boundary condition $d\theta/ds = 0$ is now imposed for $s = -\nu_p$. The second boundary condition does not change: $d\theta/ds \to 0$ at $s \to -\infty$. Eq. (26) has solutions which do not depend on s and satisfy the boundary conditions, they are: $\theta = n\pi$, where $n = 0, \pm 1, \pm 2, \ldots$. For these solutions the vector $\vec{\sigma}$ is either parallel or antiparallel to $\vec{\omega}_L$. Let us linearize Eq. (27) in the vicinity of these θ. For small deviations ϕ_0 from "even" points with the substitution $\theta = 2m\pi + \phi_0$ we arrive at

$$\frac{d^2\phi_0}{ds^2} + s\phi_0 = 0. \tag{28}$$

For "odd" points the substitution $\theta = (2m+1)\pi + \phi_1$, $\phi_1 \ll 1$ gives an equation, which differs by the sign of s:

$$\frac{d^2\phi_1}{ds^2} - s\phi_1 = 0. \tag{29}$$

As an even stationary point one can always take $\theta = 0$. The bounded solutions of Eqs. (28) and (29) are the Airy functions. For even n the solution $\phi_0 = C_0 * Ai(s)$, where C_0 is a constant, decreases when $s \to -\infty$ as $\exp[-(2/3)|s|^{3/2}]$ and oscillates for $s > 0$ (the lower curve in Fig. 2). At sufficiently small C_0, this function satisfies Eq. (27) as well. The condition $d\phi_0/ds = 0$ is met for $s = s_j$, where s_j are roots of Airy function derivative. It means that the boundary condition $d\theta/ds = 0$ can be met at $s = 0$ only for chosen values of $\nu_p = \nu_p^j = -s_j$. The function ϕ_0 in the intervals $(-\infty, s_j)$ gives a variation of θ for consecutive modes of standing waves. The eigenvalues s_j determine the frequencies of these modes in units of $\lambda\nabla\omega_L$. Several first eigenvalues of s_j are: $s_0 = -1,01188$, $s_1 = -3,2482$, $s_2 = -4,8201$. Analogously the function $\phi_1 = C_1 * Ai(-s)$ in the intervals $(-s_j, \infty)$ represents consecutive modes of spin waves, when the liquid occupies a region $s > 0$ and the magnetization is antiparallel to the field. Both functions ϕ_0 and ϕ_1 are real. It means that each of spin-wave modes corresponds to precession with a constant phase.

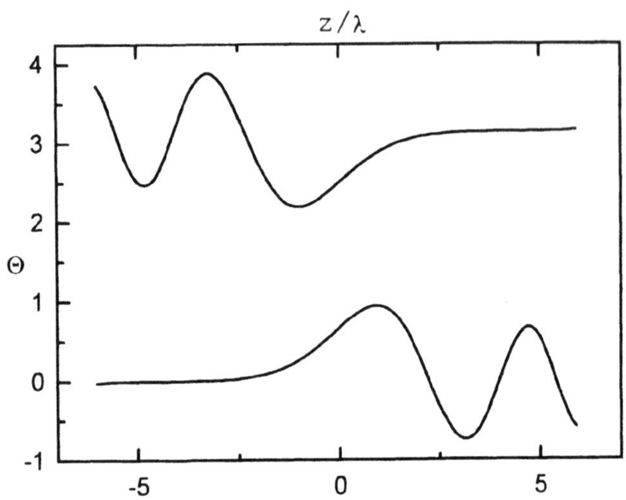

Figure 2. Bounded solutions of Eq. (15), when θ is close to π and θ is close to 0.

With the increase of amplitudes C_0 and C_1 nonlinearities become important and the functions ϕ_0 and ϕ_1 do not approximate a solution of Eq. (27) for all values of s. The nonlinearities appear first in the region $s \approx 0$ where the values of ϕ_0 and ϕ_1 are greater, while ϕ_0 and ϕ_1 can be still a good approximation of true solution in regions of s, where they assume sufficiently small values. At a certain amplitude the solution, for which $\theta \to 0$ monotonically when $s \to -\infty$ matches continuously in the region of small s, the solution, for which $\theta \to \pi$ monotonically when $s \to +\infty$. As a result of matching a domain wall is formed with an overall change of $\vec{\sigma}$ with an angle π. With the amplitude growth the nonlinear spin wave changes its frequency. It is convenient to take the frequency as a parameter, specifying a given solution. Fig. 3. demonstrates the evolution of the ground standing spin-wave mode into a domain wall.

When the wall is formed the effect of further decrease of frequency is the domain wall motion in a region of weaker field practically without a change of wall shape. Generically the two domain precessing pattern is a standing spin wave of arbitrary large amplitude.

5. SYMMETRY AND GENERAL PROPERTIES OF PRECESSING PATTERN

The existence of the domain wall solution is not an accident but a consequence of the symmetry of the problem and there are other solutions of Leggett equations, having the same symmetry and describing "multiple" domain walls. The symmetry in question can be easily demonstrated by the substitution of $\vec{\omega}_L$ in the form $\vec{\omega}_L = \vec{\omega}_p + (d\vec{\omega}_L/dz)z$ in Eqs. (14) and (15) in the limit $\tau_1 \to \infty$:

$$\frac{\partial \mathbf{j}_3}{\partial z} = \vec{\sigma} \times \frac{d\vec{\omega}_L}{dz} z, \tag{30}$$

$$u^2 \frac{\partial}{\partial z}(\vec{\sigma} - \vec{\omega}_L) = \mathbf{j}_3 \times (\frac{d\vec{\omega}_L}{dz} z + \kappa\vec{\sigma}). \tag{31}$$

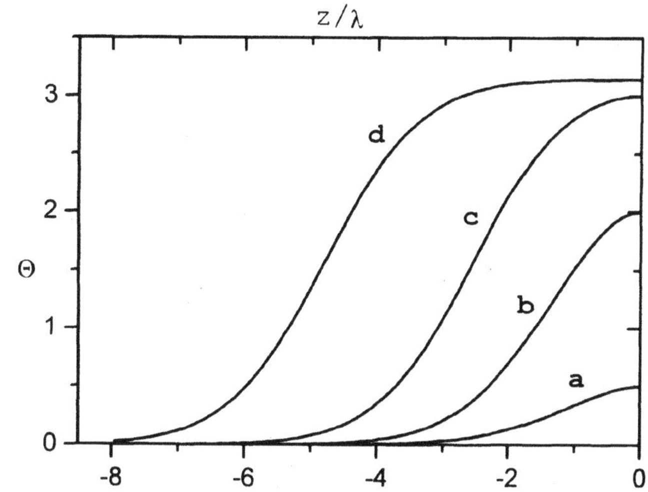

Figure 3. The transformation of "ground" spin-wave mode, confined by the magnetic field gradient and the container wall, into the two-domain pattern when increasing the amplitude of the wave, or a shift of its frequency $\varepsilon = [\omega_p - \omega_L(0)]/\lambda\nabla\omega_L$. The line a corresponds to $\varepsilon = -1,0349$, b — to $\varepsilon = -1,3540$, c — to $\varepsilon = -2,5550$ and d — to $\varepsilon = -4,8018$.

One can check directly that these equations are symmetric with respect to the combination of two transformations — the time reversal and the reflection in xy plane. This combination transforms $\sigma^x(z), \sigma^y(z), \sigma^z(z)$ into $\sigma^x(-z), \sigma^y(-z), -\sigma^z(-z)$ and $j_3^x(z)$, $j_3^y(z)$, $j_3^z(z)$ into $j_3^x(-z)$, $j_3^y(-z)$, $-j_3^z(-z)$. The symmetry transformation has to be applied to boundary conditions as well. If the container length is much greater than the characteristic scale of the problem λ, which is defined below, one can consider a solution with boundary conditions $\mathbf{j}_3 = 0$ at $z \to \pm\infty$. These conditions do not change for the given transformation and solutions specified by these conditions either appear in pairs, or they are symmetric with respect to the transformation. In the latter case x and y components of $\vec{\sigma}$ and $\mathbf{j}_3$ are even functions of z and z-components are odd. In particular $\sigma^z(0) = 0$ and $j_3^z(0) = 0$. The latter condition, combined with the z-component of Eq. (30) $dj_3^z/dz = 0$, gives $j_3^z = 0$. It has been shown before[22] that if $j_3^z = 0$ then $\vec{\sigma}$ will lie in a plane, which can be taken as the yz-plane.

For a typical situation when $\kappa\sigma \gg |\, \tilde{\omega}_L - \tilde{\omega}_p \,|$, the variation of $|\, \vec{\sigma} \,|$ is small and only the variation of the $\vec{\sigma}$ direction is of importance. The orientation of $\vec{\sigma}$ is specified by the angle θ, defined above, which satisfies Eq. (27) with boundary condition

$$d\theta/ds = 0, \qquad (32)$$

at $z \to \pm\infty$. The discussed symmetry of the original equations manifests itself in that case as a symmetry with respect to the simultaneous transformation $z \to -z$ and $\theta \to -\theta$. The discussed precessing domain wall is described by a solution of nonlinear Eq. (27) which is symmetric with respect to the transformation $z \to -z$ and $\theta \to -\theta$. The spin density $\vec{\sigma}$ changes its orientation by angle π when crossing the domain wall. Although boundary conditions have been imposed on infinities, the obtained solution will be a good approximation of finite volume problem as well, if the distance of the domain wall from container walls in the z-direction is greater than several times λ. Apart from this solution there are other symmetric solutions describing "multiple" domain walls with a change of orientation of $\vec{\sigma}$ for any odd multiple of π. The extra loops cannot be unwound without taking $\vec{\sigma}$ out of the plane. This means that there are potential barriers, separating patterns of different "multiplicities".

A very important question for the above discussion is the stability of the coherently precessing pattern. The π-wall stability has been investigated directly[25] in a standard way. A solution of Eqs. (7) and (8) has been searched in a form $\vec{\sigma} = \vec{\sigma}^0 + \vec{\psi}$. $\mathbf{j}_i = \mathbf{j}_i^0 + \mathbf{g}_i$, where $\vec{\sigma}^0$ and $\mathbf{j}_i^0$ is the steady-state solution describing the coherently precessing pattern and $\vec{\psi}$ and $\mathbf{g}_i$ are small perturbations of the pattern. The equations of motion are linearized with respect to $\vec{\psi}$ and $\mathbf{g}_i$. The obtained linear equations have oscillating solutions, which describe vibrations of two-domain pattern. The modes of vibrations appear in pairs. This is also a consequence of symmetry of Eqs. (30) and (31) discussed above. In a frame rotating with the angular velocity $\vec{\omega}_p$, the modes circularly polarized in right and left sense with respect to precessing $\vec{\sigma}$ have frequencies $\Omega_\pm$, which differ only in sign. Explicit calculations for the π-wall[25] give

$$\Omega_\pm = \pm \left[\varepsilon_n \left(\frac{u^2 (d\omega_L/dz)^2}{\kappa\sigma_0} \right)^{1/3} - \frac{1}{\kappa\sigma_0} (uk_\perp)^2 \right], \tag{33}$$

where $k_\perp$ is a wave vector in a direction, perpendicular to $\vec{\omega}_p$ and ε_n is an eigenvalue of operator $d^2/ds^2 + s\cos\theta$. Several first values of ε_n found numerically are: $\varepsilon_0 = 0,850$; $\varepsilon_1 = 2,234$; $\varepsilon_2 = 3,207$; $\varepsilon_3 = 4,040$; $\varepsilon_4 = 4,792$.

The most interesting general feature of precessing pattern is its phase coherence – the phase of precession is the same over the container volume. The physical origin of such long-range order is the broken symmetry of precessing state. The Hamiltonian of normal Fermi-liquid and the equilibrium state of such liquid in magnetic field have an axial symmetry. This symmetry is broken when the magnetization is tipped away from the equilibrium by a radio-frequency (rf) pulse (off-diagonal spin components appear). The resulting state is degenerate with the other states, having the same absolute value of magnetization and the same tipping angle, but different phases of precession. In the ideal Fermi-gas this degeneracy should be local, i.e. a variation of the phase of precession in space would not change the energy of state. It is the Fermi-liquid interaction, which lifts the degeneracy and sets up the unique phase for a given volume. The pattern rigidity is determined by the combination $\kappa/(1 + F_1^a/3)$. This combination can have any sign and be arbitrary small, provided the condition $\kappa\sigma\tau_1 \gg 1$ is met. One can observe here a close analogy with the occurrence of a long range order at phase transitions, except that in the case of precession, the symmetry is broken not spontaneously, but by the method of preparation of initial state. The conservation of spin longitudinal projection and (at $\tau_1 \to \infty$) of energy imposes the constraints which prevent spin from relaxation to the full equilibrium. The lowest energy state under constraints has a broken symmetry and the usual arguments apply. In such general form the argument applies to any system, occupying a continuously degenerate state with sufficiently long time of life.

6. RELAXATION OF TWO–DOMAIN PATTERN

The steady-state solution of Eqs. (7) and (8) is specified by two parameters. As such, one can choose the longitudinal component of total spin and the energy, or instead - the precession frequency ω_p and the constant σ_0 defined at the end of Sec. 3. At finite but sufficiently large values of τ_1, this solution becomes quasistationary, i.e.

ω_p and σ_0 begin to depend on time. To find this dependence in first order in $1/\tau_1$, let us substitute the steady-state solution in both sides of Eq. (11). If the chamber height L is large in comparison with λ, one can ignore terms of the order of λ/L in the expression for the energy. Let us place the coordinate system origin in the middle of chamber, then for the integral in the left-hand side of Eq. (11), representing the energy of pattern we have

$$E = \frac{1}{2}\left[L(\sigma_0^2 + \omega_p^2) - 2z\sigma_0\omega_p\right]. \tag{34}$$

The energy E depends both on ω_p and σ_0, so one needs one more equation. This equation depends on conditions of experiment. If the precessing pattern is formed in an isolated volume, then the total longitudinal spin component is conserved. In terms of wall coordinate z_0 this constraint reads as $z_0\sigma_0 \equiv \Sigma = \text{const}$. After the substitution of Eq. (32) into Eq. (9) and straightforward transformations with the above constraint, we find the equation

$$\frac{d}{dt}[\sigma_0(t)^{5/3}] = -\frac{5w^{4/3}(\nabla\omega_L)^{1/3}}{(3\kappa)^{5/3}\tau_1 L}J, \tag{35}$$

where $J = \lambda\int_{-\infty}^{\infty}(d\vartheta/dz)^2dz$. The value of J depends on the particular form of domain wall. For the most interesting π-wall $J_1 \simeq 2.35$. J grows rapidly with the number of turns of $\vec{\sigma}$ within the wall, for the 3π-wall $J_3 \simeq 18.1$, and for 5π-wall $J_5 \simeq 37.1$.

According to Eq. (35), $\sigma_0^{5/3}$ falls off linearly with time. The temporal dependence of the domain wall position $z_0(t)$ is found as $z_0(t) = \Sigma/\sigma_0(t)$. The z_0 variation determines the temporal variation of ω_p. As one can see from the above formula the sense of variation of ω_p depends on sign of initial total longitudinal projection of spin 2Σ. With the decrease of σ_0 the absolute value of z_0 increases, but its sign remains the same. For $\Sigma > 0$, ω_p grows with time, while for $\Sigma < 0$ it decreases. The induction signal disappears, when the domain wall reaches one of container walls. Only at $\Sigma = 0$ the domain wall does not move and ω_p remains constant. In that case σ_0 disappears in time t_f, which depends on the initial value of $\sigma_0(0)$:

$$t_f = \frac{\tau_1}{5J}\left(\frac{3\kappa\sigma_0(0)}{w}L\right)^{4/3}\left(\frac{3\kappa\sigma_0(0)}{\Delta\omega}\right)^{1/3}, \tag{36}$$

where $\Delta\omega = L(d\omega/dz)$ is the total Larmor frequency variation in the cell. For $\sigma_0(0) \cong \omega_L$ time t_f contains two large factors: $(L/\lambda)^{4/3}$ and $(\kappa\omega_L/\Delta\omega_L)^{1/3}$.

It should be kept in mind, however,that with decrease of σ, the condition $|\kappa\sigma| \gg |\omega_L - \omega_p|$ will be violated eventually. The expression for t_f should thus be regarded as approximate. In the course of relaxation, with decreasing of σ_0, the domain wall thickness increases slowly (as $\sigma_0^{-1/3}$).

For orientation, we give here numerical estimations of characteristic time t_f and characteristic length scale λ both for ^{3}He and for the saturated solutions of ^{3}He in ^{4}He for realistic conditions. For pure ^{3}He the data of D.Candela et al.[24] have been used, which are presented in terms of following parameters $u^2 = w^2/3\kappa$ and $D_0 = w^2\tau_1/3$. For pressure $P = 0$ bar and at temperature of the transition into the superfluid B-phase $u = 1.72 \cdot 10^{-6}\text{cm}^2/\text{s}$ and $D_0 = 1.16\text{cm}^2/\text{s}$. Then $\lambda \simeq 0.5$mm and $t_f \simeq 0.2$s for field of 300 Oe, field gradient of 0.1Oe/cm and for cell length $L = 1$cm. The obtained t_f in this example is about thousand times greater than the dephasing time $T_2^* = 1/\Delta w$. For the 6.4% solution of ^{3}He in ^{4}He at $P = 0$ bar and $T = 1$mK, using the data of Ishimoto et al.[40] one obtains $\lambda \simeq 2$mm and $t_f \simeq 1.8 \cdot 10^{-2}$ s for the same field. field gradient and cell length. The time t_f can be increased substantially since according to the formula (36) t_f grows as $\omega_L^{5/3}$ with the magnetic field growth and as $1/T^2$ with the temperature decrease.

The relaxation scenario will be different, if the experimental volume has a contact with a reservoir of spin-polarized liquid. In that case the longitudinal current (Eq. (17)) keeps constant the spin absolute value and the effect of relaxation for $\sigma_0 = \text{const}$ is the domain wall motion in the direction of antiparallel domain. In leading order in $1/\tau_D$ the speed of this motion is given by the formula

$$\frac{dz_0}{dt} = J\left(1 + \frac{F_1^a}{3}\right)\frac{\lambda^2}{\kappa\omega_p\tau_D}\frac{d\omega_L}{dz}. \tag{37}$$

Dmitriev et al.[26] have shown by a numerical analysis and demonstrated experimentally, that for a semi-open volume, it is possible to compensate the spin relaxation by resonant r.f. field, i.e., using the CW NMR technique. With this technique the coherently precessing pattern can be maintained stationary indefinitely long. The precession frequency ω_p in that case is constant and equal to the frequency of r.f. field.

7. EFFECT OF DEMAGNETIZING FIELD

A spin polarized liquid produces its own (demagnetizing) field. Until now the demagnetizing field has not been taken into account in the theoretical description of coherently precessing pattern. This is justified for all cases, in which the pattern has been observed. The importance of the demagnetizing field increases with polarization. In the liquid ^{3}He the demagnetizing field produces essential changes in the precessing pattern for a polarization of a few percent, when the Leggett and Rice equations still hold. In this region of parameters the demagnetizing field effect can be easily included in the theoretical description of the coherently precessing pattern by extending the description to Fermi liquids with higher magnetization.[27] The coherently precessing pattern, considered in Sec. 3 is one-dimensional, i.e., both the spin density $\vec{S}$ and the spin current tensor $\vec{J}_i$ depend on z and do not on transverse coordinates, x and y. In that case, according to Deville et al.[28] the effect of demagnetizing field is local, it is equivalent to magnetic field $\vec{H}_d$, which is determined by the magnetization $\vec{M} = \gamma\vec{S}$ at the same point :

$$\vec{H}_d = 4\pi\chi\left[M_z\hat{z} - \frac{1}{3}\vec{M}\right]. \tag{38}$$

Here χ is the magnetic susceptibility of liquid. The demagnetizing field $\vec{H}_d$ has to be added to the applied field $\vec{H}$ in the equations of spin dynamics (5) and (6) or with the shorthand notations introduced above Eqs. (7) and (8). Because the spin current can exist only in the z-direction under the assumed conditions, the subscript i in the tensor $\vec{j}_i$ in what follows is dropped. As a result Leggett and Rice equations take the following form

$$\frac{\partial\vec{\sigma}}{\partial t} + \frac{\partial\vec{j}}{\partial z} = \vec{\sigma}\times(\vec{\omega}_L + 4\pi\chi\sigma_z\hat{z}), \tag{39}$$

$$\frac{\partial \vec{j}}{\partial t} + \frac{w^2}{3}\frac{\partial}{\partial z}(\vec{\sigma} - \vec{\omega}_L - \vec{\omega}_d) = \vec{j} \times (\vec{\omega}_L + \vec{\omega}_d + \kappa\vec{\sigma}) - \frac{\vec{j}}{\tau_1}. \tag{40}$$

The conservation laws (9) and (10) will be modified:

$$\frac{\partial \sigma_z}{\partial t} + \frac{\partial j^z}{\partial z} = 0, \tag{41}$$

$$\frac{\partial}{\partial t}\left[\frac{(\sigma - \omega_L)^2}{2} + \frac{3\vec{j}^2}{2w^2} - 2\pi\chi\left(\sigma_z^2 - \frac{\sigma^2}{3}\right)\right] + \frac{\partial}{\partial z}\left[(\vec{\sigma} - \vec{\omega}_L - \vec{\omega}_d)\cdot\vec{j}\right] = -\frac{3\vec{j}^2}{w^2\tau_1}. \tag{42}$$

And the integral relation (11) now looks as

$$\frac{\partial}{\partial t}\int\left[\frac{(\sigma - \omega_L)^2}{2} + \frac{3\vec{j}^2}{2w^2} - 2\pi\chi\left(\sigma_z^2 - \frac{\sigma^2}{3}\right)\right]dV = -\frac{3}{w^2\tau_1}\int \vec{j}^2 dV. \tag{43}$$

The z-dependence of $\vec{\sigma}$ and $\vec{j}$ for steady state solutions of Eq. (12) now is determined by the following equations:

$$\frac{\partial \vec{j}}{\partial z} = \vec{\sigma} \times (\vec{\omega}_L - \vec{\omega}_p + \vec{\omega}_d), \tag{44}$$

$$\frac{w^2}{3}\frac{\partial}{\partial z}(\vec{\sigma} - \vec{\omega}_L - \vec{\omega}_d) = \vec{j} \times (\vec{\omega}_L - \vec{\omega}_p + \vec{\omega}_d + \kappa\vec{\sigma}) - \frac{\vec{j}}{\tau_1}. \tag{45}$$

Because $\chi \approx 10^{-7} - 10^{-8}$ in liquid ^{3}He, $\vec{\omega}_d$ is negligible in comparison with $\vec{\sigma}$ in Eq. (45). In Eq. (44), $\vec{\omega}_d$ has to be compared with the difference $\vec{\omega}_L - \vec{\omega}_p$, which becomes zero at the domain wall. The effect of $\vec{\omega}_d$ is important up to the distance z_d, determined by the inequality $z_d\,(d\omega_L/dz) \leq \omega_d$. The overall effect of ω_d can be neglected, if $z_d \ll \lambda$, where $\lambda = \left(w^2/3|\kappa|\sigma\nabla\omega_L\right)^{1/3}$ is the domain wall thickness for $\omega_d = 0$. Finally this criterion can be formulated as

$$4\pi\chi|\sigma| \ll \lambda\frac{d\omega_L}{dz}. \tag{46}$$

The left hand side grows, while the right hand side decreases with increasing of $|\sigma|$. For liquid ^{3}He at saturated vapor pressure and field gradient of 0.25 T/m (25 Oe/cm), both sides are approximately equal at $\sigma = 1.4266 \cdot 10^9$ s^{-1} corresponding to field $H = 7$ T (70000 Oe).

We pursue the analysis while keeping ω_d in Eq. (44). Repeating previous arguments[22, 23] we conclude that $|\sigma|$ is a function of z but remains within a plane going through the z-axis and rotating with $\vec{\omega}_p$ (the zy-plane of the rotating frame). With this restriction, $\vec{\sigma}$ is determined by two parameters, for instance $\sigma = |\vec{\sigma}|$ and the angle θ between $\vec{\sigma}$ and the z-axis positive direction: $\vec{\sigma} = (0, \sigma\sin\theta, \sigma\cos\theta)$. The spin current for the steady state solution is perpendicular both to $\vec{\omega}_L$ and $\vec{\sigma}$. In the rotating frame, specified above, it has only one component. Usually $\kappa\sigma \gg |\omega_L - \omega_p|$ then as before

$$j^x = -\frac{w^2}{3\kappa}\frac{d\theta}{dz}. \tag{47}$$

The substitution of Eq. (47) into Eq. (44) gives an equation for θ:

$$\frac{d^2\theta}{dz^2} = -\frac{3\kappa\sigma}{w^2}\sin\theta\left[\frac{d\omega_L}{dz}(z - z_p) + 4\pi\chi\sigma\cos\theta\right]. \tag{48}$$

The coefficients in front of two terms in the right hand side determine two characteristic lengths: $\lambda = (w^2/3|\kappa|\sigma(d\omega_L/dz))^{1/3}$ and $\mu = (w^2/12\pi|\kappa|\chi\sigma^2)^{1/2}$. The overall sign of the right hand side is of importance. It is determined by the sign of κ. In ^{3}He and in ^{3}He-^{4}He solutions with a concentration $x > 3.5$ %, κ is positive. In that case introducing the dimensionless coordinate $\zeta = (z - z_p)/\lambda$, we have instead of Eq. (24) :

$$\frac{d^2\theta}{d\zeta^2} = -\zeta\sin\theta - b^2\sin\theta\cos\theta, \tag{49}$$

where $b^2 = (\lambda/\mu)^2 = 4\pi\chi\sigma/\lambda\nabla\omega_L$. The effect of the demagnetizing field is small ($b^2 \ll 1$), when the condition (46) is met. When $b > 1$, the domain wall thickness is of order of λb^2. When $b^2 \to \infty$, the dependence of θ on ζ in the region $|\zeta| < b^2$ approaches that given by the equation

$$\cos\theta = -\zeta/b^2. \tag{50}$$

Formula (50) will be a good approximation, if ζ satisfies the condition $(b^2 - \zeta)(b^2 + \zeta) \gg |\zeta|$. The approximation breaks down in the region $\zeta \sim 1$ at both ends of the interval. When $b^2 \gg 1$, the situation is analogous to that in the homogeneously precessing domain in the superfluid B-phase of ^{3}He, where the Larmor frequency gradient is compensated by the dipolar frequency shift. The difference is that in the normal phase the shift starts from $\theta = 0$ and not from $\theta = \arccos(-1/4)$. With this analogy the region $|\zeta| < b^2$ at $b^2 \gg 1$ can be considered as a separate domain with thickness $2\lambda b^2$. Outside the interval $|\zeta| < b^2$ the ζ-dependent term takes over and $\sin\theta \to 0$ with the following asymptotic

$$\theta \approx -\pi + \frac{const}{(\zeta - b^2)^{1/4}}\exp\left\{-\frac{2}{3}\left(\zeta - b^2\right)^{3/2}\right\}, \tag{51}$$

when $\zeta \to \infty$ ($|\zeta - b^2| \gg 1$), and

$$\theta \approx \frac{const}{(\zeta + b^2)^{1/4}}\exp\left\{-\frac{2}{3}\left(\zeta + b^2\right)^{3/2}\right\}, \tag{52}$$

when $\zeta \to -\infty$.

Eventually, there are three domains: a parallel domain with the magnetization along the magnetic field in the region $\zeta < -b^2$ (at low ω_L), a precessing domain in the region $-b^2 < \zeta < b^2$ and an antiparallel domain with a reversed magnetization in the region $\zeta > b^2$ (at high ω_L). The transition regions, $|\zeta - b^2| \sim 1$ and $|\zeta + b^2| \sim 1$ at both ends of precessing domain play the role of domain walls.

When κ is negative, the equation for θ has the following form

$$\frac{d^2\theta}{d\zeta^2} = -\zeta \sin\theta - b^2 \sin\theta \cos\theta. \tag{53}$$

In this case the orientation of domains with respect to the $\nabla\omega_L$ direction is opposite to that considered above: the parallel domain occupies a region of relatively high ω_L. The demagnetizing field instead of compensating the difference between ω_L and ω_p effectively increases it. As a result the domain wall becomes more steep. At $b^2 \gg 1$ its shape is described by the equation

$$\cos\theta = \tanh b\zeta, \tag{54}$$

i.e. the domain wall thickness is $\sim \lambda/b = \mu$.

The relative variation of $\vec{\sigma}$ absolute value in the domain walls and within the domains is small and can be neglected for all practically interesting situations.

The precessing pattern may be of practical interest when long lived. For small dissipation the pattern becomes quasistationary, i.e. two parameters, determining the pattern: the frequency of precession – ω_p and the absolute value of $\vec{\sigma}$ in the middle of the domain wall σ_0 will be time dependent. For finite but large τ_1 the substitution of the stationary solution in both sides of Eq. (43) gives an equation determining their dependence on time. A second equation depends on a particular experimental condition. If the cell is considered closed, then it will be the conservation of the spin longitudinal projection: $\int \sigma_z dV = 0$. If the parallel domain is in contact with the unperturbed liquid, then $\sigma_0 = \text{const}$, because of the fast longitudinal spin diffusion.

We shall consider the latter case. Evaluating the time derivative in the left hand side of Eq. (43) we take into account that the Larmor frequency change over the antiparallel domain is small in comparison with ω_L itself. Then the main contribution to this derivative comes from the change of the Zeeman energy because of change of the domain wall position z_p. In the right hand side of Eq. (43) we substitute Eq. (47) for $\vec{j}$. After straightforward transformations one arrives at the expression for the domain wall velocity,

$$\frac{dz_p}{dt} = \frac{\lambda^2 \nabla\omega_L}{\kappa\omega_p\tau_1} J(b), \tag{55}$$

where the integral

$$J(b) = \int_{-\infty}^{+\infty} \left(\frac{d\theta}{d\zeta}\right)^2 d\zeta = 2\int_{-\infty}^{0} \left(\frac{d\theta}{d\zeta}\right)^2 d\zeta \tag{56}$$

is determined by the domain wall shape. In the limit $b^2 \to 0$, $J \to J(0) \approx 2.35$, as has been found before.[23] In the opposite limit $b^2 \to \infty$, the substitution of the dependence

$\theta(\zeta)$ given by Eq. (47) gives a logarithmically divergent integral. The asymptotic expression for $J(b)$ at $b^2 \to \infty$ is obtained by a separate analysis

$$J(b) = \frac{1}{b^2} \ln \frac{b^2}{Q} . \tag{57}$$

The constant Q is found numerically: $Q = 0.1278$. In comparison with the case $b^2 \leq 1$ the domain moves roughly b^2 times more slowly

$$\frac{dz_p}{dt} = \frac{\mu^2 \nabla\omega_L}{\kappa\omega_p\tau_1} \ln\frac{b^2}{Q} = \frac{w^2}{12\pi\chi\kappa^2\sigma^2} \frac{\nabla\omega_L}{\omega_p\tau_1} \ln\frac{b^2}{Q}. \tag{58}$$

In the intermediate region, $b^2 \sim 1$, a numerical evaluation of $J(b)$ is necessary. The integral $J(b)$ has been evaluated by a direct numerical integration for $b^2 < 10$. It turns out that the asymptotic formula is already a good approximation for $b^2 = 3$.

For $\kappa < 0$ and $b^2 \gg 1$ the stationary solution is given by Eq. (53) and the integral $J(b)$ can be found directly

$$J(b) = 2b.$$

In that case the domain moves much faster.

We can conclude, that the demagnetizing field can have a pronounced effect on the coherent spin precession in Fermi liquids already at modest polarizations. The effect depends on the sign of particular combination of Fermi liquid constants, entering in the spin dynamics equations. In particular, in the pure liquid ^{3}He the demagnetizing field gives rise to an increase of the domain wall width. With the increase of polarization the wall develops into a precessing domain. As a result the amplitude of induction signal and its duration will grow faster than the predicted behavior without a demagnetizing field. For example, at $T = 10$ mK, $H = 7$ T, $\nabla H = 10$ mT/m and $p = 12$ bar and κ and τ_1 measured by Candela et al.[24], we find $\kappa\tau_1\omega_L = 4.56$, $\lambda = 31.4$ μm, $\omega_d = 1014$ rad/s, $\lambda\nabla\omega_L = 64.0$ rad/s and $b^2 = 15.9$. Without the effect of demagnetizing field the domain wall would move with velocity $(dz_p/dt)_L = 1.03$ mm/s and taking this effect into account $(dz_p/dt)_d = 0.134$ mm/s. According to Eq. (58) this velocity is inversely proportional to relaxation time τ_1. The effect of demagnetizing field is well pronounced and it can be accounted for in a straightforward way within the existing theory. A more serious difficulty is the expected deviation of spin polarized liquid from predictions, based on the Fermi-liquid theory. These deviations have been discussed in the current literature.[29, 30]

8. DIFFICULTY OF SPIN DYNAMICS AT HIGH POLARIZATION

At high polarizations the Landau theory of Fermi liquids in its original form does not apply. In particular, the Leggett and Rice equations of spin hydrodynamics (Eqs. (5) and (6)) have to be reanalyzed. An indication of this demand is already the existence of two different Fermi momenta p_{F+} – for spin up, and p_{F-} – for spin-down quasiparticles. With these the meaning of Fermi liquid parameters, entering in the equations is not clearly defined. The problem cannot be solved by introduction of Fermi liquid parameters separately for each Fermi sphere. The origin of the difficulty is seen more clearly in the microscopic argument used in the Fermi liquid theory.[31] The very essential part of this argument is the existence of singularity in the equation for the vertex part $\Gamma_{\alpha\beta\gamma\delta}(p_1, p_2; k)$ at small momentum transfer $k = (\mathbf{k}, \omega)$. The singularity originates from the "loop" diagram, which contains a product of two Green functions

$$G(q)G(q+k) = \frac{2\pi i a^2}{v}\delta(\epsilon)\delta(|\,\mathbf{q}\,| - p_0)\frac{\mathbf{kv}}{\omega - \mathbf{kv}} + \phi_{reg}. \tag{59}$$

The notations are conventional - a is a residue in the pole of one-particle Green function, v - Fermi velocity, $q = (\mathbf{q}, \epsilon)$. The separation of the above product in its singular and regular (ϕ_{reg}) parts is nonambiguous up to terms, which are zero for $k = 0$. Spin indices are not specified, since for an unpolarized liquid the character of the singularity is the same for all spin components.

At finite polarization the Green functions for each spin orientation spin have poles for the Fermi momenta p_+ and p_- corresponding to the specified orientation:

$$G_\pm(\epsilon, p) = \frac{a_\pm}{\epsilon - \mu_\pm - v_\pm(p - p_\pm) + i\delta sign(p - p_\pm)} + G_{\pm reg}. \tag{60}$$

Generally the polarization is nonequilibrium, then $\mu_+ - \mu_- = \Omega - \omega_L$, where ω_L is the Larmor frequency corresponding to the real magnetic field, while Ω is the Larmor frequency corresponding to the effective field which would produce the existing polarization. The polarization introduces an essential change in the equation for "transverse" components of the vertex part Γ_{1221} and Γ_{2112}, which contain products $G_+(q)G_-(q+k)$ and $G_-(q)G_+(q+k)$. The separation of loop contribution into singular and regular parts now has a meaning up to terms which are zero for $\omega - \omega_L = 0$, $\mathbf{k} = 0$ and $\Omega = 0$, such a separation can be carried out only in the small polarization limit $\Omega \to 0$. In this limit instead of Eq. (59) we have

$$\begin{aligned} G_+(q)G_-(q+k) = &\frac{2\pi i a^2}{v}\delta(\epsilon)\delta(|\,\mathbf{q}\,| - p_0) \\ &\times \frac{\mathbf{kv} + \Omega/(1+F_0^a)}{(\omega - \omega_L) - \mathbf{kv} + \Omega F_0^a/(1+F_0^a)} + \phi_{\pm reg}. \end{aligned} \tag{61}$$

The usual procedure is valid in the principal order in polarization because the first term is a singular function of $\omega - \omega_L$, $\mathbf{k}$ and Ω. Following the procedure one introduces Γ^ω_{1221} as a limit at $\omega \to 0, \Omega/\omega \to 0, |\,\mathbf{k}\,| /\omega \to 0$. It coincides with Γ^ω_{1221} for an unpolarized liquid. The standard argument yields the equation for collective modes:

$$\nu(\mathbf{n}) = \frac{\mathbf{kn}v + \Omega/(1+F_0^a)}{(\omega - \omega_L + \Omega) - (\mathbf{kn}v + \Omega/(1+F_0^a))}\int F^a(\mathbf{n}, \mathbf{l})\nu(\mathbf{l})\frac{d\Omega_l}{4\pi}. \tag{62}$$

The notations are conventional[32] – F_0^a is the antisymmetric part of interaction function in Fermi liquid theory. In comparison with the unpolarized case, in Eq. (62) we have $\mathbf{kn}v + \Omega/(1+F_0^a)$ instead of $\mathbf{kn}v$ and $(\omega - \omega_L + \Omega)$ instead of ω. For $\Omega = \omega_L$ such equation has been derived by Silin from the kinetic equation of Landau theory. For finite Ω both terms in the r.h.s. of Eq. (61), in contrast to Eq. (59), are regular with respect to $(\omega - \omega_L)$ and $\mathbf{k}$, the separation in two terms contains an ambiguity and the usual procedure of Landau theory does not apply for the Γ transverse components. An independent argument is required for the derivation of spin dynamics equations at finite polarization.

In series of papers[34] a microscopic theory of spin dynamics has been developed. An important result of this theory is the so called zero-temperature attenuation – the damping of spin waves remains finite at $T = 0$. The transverse components of spin diffusion coefficient at a finite polarization remains finite down to the absolute zero as well. The origin of zero temperature attenuation has been explained by a simple qualitative argument. The variation of spin direction when described in terms of Fermi quasiparticles requires a change in the quasiparticles distribution in the whole interval of momenta – from p_{F-} to p_{F+}. The probability of a quasiparticle decay within the interval increases with polarization as $(p_{F+} - p_{F-})^2/p_F^2$. If the polarization is not small the inverse quasiparticle life time can become comparable with its energy and the description of nonuniformly polarized state in terms of quasiparticles fails. The above qualitative argument has a weak point, it does not take into account an essential effect of Fermi-liquid interaction on the spin motion. This interaction gives rise to a mean field, which influences the spin motion of quasiparticles. The mean field changes its direction together with the spin density. If the temporal and spatial variation of spin density is slow and smooth its effect can be treated as an adiabatic perturbation.

In what follows the equations of the transverse spin dynamics are derived at $T = 0$ with an account of discussed adiabaticity. An expansion in the inverse polarization is made and terms of principal order in frequencies and wavevectors characterizing the spin motion are taken into account. The obtained equations are nondissipative. The spin waves spectrum following from these equations in principal order in the wavevector contains no damping. This conclusion is in agreement with the results of the theory of ferromagnetic Fermi liquid[33] and in conflict with the conclusions of papers.[29, 34]

9. EQUATIONS OF SPIN DYNAMICS AT HIGH POLARIZATION

The Fermi liquid Hamiltonian includes the kinetic energy $\mathcal{H}_{kin}$, the two-particle interaction $\mathcal{H}_{int}$ and the interaction with magnetic field, directed along the z-axis $\mathcal{H}_L = -\omega_L \cdot S_z$, where S_z is the z-projection of the spin density and ω_L is the Larmor frequency for the applied field. Both in ^{3}He and in the solutions of ^{3}He in ^{4}He the spin-orbit interaction is extremely weak and we neglect it.

Let us introduce at each point a frame rotating with an angular velocity $\mathbf{\Omega}$ such that the polarization in the rotating frame is locally in equilibrium. $\mathbf{\Omega}$ depends on time and coordinates. This dependence is assumed to be slow, i.e. the characteristic frequencies ω and the wavevectors $\mathbf{k}$ of motion of $\mathbf{\Omega}$ in the frame rotating with Larmor frequency meet strong inequalities $\omega \ll \Omega$ and $k \ll \Delta p = p_{F+} - p_{F-}$. Then for the derivation of spin dynamics equations one can apply an expansion procedure in small frequencies and wavevectors which is analogous to that previously used in ^{3}He–B.[35]

Consider the second quantized Lagrangian of polarized Fermi liquid

$$\hat{\mathcal{L}} = \frac{1}{2}\int \psi\dagger_\mu(\mathbf{r})[\delta_{\mu\nu} i\frac{\partial}{\partial t} + \sigma^z_{\mu\nu}\cdot\omega_L + \sigma_{\mu\nu}\cdot\mathbf{\Omega}]\psi_\nu(\mathbf{r})d^3r$$
$$-\frac{1}{2m}\int[\nabla\psi\dagger_\mu(\mathbf{r})][\nabla\psi_\mu(\mathbf{r})]d^3r - \hat{\mathcal{H}}_{int}. \qquad (63)$$

Let us make a local rotation of spin axes $\hat{x}, \hat{y}, \hat{z}$, generated by the rotation matrix $\hat{R} = \exp(-\gamma\hat{\sigma}^z/2)\exp(-\beta\hat{\sigma}^y/2)\exp(-\alpha\hat{\sigma}^z/2)$, such that the new quantization axis $\hat{\zeta}$ is directed along $\mathbf{\Omega}$. The orientation of axes $\hat{\xi}$, $\hat{\eta}$ will be specified later. The rotation parameters α, β, γ are the Euler angles. The rotation gives rise to a transformation of the derivatives, entering in the Lagrangian, Eq. (63), according to the recipe: $\partial/\partial t \to \partial/\partial t + R_{\mu\lambda}\partial/\partial t R_{\lambda\nu}$ and an analogous transformation for the spatial derivatives. This introduces in the Lagrangian additional terms, depending on the Euler angle derivaties $\dot{\alpha}, \dot{\beta}, \dot{\gamma}$ and $\alpha_l, \beta_l, \gamma_l$, where the index denotes differentiation over the spatial coordinate x_l. The time derivatives together with $\hat{\mathcal{H}}_L$ give

$$\hat{\mathcal{L}}_1 = \frac{1}{2}\int \psi\dagger_\mu(\mathbf{r})(\sigma_{\mu\nu}\cdot\omega)\psi_\nu(\mathbf{r})d^3r, \tag{64}$$

where the angular velocity ω has the following projections on the rotating axes:

$$\begin{aligned} \omega_\xi &= -(\dot{\alpha}+\omega_L)\sin\beta\cos\gamma + \dot{\beta}\sin\gamma, \\ \omega_\eta &= (\dot{\alpha}+\omega_L)\sin\beta\sin\gamma + \dot{\beta}\cos\gamma \\ \omega_\zeta &= \dot{\gamma} + (\dot{\alpha}+\omega_L)\cos\beta. \end{aligned} \tag{65}$$

The spatial derivatives enter in the combination

$$\begin{aligned} \hat{\mathcal{L}}_2 = \frac{1}{2m}\int &\{\partial_l\psi\dagger_\mu(\mathbf{r})(A^l_a\sigma^a_{\mu\nu})\psi_\nu(\mathbf{r}) - \psi\dagger_\mu(\mathbf{r})(A^l_a\sigma^a_{\mu\nu})\partial_l\psi_\nu(\mathbf{r}) \\ &+\psi\dagger_\mu(\mathbf{r})\psi_\mu(\mathbf{r})A^l_aA^l_a\}d^3r, \end{aligned} \tag{66}$$

where A^l_a are spin or "chiral" velocities:

$$\begin{aligned} A^l_\xi &= -\alpha_l\sin\beta\cos\gamma + \beta_l\sin\gamma, \\ A^l_\eta &= \alpha_l\sin\beta\sin\gamma + \beta_l\cos\gamma, \\ A^l_\zeta &= \gamma_l + \alpha_l\cos\beta. \end{aligned} \tag{67}$$

Now we can specify the orientation of axes ξ and η so that the velocity A^l_ζ corresponding to transport of longitudinal spin component to be zero:

$$\gamma_l + \alpha_l\cos\beta = 0. \tag{68}$$

The spin current operator $\hat{j}^l_a(\mathbf{r})$ is obtained by variation of $\hat{\mathcal{L}}_2$ with respect to A^l_a following the definition $\delta\hat{\mathcal{L}}_2 = \int \hat{j}^l_a\delta A^l_a d^3r$. After the transformation, the Lagrangian $\hat{\mathcal{L}}$ consists of three parts

$$\hat{\mathcal{L}} = \hat{\mathcal{L}}_0 + \hat{\mathcal{L}}_1 + \hat{\mathcal{L}}_2, \tag{69}$$

where $\hat{\mathcal{L}}_0$ is the Fermi liquid Lagrangian in a uniform field $\mathbf{\Omega}$ oriented along the z-axis. Its ground stationary state corresponds to $\langle\hat{S}_\zeta\rangle = S$, $\langle\hat{S}_\xi\rangle = 0$, $\langle\hat{S}_\eta\rangle = 0$.

The derivatives $\alpha_l, \beta_l, \gamma_l, \dot\beta, \dot\gamma$, and the combination $(\dot\alpha+\omega_L)$ are assumed to be small and the sum $(\hat{\mathcal{L}}_1 + \hat{\mathcal{L}}_2)$ is an adiabatic perturbation. Such perturbation does not give rise to transitions to excited states. A change of state occurs only via the variation of spin orientation with respect to the nonrotating frame. This orientation is determined by the angles α, β according to $S_x = S\sin\beta\cos\alpha, S_y = S\sin\beta\sin\alpha, S_z = S\cos\beta$. The first order correction to the averaged Lagrangian contains only time derivatives of angles. For the account of spatial derivatives the second order corrections have to be calculated as in Ref.[35] The sum of both corrections forms a Lagrangian which describes the motion of spin $\mathbf{S}$ in the principal order in ω and k:

$$\mathcal{L} = S[(\dot\alpha + \omega_L)\cos\beta + \dot\gamma] + \frac{1}{2}\chi_\perp^J(\alpha_l^2 \sin^2\beta + \beta_l^2). \tag{70}$$

The coefficient $\chi_\perp^J$ in this formula is the transverse component of static current susceptibility. The current susceptibility $(\chi^J)_{ab}^{lm}(\omega, \mathbf{k})$ is a tensor coefficient in the linear relation between the Fourier components of spin current and the chiral velocity $j_a^l = \chi_{ab}^{lm} A_b^m$. The susceptibility $\chi_\perp^J$ enters Eq. (70) as a coefficient in front of terms of second order in k, so it can be taken for $\omega = 0$ and $k = 0$. Since the spin-orbit interaction is neglected the susceptibility will be isotropic tensor with respect to spatial indices l, m. With respect to the spin indices it is uniaxial tensor. Its longitudinal component does not enter in the expression for $\mathcal{L}$ because of the constraint (68). The transverse component $\chi_\perp^J(0,0)$ is a real number at the strength of general properties of linear response. All components of tensor $(\chi^J)_{ab}^{lm}(\omega, \mathbf{k})$ are expressed in terms of Fourier components of retarded commutators of corresponding components of the spin current operator (cf. Ref.[35]). These commutators are not calculated in the general case. In the weak polarization limit according to the Fermi liquid theory $\chi_\perp^J(0,0) = (\hbar/2)^2 N(0) v_f^2/3\Lambda$, where Λ is the combination of Fermi liquid parameters: $\Lambda = 1/(1+F_0^a) - 1/(1+F_1^a/3)$.

The Lagrangian (70) gives rise to the following equations of motion:

$$\frac{\partial S}{\partial t} = 0, \tag{71}$$

$$\frac{\partial}{\partial t}(S\cos\beta) + \chi_\perp^J \frac{\partial}{\partial x_l}(\alpha_l \sin^2\beta) = 0, \tag{72}$$

$$S\sin\beta(\frac{\partial\alpha}{\partial t} + \omega_L) - \chi_\perp^J[(\nabla\alpha)^2 \sin\beta\cos\beta - \Delta\beta] = 0, \tag{73}$$

where Δ is the Laplace operator. Eqs. (71)–(73) coincide with the spin dynamics equations for an isotropic ferromagnet[12] expressed in terms of Euler angles. According to Eq. (71) S=const. The variation of orientation of $\mathbf{S}$ is determined by equations (72) and (73).

Let us consider two simple stationary solutions of these equations. In static and uniform magnetic field there exists a solution of the form: β =const, $\nabla\alpha = \mathbf{k}$, $\dot\alpha = -\omega_p$. By substitution of this solution in Eq. (73), we arrive at

$$\omega_p = \omega_L - \frac{\chi_\perp^J}{S} k^2 \cos\beta. \tag{74}$$

When β is small $\cos\beta \approx 1$ and one obtains a usual spin wave with a quadratic dispersion. In the principal order in k the wave does not attenuate. The region of applicability of the present approach is limited by the adiabaticity condition $(\omega_p - \omega_L) \ll \Omega$. Then from Eq. (74) one obtains $(\chi_\perp^J/S)k^2 \ll \Omega$. For a weak polarization, $\chi_\perp^J$ is expressed in terms of Fermi liquid parameters $(kv_F)/\Omega \ll \sqrt{3\Lambda/(1+F_0^a)}$. In pure ^{3}He the r.h.s. is ~ 1 and the criterion $(kv_F)/\Omega \ll 1$ is recovered. In solutions $\Lambda \ll 1$ and the condition is more restrictive $(kv_F)/\Omega \ll \sqrt{3\Lambda}$. For $(kv_F)/\Omega \sim \sqrt{\Lambda}$ the spin waves effectively interact with the longitudinal degree of freedom and the separation in the longitudinal and transverse dynamics fails.

In a weakly nonuniform field $\omega_L = \omega_L(z)$ there exists another solution of Eqs. (72) and (73): $\dot{\beta} = 0$, $\dot{\alpha} = -\omega_p$, $\alpha_l = 0$. The variation of β is described by Eq. (73) which for given values of $\beta, \dot{\alpha}, \alpha_l = 0$ takes the form

$$\chi_\perp^J \frac{\partial^2 \beta}{\partial z^2} + S\frac{d\omega_L}{dz}(z - z_0)\sin\beta = 0. \tag{75}$$

Eq.(75) coincides with the corresponding equation from Ref.[22] which describes the coherently precessing pattern of weakly polarized Fermi liquid. It means that such a pattern must exist in a strongly polarized Fermi liquid as well. The possible effect of the demagnetizing field has not been taken into account in Eq. (75). The role of this field increases with the polarization and its effect can be accounted for as in Section 7.

The equivalence of spin dynamics equations of spin-polarized Fermi-liquid to that of isotropic ferromagnet reflects the equivalence of their symmetries. One can arrive at the same equations taking the limit $\Omega\tau \to \infty$ in the Leggett equations and redefining coefficients in these equations. This is not true for the dissipative terms. The dissipation occurs in the higher order terms in frequencies and wavevectors. An account of dissipation cannot resolve the controversy with the discussed zero-temperature attenuation, since the disagreement is present already in terms of order of k^2. The results of spin diffusion measurement in the pure spin-polarized ^{3}He[10] and in solutions of ^{3}He in ^{4}He[11] demonstrate a substantial decrease of transverse component of the spin diffusion coefficient with respect to its longitudinal component at temperatures $T \leq \Omega$. This tendency is in a qualitative agreement with the idea of the zero-temperature attenuation. The available data still cannot be considered as a direct proof of the existence of such attenuation. The data interpretation requires a proper calculation of the diffusion coefficient and spin waves attenuation. In these calculations all elementary excitations which are involved in the spin transport have to be taken into account. These are spin waves and single particle excitations. The existence of two types of excitations may give rise to the existence of regions with different temperature dependences of dissipation rate. A change of character of transverse diffusion can be expected in the region $T \sim \Omega$ for a different scenario of process. A final conclusion about the experimental verification of zero-temperature attenuation can be made only when the results of proper calculations are available.

10. CONCLUSIONS

A very essential property of spin polarized Fermi liquid is its broken symmetry with respect to spin rotations. Unlike ferromagnet this is not a spontaneous symmetry breaking, but induced, or "frozen". Nevertheless it gives rise in a usual way to a soft dynamics, which is the spin dynamics. The Fermi liquid interaction is assumed to be not strong enough for producing a spontaneous polarization, which is true both for solutions of ^{3}He in ^{4}He and for pure liquid ^{3}He. Once the symmetry is broken even a small interaction can have a pronounced effect, when lifting the degeneracy. In particular, the exchange part of Fermi liquid interaction is responsible for the existence of spin waves and coherently precessing patterns, discussed in Sec. 3. The spin waves present a new branch of excitations, which is absent in the unpolarized liquid. With the growth of polarization the contribution of these excitations to thermodynamics and kinetics of spin polarized liquid increases. A proper theory describing the kinetics of Fermi liquid at high polarizations is still a challenging problem. The possibility to vary the "frozen" polarization provides an additional parameter for the test of theories. In the solutions of ^{3}He in ^{4}He one can vary the concentration as well. All that makes the spin polarized Fermi liquid a promising object of investigation and one can expect a new interesting physics as a result of these investigations.

Acknowledgements

I am grateful to the Bulgarian Academy of Sciences and personally to Prof. D. I. Uzunov for the kind hospitality during the Workshop. I am also grateful to V. V. Dmitriev for stimulating discussions and to I. V. Kosarev for the help in preparation of the figures. This work was supported in part by CRDF grant RP1–249. and by INTAS grant 96 – 0610.

REFERENCES

1. A. E. Meyerovich, *in:* "Helium Three", W. P. Halperin and L. P. Pitaevskii ed., North-Holland, Amsterdam, (1990), Chap. 13; A. E. Meyerovich, *in:* "Progress in Low Temperature Physics", D. F. Brewer, ed., North-Holland, Amsterdam, (1987), Vol. XI.
2. G. Nunes, Jr., C. Jin, D. L. Hawthorne, A. M. Putnam, and D. M. Lee, *Phys. Rev. B* 46:9082 (1992).
3. B. Castaing and P. Nozieres, *J.Phys. (Paris)* 40:257 (1979).
4. M. Bravin, S. A. J. Wiegers, L. Puech, and P. E. Wolf, *Physica.* B197:410 (1994).
5. O.Buu, A. C. Forbes, A. C. van Steenbergen, S. A. J. Wiegers, G. Rementyi, L. Puech and P. E. Wolf, *J. Low. Temp.Phys.*, 110:311 (1998).
6. A. Rodrigues, and G. Vermeulen, *J. Low Temp. Phys.* 101:151 (1995).
7. A. Rodrigues, and G. Vermeulen, *J. Low. Temp.Phys.* 108:103 (1997).
8. P.-J. Nacher, E. Stolz, and G. Tastevin, *Czechoslovak Journal of Physics.* 46 Suppl. S6:3025 (1966).
9. D. M. Lee, *preprint.*
10. L.-J. Wei, N. Kalenchofsky and D. Candela, *Phys. Rev. Lett.* 71:879 (1993).
11. J. H. Ager, A. Child, R. Konig, J. R. Owers-Bradley and R. M. Bowley, *J. Low Temp. Phys.* 99:683 (1995).
12. L. D. Landau and E. M. Lifshitz, *Phys. Zs. Sovjet.* 8:153 (1935).
13. V. P. Silin, *Zh. Exp. Theor. Fiz.* 33:1227 (1957).
14. A. J. Leggett, *J. Phys. C* 3:448 (1970).
15. A. J. Leggett and M. J. Rice, *Phys.Rev.Lett.* 20:586 (1968).
16. D. Einzel, P. Wölfle, H. H. Jensen, and H. Smith, *Phys. Rev. Lett.* 51:2321 (1984).
17. L. R. Corruccini and D. D. Osheroff, *Phys.Rev.* B17: 126 (1978).
18. A. S. Borovik-Romanov, Yu. M. Bunkov, V. V. Dmitriev, and Yu. M. Mukharskii, *JETP Lett.* 40:1033 (1984).

19. I. A. Fomin, *JETP Lett.* 40:1037 (1984).
20. H. Akimoto, O. Ishikawa, Gong-Hun Oh, M. Nakagawa, T. Hata, and T. Kodama. *J. Low Temp. Phys.* 82:292 (1991).
21. V. V. Dmitriev, V. V. Moroz, A. S. Visotskiy, and S. R. Zakazov, *Physica* B 210:366 (1995).
22. V. V. Dmitriev, and I. A. Fomin, *JETP Lett.* 59:378 (1994).
23. I. A. Fomin, *Physica.* B 210:373 (1995).
24. D. Candela, N. Masuhara, D. S. Sherill, and D. D. Edwards, *J. Low Temp. Phys.* 63:369 (1986).
25. I. A. Fomin, *JETP.* 81:347 (1995)
26. V. V. Dmitriev, *Czechoslovak Journal of Physics.* 46, Suppl. S6:3011 (1966).
27. I. A. Fomin, and G. A. Vermeulen, *J. Low. Temp.Phys.* 106:133 (1997).
28. G. Deville, M. Bernier, and J. M. Delrieu, *Phys. Rev.* B19:5666 (1979).
29. A. E. Meyerovich, *Phys. Lett.* A107:177 (1985).
30. J. W. Jeon, and W. J. Mullin, *J. Low Temp. Phys.* 49:421 (1987); *Phys. Rev. Lett.* 62:2691 (1989); *J. Low Temp. Phys.* 88:433 (1992).
31. L. D. Landau, *Zh. Eksp. Teor. Fiz.* 35:97 (1958).
32. A. Abrikosov, L. P. Gorkov, and I. E. Dzyaloshinski. "Methods of Quantum Field Theory in Statistical Physics," Dover, NY (1975).
33. P. S. Kondratenko, *Sov.Phys.-JETP.* 20:1032 (1965); 23:509 (1966).
34. A. E. Meyerovich, and K. A. Musaelian, *J. Low Temp. Phys.* 89:781 (1992); *J. Low Temp. Phys.* 94:249 (1994); *J. Low Temp. Phys.* 95:789 (1994); *Phys. Rev. Lett.* 72:1710 (1994).
35. K. Maki, *Phys. Rev.* B 11:4264 (1975).
36. V. V. Dmitriev, V. V. Moroz, A. S. Visotskiy, and S. R. Zakazov, *Physica* B210:366 (1995).
37. V.V. Dmitriev, S.R. Zakazov, and V.V. Moroz, *Pis'ma Zh. Eksp. Teor. Fiz.* 61:309 (1995).
38. B. Castaing, *Physica* B126:212 (1984).
39. L. P. Levy, and A. E. Ruckenstein, *Phys. Rev. Lett.* 52:1512 (1984).
40. H. Ishimoto, H. Fukuyama, T. Fukuda, T. Tazaki, and S. Ogava, *Phys. Rev.* B 38:6422 (1988).

CHAOTIC VORTICES IN HE II AS AN EXAMPLE OF HIGHLY DISORDERED STATE OF ONE DIMENSIONAL SINGULARITIES

Sergey K. Nemirovskii

Institute of Thermophysics, Prospect Lavrentyeva 1,
630090, Novosibirsk, Russia

This lecture is devoted to chaotic vortex filaments in superfluid turbulent He II. The interest in this system extends beyond the theory of superfluidity to include the field of statistical physics of extended objects. In spite of the large amount of works devoted to this topic there is practically no advanced theory of this phenomenon. Formally, this is due to the extreme complexity of the vortex line dynamics. At the same time there are several approaches such as the phenomenological theory of superfluid turbulence, direct numerical simulations of the vortex line dynamics and Gaussian model of the vortex tangle giving some notions of the vortex tangle structure and allowing to describe a number of physical phenomena. These approaches as well as their numerous applications are reviewed in the lecture.

1. INTRODUCTION AND SCIENTIFIC BACKGROUND

The lecture is concerned with the pretty old but still an open chapter in the theory of superfluidity – the superfluid turbulence. The conception of superfluid turbulence was introduced by Feynman[1] in 1955. He describes the superfluid turbulence as a disordered set of quantized vortex lines or vortex tangle (VT), which appears in He II flows (or counterflows) whenever the velocity (or the relative velocity) exceeds a critical value.

The most standard scheme to study the superfluid turbulence is depicted in Fig. 1. The counterflow is created by applying a heat load q to the end of channel filled with He II. When the heat load is small, the counterflow is supported by an extremely small drop of temperature ($\Delta T \propto q$) along the channel necessary to overcome the normal component viscous flow. After exceeding some critical value of heat flux (of order of 10^{-3} W/cm^2) the temperature drop increases rapidly ($\Delta T \propto q^3$), which indicates that

Correlations, Coherence, and Order. Edited by Shopova and Uzunov
Kluwer Academic/Plenum Publishers, 1999

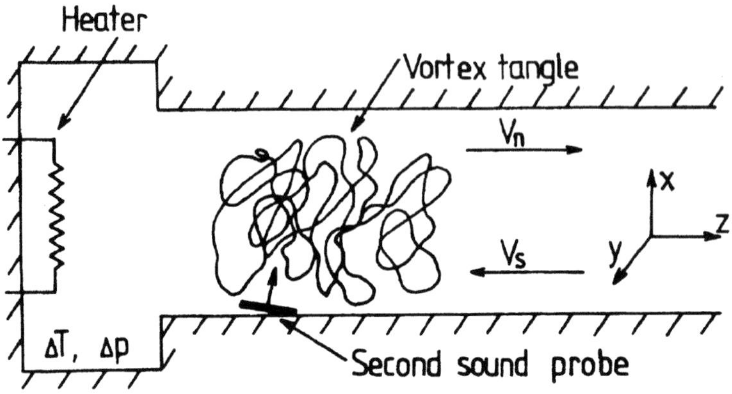

Figure 1. The turbulent counterflow in He II. The normal component flows from the heater carrying heat flux $q = ST\mathbf{V}_n$; the superfluid component flows towards the heater. The total mass current is $\mathbf{J} = \rho_n \mathbf{V}_n + \rho_s \mathbf{V}_s = 0$. Usual measured quantities are the drop of temperature ΔT or/and pressure Δp, the attenuation and the velocity of the second sound propagating at different angles through the counterflow, the shape of heat pulses etc. The axes used in the present paper are also depicted, the z-axis is directed along the relative velocity $\mathbf{V}_{ns}$, axes x and y are arbitrary, however a symmetry between x and y is assumed.

an additional strong dissipative mechanism appears. Due to the smallness of the critical value this situation is almost always realized, therefore one has to take into account the presence of vortex tangle and its impact on the dynamics of He II while treating one or other experiment. This is one of the reasons why the study of superfluid turbulence is important for the general theory of superfluidity.

Furthermore, the theory of the superfluid turbulence is tightly connected with such particular but very important branch of the theory of superfluidity as the theory of quantum vortices. Together with this field the theory of superfluid turbulence meets the common problems such as the problem of vortices initial creation, interaction between very closely spaced vortex lines and hence their reconnection, the problem of critical velocities, etc. In addition, the presence of vortex tangle influences greatly thermo- and hydrodynamic processes in superfluid Helium II. The latter is of a crucial importance for various applied problems connected with, say, cooling of superconducting magnets, designing of low temperature refrigerators, etc. These and many other aspects of superfluid turbulence have been reviewed in handbook[2] and in reviews.[3, 4]

We would like, however, to shift the accents and consider this phenomenon from another point of view. Indeed, the theory of stochastic vortex tangle in He II is of great interest and importance from the point of view of general physics. This is justified by the existence of similar systems in many physical fields, in which highly disordered sets of one-dimensional (1-D) singularities are present, e.g., chaotic vortices in ^{3}He or in superconductors, as they determine many of their physical properties. Polymer chains, linear defects in solids, strings are other examples of disordered 1-D singularities. The possibility to use the theory of stochastic vortex lines for clarifying the perennial problem of classical turbulence would be of great value.

Some physicists express the view[5-9], that the classical turbulence can be treated using the model of interacting vortex tubes. This follows from the assumption that the Kolmogorov spectrum is formed at crossings of filaments, i.e., positions of strong singularities. It is also likely that the kinetic energy dissipation is not uniformly distributed in space but occurs only at instances of quantum vortex lines collisions and reconnections. The connection between the observed coherent structures in the classical turbulence and the formation of turbulence lines is also discussed. An additional asset of this model is the possibility of using effective numerical methods because the dynamics of such systems is described by a set of one-dimensional equations. The study of stochastic properties of similar objects (and also, e.g., chaotic sets of 2–D singularities, such as membranes, triangular surfaces, etc.) belongs to the field of stochastic physics of extended objects. The vortex tangle in He II, which is formed by a stochastic complex of one-dimensional singularities conforming to a non-trivial nonlinear equation of motion with random changes of its topological structure belongs undoubtedly to this class of objects.

Therefore, the study of vortex tangle in He II may significantly contribute to the general theory of extended objects. The case of He II has a number of advantages. First of all, as we shall see later the vortex line dynamics strongly varies with temperature. For the low temperature region, where coupling to the normal component is negligibly small, the motion of lines is almost free, that coincides with the ideal fluid case. On the contrary, near T_λ the lines are almost frozen in the normal component and their dynamics is closer to the pure relaxation. Thus He II covers many practically important cases. The second advantage of He II is connected with the possibility to fullfil precise experiments, therefore, there is an outstanding opportunity to study experimentally various problems of stochastic dynamics of 1-D singularities.

The study of the vortex tangle in He II carried out for many years has yielded many important results leading to an understanding of the stochastic behavior of chaotic 1-D singularities. One of the main aims of the lecture, resulting from what we have stated above, is to review the properties of stochastic vortex tangle in He II gained from the existing interconnection between different elements appearing in the investigations of superfluid turbulence in He II such as theoretical models, numerical simulations and experimental findings.

The plan of the lecture is following. In Section 2 we shall present the general statement of the problem of stochastic dynamics of the vortex filaments in He II. The Section 3 is devoted to the classical Feynman–Vinen theory of superfluid turbulence. In Section 4 we shall reveal recent numerical simulations on the vortex lines (VL) dynamics. We will also briefly describe an analytical investigation of one model problem connected with the chaotic dynamics of VL.

The second part of this lecture is devoted to a new approach which is called the Gaussian model of vortex tangle. The main goal of this approach is to build a trial distribution function in space of vortex loop configuration satisfying all properties of the superfluid turbulence known both from the experiment and numerical simulations of the vortex tangle. A number of GM applications will be presented in Section 6.

2. STATEMENT OF THE PROBLEM

In this Section we shall formulate a general statement of the problem of the stochastic dynamics of the vortex filaments in He II. It comprises the description of equations of vortex line motion, including the interaction between vortkces and boundaries, if any, and their interaction with the normal component. The latter process is specific to superfluids. The considered problem includes also the reconnection process when two lines cross and reconnect thus changing the system topology. Finally, the stochastic approach requires also a consideration of its origin, hence, of possible external random forces, instabilities and so on.

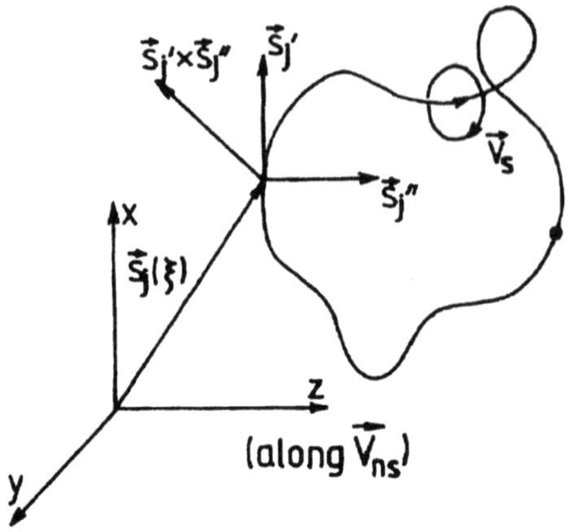

Figure 2. The space curve represents a closed vortex loop. The loop points are described as $\mathbf{s}(\xi, t)$, where ξ is the arclength, $\mathbf{s}' = d\mathbf{s}/d\xi$ is a unit vector along the vortex line, $\mathbf{s}'' = d^2\mathbf{s}/d\xi^2$ is the local curvature vector (whose magnitude is $1/R$), and $\mathbf{s}' \times \mathbf{s}''$ is the binormal with a magnitude $1/R$. Further in the text it is supposed that the axis z is directed along the applied counterflow $\mathbf{V}_{ns}$.

2.1 Dynamics of Free Lines

Onsager[10] and Feynman[1] have suggested that the Landau assumption of rotationless superfluid component flow, $\omega(\mathbf{r})$ =rot $\mathbf{v}_s = 0$, is violated on one-dimensional singularities $\mathbf{s}(\xi, \mathbf{t})$ which depend on the position parameter ξ and time t. This line singularity is shown in Fig. 2, where ξ denotes the arc length parameter. On these singularities rot $\mathbf{v}_s \to \infty$, the velocity increases also to infinity so that the circulation, $\tilde{\kappa}$, of superfluid velocity about these lines remains constant, $\kappa = h/m_{He} = 9.97 \cdot 10^{-4}$ cm^2/s; here h is Planck constant, and m_{He} is the mass of helium atom. Formally this assertion can be written as follows:

$$\omega(\mathbf{r}) \;=\; \text{rot } \mathbf{v}_s \;=\; \tilde{\kappa} \oint ds\, \delta(\mathbf{r} - \mathbf{s}(\xi, t)), \tag{1}$$

where the integration is along the line singularity $\mathbf{s}(\xi, \mathbf{t})$. At distances not very close to the vortex line the equation of vortex line motion resulting from (1) and the condition div $\mathbf{v}_s = 0$, following the Biot–Savart law will be[11]

$$\dot{\mathbf{s}}_i(\xi, t) = \frac{\tilde{\kappa}}{4\pi} \oint \frac{[\mathbf{s}(\xi', t) - \mathbf{s}(\xi, t)]}{|\mathbf{s}(\xi', t) - \mathbf{s}(\xi, t)|^3} \times \mathbf{s}'(\xi', t) d\xi', \tag{2}$$

where the integral should be taken along the line $\mathbf{s}(\xi', \mathbf{t})$ and the first argument is an arbitrary parameter not necessarily the arc length.

The above integral diverges when $\mathbf{s}(\xi', \mathbf{t}) \to \mathbf{s}(\xi, \mathbf{t})$ which requires additional conditions to be imposed on $\dot{\mathbf{s}}_i(\xi, t)$. This problem is closely connected with the small scale structure of vortex line. As the integral (2) diverges logarithmically one of the simplest, widely used approaches is to introduce a finite core radius r_0 and to replace the denominator in (2) by

$$|\mathbf{s}(\xi', t) - \mathbf{s}(\xi, t)| \to \sqrt{(\mathbf{s}(\xi', t)^2 - \mathbf{s}(\xi, t)^2 + r_0^2}. \tag{3}$$

Frequently, the so called local approach (instead of Biot–Savart law) is used. It is obtained by developing the vector function $\mathbf{s}(\xi', t)$ close to $\mathbf{s}(\xi, t)$ and by retaining the first term, which is logarithmically large with respect to other terms of the series[12]:

$$\dot{\mathbf{s}}_i(\xi, t) \;=\; \frac{\tilde{\kappa}}{4\pi}\,\frac{\mathbf{s}' \times \mathbf{s}''}{|\mathbf{s}'|^3} \int \frac{d(\xi' - \xi)}{(\xi' - \xi)} + \text{nonlocal terms.} \tag{4}$$

This integral diverges logarithmically at both the upper and the lower limits of integration. As far as the lower limit is concerned the divergence can be dealt with by introducing the radius r_0 of vortex core in the full Biot–Savart law (2). As to the upper limit the average radius of vortex line curvature $\langle R \rangle$ seems to be the appropriate limit. Finally we have

$$\dot{\mathbf{s}}_i(\xi, t) \;=\; \beta\,\frac{\mathbf{s}' \times \mathbf{s}''}{|\mathbf{s}'|^3} \;+\; \text{nonlocal terms.} \tag{5}$$

The coefficient β playing the role of a coupling constant in the nonlinear term is

$$\beta = \frac{\tilde{\kappa}}{4\pi}\,\ln\frac{\langle R \rangle}{r_0}. \tag{6}$$

Let us recall that $(\mathbf{s}' \times \mathbf{s}'')/\,|\mathbf{s}'|^3$ is directed along the binormal and its value is equal to the curvature at the considered point (see Fig. 2). As to nonlocal terms, according to the procedure used they are by an order of magnitude of $\ln(\langle R \rangle / r_0)$ smaller than the local term. Since for a vortex tangle in the superfluid Helium II, $\ln(\langle R \rangle / r_0)$ is typically of order of 10, Schwarz[13] has noted that, certainly within (5) the local approximation is valid, except in cases when two lines are very close to each other or a line is close to a boundary, considered separately. However, the problem is more involved because the neglect of small nonlocal terms means to discard the very important process of stretching of vortex lines due to nonlocal effects.

The vortex line velocity, $\dot{\mathbf{s}}_i$, can vary under the influence of external flow $\mathbf{V}_s$. Another obvious correction must be made when the vortex line approaches a boundary. A correction due to the induced velocity $\mathbf{v}_{s,b}$, which depends on the shape of boundary and, in particular, on its roughness appears and must be taken into account. Finally the free vortex line velocity $\dot{\mathbf{s}}_{free}$, taking into account these corrections will be

$$\dot{\mathbf{s}}_{free} \;=\; \dot{\mathbf{s}}_i + \mathbf{V}_s + \mathbf{v}_{s,b}. \tag{7}$$

2.2 Interaction with the Normal Component

The next factor determining the vortex line dynamics, which we will now consider, is the interaction between quantum vortices and normal component. This is specific for He II and there is no analogy in the theory of vortex tubes in classical fluids. As is well known[14], the normal component motion with a velocity $\mathbf{V}_n$ is equivalent to a drift of quasiparticles — phonons and rotons which form this component. The energy of these quasiparticles is a function of local value of superfluid velocity $\mathbf{v}_s(\mathbf{r})$ and, therefore, it is a strongly varying function near the vortex line. In other words, there exists an effective potential describing the interaction between the quasiparticles and the vortex line. Hence, it results in a momentum transfer between the quasiparticles and the vortex line during the relative motion. Thus an interaction force, called a mutual friction, appears. The corresponding theory is fully described in many reviews and handbooks, e.g., in Donnelly book[2]. We will give here only the result, important for the dynamics of vortex lines, and make some comments. The force, $\mathbf{f}_D$, acting on a vortex line unit length is

$$\mathbf{f}_D = D_1\,\frac{\mathbf{s}'}{|\mathbf{s}'|} \times \left[\frac{\mathbf{s}'}{|\mathbf{s}'|} \times (\mathbf{V}_n - \dot{\mathbf{s}})\right] + D_2\,\frac{\mathbf{s}'}{|\mathbf{s}'|} \times (\mathbf{V}_n - \dot{\mathbf{s}}), \tag{8}$$

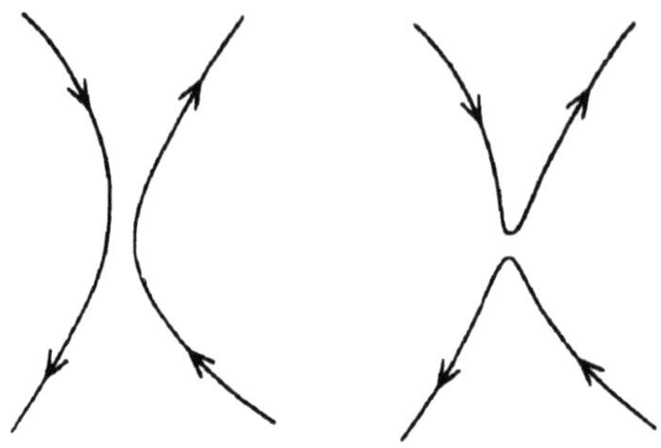

Figure 3. A schematic illustration of reconnection process.

where $\dot{\mathbf{s}}$ is the vortex line actual velocity. Many papers have been published concerning the calculation of coefficients D_1, D_2. There are also many works concerning different methods of their experimental determination[2]. The vortex line actual velocity $\dot{\mathbf{s}}$, entering in this relation, is not identical to the previously introduced velocity $\dot{\mathbf{s}}_{free}$ (see Eq. (7)) precisely, because of the existence of force $\mathbf{f}_D$. But it is known that if a vortex slips with respect to the local value of velocity, the Magnus force $\mathbf{f}_M$ equal to

$$\mathbf{f}_M = \rho_s \kappa \, \frac{\mathbf{s}'}{|\mathbf{s}'|} \times (\mathbf{s} - \dot{\mathbf{s}}_{free}) \tag{9}$$

will occur. From the comparison of $\mathbf{f}_M$ and $\mathbf{f}_D$ which is justified when there are no mass effects during the vortex motion, it follows that

$$\dot{\mathbf{s}} \; = \; \dot{\mathbf{s}}_i \; + \; \mathbf{V}_s + \mathbf{v}_{s,b} \; + \; \alpha \, \frac{\mathbf{s}'}{|\mathbf{s}'|} \times (\mathbf{V}_{ns} - \dot{\mathbf{s}}_i) \; - \; \alpha' \, \frac{\mathbf{s}'}{|\mathbf{s}'|} \times \frac{\mathbf{s}'}{|\mathbf{s}'|} \times (\mathbf{V}_{ns} - \dot{\mathbf{s}}_i) \, . \tag{10}$$

The coupling constants α and α' can be expressed by previously used coefficients D_1 and D_2 (see, e.g. paper[13]). They are also related to Hall and Vinen B, B' coefficients by following relations $\alpha = \rho_n B/2\rho$, $\alpha' = \rho_n B'/2\rho$. Thus we have obtained the master equation governing the vortex line motion in He II. Let us note that the self-induced velocity $\dot{\mathbf{s}}_i$, in the r.h.s. of Eq. (10) can be expressed in terms of the complete Biot–Savart law (2) or by the singled-out local part (5).

2.3 Reconnection of Lines

The process of greatest influence on the vortex tangle evolution to be considered also is the reconnection of vortex lines. This remarkable and only scarcely studied process occurs in other extended objects such as polymers, linear defects in solids, etc. During their motion the vortex lines unavoidably cross each other and the problem of what happens when this occurs is of a crucial importance for structure and dynamics of vortex tangle. Feynman[1] in his famous, fundamental paper has predicted that crossing of two vortex lines is accompanied by a reconnection process. This is shown schematically in Fig. 3. Lines at crossing break and reconnect with their nearest neighbors thus changing the flow topology. A number of papers, according to the Sethian review[5], deal with this problem for the case of vortex tubes in classical fluids. These results can be partly used to describe or model the vortex reconnection in the superfluid Helium II.

The reconnection problem can be divided into two parts. One of them describes the line motions in their evolution process when they approach each other up to the point where their mutual influence on the velocity of their motion becomes larger than the self-induced velocity due to the local curvature. But at the same time they are not close enough to influence the flow inside the vortex cores. The quantitative criterion is $r_0 < \Delta \leq R/\ln(R/r_0)$, where Δ is the distance between the lines. In these cases the evolution of approaching segments follows the Biot–Savart law (2) where the cut-off radius r_0 is either a constant (see Schwarz[15], Tsubota et al.[16]), or depends on the label ξ (Siggia[6]). Qualitatively the results of these investigations are very similar and can be described according to the Siggia model as follows. Due to the long range interaction in the Biot–Savart integral, cusps may appear when approximately antiparallel segments of two vortex lines approach. The curvature of these cusps may be so large that the self-induced velocity (see (5)) of each perturbation overcomes the repulsion from the adjoining vortex line. Further the cusps grow, approach each other closer increasing their curvature and correspondingly their self-induced velocities and this process is repeated faster and faster. It is important that this process grows explosively, since the distance between two perturbed segments, Δ, decreases according to the relation $\Delta \sim (t^* - t)^{1/2}$, where t^* is some quantity depending on relevant parameters and initial conditions. Thus after a finite time the vortex lines collapse. It is very important that time of collapse is much smaller than the characteristic time of motion of other vortex line elements. Schwarz[15] has described a similar behavior of vortex lines and of a vortex line and its image close to a boundary. In this way antiparallel vortex lines (a vortex line and its image) whenever they are at a distance $\Delta \leq R/\ln(R/r_0)$ suffer a rapid approach and collapse. Initially arbitrarily oriented vortices, as shown also by Schwarz, on approaching each other closer than some critical distance, start reorienting their close segments so as to bring them into an antiparallel position which is followed by a collapse as described above.

The second part of the process starts when the distance between vortex lines Δ is comparable with the core radius r_0. In this case the induced velocities calculated from the Biot–Savart integral distort the flow inside the core, which imposes the necessity to solve full Navier–Stokes or Euler equations. The full investigation of classical vortex tube shows that there exists an extremely complicated picture of vortex interaction. Instead of a full annihilation some very involved structures appear where strong dissipation effects take place. As far as Helium II is concerned where atomic scales come into play it is necessary to find solutions using either the weakly interacting Bose gas model or the Hills and Roberts theory (see, e.g., Donnelly book[2]). We are interested to know first whether a full annihilation of two antiparallel line segments will occur because this is necessary for the reconnection and second, how much time this process will require if it takes place at all. The answer to the first question is positive as shown by Nakajima et al.,[17] who using a numerical modelling of Gross–Pitaevskii equation, have proven that the approaching antiparallel quantized vortices are completely annihilated. Frisch et al.[18] have obtained a similar result for the collapse and the annihilation of vortex rings. It has been also shown that the duration of this process is very short in comparison with the characteristic time of vortex tangle dynamics. This information about the small duration of reconnection process is very useful because it justifies the assumption used by Schwarz[20] and Buttke[19] that the reconnection process is instantaneous, just as a collision of particles in the gas dynamics problem.

The instantaneous reconnection is, of course, a great simplification of the problem. But even so the problem remains extremely complex in spite of simplifications made in obtaining (10). Indeed this equation describing the dynamics of the vortex line motion is substantially nonlinear with several kinds of couplings and the nonlinearity is not polynomial due to the existence of denominators of type $1/|\mathbf{s}'|$. The equation contains also nonlocal terms expressed by the Biot–Savart law. The presence of mutual friction terms leads to the violation of some conservation laws, e.g. the conservation of energy. Finally the reconnection process changes the system topology, hence the quantity $\mathbf{s}(\xi, t)$ as a function of parameter ξ during the collision process receives and stores discontinuities which accumulate during the stochastic process of vortex structure development. To appreciate the complexity of problem we would like to point out that as shown by Hasimoto[21], if in (10) all terms except the first one are omitted, the local approach can be used and finally the local length can be fixed, $|\partial \mathbf{s}/\partial \xi| = 1$; then Eq. (10) can be reduced to the nonlinear Schrödinger equation. The stochastic behavior of nonlinear Schrödinger equation is a quite nontrivial problem which is intensively studied at present.

Resuming results exposed in this section it can be concluded that the vortex line dynamics is governed by the master equation of motion (10) plus the instantaneous reconnection. This approach is referred to as the reconnection anzatz, and it should be considered as a basis for the study of the chaotic vortex filament dynamics. To do so, however, we have to introduce some source of chaos, for instance, to introduce in r.h.s. of the master equation (10) some random force, or Langevin force. Resuming again it can be concluded that the problem of a stochastic vortex filaments motion is deadly involved. For this reason there is practically no adequate theory concerning the stochastic vortex line dynamics in He II. There are, however, some approaches based on both experimental data and numerical simulations, which allow to describe the vortex tangle properties as well as its dynamics. We shall describe them in the following sections.

3. FEYNMAN–VINEN PHENOMENOLOGICAL THEORY OF SUPERFLUID TURBULENCE

3.1 Feynman Qualitative Scenario

The greatest success has been gained by the phenomenological theory of superfluid turbulence. This theory in its original form based on Feynman[1] qualitative considerations has been developed by Vinen.[22]

Feynman has assumed the following scenario for the vortex structure evolution. As appears from the equation of vortex lines motion, segments of a vortex line are exposed to a Magnus force, directed perpendicularly to the tangent vector. Then a variation of the length of the considered segment occurs. Depending on the orientation and the curvature and also on the counterflow velocity $\mathbf{V}_{ns}$, stretching or shrinking of the segment is possible. Feynman has assumed that stretching of lines prevails, i.e. the length of evolving vortices *on the average* grows. While increasing their length the lines fill in more densely the liquid volume, and the processes of lines crossing and reconnection come into play. As a result of the reconnection a fusion of small vortex rings into larger ones as well as a break up into smaller ones are possible.
Assuming again that the last property dominates, Feynman has concluded that *on the average* a break up of vortex loops takes place. This leads to a cascade like process of a formation of smaller and smaller loops, as is schematically depicted in Fig. 4. When the scale of small rings becomes of order of interatomic distances, which is the final stage of the cascade, the vortex motion is degenerated into thermal excitations.

Thus the reduction of the total vortex lines length and the transformation of the vortex energy, which has been initially drained from the main flow into thermal excitations, take place. This total length decrease, at a sufficiently high density of the

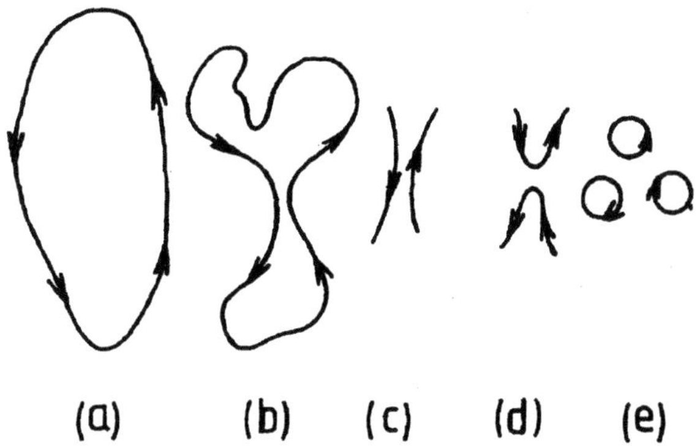

Figure 4. A cascade-like process of vortex ring breaking due to the reconnection: (a) the initial stage; (b), (c) the stage of approaching the line elements; (d) the stage of collapse and reconnection; (e) the stage of cascade-like degeneration of vortex loops into thermal excitations.

vortex tangle, compensates the growth process due to the mutual interaction with the normal component. Thus an equilibrium state, characterized by the total vortex line length which is a function of thermodynamic variables and the quantity $\mathbf{V}_{ns}$, is reached.

3.2 Vinen Phenomenological Theory

The Feynman qualitative model has been further developed in the classical works of Vinen.[22] He has formulated these ideas in quantitative relations and, in particular, obtained the equation, bearing his name, which gives a quantitative description of the macroscopic vortex tangle dynamics, i.e., the evolution of total vortex line length per unit volume, or the vortex line density (VLD) $\mathcal{L}(t)$.

The time derivative $d\mathcal{L}/dt$ has been assumed to be composed of two terms corresponding exactly to the Feynman qualitative model

$$\frac{\partial \mathcal{L}}{\partial t} = \left(\frac{\partial \mathcal{L}}{\partial t}\right)_{growth} - \left(\frac{\partial \mathcal{L}}{\partial t}\right)_{decay}. \tag{11}$$

To find the form of two components Vinen has used dimensional considerations, the familiar results concerning the dynamics of single vortex rings and the analogy with the classical turbulence. He has also used widely the results of his experiments on the temperature drop and the additional second sound attenuation. In these considerations Vinen has made a very important assumption that $d\mathcal{L}/dt$ is a function only of the instantaneous value $\mathcal{L}$, the friction force $\mathbf{f}$ and the circulation $\tilde{\kappa}$. This property is called the self-preserving state. In this way he has derived the equation for the evolution of vortex line density

$$\frac{\partial \mathcal{L}}{\partial t} = \alpha_V\,|\mathbf{V}_{ns}|\,\mathcal{L}^{3/2} - \beta_V\,\mathcal{L}^2. \tag{12}$$

In the stationary case, the Vinen equation yields the relation

$$\mathcal{L}_\infty = \frac{\alpha_V^2}{\beta_V^2}\,\mathbf{V}_{ns}^2 = \gamma^2\,\mathbf{V}_{ns}^2. \tag{13}$$

Having the total length and remembering the expression for force per unit length it is possible to obtain an expression for the total friction force. It is easy to see, however, that some additional supposition about the vortex tangle arrangement is required. In other words, we need some rule to fulfil the averaging over various vortex tangle configurations $\langle\rangle_{VT}$. As a matter of fact this is the main problem, which of course, could not be solved in the frame of the Vinen–Feynman model operating with the macroscopic variable $\mathcal{L}(t)$ *only*. The problem of averaging has frequently arisen while physicists try to calculate some quantities due to the vortex tangle presence. Usually they make some additional suppositions concerning the tangle arrangements.

In particular, Vinen himself assumed that there is a full isotropy in the distribution of tangent vector $\mathbf{s}'(\xi)$ and obtained

$$\mathbf{F}_{ns} \;=\; (2\alpha\,\rho_s\,\tilde{\kappa}/3)\,\mathcal{L}\,\mathbf{V}_{ns} \;=\; A(T)\,\rho_s\,\rho_n\,\mathbf{V}_{ns}^3. \tag{14}$$

Here $A(T)$ is the so-called Gorter–Mellink constant obtained earlier in experiments on the anomalous heat conductivity of He II. Introducing this relation in superfluid hydrodynamics equations one immediately obtains that the temperature gradient is proportional to the cubed heat load, or to the Gorter–Mellink formula. In a similar way the other characteristics, e.g., the extra-attenuation of second sound can be evaluated.

Behind the very simple relations introduced by (12)–(14) there is a huge variety of phenomena studied by many scientific groups for may years. Good reviews can be found in a book by Donnelly[2] and a review article by Tough[3]. Let us list some important results. One of them is the temperature drop (the Gorter–Mellink result) discussed above. The other one is the additional sound attenuation due to the vortex tangle and the friction force. This extra-attenuation turns out to be anisotropic (with respect to the counterflow direction), which indicates that the vortex tangle is actually anisotropic. Besides, it has been revealed that there is an additional pressure drop, probably connected with the phase slippage of vortices. Interesting observations have been made by several groups who notice the appearance of organized structures – "plugs" of vortex line density. Furthermore, the Vinen equation can be incorporated[23-25] into classical hydrodynamics of He II. This unified hydrodynamics of superfluid turbulence describes nonstationary processes in the superfluid turbulent Helium II. As an example we can give the problem of propagation of strong second sound pulses which generate quantum vortices and interact with these "own" vortices.[26-28] Another example is the practically important case of developing a temperature field under a varying heat load. Many of these and other phenomena such as the vortex tangle free decay or the vortex line density fluctuation are described in a review[4] by Nemirovskii and Fiszdon.

The approach developed by Feynman and Vinen is frequently referred to as the phenomenological theory of superfluid turbulence, for the main constituents of this theory have been the Feynman acute conjecture as well as the Vinen experimental data.

4. MODERN NOTION OF CHAOTIC VORTEX LINE DYNAMICS

4.1 Numerical Simulations in the Vortex Dynamics

The Vinen phenomenological theory of the superfluid turbulence described in the previous chapter, was derived in the early sixties. During the following two decades the superfluid turbulence has been intensely studied mainly experimentally. It should be said that all of these experimental and semi-theoretical investigations operate only with "macroscopic" quantities such as the vortex line density $\mathcal{L}(t)$, the total friction force, the sound attenuation, etc. Almost no attempts has been made to obtain further insight into the subtle vortex tangle structure. Moreover, no attempts have been made to substantiate the fundamental conception of this theory and rigorously to connect the vortex line dynamics with the vortex tangle properties.

The break-through in the understanding of superfluid turbulence nature and its relation to the vortex line dynamics has been reached in the key series of works[13,5,20] made by Schwarz. In the striking paper[20] of this series Schwarz has reported the results from direct numerical simulations of the vortex filament dynamics. Starting from the equation of vortex line elements motion in He II (in the local form, see (5)) and assuming that the vortex lines reconnect while approaching each other, Schwarz has shown that initially smooth vortex rings develop into a chaotic vortex tangle. He has calculated some of characteristics of this vortex tangle, which he calls the structure parameters of vortex tangle. Pursuing a goal to compare his results with experimental data Schwarz primarily concentrated on phenomena and effects studied before in the phenomenological theory. For instance, the friction force, which has been derived by Vinen practically qualitatively (see Eq. (14)), can be rigorously written as

$$\mathbf{F}_{ns} = \frac{\rho_s \tilde{\kappa} \alpha}{\Omega} \int \langle \mathbf{s}' \times [\mathbf{s}' \times (\mathbf{V}_{ns} - \dot{\mathbf{s}}_i)] \rangle \, d\xi \, . \tag{15}$$

Here the integration is fulfilled over the whole vortex tangle configuration. The quantity $\dot{\mathbf{s}}_i$ is the self-induced velocity of line elements given by the local approach (5). Similarly, the drift velocity $\mathbf{v}_L$ of the vortex tangle as a whole in the superfluid velocity reference frame can be expressed as follows:

$$\mathbf{V}_L = \frac{1}{\Omega \, L} \int \langle \dot{\mathbf{s}} \rangle \; d\xi \; - \; \mathbf{V}_s \, . \tag{16}$$

The expressions for second sound damping, temperature drop or chemical potential drop can be formulated in a similar way. On the basis of the vortex line dynamics Schwarz has even managed to obtain an equation for the rate of change of vortex line density $\mathcal{L}(t)$. Using that the line element length is $\Delta l = |\mathbf{s}'| \; d\xi$ and applying the equation of motion (10) one obtains that the rate of change of quantity Δl will be:

$$\frac{\partial \Delta l}{\partial t} = \frac{\alpha}{\beta} \, (\mathbf{V}_{ns} - \dot{\mathbf{s}}_i) \, \dot{\mathbf{s}}_i \, \Delta l. \tag{17}$$

Integrating Eq. (17) over the whole vortex tangle and averaging one arrives at the following relation for the vortex line density evolution:

$$\frac{\partial \mathcal{L}}{\partial t} = \frac{\alpha}{\Omega} \mathbf{V}_{ns} \int \langle \mathbf{s}' \times \mathbf{s}'' \rangle \; d\xi \; - \frac{\alpha}{\Omega} \int \left\langle |\mathbf{s}''|^2 \right\rangle \; d\xi \, . \tag{18}$$

Relations like (15)–(18) are exact relations between the vortex tangle structure and measured quantities, they replace phenomenological relations like (14) or (12). Of course, these relations will be nothing but some "useless" identities unless one has some knowledge of vortex line distribution in space or the statistical properties of this distribution. Put in other way, we again face the problem how to fulfil the $\langle \rangle_{VT}$ procedure. Schwarz has solved this most important and most difficult part of the problem using the results from his direct numerical simulation of the vortex line dynamics. Fulfilling this task he used the reconnection anzatz, described in the Section 2, namely he used the master equation of vortex line motion (10) with the local from of the Biot–Savart law (5) plus an instantaneous reconnection.

The extreme complexity of the dynamics of He II vortex lines can be seen in Fig. 5 illustrating the evolution of six initially smooth vortex rings. Using the result of his

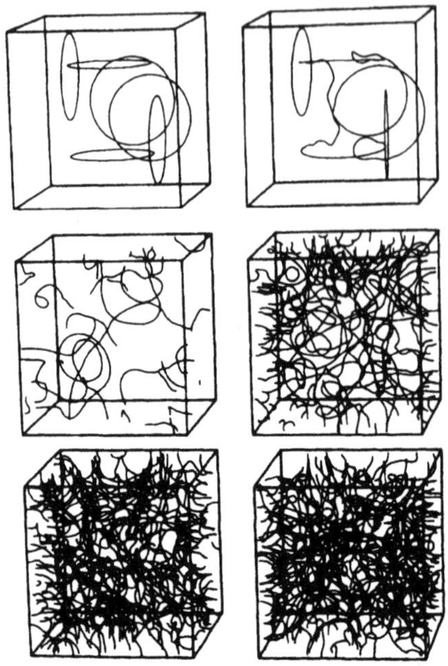

Figure 5. Case study of the development of vortex tangle in a real channel, (see Schwarz[20], Fig. 4). Here the temperature is about 1.6 K, and $|\mathbf{V}_{ns}| = 0.47$ cm/s in the front face of channel section is shown. Upper left: $t = 0$, no reconnections; upper right: $t = 2.8$ s, three reconnections; middle left: $t = 50$ s, 18 reconnections; middle right: $t = 200$ s, 844 reconnections; lower left: $t = 550$, 12128 reconnections; lower right: $t = 2750$ s, 124781 reconnections.

numerical simulations Schwarz calculated some quantities, namely the averages, entering in relations (15)–(18), which he has named structure parameters of vortex tangle and which allowed him to calculate the friction force, the Gorter–Mellink constant, longitudinal and transverse attenuations of second sound travelling in turbulent He II. He also calculated the coefficients of Vinen equations. The calculated quantities are in a good agreement with the large body of experimental data obtained earlier by many experimentalists (see, for details, the original paper[20], and also the review[4]).

Undoubtedly the main results of Schwarz work is the direct confirmation of the Feynman conjecture. Indeed, before his numerical calculations the Feynman picture has been just a conjecture, confirmed only by indirect measurements. It can be asserted, that Schwarz has started an activity and made first steps to change the phenomenological and qualitative theory of superfluid turbulence into a more or less rigorous theory connecting the deterministic dynamics of vortex filaments with stochastic properties of vortex tangle.

In spite of the obvious progress in understanding of superfluid turbulence and the good agreement with experimental results the Schwarz theory is sometimes criticized. Buttke[19] asserts, after repeating the calculations with a different algorithm, that the main results are confirmed only for rough mesh sizes, but the reducing of mesh size gives very different results. Obviously the recent results of Aarts and de Waele[29] who not only have confirmed Schwarz numerical findings but have also obtained new results for the non-uniform, Poiseuille like $\mathbf{V}_{ns}$ velocity distribution clarify somewhat this problem. We think that a full clarification of this problem is of special importance because this controversy has led to some doubts concerning the formulation of problem, especially the use of local induction approximation (5) instead of the full Biot–Savart law (2). This is the second critical remark concerning Schwarz results.

Doubts concerning the use of local induction approximation for the adequate description of stochastic vortex line evolution comes from scientists investigating vortex tubes in classical fluids. From their point of view the local approximation which in classical fluids does not lead to a change in the vortex tube length, i.e. does not include stretching, is not suitable to describe satisfactorily the vortex dynamics. Moreover, there are some statements that ignoring the stretching one cannot obtain the stochastization.

Hansen and Nelkin[30] have supported Schwarz results indicating the difference between classical vortices and quantized vortices. The stretching of classical vortices is accompanied by a reduction of their core radius whereas the core radius of quantized vortices is fixed. This difference according to Hansen and Nelkin in the vortex properties removes the problem of importance of vortex stretching due to nonlocal processes in the vortex dynamics of He II. Agistein and Migdal[9] on the other hand have criticized the introduction of cut off radius, which depends on the label ξ. In their calculations they have fixed the core radius (which is the case of He II) and observed the stochastization of initially smooth vortex lines.

We think that a large degree of uncertainties concerning the relevant questions is connected with the rapid development of numerical methods used in the investigation of stochastic vortices in Helium II and in classical fluids which are very much ahead in corresponding theoretical analysis. This results in the introduction of different vague procedures, which permit to make different unsubstantiated conclusions.

4.2 Analytical Study

We should mention an attempt[31] of simultaneous theoretical and analytical investigation of the stochastic free vortex line dynamics in the local approximation. In addition to the listed simplifications, the processes of reconnection is omitted. Of course, this model is very far from the physical reality, nevertheless the good agreement between numerical results and analytical considerations allows to draw some conclusions concerning the stochastic vortex line dynamics and even to apply them to the superfluid turbulence problem. Fig. 6 illustrates the evolution of vortex ring satisfying the following dynamic equation

$$\frac{\partial \mathbf{s}}{\partial t} = \beta\ \mathbf{s}' \times \mathbf{s}'' + \mathbf{f}(\xi, t); \tag{19}$$

$\mathbf{f}(\xi, t)$ is a random "Langevin" force with a correlator

$$< f_\kappa^\alpha f_\kappa^\beta >= \delta_{\alpha\beta} F_\kappa \delta(t - t'),$$

where κ is the wave vector of the one dimensional Fourier transformation along the line (conjugate to the variable ξ). The Langevin force models the random long range interaction of different line elements.

The nonlinear interaction described by the first r.h.s. term in (19) leads to a generation of higher harmonics (larger κ) which in their turn generate still higher harmonics and so on. This cascade process induces the creation of segments of a large curvature as shown in Fig. 6. An additional curvature generated by the force $\mathbf{f}(\xi, t)$ stretches into the region of large κ. Thus a flux of the curvature P appears in κ space. At very large $|\kappa|$, that corresponds to very kinked segments of vortex line, a dissipation mechanism, e.g., a collapse and vortex annihilation appears.

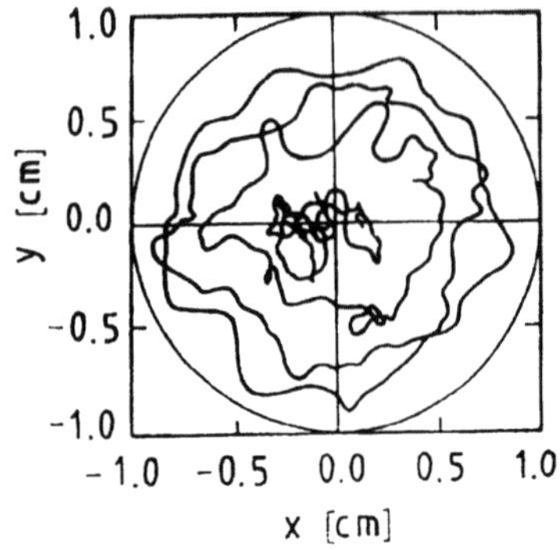

Figure 6. The evolution of vortex ring under the influence of an external random force in the local approach. Projections of vortex filament on the plane of the initial ring at different instants are shown.

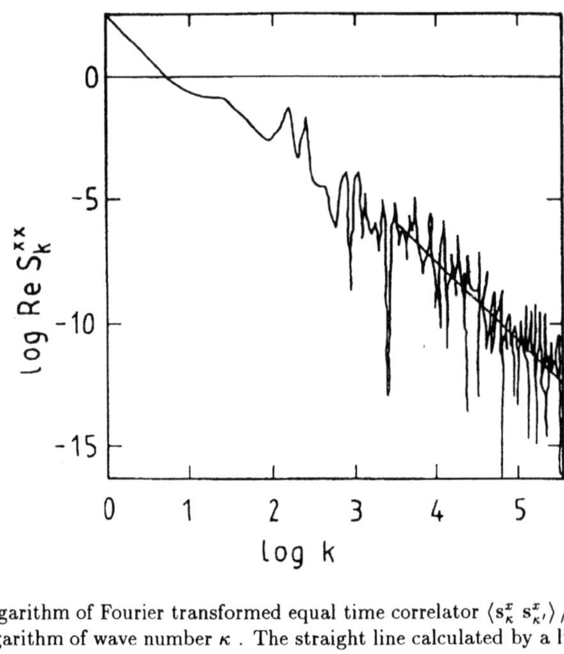

Figure 7. The logarithm of Fourier transformed equal time correlator $\langle s_\kappa^x \, s_{\kappa'}^x \rangle / \delta(\kappa + \kappa')$ as a function of the logarithm of wave number κ . The straight line calculated by a linear regression has a slope −3.13.

The above processes lead to the dissipation of vortex lines curvature. The competition between the non-linear generation of high harmonics and dissipative mechanisms leads to a stationary picture with a spectral distribution of vortex curvature. A vortex line in this case forms a small tangle shown in Fig. 6. This statement of the problem is fully analogous to the Kolmogorov description of classical turbulence. The spectral distribution has been obtained in the above cited paper with the help of field theoretical methods. It was found that the Fourier image (spectrum) of the equal time correlator of vortex line follows the relation

$$\left\langle \mathbf{s}_{\kappa}^{\alpha}\, \mathbf{s}_{\kappa'}^{\beta} \right\rangle = \mathrm{const}\ \delta(\kappa + \kappa')\ \frac{P^{2/3}}{\kappa^3}. \tag{20}$$

The spectrum obtained from an exact numerical solution of problem (19) is illustrated in Fig. 7. It has been found that the spectrum varies depending on the external force intensity although it remains close to theoretical predictions (20).

We would like to make a few remarks concerning Schwarz results with reference to the solutions of the simple model described above. The first one is concerned with the scaling properties. In study of these properties it has been assumed that the stochastic vortex tangle dynamics is described by $\mathbf{V}_{ns}$ and $\mathcal{L}$. Correspondingly, in the stationary case, when these quantities are connected by the relation $\mathcal{L}^{1/2} = \gamma \mathbf{V}_{ns}$ everything is determined by one of them (Schwarz has selected $\mathcal{L}$). However, the solution of the model problem shows that there exists one more parameter in the problem, the curvature flux P in κ-space. The presence of an additional dimensional parameter implies that the Schwarz scaling analysis appears at least not complete.

Furthermore, the solution of the model problem shows the importance of random forces which model the long range interaction in a vortex tangle. Since in the Schwarz local approximation this long range contribution is missing, the corresponding analysis is apparently not complete. If we add here also the absence of stretching as a nonlocal effect it may be concluded that some processes remain outside the Schwarz model. Agreeing generally to this incompleteness Schwarz[20] states that omitted processes have a small influence as compared to those left in his approximation which is confirmed by the good agreement of his calculations with experimental observations. Taking into account the absence of adjustable parameters in his calculations such an argumentation is acceptable. However, one should remember that the experimental data also have a large scatter.

We have exposed the current state of art in the theory of chaotic vortex lines in the superfluid Helium II. It is apparent that this problem is sooner open than essentially advanced. Nevertheless results described above enable us to obtain an insight into the vortex tangle structure.

4.3 The Vortex Tangle Structure

This subsection is devoted to the summary of our knowledge on the vortex tangle arrangement obtained from investigations of superfluid turbulence. Primarily this knowledge has been accumulated from experimental works, however, while interpreting one or other experiment, investigators have used a conception of the superfluid turbulence as a set of vortex filaments chaotically distributed in space. Fetching various semi-quantitative speculations investigators have drawn a number of conclusions concerning the vortex tangle structure. As has been already explained these pure phenomenological results are confirmed in numerical simulations of vortex lines dynamics made by Schwarz.[20] The numerical modelling not only establishes numerical values for a number of vortex tangle characteristics, but also allows to determine them in temperature regions, where the experimental data are absent. For this reason we widely use the structure parameters calculated by Schwarz as a basis, although, in principle, we would rely only on experimental results. Therefore, besides the simple introduction of quantities characterizing the vortex tangle structure, we will briefly discuss what experimental results have led to them.

The current view on the vortex tangle arrangement can be summarized as follows: the vortex tangle developed in counterflowing He II consists of a set of closed lines labelled by an index j. They can be described by a set of functions $\mathbf{s}_j(\xi_j)$, where $\mathbf{s}_j(\xi_j)$ is the radius-vector of points resting on a j-loop. The variable ξ_j labels the points of the j-loop. It is convenient to choose the variable ξ_j to be equal to the arc length ξ_j, $(0 \leq \xi_j \leq L_j)$, (see Fig. 2). The phenomenological theory gives an instantaneous picture of vortex tangle, therefore, the dependence on time is omitted. The whole configuration of vortex tangle $\{\mathbf{s}_j(\xi_j)\}$ is a unification of all curves $\{\mathbf{s}_j(\xi_j)\} = \bigcup_j \mathbf{s}_j(\xi_j)$. Due to frequent reconnections, both the number of loops and their lengths L_j are arbitrary quantities. In addition, each of the loops can take any arbitrary shape $\mathbf{s}_j(\xi_j)$. It should be understood, however, that despite the arbitrariness of these quantities the whole configuration should meet a number of requirements.

For instance, the total length of loops per unit volume – the vortex line density $\mathcal{L}_v$ is a well determined quantity, satisfying the relation:

$$\left\langle \frac{1}{\mathcal{V}} \sum_j \int_0^{L_j} |\mathbf{s}'_j(\xi_j)| \, d\xi_j \right\rangle = \mathcal{L}_v. \tag{21}$$

Here $\mathcal{V}$ is the volume, the prime denotes the derivative with respect to the arc length ξ_j. The brackets $\langle\rangle$ denote an overall averaging over vortex loop configurations $\{\mathbf{s}_j(\xi_j)\}$. Since the variable ξ_j is chosen to be the arc length, the tangent vector absolute value is just a unity

$$|\mathbf{s}'_j(\xi_j)| = 1, \tag{22}$$

which leads to the relation

$$\left\langle \sum_j \int_0^{L_j} d\xi_j \right\rangle = \mathcal{V}\mathcal{L}_v. \tag{23}$$

The filaments comprising the vortex tangle are distributed in space in an anisotropic manner. There are two kinds of anisotropy. The first one is connected to the orientations of line elements and it has been discovered in experiments on the attenuation of transverse (with respect to the counterflow, see Fig.1) and longitudinal second sound.[32] The measure of this anisotropy can be defined with the use of structure parameters $I_{\|}$ and $I_{\perp}$ introduced by Schwarz[20] who has confirmed the vortex tangle anisotropy in the numerical modelling. For our purposes it is more convenient to use other parameters $I_{\alpha\alpha}$ $(\alpha = x, y, z)$ simply connected with those introduced by Schwarz. The fraction of vortex line elements orientated along the z–axis is

$$\left\langle \frac{1}{\mathcal{V}\mathcal{L}_v} \sum_j \int_0^{L_j} \mathbf{s}'_{jz}(\xi_j)\mathbf{s}'_{jz}(\xi_j) \, d\xi_j \right\rangle = 1 - I_{\|} = I_{zz}. \tag{24}$$

To move further we have to discuss what we mean by the term "an overall averaging over the vortex loop configuration". Based on what has been said above, we state that the overall averaging includes an averaging: (i) over the shape of each of the loops, (ii) over the number of loops and their lengths, and (iii) over initial points of each of the loops. As far as second and third items are concerned, unfortunately neither the experiment nor the theory give any idea what are the respective distributions. To overcome this problem we accept the supposition of the full uniformity of the vortex tangle made by many investigators and likely confirmed in numerical simulations. Because of the uniformity any overall *local* average such as $\langle \mathbf{s}_{j\alpha}(\xi_j) \rangle$, $\langle \mathbf{s}_{j\alpha}(\xi_j) \mathbf{s}'_{j\beta}(\xi_j) \rangle$, etc. should not depend on both ξ_j and j. Therefore, the relation (24) can be factorized

$$\left\langle \mathbf{s}'_{jz}(\xi_j)\mathbf{s}'_{jz}(\xi_j)\right\rangle \left\langle \frac{1}{\mathcal{V}\mathcal{L}_v} \sum_j \int_0^{L_j} d\xi_j \right\rangle = I_{zz},$$

and, in combination with (23), can be rewritten in a local form

$$\left\langle \mathbf{s}'_{jz}(\xi_j)\mathbf{s}'_{jz}(\xi_j)\right\rangle = I_{zz}. \tag{25}$$

The meaning of performed procedure is that we have separated the overall averaging into one over shapes of loops and other averaging. Therefore, accomplishing pre-averaging over loop lengths, over number of loops and their initial points, which is trivial due to the full uniformity we are left with an averaging over the shape of some "averaged" loop having the same structure parameters as the whole vortex tangle. Thus the problem is reduced to construct a distribution function in the configuration space of "averaged" curve which takes various shapes. We, however, retain the index j for a consistency of presentation.

Relations analogous to (25) can be written for x and y components:

$$\left\langle \mathbf{s}'_{jx}(\xi_j)\mathbf{s}'_{jx}(\xi_j)\right\rangle = 1 - I_{\perp} = I_{xx}, \tag{26}$$

$$\left\langle \mathbf{s}'_{jy}(\xi_j)\mathbf{s}'_{jy}(\xi_j)\right\rangle = 1 - I_{\perp} = I_{yy}. \tag{27}$$

Since $|\mathbf{s}'_j(\xi_j)| = 1$ (see Eq. (22)) parameters I_{xx}, I_{yy}, I_{zz} will obey the obvious identity

$$I_{xx} + I_{yy} + I_{zz} = 1. \tag{28}$$

The second kind of anisotropy, the so-called polarization is connected with the mutual orientation between the tangent vector $\mathbf{s}'_j(\xi_j)$ of filament segments and the curvature vector $\mathbf{s}''_j(\xi_j)$. A measure of polarization I_l has been introduced quantitatively by Schwarz with the help of the relation

$$\left\langle \frac{1}{\mathcal{V}\mathcal{L}_v^{3/2}} \sum_j \int_0^{L_j} \mathbf{s}'_j(\xi_j) \times \mathbf{s}''_j(\xi_j)\, d\xi_j \right\rangle = I_l \mathbf{e}_z, \tag{29}$$

where $\mathbf{e}_z$ stands for the unit vector in the z-direction (along the counterflow $\mathbf{V}_{ns}$). On the strength of arguments discussed above, the condition (29) can be also brought into a local form

$$\left\langle \mathbf{s}'_j(\xi_j) \times \mathbf{s}''_j(\xi_j)\right\rangle = I_l \mathcal{L}_v^{1/2} \mathbf{e}_z. \tag{30}$$

Though the relation (29) has been taken from a numerical simulation it can be readily obtained from the experimental data. Indeed, the combination $\left\langle \mathbf{s}'_j(\xi_j) \times \mathbf{s}''_j(\xi_j)\right\rangle$ appears as a positive term in the equation for the rate of change of the line element length. Therefore, it can be extracted by comparing this equation with the first term in the right-hand side of Vinen equation (12). This procedure has been carried out in papers.[25, 20]

The next property of the vortex tangle concerns the mean line curvature. The idea that the mean vortex tangle curvature should be of order of interline space goes back to Hall works on the superfluid turbulence.[33] Later this view was discussed by many authors and has been rigorously confirmed in the Schwarz numerical simulation. He calculated the coefficient c_2^2 connecting the averaged squared curvature with the quantity $\mathcal{L}_v$. In our notation this property reads

$$\left\langle \frac{1}{\mathcal{V}\mathcal{L}_v} \sum_j \int_0^{L_j} \mathbf{s}_j''(\xi_j)\mathbf{s}_j''(\xi_j)\, d\xi_j \right\rangle = c_2^2\, \mathcal{L}_v, \tag{31}$$

or, in a local form

$$\left\langle \mathbf{s}_j''(\xi_j)\mathbf{s}_j''(\xi_j) \right\rangle = c_2^2\, \mathcal{L}_v. \tag{32}$$

The uniform superfluid turbulence which we are interested in, is realized in wide channels. More rigorously it implies that the interline space $\delta = \mathcal{L}_v^{-1/2}$ should be much smaller than the channel size. In this case we can disregard lines ending on surfaces and consider all lines as closed loops. The condition of line closeness can be written as follows:

$$\int_0^{L_j} \mathbf{s}_j' d\xi = 0. \tag{33}$$

In addition we suppose that the length of each of the loops is greater than the mean curvature radius, as has been observed by Schwarz[20]

$$L_j \gg \left\langle |\mathbf{s}_j''(\xi_j)| \right\rangle^{-1}. \tag{34}$$

Apart from the direct evidence following from the Schwarz work we can bring forward a following argument in favor of condition (34). The vortex filament length is changed firstly due to the deterministic process of ballooning or shrinking of loops, and, secondly due to reconnection processes. In the process of reconnection, the loops can decay into two smaller loops in case of a self-collision or can fuse into a larger loop in case of a collision with other loops. It is clear that the loops collide more frequently with other loops than are subjected to the self-collision. Therefore, it seems plausible that long loops prevail due to reconnection processes.

The properties expressed by relations (21) – (34) are almost all we know about the vortex tangle arrangement in the superfluid turbulent Helium II. Following Schwarz we shall call the quantities $I_\perp$, $I_\|$, (and I_{xx}, I_{yy}, I_{zz}), I_l, c_2, the structure parameters of vortex tangle. They depend on temperature T and pressure p and do not depend on the applied counterflow velocity $\mathbf{V}_{ns}$. Numerical values of structure parameters as functions of temperature are given in the original work of Schwarz[20] and in papers.[2, 4]

5. GAUSSIAN MODEL OF VORTEX TANGLE

In the previous sections we have exposed the current state of art in the description of chaotic vortex filaments or the superfluid turbulence. It has been shown how using the experimental results interpreted in terms of phenomenological theory as well as results from numerical simulations one can draw some conclusions about the vortex tangle structure. It is clear that the bigger part of respective relations (21)–(34) is expressed via structure parameters obtained by Schwarz in numerical simulations. Although the Schwarz numerical modelling expands considerably the limits of phenomenological theory it, at the same time, has restricted possibilities for the study of effects connected with the fine vortex tangle structure. Indeed, as is often in case of numerical simulations, structure parameters calculated by Schwarz can hardly be used to evaluate other quantities than those he has calculated. It should be understood that the above mentioned statement does not concern the quantities which are directly expressed via the structure parameters. Before we formulate the purpose of this section we would like to demonstrate that there exist many other physical quantities related with other physical phenomena which cannot be expressed in terms of Schwarz theory.

5.1 Examples of Other Quantities of Interest

As a first example we take the average vorticity of superfluid velocity $\mathbf{\Omega}(\mathbf{r})$ and its Fourier transform $\mathbf{\Omega}_{\mathbf{k}}$. These quantities may have nonzero values and should be taken into account, when more complicated flows than a one-dimensional counterflow is considered. They have to be introduced as averages over vortex lines configurations:

$$\mathbf{\Omega}(\mathbf{r}) = \left\langle \tilde{\kappa} \sum_j \oint d\mathbf{s}_j \delta(\mathbf{r} - \mathbf{s}_j(\xi_j)) \right\rangle, \tag{35}$$

and

$$\begin{aligned} \mathbf{\Omega}_{\mathbf{k}} &= \left\langle \frac{\tilde{\kappa}}{(2\pi)^{3/2}} \int d^3\mathbf{r}\, e^{-i\mathbf{k}\mathbf{r}} \sum_j \oint d\mathbf{s}_j \delta(\mathbf{r} - \mathbf{s}_j(\xi_j)) \right\rangle \\ &= \left\langle \frac{\tilde{\kappa}}{(2\pi)^{3/2}} \sum_j \oint d\mathbf{s}_j e^{-i\mathbf{k}\mathbf{s}_j(\xi_j)} \right\rangle, \end{aligned} \tag{36}$$

where $\tilde{\kappa}$ is the quantum of circulation. The closest example of a flow of superfluid component with a nonzero averaged vorticity created by a set of vortex lines will be the rotating He II. In this case the superfluid component simulates (in average) the solid-body rotation and the averaged vorticity is just the real density of vortex lines multiplied by $\tilde{\kappa}$ as follows from relation (35).

Another quantity which influences the hydrodynamic properties of superfluid turbulence is the mean superfluid velocity created by the vortex tangle $\mathbf{V}_s^V(\mathbf{r})$. It is obtained by averaging the Biot–Savart law

$$\mathbf{V}_s^V(\mathbf{r}) = \left\langle \frac{\tilde{\kappa}}{4\pi} \sum_j \int_0^{L_j} \frac{\mathbf{s}_j'(\xi_j) \times (\mathbf{r} - \mathbf{s}_j(\xi_j))}{|\mathbf{r} - \mathbf{s}_j(\xi_j)|^3} d\xi_j \right\rangle. \tag{37}$$

Accordingly the full momentum $\mathbf{P}_V$ of additional superfluid motion connected with the vortex tangle presence will be:

$$\mathbf{P}_V = \rho_s \int \mathbf{V}_s^V(\mathbf{r})\, d^3\mathbf{r}\, . \tag{38}$$

The direct use of relations (37) and (38) faces a problem typical of vortex flows. The integral in the relation (38) diverges both for small and large $|\mathbf{r} - \mathbf{s}_j(\xi_j)|$; see Refs.[11,12] Therefore, the question of the averaged velocity or the full momentum generated by vortices cannot be resolved in a straightforward way. On the other hand it is known that in many respects the so-called Lamb impulse plays the role of momentum. In general, the Lamb impulse density is defined by

$$\mathbf{J}_V = \frac{\rho_s}{2\mathcal{V}} \int \mathbf{r} \times \omega(\mathbf{r})\, \mathbf{d}^3\mathbf{r}, \tag{39}$$

where $\omega(\mathbf{r})$ is the distribution (microscopic, not averaged $\mathbf{\Omega}(\mathbf{r})$) of the vorticity. For the singular vorticity distribution, viz. for chaotic vortex filaments the Eq. (39) can be rewritten as

$$\mathbf{J}_V = \left\langle \frac{\rho_s \widetilde{\kappa}}{2\mathcal{V}} \sum_j \int_0^{L_j} \mathbf{s}_j(\xi_j) \times \mathbf{s}_j'(\xi_j)\, d\xi_j \right\rangle . \tag{40}$$

Because of the interaction between vortices the vortex tangle should display some kind of elasticity. As a result it should be expected that the long-range interaction between different vortex line elements will lead to new macroscopic modes, e.g., waves of vortex line density, a kind of 3D analog to Tkachenko waves.[2] The measure of the elasticity is determined by the interaction energy. The vortex tangle energy is defined by

$$E = \left\langle \frac{1}{2} \int \rho_s \mathbf{V}_s^2\, d^3\mathbf{r} \right\rangle = \left\langle \frac{\rho_s \kappa^2}{8\pi} \sum_{j,i} \int_0^{L_i} \int_0^{L_j} \frac{\mathbf{s}_i'(\xi_i)\mathbf{s}_j'(\xi_j)}{|\mathbf{s}_i(\xi_i) - \mathbf{s}_j(\xi_j)|} d\xi_i d\xi_j \right\rangle . \tag{41}$$

By use of the well-known formula

$$\frac{1}{|\mathbf{r}|} = \int_{\mathbf{k}} \frac{4\pi d^3\mathbf{k}}{\mathbf{k}^2} e^{i\mathbf{k}\mathbf{r}}, \tag{42}$$

the average energy E (41) can be rewritten as

$$E = \left\langle \frac{\rho_s \kappa^2}{2} \sum_{i,j} \int_{\mathbf{k}} \frac{d^3\mathbf{k}}{\mathbf{k}^2} \int_0^{L_i} \int_0^{L_j} \mathbf{s}_j'(\xi_i)\mathbf{s}_j'(\xi_j) d\xi_i d\xi_j\, e^{i\mathbf{k}(\mathbf{s}_i(\xi_i) - \mathbf{s}_j(\xi_j))} \right\rangle . \tag{43}$$

The next example, we would like to mention, is the proper vortex tangle entropy

$$S^V = k_B \left\langle \ln \Gamma(\{\mathbf{s}_j(\xi_j)\}) \right\rangle , \tag{44}$$

where $\Gamma(\{\mathbf{s}_j(\xi_j)\})$ is the number of vortex loop configurations (see the next Section). The quantity S^V enters the equations of the hydrodynamics of superfluid turbulence[23, 4] and its knowledge is necessary for the correct study of unsteady hydrodynamic processes.

Let us consider the average

$$\mathcal{A} = \left\langle \sum_j \int_0^{L_j} \left(\dot{\mathbf{s}}_j\,(\xi_j) \times \mathbf{s}_j'(\xi_j) \right) d\xi_i \right\rangle , \tag{45}$$

where $\dot{\mathbf{s}}_j\,(\xi_j)$ is the line element velocity. The quantity $\dot{\mathbf{s}}_j\,(\xi_j)$ is expressed (in general as a functional) via the instantaneous configuration of vortex tangle $\{\mathbf{s}_j(\xi_j)\}$ with the help of the equation of motion (10). The right-hand side of Eq. (45) is an averaged net area swept out by the motion of lines. The quantity $\mathcal{A}$ bears a manifold physical interest. For example, the equation of vortex line motion can be derived from the variational principle, and the contribution of lines into the action is proportional to the area swept out by the moving lines[34]. Furthermore, the z-component of $\mathcal{A}$ is just the phase slippage rate caused by the motion of vortex lines transverse to the counterflow $\mathbf{V}_{ns}$ (see Ref.[35]). Finally the integrand in Eq.(45) is the discrete variant of the quantity $\mathbf{V}_s \times (\nabla \times \mathbf{V}_s)$, which is called a vortex force and which plays a significant role in the vortex dynamics.[12] Of course, all discussed above properties of $\mathcal{A}$ are not independent and we have listed them just to stress the importance of $\mathcal{A}$ for the applications.

Using the distribution function it is even possible to describe the dynamics of various quantities $A(\{\mathbf{s}_j(\xi_j)\})$ averaged over loop configurations $\langle A(\{\mathbf{s}_j(\xi_j)\})\rangle$. Indeed, reverting the time dependence for line element positions $\mathbf{s}_j(\xi_j) \to \mathbf{s}_j(\xi_j, t)$ and using a chain rule, we can write down the rate of change of the quantity $\langle A(\{\mathbf{s}_j(\xi_j,t)\})\rangle$ in the form

$$\frac{\partial \langle A(\{\mathbf{s}_j(\xi_j,t)\})\rangle}{\partial t} = \left\langle \sum_i \int_0^{L_j} \frac{\delta A(\{\mathbf{s}_j(\xi,t)\})}{\delta \mathbf{s}_i(\xi_i',t)} \frac{\partial \mathbf{s}_i(\xi_i',t)}{\partial t} d\xi_i' \right\rangle . \tag{46}$$

Expressing the velocity of line element $\dot{\mathbf{s}}_j(\xi_j)$ with the help of the equation of motion[20] we obtain that the right-hand side of (46) is an average of some functional of the vortex line configuration. Thus we have got a rule how to calculate the evolution of the quantity $\langle A(\{\mathbf{s}_j(\xi_j,t)\})\rangle$.

A word of precaution should be said. The future distribution function will correspond to the instantaneous picture of vortex tangle, i.e., to the "equilibrium" state of vortex tangle. Therefore, when we say dynamics, we have in mind only small deviations from the "equilibrium". Nevertheless the possibility to introduce a macroscopic vortex tangle dynamics in a regular way seems to be very important. For example, in the process of the derivation of govern equation (12) for the vortex line density evolution Vinen has made a very important assumption that the rate of change of vortex line density $d\mathcal{L}_v/dt$ is a function only of the instantaneous value of $\mathcal{L}_v$. This property has been called the self-preserving assumption. As has been discussed[4] this assumption is justified, if and only if the other characteristics of vortex tangle different from the quantity $\mathcal{L}_v(t)$ relax to the "equilibrium" state much faster than $\mathcal{L}_v(t)$ itself. The self-preserving assumption can be confirmed or refuted inspecting relations similar to Eq. (46).

We have given several examples of the averages responsible for new interesting phenomena in the turbulent counterflowing He II. Apart from the independent interest the study of these phenomena can supply a new important information concerning the vortex tangle fine structure. It is easy to see that none of introduced quantities can be expressed via structure parameters of the phenomenological theory of the superfluid turbulence. To evaluate them one should have some theory yielding the rules how to fulfil the averaging.

Of course, the most honest way to develop such a theory is to resolve the problem about stochastic vortex filaments dynamics on the basis of equations of motion. However, as it has been already discussed, this way seems to be almost hopeless because of the extremely involved vortex line dynamics.

We would like to offer another, far more moderate approach.[36] The master idea of our proposal is to construct a trial distribution function in space of vortex loop configurations in the most general form, which satisfies all properties of vortex tangle established earlier in the phenomenological theory and numerical simulations and expressed by relations (21)–(34). We assume that this trial distribution function will enable us to calculate any physical quantities owing to the vortex tangle presence. Now we will describe the way of constructing this function.

5.2 Trial Distribution Function

We make the usual in statistical physics supposition that all configurations corresponding to the same macroscopic state have equal probabilities. Thus the probability $\mathcal{P}(\{\mathbf{s}_j(\xi_j)\})$ for a vortex tangle to have a particular configuration $\{\mathbf{s}_j(\xi_j)\}$ should be proportional to $1/N_{allowed}$, where $N_{allowed}$ is the number of allowed configurations, which of course, are infinite

$$\mathcal{P}(\{\mathbf{s}_j(\xi_j)\}) \propto \frac{1}{N_{allowed}} \,. \tag{47}$$

Under the term "allowed configurations" $N_{allowed}$ we mean only the configurations that will lead to the correct values for all average quantities known from the experiment and given by relations (21)–(34).

The number of allowed configurations, or the curve number is expressed by a path integral in the space of 3D curves supplemented by some constraints which follow from conditions (21)–(34), or, to be precise, which will lead to these conditions

$$N_{allowed} \propto \int \mathcal{D}\{\mathbf{s}_j(\xi_j)\} \times constraints\ \{\mathbf{s}_j(\xi_j)\}. \tag{48}$$

In this subsection we will introduce the constraints dictated by conditions (21)–(34), and modify the expression (48) into a standard and tractable form.

Let us begin with the first condition (21) concerning the total vortex line length. It implies that among the possible curves, labelled by ξ_j we have to choose only curves whose lengths are fixed and equal to L_j. Taking the local form (22) of this condition we impose the corresponding constraint into the integrand of equation (48) as a delta function

$$N^{(j)}_{allowed} \propto \int \mathcal{D}\mathbf{s}_j(\xi_j) \times \delta(|\mathbf{s}_j{}'(\xi_j)| - 1) \,. \tag{49}$$

Because the absolute value of $\mathbf{s}_j{}'(\xi_j)$ is a nonanalytical function this expression will lead to a not tractable theory.

We will use here a trick known from the theory of polymer chains, where a similar problem appears. Let us divide the j vortex line into a set of discrete points and change the path integral by a product of usual ones

$$\int \mathcal{D}\mathbf{s}_j(\xi_j) = \int J\mathcal{D}\mathbf{s}'_j(\xi_j) \to \prod_n \int J\, d(\mathbf{s}_{j(n+1)} - \mathbf{s}_{jn}). \tag{50}$$

Here J is the Jacobian corresponding to the change from the variable $\mathbf{s}_j(\xi_j)$ to the derivative $\mathbf{s}'_j(\xi_j)$. The explicit shape of the Jacobian J will be not significant since it is usually cancelled against the one in the normalizing factor for the probability $\mathcal{P}(\mathbf{s}_j(\xi_j))$ in equation (47). Then the integration for each of the links $\mathbf{s}_{j(n+1)} - \mathbf{s}_{jn} = \mathbf{l}_n$ in the relation (50) should be accomplished with delta function constraints $\delta(|\mathbf{l}_n| - 1)$ corresponding to a fixed link length. In the theory of polymer chains they offer to relax the rigorous condition $|\mathbf{l}_n| = l_n$ and to change it by a smeared-out (Gaussian) distribution of the link length with the same value of integral[37]

$$\int d^3\mathbf{l}_n\, \delta(|\mathbf{l}_n| - l_n) \Rightarrow \int d^3\mathbf{l}_n \left(\frac{4}{\sqrt{\pi} l_n}\right) e^{-\mathbf{l}_n^2/l_n^2}. \tag{51}$$

The parameter l_n here, the so-called effective bond length, is usually established in the experiment. Gathering contributions from all links and coming back to the continuous case

$$\prod_n e^{-\mathbf{l}_n^2/a^2} = \exp(-\sum_n (\mathbf{s}_{jn+1} - \mathbf{s}_{jn})^2/a^2) \Rightarrow \exp(-\lambda_1 \int_0^{L_j} |\mathbf{s}_j{}'|^2 d\xi_j), \tag{52}$$

one obtains that the number of configurations $N^{(j)}_{allowed}$ of j vortex loop supplemented with the constraint $\delta(|\mathbf{s}'_j(\xi_j)| - 1)$ can be written as

$$N^{(j)}_{allowed} \propto \int J\mathcal{D}\mathbf{s}_j(\xi_j) \times \exp(-\,\lambda_1 \int_0^{L_j} |\mathbf{s}_j{}'|^2 d\xi_j) \,. \tag{53}$$

Here the Jacobian J differs from the one used in Eq. (50), however this has no effect because of the remark made after (50). The quantity λ_1 is the parameter of our theory which is to be determined later.

Before we move further it is worth to discuss once more the meaning of the fulfilled procedure. Concerning the polymer theory, the introduction of Gaussian chains instead of real polymers does not influence significantly most of the important physical applications.[37] One can say that a loss of some rigorousness is a good price for the obvious simplicity. This conclusion concerns our model to a greater extent than the polymer theory. Indeed, the choice of the arc length ξ_j as a label for the vortex filament is a question of convenience. The same treatment can be applied to another label, say, the arc length $\tilde{\xi}_j$ at some moment of time. During the line motion some parts of the curve shrink whereas other parts stretch. Therefore, the real value of $|\mathbf{s}_j{}'(\xi_j)|$ must not be equal to a unity exactly but should be smeared out, which we have used.

Returning to the relation (53) we resume that the use of constraint (22) leads to the familiar form for the number of allowed configurations of the j-loop as a path integral over all configurations with a weight proportional to $\exp(-\int \mathcal{L}_j\,)$, where the effective Lagrangian is determined by

$$\int \mathcal{L}_j \;=\; \lambda_1 \int_0^{L_j} |\mathbf{s}_j{}'|^2 d\xi_j. \tag{54}$$

Using the formula (47) we conclude that the probability $\mathcal{P}_j(\mathbf{s}_j(\xi_j))$ for the j-loop to have a configuration $\mathbf{s}_j(\xi_j)$ is the Wiener distribution[37]

$$\mathcal{P}_j(\mathbf{s}_j(\xi_j)) = \mathcal{N}_j \exp\left(-\lambda_1 \int_0^{L_j} |\mathbf{s}_j{}'|^2 d\xi_j\right), \tag{55}$$

where $\mathcal{N}_j$ is the corresponding normalizing factor. In a similar way we could take into account other loops so the full Lagrangian corresponding to the constraint $|\mathbf{s}_j'(\xi_j)| = 1$ will be given by

$$\int \mathcal{L} \;=\; \lambda_1 \sum_j \int_0^{L_j} |\mathbf{s}_j{}'|^2 d\xi_j. \tag{56}$$

The constant λ_1 factors out of the sum otherwise the average will depend on the index j what will contradict to the full uniformity supposition.

The next constraint to be discussed is connected with the line curvature. It is known that although the resulting curve averaged according to Eq. (55) has a correct length L_j, it is not smooth but is wiggly at each point. Indeed, the distance between points $\mathbf{s}_j(\xi_{j1})$ and $\mathbf{s}_j(\xi_{j2})$ of the curve is expressed by the relation[37]

$$\langle|\mathbf{s}_j(\xi_{j1}) - \mathbf{s}_j(\xi_{j2})|\rangle = \sqrt{(\xi_{j1} - \xi_{j2})/\lambda_1}, \tag{57}$$

which is nonanalytical when $\xi_{j1} \longrightarrow \xi_{j2}$. In other words the average curve $\langle \mathbf{s}_j(\xi_j)\rangle$ does not possess even a first derivative. To make it smooth we have to introduce into the effective Lagrangian $\mathcal{L}_j$ a term with a second derivative and, further; to make it have a finite curvature we have to introduce a term with a third derivative, etc., namely,

$$\int \mathcal{L}_j \;=\; \lambda_1 \int_0^{L_j} |\mathbf{s}_j{}'|^2 d\xi_j \;+\; \lambda_2 \int_0^{L_j} |\mathbf{s}_j{}''|^2 d\xi_j \;+\; \lambda_3 \int_0^{L_j} |\mathbf{s}_j{}'''|^2 d\xi_j \;+\; \ldots \tag{58}$$

The effective Lagrangian (58) is isotropic, therefore, it will give an isotropic line distribution which is wrong. To improve the situation we have to impose the coefficients λ to be matrices (so far diagonal). For instance, the relation (54) should be changed to the expression

$$\lambda_{1x}\int_0^{L_j} \mathbf{s}'_{jx}\mathbf{s}'_{jx}d\xi_j + \lambda_{1y}\int_0^{L_j} \mathbf{s}'_{jy}\mathbf{s}'_{jy}d\xi_j + \lambda_{1z}\int_0^{L_j} \mathbf{s}'_{jz}\mathbf{s}'_{jz}d\xi_j = \int_0^{L_j} \mathbf{s}'_{j\alpha}\, \Lambda^{\alpha\beta}\mathbf{s}'_{j\beta}\, d\xi_j\,. \tag{59}$$

The next step is to take into account the vortex tangle polarization. To have a nonzero value for the averaged polarity we have to add nondiagonal terms into the matrix $\Lambda^{\alpha\beta}$. The respective correction to the Lagrangian will be

$$\int \delta\mathcal{L}_j \;=\; \lambda_p \int_0^{L_j} (\mathbf{s}'_j \times \mathbf{s}''_j)_z d\xi_j. \tag{60}$$

The index z implies that we have to take only the z coordinate in the vector product $\mathbf{s}'_j \times \mathbf{s}''_j$. Finally, since the effective Lagrangian includes derivatives of different orders ($\mathbf{s}'_j, \mathbf{s}''_j$ and $\mathbf{s}'''_j$) it is convenient to perform the one-dimensional Fourier transform along ξ_j:

$$\mathbf{s}_j(\xi_j) \;=\; \sum_\kappa \mathbf{s}_j(\kappa)e^{i\kappa\xi_j}, \qquad \kappa = 2\pi n/L_j\;. \tag{61}$$

Because of the closure condition (33)

$$\frac{1}{L_j}\int_0^{L_j} \mathbf{s}_j{}'d\xi_j \;=\; \mathbf{s}'_j(\kappa)\Big|_{\kappa=0} \;=\; 0, \tag{62}$$

the zero harmonic of derivative $\mathbf{s}'_j(\xi_j)$ is zero, therefore, we will further exclude the harmonic with $\kappa = 0$ in the summation in 1D Fourier space. Correspondingly the evaluation of the path integral in the κ-representation should be accomplished according to the following rule:

$$\int \mathcal{D}\{\mathbf{s}_j(\kappa)\} \;= \prod_j \prod_{\kappa\neq 0} \int d\mathbf{s}_j(\kappa). \tag{63}$$

Further we will use both sides of this relation interchangably.

Summarizing everything concerning the effective Lagrangian and using the relation (47) it can be inferred that the probability $\mathcal{P}(\{\mathbf{s}_j(\kappa)\})$ for the vortex tangle to have a particular configuration $\{\mathbf{s}_j(\kappa)\}$ will be

$$\mathcal{P}(\{\mathbf{s}_j(\kappa)\}) \;= \mathcal{N}\exp\left(-\sum_{\kappa\neq 0}\mathcal{L}(\{\mathbf{s}_j(\kappa)\})\right), \tag{64}$$

where $\mathcal{N}$ is an overall normalization. The density of the Lagrangian $\mathcal{L}\{\mathbf{s}_j(\kappa)\}$ in κ-space is

$$\mathcal{L}\{\mathbf{s}_j(\kappa)\} \;= \sum_j \begin{pmatrix} \mathbf{s}_{jx}(\kappa) \\ \mathbf{s}_{jy}(\kappa) \\ \mathbf{s}_{jz}(\kappa) \end{pmatrix} \begin{pmatrix} \Lambda^{xx}(\kappa) & \Lambda^{xy}(\kappa) & \Lambda^{xz}(\kappa) \\ \Lambda^{yx}(\kappa) & \Lambda^{yy}(\kappa) & \Lambda^{yz}(\kappa) \\ \Lambda^{zx}(\kappa) & \Lambda^{zy}(\kappa) & \Lambda^{zz}(\kappa) \end{pmatrix} \begin{pmatrix} \mathbf{s}_{jx}(-\kappa) \\ \mathbf{s}_{jy}(-\kappa) \\ \mathbf{s}_{jz}(-\kappa) \end{pmatrix}. \tag{65}$$

The diagonal terms of matrix $\Lambda^{\alpha\beta}$ have the following structure:

$$\Lambda^{\alpha\alpha} \;=\; \lambda_{1\alpha}\kappa^2 \;+\; \lambda_{2\alpha}\kappa^4 \;+\; \lambda_{3\alpha}\kappa^6 \;+\; ...., \qquad \alpha = x, y, z. \tag{66}$$

From the definition of the nondiagonal part of Lagrangian (60) and from the assumed symmetry in the plain x, y it follows that

$$\Lambda^{xy} = (i\kappa)^3 \lambda_p, \qquad \Lambda^{yx} = -(i\kappa)^3 \lambda_p. \tag{67}$$

Expressions (64)-(67) determine the probability of allowed vortex loop configuration of most general form satisfying all known vortex tangle properties. Thus they can be considered as a trial distribution function which we are looking for. The effective Lagrangian has a bilinear form over the "field" variables $\mathbf{s}_j(\kappa)$, therefore, all path integrals, we have considered, are Gaussian, hence the developed approach is referred to as the Gaussian model of a vortex tangle. Of course, to use this function one has to specify all parameters λ and cancel the uncertainty in the expansion (66). The most convenient way to do this is to calculate the characteristic functional and studying its properties to specify the explicit form of parameters which enter in the trial distribution function. The following subsection is devoted to this procedure.

5.3 Characteristic Functional

Let us consider the following averaged quantity, the so-called characteristic functional

$$W(\{\mathbf{P}_j(\xi_j)\}) = \left\langle \exp\left(i \sum_j \int_0^{L_j} \mathbf{P}_j(\xi_j)\mathbf{s}'_j(\xi_j)\, d\,\xi_j \right) \right\rangle . \tag{68}$$

The characteristic functional is of special interest. The point is that one is able to calculate any averages depending on the vortex line configuration by a simple functional differentiation. For instance, the average tangent vector $\left\langle \mathbf{s}'_{j\alpha}(\xi_j)\right\rangle$, the average curvature vector $\left\langle \mathbf{s}''_{j\alpha}(\xi_j)\right\rangle$, or the correlation function between the orientation of the different elements of vortex filaments $\left\langle \mathbf{s}'_{j\alpha}(\xi_j)\mathbf{s}'_{j\beta}(\xi_j)\right\rangle$ are readily expressed by the characteristic functional according to the rules

$$\left\langle \mathbf{s}'_{j\alpha}(\xi_j)\right\rangle = \left.\frac{\delta W}{i\delta \mathbf{P}^{\alpha}_j(\xi_j)}\right|_{all\ \mathbf{P}=0}, \tag{69}$$

$$\left\langle \mathbf{s}''_{j\alpha}(\xi_j)\right\rangle = \left.\frac{\partial}{\partial \xi_j}\left(\frac{\delta W}{i\delta \mathbf{P}_j(\xi_j)}\right)\right|_{all\ \mathbf{P}=0}, \tag{70}$$

$$\left\langle \mathbf{s}'_{j\alpha}(\xi_j)\mathbf{s}'_{j\beta}(\xi_j)\right\rangle = \left.\frac{\delta^2 W}{i\delta \mathbf{P}^{\alpha}_j(\xi_{j1})\ i\delta \mathbf{P}^{\beta}_j(\xi_{j2})}\right|_{all\ \mathbf{P}=0}. \tag{71}$$

The other quantities are expressed by the characteristic functional in a bit more sophisticated way. For instance, the Fourier transform of averaged vorticity $\mathbf{\Omega_k}$ (36) can be evaluated with the help of the characteristic functional by use of the following procedure:

$$\mathbf{\Omega_k} = \sum_j \int_0^{L_j} d\xi_j e^{-i\mathbf{k}\mathbf{s}_j(0)} \left.\frac{\delta W}{i\delta \mathbf{P}_j(\xi_j)}\right|_{\{\mathbf{P}_j(\xi'_j)\} = -\mathbf{k}\theta(\xi'_j)\theta(\xi_j-\xi'_j)}, \tag{72}$$

where $\theta(\xi'_j)$ is the unit step-wise function. The product $\theta(\xi'_j)\theta(\xi_j - \xi'_j)$ selects out points lying in the range $0 \le \xi'_j \le \xi_j$ on the j-curve. This choice assures the appearance of the correct quantity $\exp(i\mathbf{k}\mathbf{s}_j(\xi_j))$ after the integration in the exponent in the relation (68) is done. The quantity $\mathbf{s}_j(0)$ is the initial point of the j-curve which is chosen arbitrary.

In a similar way the mean energy $\langle E\rangle$ (43) can be calculated as

$$\langle E\rangle = \frac{\rho_s\kappa^2}{2}\sum_{i,j}\int_{\mathbf{k}}\frac{d^3\mathbf{k}}{\mathbf{k}^2}\int_0^{L_i}\int_0^{L_j} d\xi_i d\xi_j\, e^{-i\mathbf{k}(\mathbf{s}_i(0)\,-\,\mathbf{s}_j(0))}\,\frac{\delta^2 W}{i\delta\mathbf{P}_i^\alpha(\xi_i)\, i\delta\mathbf{P}_j^\alpha(\xi_j)}. \tag{73}$$

Here the set $\mathbf{P}_n(\xi_n')$ in the characteristic functional $W(\{\mathbf{P}_n(\xi_n')\})$ is again determined with the help of θ-functions:

$$\begin{aligned}
\mathbf{P}_i(\xi_i') &= -\mathbf{k}\theta(\xi_i')\theta(\xi_i-\xi_i'),\\
\mathbf{P}_j(\xi_j') &= -\mathbf{k}\theta(\xi_j')\theta(\xi_j-\xi_j'),\\
\mathbf{P}_n(\xi_n) &= 0, \qquad n\neq i,j.
\end{aligned} \tag{74}$$

The Eq. (74) implies that we choose in the integrand of characteristic functional (68) only points lying in the interval from 0 to ξ_i on an i-curve and from 0 to ξ_j on a j-curve. During the evaluation of the self-energy of the same loop, $i=j$, one has to distinguish between points ξ_i and to put them to be, e.g., ξ_i' and ξ_i''.

Although the characteristic functional is defined via the distribution function as some auxiliary quantity, it plays a significant independent role in the stochastic theory. For instance, in the many body problem the use of a characteristic functional (in this theory it is usually called a generating functional) allows to get a shorten description of statistical properties in terms of Green functions and equations for them. Another example is the case of classical turbulence, where they derive the master equation for the characteristic functional directly from the equation of fluid motion escaping the use of distribution function, which, in addition, is unknown (see, e.g., book[38]). Similarly in our work the characteristic functional is used not only for the calculation of different averages but it plays a key role in the derivation of trial distribution function.

Our first step is to calculate the characteristic functional defined by the relation (68). It is convenient to do this in κ - space. In the κ-space the characteristic functional can be obtained by accomplishing the 1D Fourier transform (61) in Eq. (68):

$$W(\{\mathbf{P}_j(\kappa)\}) = \left\langle \exp\left(i\sum_j\sum_{\kappa\neq 0} L_j\mathbf{P}_j(\kappa)\mathbf{s}_j'(-\kappa)\right)\right\rangle. \tag{75}$$

The various averages are readily obtained using this definition. For instance, the averaged values of tangent vector (69) and curvature vector (70) can be evaluated with the help of the rules

$$\langle \mathbf{s}_{j\alpha}'(\xi_j)\rangle = \sum_{\kappa\neq 0} e^{-i\kappa\xi_j}\left.\frac{\delta W}{iL_j\delta\mathbf{P}_j^\alpha(\kappa)}\right|_{All\quad \mathbf{P}(\kappa)=0}, \tag{76}$$

$$\langle \mathbf{s}_{j\alpha}''(\xi_j)\rangle = \sum_{\kappa\neq 0}(-i\kappa)\, e^{-i\kappa\xi_j}\left.\frac{\delta W}{iL_j\delta\mathbf{P}_j^\alpha(\kappa)}\right|_{All\quad \mathbf{P}(\kappa)=0}. \tag{77}$$

Likewise the two point correlation function (relation (71)) is expressed via the characteristic functional in κ-space as follows:

$$\langle \mathbf{s}_{j\alpha}'(\xi_j)\mathbf{s}_{j\beta}'(\xi_j)\rangle = \sum_{\kappa_1,\,\kappa_2\neq 0} e^{-i\kappa_1\xi_{j1}}e^{-i\kappa_2\xi_{j2}}\left.\frac{\delta^2 W}{iL_j\delta\mathbf{P}_j^\alpha(\kappa_1)\, iL_j\delta\mathbf{P}_j^\beta(\kappa_2)}\right|_{All\quad \mathbf{P}(\kappa)=0}. \tag{78}$$

To calculate the characteristic functional in κ-space (Eq. (75)) we shall employ the trial distribution function introduced in the previous subsection. Using relations

(64)–(67) one can rewrite the expression (75) in the form

$$W(\{\mathbf{P}_j(\kappa)\}) = \mathcal{N}\int J\mathcal{D}\{\mathbf{s}_j(\kappa)\}\exp\left(-\sum_j\sum_{\kappa\neq 0}\mathbf{s}_{j\alpha}(\kappa)\Lambda^{\alpha\beta}(\kappa)\mathbf{s}_{j\beta}(-\kappa)\right)$$
$$\times\exp\left(i\sum_j\sum_{\kappa\neq 0}\left(L_j\frac{1}{2}(\mathbf{P}_j(\kappa)\mathbf{s}_j'(-\kappa) + \mathbf{P}_j(-\kappa)\mathbf{s}_j'(\kappa))\right)\right), \quad (79)$$

where

$$\mathcal{N} = \left\{\int J\mathcal{D}\mathbf{s}_j(\kappa)\exp\left(-\sum_j\sum_{\kappa\neq 0}\mathbf{s}_j^{\alpha}(\kappa)\Lambda^{\alpha\beta}(\kappa)\mathbf{s}_j^{\beta}(-\kappa)\right)\right\}^{-1}$$

is the overall normalization. The r. h. s. of Eq. (79) is evaluated by the standard "full square procedure" expressed by the identity

$$\int\prod_k dz_k\exp\left(-\sum_{nm}z_nA_{nm}z_m^* + \sum_n(z_nu_n^*z_n^*u_n)\right)$$
$$= \exp\left(-\sum_{nm}u_nA_{nm}^{-1}u_n^*\right) * \int\prod_k dz_k\exp\left(-\sum_{nm}z_nA_{nm}z_m^*\right). \quad (80)$$

The integral in (80) is taken over a set of complex variables z_k , $\int dz_k = \int d\mathrm{Re}z_k d\mathrm{Im}z_k$. The matrix A_{nm} is supposed to be Hermitian, and the matrix A_{nm}^{-1} is inverse to the matrix A_{nm}.

Using this rule for each of Fourier harmonics in (79) we get

$$W(\{\mathbf{P}_j(\kappa)\}) = \exp\left(-\sum_j\sum_{\kappa\neq 0}L_j^2\mathbf{P}_j^{\alpha}(\kappa)N^{\alpha\beta}(\kappa)\mathbf{P}_j^{\beta}(-\kappa)\right), \quad (81)$$

where the matrix $N^{\alpha\beta}(\kappa)$ is equal to $[1/4\kappa^2\ (\Lambda^{\alpha\beta}(\kappa))^{-1}]$. Elements of both matrices, $N^{\alpha\beta}(\kappa)$ and $\Lambda^{\alpha\beta}(\kappa)$, do not depend on the index j otherwise the local averages will depend on j which contradicts the full uniformity supposition.

The second step in realizing the scheme outlined at the end of previous subsection is to study general properties of characteristic functional (81) beginning with the method it has been derived. Before doing it let us connect the matrix $N^{\alpha\beta}(\kappa)$ (so far not determined explicitly) with the vortex tangle characteristics expressed by formulas (21)–(34). The functional derivatives entering in relations (76)–(78) as applied to the characteristic functional (81) can be evaluated according to the rules

$$\frac{\delta W}{iL_j\delta\mathbf{P}_j^{\alpha}(\kappa)}$$
$$= \frac{L}{i}2\ N^{\alpha\nu}(\kappa_1)\ \mathbf{P}_j^{\nu}(-\kappa_1)\exp\left(-\sum_j\sum_{\kappa\neq 0}L_j^2\mathbf{P}_j^{\mu}(\kappa)N^{\mu\nu}(\kappa)\mathbf{P}_j^{\nu}(-\kappa)\right). \quad (82)$$

Here it has been taken into account that $N^{\mu\nu}(\kappa)$ is a Hermitian matrix, $N^{\alpha\nu}(\kappa) = N^{\nu\alpha}(-\kappa)$.

Likewise the second derivative will be

$$\delta^2W/\left[iL_j\delta\mathbf{P}_j^{\alpha}(\kappa_1)\ iL_j\delta\mathbf{P}_j^{\beta}(\kappa_2)\right]$$
$$= \left\{2N^{\alpha\beta}(\kappa_1)\ \delta_{-\kappa_1,\kappa_2} - 4N^{\alpha\nu}(\kappa_1)\mathbf{P}_j^{\nu}(-\kappa_1)N^{\beta\gamma}(\kappa_2)\mathbf{P}_j^{\gamma}(-\kappa_2)\right\}$$
$$\times\exp\left(-\sum_j\sum_{\kappa\neq 0}L_j^2\mathbf{P}_j^{\mu}(\kappa)N^{\mu\nu}(\kappa)\mathbf{P}_j^{\nu}(-\kappa)\right). \quad (83)$$

Using (82) and (83) in Eq. (76)–(78) we can conclude that

$$\langle \mathbf{s}'_{j\alpha}(\xi_{j1})\mathbf{s}'_{j\beta}(\xi_{j2})\rangle = \sum_{\kappa_1\neq 0} e^{-i\kappa_1(\xi_{j1}-\xi_{j2})}2\, N^{\alpha\beta}(\kappa_1)\,, \tag{84}$$

hence the averaged squared tangent vector $\langle \mathbf{s}'_{j\alpha}(\xi_j)\mathbf{s}'_{j\alpha}(\xi_j)\rangle$ is just

$$\langle \mathbf{s}'_{j\alpha}(\xi_j)\mathbf{s}'_{j\alpha}(\xi_j)\rangle = \sum_{\kappa\neq 0} 2\, N^{\alpha\alpha}(\kappa). \tag{85}$$

Accordingly the averaged squared curvature vector $\langle \mathbf{s}''_{j\alpha}(\xi_j)\mathbf{s}''_{j\alpha}(\xi_j)\rangle$ will be

$$\langle \mathbf{s}''_{j\alpha}(\xi_j)\mathbf{s}''_{j\alpha}(\xi_j)\rangle = \sum_{\kappa\neq 0} 2\,\kappa^2\, N^{\alpha\alpha}(\kappa). \tag{86}$$

Note that formulas (85)–(86) are valid both for each of components α and for sum over α.

As far as the average vortex tangle polarization is concerned it is expressed via the matrix $N^{\alpha\beta}(\kappa)$ as follows:

$$\langle (\mathbf{s}'_{jx}\mathbf{s}''_{jy} - \mathbf{s}'_{jy}\mathbf{s}''_{jx})\rangle = \sum_{\kappa_1,\,\kappa_2\neq 0} e^{-i(\kappa_1+\kappa_2)\xi_j}$$
$$\times\left[(\,i\kappa_2)\;N^{xy}(\kappa_1)\,\delta_{-\kappa_1,\,\kappa_2} - (i\kappa_2)\;N^{yx}(\kappa_2)\delta_{-\kappa_1,\,\kappa_2}\right]. \tag{87}$$

By their construction the nondiagonal elements of Hermitian matrices $\Lambda^{xy}(\kappa)$ and $\Lambda^{yx}(\kappa)$ are odd functions of the argument κ. It is obvious that the inverse matrix $N^{\alpha\beta}(\kappa)$ satisfies same conditions. Therefore

$$N^{yx}(\kappa) = N^{xy}(-\kappa) = -N^{xy}(\kappa). \tag{88}$$

Using this chain of relations we finally arrive at

$$\langle (\mathbf{s}'_{jx}\mathbf{s}''_{jy} - \mathbf{s}'_{jy}\mathbf{s}''_{jx})\rangle = \sum_{\kappa\neq 0} 2\,(i\kappa)N^{xy}(\kappa). \tag{89}$$

Inspecting the method of the trial distribution function construction as well as the way of characteristic functional derivation we can deduce several quite general properties of matrix $N^{\alpha\beta}(\kappa)$. Let us resume them:

(i) The matrix $N^{\alpha\beta}(\kappa) = N^{\beta\alpha}(-\kappa)$ is Hermitian.

(ii) The diagonal terms $N^{\alpha\alpha}(\kappa)$ should be even functions of κ.

(iii) The nondiagonal terms $N^{xy}(\kappa)$ and $N^{xy}(\kappa)$ should be odd functions of κ.

(iv) To guarantee an existence of any $\langle \mathbf{s}_j^{(n)}(\xi_j)\mathbf{s}_j^{(n)}(\xi_j)\rangle$ for any n one has to require the elements of matrix $N^{\alpha\beta}(\kappa)$ to decrease faster than any power function κ^{n+2}.

Besides of the listed properties, the matrix $N^{\alpha\beta}(\kappa)$ should give correct values for mean tangent vector $\langle \mathbf{s}'_{j\alpha}(\xi_j)\mathbf{s}'_{j\alpha}(\xi_j)\rangle$, mean squared curvature $\langle \mathbf{s}''_{j\alpha}(\xi_j)\mathbf{s}''_{j\alpha}(\xi_j)\rangle$ and polarization $\langle (\mathbf{s}'_{jx}\mathbf{s}''_{jy} - \mathbf{s}'_{jy}\mathbf{s}''_{jx})\rangle$. Furthermore, since only few characteristics of vortex tangle are known, the matrix $N^{\alpha\beta}(\kappa)$ should not include too many parameters. And, finally, it should be simple enough and tractable otherwise the whole method would be meaningless.

As a suitable candidate satisfying all listed properties we propose the matrix $N^{\alpha\beta}(\kappa)$ *of the following form:*

$$\begin{pmatrix} N^{xx}\exp(-\,\kappa^2\xi_0^2) & (i\,\kappa)\,N^{xy}\,\exp(-\,\kappa^2\xi_0^2) & 0 \\ -(i\,\kappa)\,N^{yx}\,\exp(-\,\kappa^2\xi_0^2) & N^{yy}\exp(-\,\kappa^2\xi_0^2) & 0 \\ 0 & 0 & N^{zz}\exp(-\,\kappa^2\xi_0^2) \end{pmatrix}. \tag{90}$$

It will be shown later that the quantity ξ_0 is nothing but the correlation length. It will be also shown that ξ_0 is of order of the mean curvature, or of order of the interline space. Thus, besides the listed above conditions (i)-(iv) we have made here one more strong supposition that all correlation functions $\langle \mathbf{s}_{j\alpha}^{(n)}(\xi_{j1})\mathbf{s}_{j\beta}^{(m)}(\xi_{j2})\rangle$ have the same correlation length ξ_0 of order of the interline space. A semi-quantitative proof of that fact based on the consideration of kinematic relations of the kind $\mathbf{s}'\mathbf{s}'' = 0$, $\mathbf{s}'\mathbf{s}''' + \mathbf{s}''\mathbf{s}'' = 0$ etc. has been given by Schwarz.[13]

The final step in constructing of the characteristic functional is to specify the coefficients $N^{\alpha\beta}$ as well as the quantity ξ_0. These five quantities are to be obtained comparing relations (85)–(89), where the matrix $N^{\alpha\beta}(\kappa)$ is taken from (90), with relations (21)–(34).

Let us start with the calculation of the coefficient N^{xx}. It can be found from the comparison of Eq. (26) for the mean x-fraction of tangent vector obtained in the experiment with Eq. (85) for the same quantity expressed via the characteristic functional (81) with the matrix (90):

$$I_{xx} = 2\, N^{xx} \left\{ \sum_{all\ n} \exp\left[-\, n^2 \left(\frac{2\pi\xi_0}{L_j}\right)^2 \right] - 1 \right\}. \tag{91}$$

Employing the condition $\xi_0 \ll L_j$ and using

$$\sum_{all\ n} \to \int dn$$

we arrive at

$$N^{xx} = I_{xx} \frac{\xi_0\sqrt{\pi}}{L_j} \left(1 - \frac{2\xi_0\sqrt{\pi}}{L_j}\right)^{-1}. \tag{92}$$

Of course, this result is valid for each of the components with corresponding I_{yy} and I_{zz}. Note that we have retained the small term $2\xi_0\sqrt{\pi}/L_j$ in the denominator of the right-hand side of Eq. (92). Its origin is from the vortex line closeness and it plays a significant role in questions, where the vortex line closeness is relevant (see later).

To specify the quantity ξ_0 one has to use the relations for the averaged squared curvature vector. Comparing (31) and (86) and using the expression for $N^{\alpha\alpha}$ obtained above one concludes that

$$c_2^2\, \mathcal{L}_v = \frac{\xi_0\sqrt{\pi}}{L_j} \sum_{all\ n} 2\, n^2 \left(\frac{2\pi}{L_j}\right)^2 \exp\left[-\, n^2 \left(\frac{2\pi\,\xi_0}{L_j}\right)^2 \right]. \tag{93}$$

Passing to the integration

$$\sum_{all\ n} \to \int dn$$

we obtain

$$\xi_0^2 = \frac{1}{2\, c_2^2\, \mathcal{L}_v}. \tag{94}$$

Since c_2^2 is of order of unity[20], the quantity ξ_0 is of order of the interline space $\mathcal{L}_v^{-1/2}$. Analogous calculations for the vortex tangle polarization (29) allow to determine the coefficient N^{xy} in nondiagonal terms of the matrix $N^{\alpha\beta}(\kappa)$:

$$N^{xy} = 2\sqrt{\pi}\, I_l \frac{\xi_0^3\, \mathcal{L}_v^{1/2}}{L_j} = \sqrt{\frac{\pi}{2}} \frac{I_l}{L_j c_2^3 \mathcal{L}_v}. \tag{95}$$

The evaluation of pre-exponent factors $N^{\alpha\beta}$ as given by Eqs. (92) and (95) and the quantity ξ_0 in the matrix $N^{\alpha\beta}(\kappa)$, see Eq. (90), has completed the calculation of characteristic functional and, consequently, of trial distribution function. They are, of course, expressed by combinations composed of vortex tangle structure parameters.

Thus we have reached the intended purpose and have obtained the characteristic functional in an explicit form. Therefore, now we are ready to calculate any averaged over the vortex filament configuration.

6. APPLICATIONS OF GAUSSIAN MODEL

6.1 Arrangement of Average Vortex Loop

Correlation Functions. Among these averages we are able to calculate the correlations functions between distinct points of the line, or between tangent vectors, or the curvature and so on. Without entering in details of calculations[36] we shall give the results. For instance, the two-point correlation function between tangent vectors $\left\langle s'_{j\alpha}(\xi_{j1})s'_{j\alpha}(\xi_{j2})\right\rangle$ will be

$$\left\langle s'_{j\alpha}(\xi_{j1})s'_{j\alpha}(\xi_{j2})\right\rangle = I_{\alpha\alpha}\left(1-(2\xi_0\sqrt{\pi})/L\right)^{-1}$$
$$\times(\exp\left[-\tfrac{(\xi_{j1}-\xi_{j2})^2}{4\xi_0^2}\right]+\exp\left[-\tfrac{(L_j-(\xi_{j1}-\xi_{j2}))^2}{4\xi_0^2}\right]-(2\xi_0\sqrt{\pi})/L_j). \tag{96}$$

Inspecting the relation (96) one concludes that the close points $(\xi_{j1}-\xi_{j2})\le\xi_0$ and points satisfying the condition $[L_j-(\xi_{j1}-\xi_{j2})]\le\xi_0$ are strongly correlated (the latter condition appears because of the loop closeness). Then this correlation weakens as $\exp[-(\xi_{j1}-\xi_{j2})^2/4\xi_0^2]$ turning into a delta-correlated structure as

$$\exp(-(\xi_{j1}-\xi_{j2})^2/4\xi_0^2)\sim 2\sqrt{\pi}\xi_0\delta(\xi_{j1}-\xi_{j2}). \tag{97}$$

Thus we arrive at a very important conclusion. The correlation length of orientations of curve different parts is of order of mean curvature radius, or, in accordance with (94), of order of interline space. This view corresponds to current notions of vortex tangle and it has been discussed previously by Schwarz.[13, 20]

It is worth noting that there is a small negative correlation between distant points due to the term $(-2\xi_0\sqrt{\pi}/L_j)$ in the r.h.s. of Eq. (96). The origin of this term is connected with the line closeness that is why each of the line elements "remembers" that the whole line should return to the initial point. Discarding this effect, the correlations between remote (along the curve, $\xi_0\ll\xi_{j1}-\xi_{j2}$) points vanish and the line takes a random walk structure. The correlation between different curvature vectors $\langle s''_{j\alpha}(\xi_{j1})s''_{j\alpha}(\xi_{j2})\rangle$ behaves in a similar way. The only exception is that the small negative correlation disappears because of the differentiation over ξ_j.

In a similar way the correlation between derivatives of different orders can be evaluated. Let us consider, e.g., the correlation between the tangent vector and the curvature vector $\langle s'^{\alpha}_j(\xi_{j1})s''^{\alpha}_j(\xi_{j2})\rangle$. It has to be evaluated as

$$\langle s'_{j\alpha}(\xi_{j1})s''_{j\alpha}(\xi_{j2})\rangle = \frac{\partial}{\partial\xi_{j2}}\langle s'_{j\alpha}(\xi_{j1})s'_{j\alpha}(\xi_{j2})\rangle\ .$$

The interesting feature of this quantity is that it is zero when $\xi_{j1}=\xi_{j2}$, then it grows reaching its maximum value at the point $(\xi_{j1}-\xi_{j2})\sim\xi_0/2$, and after that the growth is changed with the usual exponential decay.

Average Size of The Chaotic Loop. Let us now calculate the quantity $\langle (\mathbf{s}^{\alpha}(\xi_j) - \mathbf{s}^{\alpha}(0))^2 \rangle$ (here a summation over α is assumed) which is the averaged squared distance between the initial point of the curve $\mathbf{s}(0)$ and points $\mathbf{s}(\xi_j)$. Note that we deal with the real distance in the usual space (not along the curve!), therefore, this consideration concerns the real vortex loop size embedded in 3D space. Note also that if one does not accomplish the summation over α, this quantity will describe the loop size along the α-axes. Using the ξ-presentation of characteristic functional the quantity $\langle (\mathbf{s}^{\alpha}(\xi_j) - \mathbf{s}^{\alpha}(0))^2 \rangle$ can be rewritten in the form

$$\begin{aligned} \langle (\mathbf{s}_{\alpha}(\xi_j) - \mathbf{s}_{\alpha}(0))^2 \rangle &= \int_0^{\xi_j}\int_0^{\xi_j} d\xi_{j1}\, d\xi_{j2} \langle \mathbf{s}'_{j\alpha}(\xi_{j1}) \mathbf{s}'_{j\alpha}(\xi_{j2}) \rangle \\ = \int_0^{\xi_j}\int_0^{\xi_j} & d\xi_{j1}\, d\xi_{j2} I_{\alpha\alpha} \left(1 - 2\xi_0 \sqrt{\pi}/L\right)^{-1} (\exp\left[-\frac{(\xi_{j1} - \xi_{j2})^2}{4\xi_0^2}\right] \\ &+ \exp\left[-\frac{(L_j - (\xi_{j1} - \xi_{j2}))^2}{4\xi_0^2}\right] - 2\xi_0 \sqrt{\pi}/L_j). \end{aligned} \tag{98}$$

For $\xi_j \leq \xi_0$ the exponent is close to unity and with an accuracy $2\xi_0\sqrt{\pi}/L_j$ we conclude that the averaged squared distance in the α-direction will be

$$\langle (\mathbf{s}_{\alpha}(\xi_j) - \mathbf{s}_{\alpha}(0))^2 \rangle = \xi_j^2 \, I_{\alpha\alpha}, \tag{99}$$

or the full distance will be

$$\langle (\mathbf{s}(\xi_j) - \mathbf{s}(0))^2 \rangle = \xi_j^2. \tag{100}$$

In the intermediate region of argument ξ, $\xi_0 \leq \xi_j \leq (L_j - \xi_0)$, the exponent can be approximately replaced by a δ-function (see Eq. (97)), which together with (98) gives the following result (with an accuracy $2\xi_0\sqrt{\pi}/L_j$):

$$\langle (\mathbf{s}_{\alpha}(\xi_j) - \mathbf{s}_{\alpha}(0))^2 \rangle \sim 2\xi_0 \, I_{\alpha\alpha} \sqrt{\pi} \, (\xi_j \; - \; \xi_j^2 \; / \; L) \, . \tag{101}$$

Note that the quantity $(-2\xi_0\sqrt{\pi}/L_j)$ has been disregarded only in the denominator of Eq. (98) while it has been retained in the numerator, where its contribution is comparable with the one from exponential terms. The reason is that for ξ_j larger than ξ_0 (but smaller than L_j) the vortex line has a random walk structure and the distance between the initial point $\mathbf{s}^{\alpha}(0)$ and the point $\mathbf{s}^{\alpha}(\xi_j)$ increases as $\sqrt{\xi_j}$.

The role of the term $(-2\xi_0\sqrt{\pi}/L_j)$ is to force the line go back in order to assure that $(\mathbf{s}_{\alpha}(\xi_j) - \mathbf{s}_{\alpha}(0)) \to 0$, when $\xi_j \to L_j$. Eq. (101) should be, however, corrected in the region near the line end $(L_j - \xi_0) \leq \xi_j$. In this region the main contribution will appear from the second exponent in the right-hand side of Eq. (98). This contribution will prevail the quantity $(\xi_j \; - \; \xi_j^2 \; / \; L)$ and the final result will be

$$\langle (\mathbf{s}_{\alpha}(\xi_j) - \mathbf{s}_{\alpha}(0))^2 \rangle = (L_j - \xi_j)^2,$$

which is obvious due to the periodicity.

Typical Shape of Averaged Curve. Summarizing the results obtained in this section we can conclude that the vortex loop behaves as a flexible polymer.[37] The small parts of the line behave as rod-like polymers whose lengths are exactly equal to the distance $(\xi' \; - \; \xi'')$ along the curve (see Fig. 8). At larger distance the filament has a random walk structure with an effective bond length of order of the correlation length

ξ_0 or, which is the same, of order of the mean curvature radius, or of order of the interline space. Because of the closeness condition the pure random walk structure $|\mathbf{s}_\alpha(\xi_j) - \mathbf{s}_\alpha(0)| \propto \sqrt{\xi_j}$ is violated and changes according to Eq. (101). Besides, due to the anisotropy the whole average loop has a "pancake" form in the z-direction. In addition, since the vortex filaments unlike polymer chains are oriented, there is an anisotropy related with the mutual orientation of vectors $\mathbf{s}'$ and $\mathbf{s}''$. Thus the vortex loop as a whole has a nonzero polarization $\langle \int_0^{L_j} \mathbf{s}_j'(\xi_j) \times \mathbf{s}_j''(\xi_j)\ d\xi_j \rangle$, and as a result it should have some drift velocity as well as it should induce a nonzero mean superfluid velocity.

6.2 Hydrodynamic Impulse of Vortex Tangle

As a more extended illustration of developed theory we shall study the hydrodynamic impulse (or the Lamb impulse) of vortex tangle $\mathbf{J_V}$ which is defined as

$$\mathbf{J}_V = \langle \frac{\rho_s \tilde{\kappa}}{2} \sum_j \int \mathbf{s}_j(\xi_j) \times \mathbf{s}_j'(\xi_j)\ d\xi_j \rangle. \tag{102}$$

In many respects the Lamb impulse plays the role of a momentum [11]. The average $\langle \mathbf{s}_j(\xi_j) \times \mathbf{s}_j'(\xi_j) \rangle$ entering in (102) is immediately evaluated by use of the characteristic functional (see for details Ref.[39]) to give the following result:

$$\mathbf{J}_V^z = - \left[\frac{\rho \tilde{\kappa} I_l \alpha_v}{\rho_n c_2^2 \beta_v} \right] \rho_s \mathbf{V}_s. \tag{103}$$

Note that the coefficient includes no fitting parameters but only characteristics known from the phenomenological theory and from numerical simulations. Eq. (103) shows that the vortex tangle induces a superfluid current directed against the external superfluid current. This should be expected since there is some preferable polarization of vortex loops.

From a macroscopic point of view, in particular, in hydrodynamic experiments, an additional superfluid mass current which partly cancels the external superfluid mass current should display itself as a superfluid density suppression. It can be said that this effect is a 3D analog of the Kosterlitz–Thouless effect. The superfluid density suppression $\Delta\rho_s$ being defined as a system response to the applied infinitesimal superfluid velocity $\delta\mathbf{V}_s$

$$\Delta\rho_s = \frac{\delta \mathbf{J}_V}{\delta(\delta \mathbf{V}_s)}, \tag{104}$$

has a tensor nature. Applying (104) to Eq. (103) one concludes that the longitudinal relative suppression $\Delta\rho_s/\rho_s$ will be

$$\frac{\Delta\rho_s}{\rho_s} = - \left[\frac{\rho I_l^2}{2\rho_n c_2^4} \right]. \tag{105}$$

Thus we have expressed the superfluid density suppression $\Delta\rho_s/\rho_s$ via vortex tangle structure parameters. Using the known values for structure parameters[20] one can evaluate the suppression $\Delta\rho_s/\rho_s$ as a function of temperature. The result depicted in Fig. 9 shows that the superfluid component suppression $\Delta\rho_s/\rho_s$ as a function of temperature is of order of several percents. We think it is a pretty large effect deserving an experimental study.

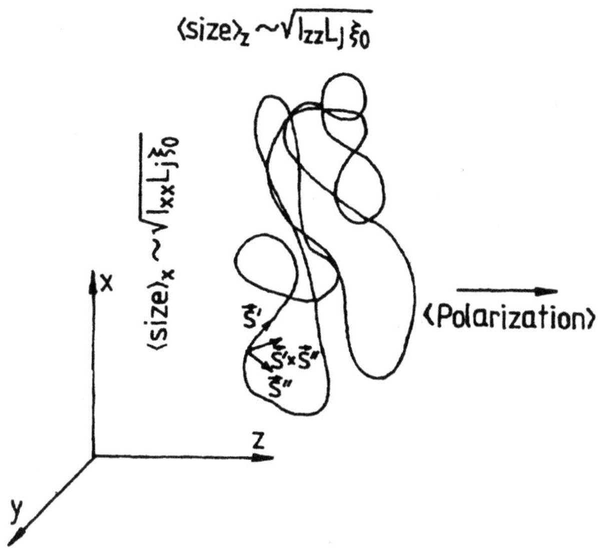

Figure 8. A snapshot of the "average" vortex loop obtained from the analysis of statistical properties. Close ($\Delta\xi \ll R$) parts of line are separated in $3D$ space by a distance $\Delta\xi$. The distant parts ($R \ll \Delta\xi$) are separated in $3D$ space by the distance $\sqrt{2\pi R\Delta\xi}$ (with a correction due to closeness). Thus, for large scales ($R \ll \xi$) the loop has a random walk structure with an effective bond equal to the mean curvature radius. As a whole the loop is not isotropic having a "pancake" shape with different sizes in longitudinal and transverse directions. In addition, the loop has a total average polarization $\langle \int \mathbf{s}_j'(\xi_j) \times \mathbf{s}_j''(\xi_j) d\xi_j \rangle$ forcing the loop to drift along the vector $\mathbf{V}_n$.

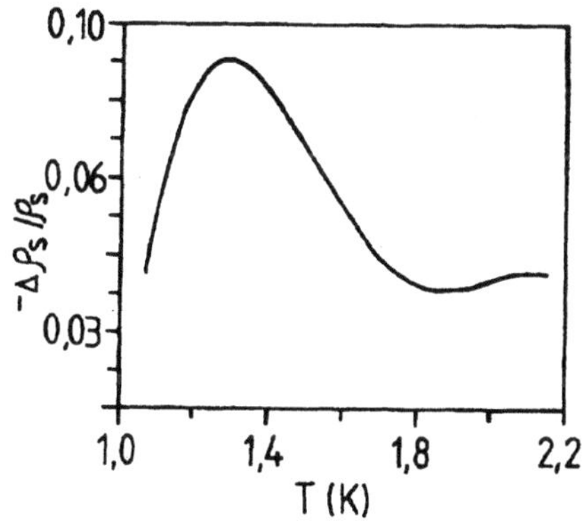

Figure 9. The predicted behavior of superfluid density suppression $\Delta\rho_s/\rho_s$ as a function of temperature.

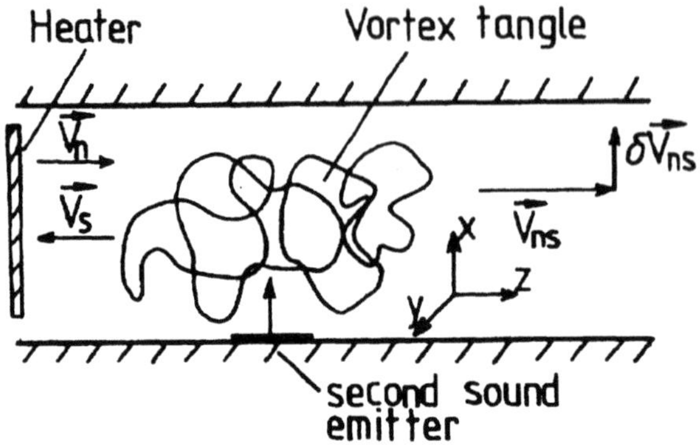

Figure 10. An illustration to the transverse experiment. A small perpendicular deviation $\delta\mathbf{V}_{ns}$ of the counterflow velocity $\mathbf{V}_{ns}$ changes the orientation of vortex tangle polarization whereas the vortex line density and the mean curvature do not change to the first order in $\delta\mathbf{V}_{ns}$. The transverse change of the superfluid density leads to decrease of second sound velocity propagating in the perpendicular (with respect to the main counterflow) direction.

Since the superfluid density enters in the expression for the second sound velocity u_2 it seems attractive to detect the effect obtained using the transverse second sound testing (Fig. 10). To do this we have firstly to evaluate the transverse change of ρ_s and, secondly, to develop a theory to match it to the nonstationary case. The general theory asserts that when applying a harmonic external second sound, the superfluid density suppression will become a function of frequency ω of the following form:

$$\Delta\rho_s^x(\omega) = \left(\frac{\delta \mathbf{J}_V^x}{\delta \mathbf{V}_s^x}\right)_{transv} \frac{1}{1+i\omega\tau_J}. \tag{106}$$

Here the transverse $(\delta\mathbf{J}_V^x/\delta\mathbf{V}_s^x)_{transv}$ is half of that given[39] by Eq. (103). The quantity τ_J is the relaxation time of superfluid current J_V which is to be found from dynamical consideration. Expressing $d\mathbf{J}_V^x/dt$ with the help of equation for vortex line element motion and then evaluating various averages by our trial distribution function we can find the change of second sound velocity. Performing all described procedures (see paper[39] for details) one obtains that the relative change $\Delta u_2/u_2$ will be

$$\frac{\Delta u_2}{u_2} = -f(T)\frac{\mathbf{V}_{ns}^4}{\omega^2}. \tag{107}$$

Here the function $f(T)$ is composed of vortex tangle structure parameters

$$f(T) = \frac{4\rho\tilde{\kappa} I_l^2 \alpha^2 (1 - I_{xx})^2}{\rho_n c_2^4 \beta^3}. \tag{108}$$

The strong $\mathbf{V}_{ns}$-dependent decrease of second sound velocity in the counterflowing He II was really observed about two decades ago by Vidal with coauthors.[40] To our knowledge there has been only an attempt to consider this effect theoretically made by Mehl[41], who explained the change of second sound velocity by introducing an imaginary part into the Hall-Vinen constant. However, besides some numerical disagreements the Mehl theory does explain the strong $\mathbf{V}_{ns}$-dependence in experimental data.

Let us compare our result (107) with the Vidal experiment. Using the data on structure parameters one obtains that, e.g., for temperatures $1.44K$ the value of the function $f(T)$ is about 620 s^2/cm^4. Taking frequency[40] $\omega = 4.3\times10^3$ rad/s, and $V_{ns} = 2$ cm/s one obtains that $\Delta u_2/u_2 \approx 4\times 10^{-4}$, which is very close to the observed value.

6.3 Energy of the Chaotic Vortex Loop

In this subsection we shall calculate the averaged energy of stochastic vortex loop (of length L) distributed according to the trial distribution function (64). For a single loop the general expression (41) can be written as

$$E = \left\langle \frac{1}{2} \int \rho_s \mathbf{V}_s^2 \, d^3\mathbf{r} \right\rangle = \left\langle \frac{\rho_s \kappa^2}{8\pi} \int_0^L \int_0^L \frac{\mathbf{s}'(\xi')\mathbf{s}'(\xi'')}{|\mathbf{s}(\xi') - \mathbf{s}(\xi'')|} d\xi' d\xi'' \right\rangle. \tag{109}$$

Accordingly in 3D Fourier space the average energy E given by Eq. (109) will be

$$E = \left\langle \frac{\rho_s \kappa^2}{2} \int_{\mathbf{k}} \frac{d^3\mathbf{k}}{(2\pi)^3 \mathbf{k}^2} \int_0^L \int_0^L \mathbf{s}'(\xi')\mathbf{s}'(\xi'') d\xi' d\xi'' \, e^{i\mathbf{k}(\mathbf{s}(\xi') - \mathbf{s}(\xi''))} \right\rangle. \tag{110}$$

In terms of characteristic functional the energy (110) is given by

$$E = \frac{\rho_s \kappa^2}{2} \int_{\mathbf{k}} \frac{d^3\mathbf{k}}{(2\pi)^3 \mathbf{k}^2} \int_0^L \int_0^L d\xi' d\xi'' \, \frac{\delta^2 W}{i\delta \mathbf{P}^\alpha(\xi') \, i\delta \mathbf{P}^\alpha(\xi'')}. \tag{111}$$

After the functional differentiation we have to take the argument of characteristic functional $\mathbf{P}(\xi)$ to be the following function:

$$\mathbf{P}(\xi) \;=\; \mathbf{k}\theta(\xi - \xi')\theta(\xi'' - \xi). \tag{112}$$

Eq. (112) implies that we take in the exponent integrand of characteristic functional only points lying in the interval from ξ' to ξ''. Calculations of the averaged energy E on the basis of Eq. (111) is straightforward and will be published elsewhere. Omitting these tremendous calculations we shall write down the final result

$$E = \frac{\rho\kappa^2 L}{4\pi} \ln \frac{R}{a_0} \frac{\rho\kappa^2 L}{4\pi} \left(1 - \frac{2}{\sqrt{\pi}}(f_2 - f_1)\right) \ln \frac{R}{a_0} \tag{113}$$

$$+ \quad \frac{\rho\kappa^2 L}{4\pi} \left[\frac{1}{\left(\sqrt{\pi} - 1\right)^{1/2}} \frac{2 f_3}{\pi^{5/2} c_{2\,3}^2} I_l^2 + \frac{f_2}{\pi^{3/2}\left(\sqrt{\pi} - 1\right)^{1/2}} \right], \tag{114}$$

where the quantities f are expressed via vortex tangle structure parameters as

$$f_1(\gamma) = \sqrt{2\left(3 - \gamma^2\right)} \frac{\arcsin(\gamma)}{\gamma}, \tag{115}$$

$$f_2(\gamma) = \sqrt{\frac{1}{2\left(3 - \gamma^2\right)}} \left(-\sqrt{1 - \gamma^2}\left(2 - \gamma^2\right) \frac{\arcsin(\gamma)}{\gamma} \right), \tag{116}$$

$$f_3(\gamma) = \left(2\left(3 - \gamma^2\right)\right)^{3/2} \left(\frac{\sqrt{1 - \gamma^2}}{\gamma^2} - \frac{\arcsin(\gamma)}{\gamma^3} \right). \tag{117}$$

Here we have introduced the parameter

$$\gamma = \sqrt{\frac{I_x - I_z}{I_x}}.$$

Let us comment the expression (114). The first term in the r.h.s. of (114) is just the energy per unit length of a straight vortex filament[2] multiplied by its length. In this form it is frequently used in the superfluid turbulence theory.[4] But there are additional terms. The second one is also logarithmically large. We consider that this contribution appears from an accidental self-crossing of remote (along the line) parts of random walking vortex filament. The third and forth terms appear from the long-range interaction, they are smaller than logarithmic ones (about ten percents). The third term is of special interest. It appears due to the vortex loop polarization and its presence implies that some elasticity of vortex tangle in $\mathbf{V}_{ns}$ direction is present. Results of the previous section assert that the vortex tangle presence leads to the appearance of additional superfluid mass current induced by vortices. The longitudinal elasticity combined with the inertia of the "joint" mass should lead to the appearance of elastic waves, the 3D analog of Tkachenko waves.

7. CONCLUSION

The present state of research on chaotic quantized vortice processes in the superfluid Helium II is exposed. In the introduction we have claimed that the study of superfluid turbulence even using macroscopic methods can supply a lot of information on the stochastic vortex line dynamics. We have also claimed that this information can be used in the study of some problems in the superfluidity theory and even in the study of more general problems of the stochastic physics of extended objects and classical turbulence. Following consistently these aims we would like to list some of the respective results.

The self preservation assumption that $d\mathcal{L}/dt$ is a function only of the instantaneous value of $\mathcal{L}$ (Sec. 3) is fairly well confirmed by the agreement with the bigger part of experimental observations. This assumption points out the presence of a hierarchy of relaxation times in the vortex dynamics. Therefore, there is some justification to introduce a reduced stochastic description similar to the kinetic or Fokker–Planck equation. Furthermore, the success of Schwarz theory demonstrates the possibility to use the local self-induced approximation for the calculation of vortex line dynamics. Another important conclusion from the comparison of theory and experiments concerns the scaling properties of vortex tangle. These results are very important for future analytical investigations as they show possibilities to simplify the basic equations of vortex line motion and an adequate formulation of the problem of vortex tangle dynamics.

Furthermore Schwarz and Feynman–Vinen conceptions about the decay of the vortex line density $\mathcal{L}$ are quite different. Indeed, in the Feynman–Vinen model the vortex tangle decay is a consequence of breaking down of vortex rings transformed into thermal excitations. Therefore, there is a flux of some physical quantity, e.g., the local curvature in space of vortex ring sizes. This model reminds the Kolmogorov cascade in the classical turbulence and the stochastic behavior is essentially nonequilibrium. In the Schwarz interpretation only the friction force is responsible for the vortex line length decrease and there is no flux in space of vortex rings. Thus the stochastic process is close to equilibrium. This conclusion may be of the crucial importance for the development of theory.

Let us discuss some questions related to the second part of the lecture, namely, the Gaussian model of the vortex tangle and its applications. The main result can be formulated as follows: *based on the well established experimental data of the vortex tangle structure in He II we have constructed a trial distribution function in space of the vortex loop configurations of most general form compatible with these data.* We assume further that the trial distribution function obtained in this way will enable us to calculate various averages over vortex loops configurations.

The use of the characteristic functional simplifies the calculation of these quantities. Let us discuss once more the assumptions made while developing the whole procedure and outline the class of problems which can be resolved with the trial distribution function method. The main premise of our approach is the relation (47) expressing that the allowed configurations corresponding to the same macroscopic state have equal probabilities. This assumption is widely used in problems of equilibrium states and it seems quite reasonable for our problem. We can refer to the work of Polyakov[42] on classical turbulence who has noted that "One can say that while Gibbs distributions are uniform on surfaces of fixed values of conserved quantities, the turbulent distributions are located on surfaces of constant fluxes of the corresponding quantities."

The next questions, which we would like to discuss concern the constraints imposed by relations (21)–(34). Of course, these few properties give by no means the full description of vortex tangle structure and a question is whether the trial distribution function, satisfying few selected conditions, is adequate enough to evaluate correctly other quantities. One more question is: what will be the possible restrictions on the class of these quantities? We can give the following answer to these questions. In spite of the small number of input conditions (21)–(34) they, as a matter of fact, include almost all requested information concerning the orientation of vortex line elements and their curvatures. In other words these input conditions involve almost all information concerning first and second derivatives of functions $\mathbf{s}_j(\xi_j)$, $\mathbf{s}_j'(\xi_j)$ and $\mathbf{s}_j''(\xi_j)$, respectively. That in turn implies that the quantities of interest containing derivatives of not too high order can be evaluated correctly by use of the trial distribution function. But it also implies that expressions containing derivatives of higher orders hardly can be calculated correctly in this way. In any case the reliability of respective calculations will be not too high, although they can be considered as a rough estimation. However, we do not know an example of quantities expressed via the high order derivatives, which bear any physical interest. On the contrary the quantities of physical interest are expressed via derivatives of first and second order. There is a wide class of such quantities and corresponding effects.

The trial distribution function has been derived based on the instantaneous picture of vortex tangle and, as a consequence, the dynamic properties (except those which deal with a small deviation from the "equilibrium" state) drop out of consideration. In particular, we are not able to answer how the vortex tangle structure develops as well as we are not able to answer how it has appeared. We suppose that this structure is a result of very subtle and very involved dynamic nonequilibrium processes which unfortunately cannot be described analytically because of the complexity of the problem. In this connection we would like to note that attempts to describe stochastic properties of classical vortex filaments based on principles of equilibrium Boltzman statistics seem incorrect. In contrast, our model is based mainly on experimental data, therefore it is rather phenomenological one. It does not explain how a certain arrangement of vortex tangle appears, instead it is assigned to calculate various averages over vortex loop configurations. In other words, the developed approach can be considered as some convenient and simple enough "tool" for the evaluation or the estimation of various effects due to the vortex tangle presence in the turbulent superfluid Helium II.

In particular, we have demonstrated how a hydrodynamic impulse of vortex tangle can be calculated with the help of the trial distribution function. Some dynamic properties of the turbulent He II due to impulse and energy of vortex configuration such as the superfluid density suppression or decrease of second sound velocity propagating in the superfluid turbulent He II have been discussed.

We have also calculated the kinetic energy E associated with chaotic vortex filaments. The calculation shows that the energy is schematically composed of several contributions. Besides the local contribution usually used in many applications, the quantity E includes additional terms due to the long-range interaction. Though they are logarithmically small, they are responsible for a number of new effects. For instance, they describe the longitudinal elasticity, which can give some new wave mode – the 3D analog of Tkachenko waves. Besides there is an additional contribution of order of the local one. It appears from an accidental self-crossing of remote (along the line) parts of vortex filament and is connected to the fractal structure of vortex loop.

We hope that we have succeeded to show that the simultaneous use of theoretical, numerical and experimental methods for the study of superfluid turbulence appears very useful for investigations of vortex filament stochastic dynamics.

Acknowledgments

The different parts of the work have been presented at various scientific meetings. I would like to thank participants of these meetings for very fruitful discussions. I am particularly grateful to Prof. G. Ahlers, Prof. M. Tsubota, Prof. W. Glaberson, Prof. C. Barenghi, Prof. D. Samuels, Prof. G. Williams, and Prof. M. Murakami for their stimulating questions and discussions. This work was partly supported under grant N 96–02–19414 from the Russian Foundation of Basic research.

REFERENCES

1. R. P. Feynman, *in*: "Progress in Low Temperature Physics", C. J. Gorter, ed., North Holland, I:17 (1955) .
2. R. J. Donnelly. "Quantized Vortices in Helium II", Cambridge University Press, Cambridge, England (1991).
3. J. T. Tough, *in*: "Progress in Low Temperature Physics", D. F. Brewer, ed., North Holland, XIII:133 (1982).
4. S. K. Nemirovskii, and W. Fiszdon, *Rev. Mod. Phys.* 67:37 (1995).
5. J. A. Sethian, *in*: "Vortex Methods and Vortex Motion", K. E. Gustafson, and J. A. Sethian ed., SIAM, Philadelphia, PA, 1 (1991).
6. E. D. Siggia, *Phys. Fluids* 28:794 (1985).
7. A. K. M. F. Hussain, *J. Fluid Mech.* 173:303 (1986).
8. A. J. Chorin, and J. Marsden. "A Mathematical Introduction to Fluid Mechanics", Springer, Berlin/New-York (1979).
9. M. E. Agistein, and A. A. Migdal, *Mod. Phys. Lett.* A 1:221 (1986).
10. L. Onsager, *Nuovo Cimento* 6:249 (1949).
11. G. K. Batchelor. "An introduction to fluid mechanics", Cambridge University Press (1967).
12. P. G. Saffman. "Vortex Dynamics", Cambridge University Press (1992).
13. K. W. Schwarz, *Phys. Rev.* B18:245 (1978).
14. I. M. Khalatnikov. "An Introduction to the Theory of Superfluidity", W. A. Benjamin, New York/Amsterdam (1965).
15. K. W. Schwarz, *Phys. Rev.* B31:5782 (1985).
16. M. Tsubota and S. Maekawa, *J. Phys. Soc. Jpn.* 61:2007 (1992).
17. K. Nakajima, Y. Sawada, and Y. Onodera, *Phys. Rev.* B17:170 (1978).
18. T. Frisch, Y. Pomeau, and S. Rica, *Phys. Rev. Lett.* 69:1644 (1992).
19. T. F. Buttke, *J. Comput. Phys.* 76:301 (1988).
20. K. W. Schwarz, *Phys. Rev.* B38:2398 (1988).
21. H. Hasimoto, *J. Fluid Mech.* 51:477 (1972).
22. W. F. Vinen, *Proc. R. Soc. London* A243:400 (1958).
23. S. K. Nemirovskii, and V. V. Lebedev, *Sov. Phys. JETP* 57:1009 (1983).
24. J. A. Geurst, *Physica* B154:327 (1989).
25. K. Yamada, S. Kashiwamura, and K. Miyake, *Physica* B154:318 (1989).
26. S. K. Nemirovskii, *Sov. Phys. JETP*, 64:803 (1986).
27. W. M. Fiszdon, V. Schwerdtner, G. Stamm and W. Poppe, *J.Fluid Mech.* 212:663 (1990).
28. S. K. Nemirovskii, L. P. Kondaurova, and A. Ya. Baltsevich, *Cryogenics* 32:47 (1992).

29. R. G. K. M. Aarts and A. T. A. M. Waele, *Phys. Rev.* B50:10069 (1994).
30. A. Hansen and M. Nelkin, *Phys. Rev.* B34:4894 (1986).
31. S. K. Nemirovskii, J. Pakleza, and W. Poppe, *Russian J. of Eng. Thermophysics*, 3:369 (1993).
32. R. T. Wang, C. E. Swanson, and R.J. Donnelly, *Phys. Rev.* B36:5240 (1987).
33. H. E. Hall, *Adv. Phys.* 9:89 (1960).
34. M. Rasetti and T. Regge, *Physica* A80:217 (1975).
35. C. E. Swanson and R. J. Donnelly, *J. Low Temp. Phys.* 61:363 (1985).
36. S. K. Nemirovskii, *Phys. Rev.* B57:5972 (1998).
37. M. Doi and S. F. Edwards. "The theory of polymer dynamics", Clarendon Press, Oxford (1986).
38. A. S. Monin and A. M. Yaglom. "Statistical Fluid Dynamics, Part II", MIT, Cambridge (1975).
39. S. K. Nemirovskii, *Phys. Rev.* B57:5986 (1998).
40. F. C. Vidal, *C.R. Acad Sci.* B 275:609 (1972).
41. J. B. Mehl, *Phys. Rev.* A10:601 (1974).
42. A. M. Polyakov, *Nuclear Physics* B396:367 (1993).

ORIENTATIONAL DISORDER AND ORDER IN C_{60}–FULLERITE AND IN MC_{60}–ALKALI METAL FULLERIDES

A. V. Nikolaev,[1*] K. H. Michel,[1] and J. R. D. Copley[2]

[1]Department of Physics, University of Antwerp,
UIA, 2610 Antwerpen, Belgium
[2]NIST Center for Neutron Research,
National Institute of Standards and Technology,
Gaithersburg, MD 20899, USA

We start with a review of the formalism of symmetry adapted functions (SAF) for the description of orientation dependent properties of solids. For nonlinear molecules SAF's are rotator functions which take into account the symmetry of the molecule and the symmetry of the crystalline site. We apply these concepts to the description of the phase transition from the orientationally disordered phase (crystal structure $Fm\bar{3}m$) to the ordered phase (crystal structure $Pa\bar{3}$) in solid C_{60}. Due to the unusually high symmetry (I_h) of the molecule, the first order phase transition is characterized by the simultaneous condensation of a multitude of order parameter components belonging to the irreducible representation X_5^+ ($\hat{\tau}^{(9)}$) at the X point of the Brillouin zone. Theoretical results are compared with recent neutron and X-ray diffraction experiments.
In the second part we extend the theory to the description of charge transfer and polymer phases in MC_{60} (M=Rb, Cs). The transfer of electronic charge is described by the t_{1u} molecular orbitals of C_{60}^-. The resulting Coulomb interactions lead to additional orientation dependent intermolecular potentials. By studying the orientation dependence of the crystal field and the molecular field, we find that, in comparison with the phase transition $Fm\bar{3}m \rightarrow Pa\bar{3}$ known from C_{60}-fullerite, additional channels to an orthorhombic $Pmnn$ structure are opened. We suggest a phase transition scenario where the orientational ordering is a precursor to polymerization.

1. INTRODUCTION

A crystal that contains molecules or molecular ions is more complex than one containing only atoms or atomic ions because the individual species have orientation as well as position. In a normal crystal both the positions and the orientations of the molecules from which the crystal is made are ordered. In a liquid there is neither translational nor rotational order. Between the crystal and liquid phases it is possible to find partially

*Permanent address: Institute of Physical Chemistry of RAS, 117915, Moscow, Leninskii prospect 31, Russia.

Correlations, Coherence, and Order. Edited by Shopova and Uzunov
Kluwer Academic/Plenum Publishers, 1999

ordered phases, either liquid crystals in which most or all of the translational order is destroyed but orientational order remains, or orientationally disordered crystals in which there is translational order with orientational disorder. Which (if either) of these intermediate phases occurs, depends on the shapes of the molecules. Orientationally disordered phases are to be expected when the molecules are small or nearly spherical. A general account is given in a book by Parsonage and Stavely.[1] Classical examples of non ionic molecular crystals with orientationally disordered phases (also called plastic crystals) are CH_4, CCl_4, SF_6, and adamantane. The most remarkable example is solid C_{60},[2] with almost spherical molecules of icosahedral symmetry,[3] which has an orientationally disordered phase at room temperature. The space group is $Fm\bar{3}m$, and below $\approx$260 K the molecules orientationally order with a $Pa\bar{3}$ structure.[4-6] The discovery of fullerites (solid C_{60} and C_{70}) and fullerides (M_xC_{60}, $x = 1 - 6$, where M is an alkali metal) has stimulated interdisciplinary research in the solid state physics and chemistry of these compounds[7, 8].

In the present paper we will treat molecular orientation dependent properties of C_{60} fullerite and fullerides, with particular emphasis on structural phase transitions. Orientational disorder is described using molecular and site symmetry adapted rotator functions belonging to the manifolds $l = 6, 10, 12$. The fact that the only orientational modes that occur are those with $l = 6, 10, \ldots$ is a consequence of the very high symmetry of the C_{60} molecule. Experimentally this fact manifests itself in the unusually rich structure of the diffuse scattering spectra in the disordered phase.[9-12]

In studying alkali doped fullerides with MC_{60} (M=Rb,Cs) stoichiometry[13-15] we have to take into account the charge transfer of one electron from the alkali atoms to the C_{60} molecules, giving rise to $M^+C_{60}^-$. The charge transfer changes the electronic structure of the molecule, leading to occupation of the lowest unoccupied molecular orbital (LUMO) levels of the C_{60} molecule, which are of t_{1u} symmetry.[16, 17] We will show that this change in electronic structure affects the orientation dependent molecular interactions, in particular the crystal field. We will find that these effects play an essential role in driving the transitions from the orientationally disordered $Fm\bar{3}m$ phase of MC_{60} to the polymer[18-20] and dimer[21, 22] phases.

2. SYMMETRY ADAPTED FUNCTIONS

The appropriate variables with which to describe molecular orientations in orientationally disordered crystals are symmetry adapted functions (SAFs). These variables take into account the symmetry of the molecule and of the site in the crystal.

The orientation of a linear molecule, considered as a rigid body, relative to crystal axes may be described by two polar angles $\Omega \equiv (\Theta, \phi)$. Any function of the orientation of a linear molecule may be described by expanding in terms of spherical harmonics $Y_l^m(\Theta, \phi)$. If the orientation is fairly uniform, a few terms will suffice. Moreover, we can simplify the expansion and reduce the number of terms by using the symmetry of the system. If the molecules are centrosymmetric, as for instance O_2, only terms with even angular momentum number l occur, otherwise odd values of l are also allowed. Further restrictions are imposed by the symmetry of the site. It is convenient to introduce symmetry adapted functions

$$S_l^\tau(\Omega) = \sum_{m=-l}^{l} \alpha_l^{m\,\tau}\, Y_l^m(\Omega), \tag{1}$$

that transform according to the various symmetry species of the crystal site symmetry group. Here the index $\tau = (\Gamma, \rho, i)$ stands for the irreducible representation Γ of the

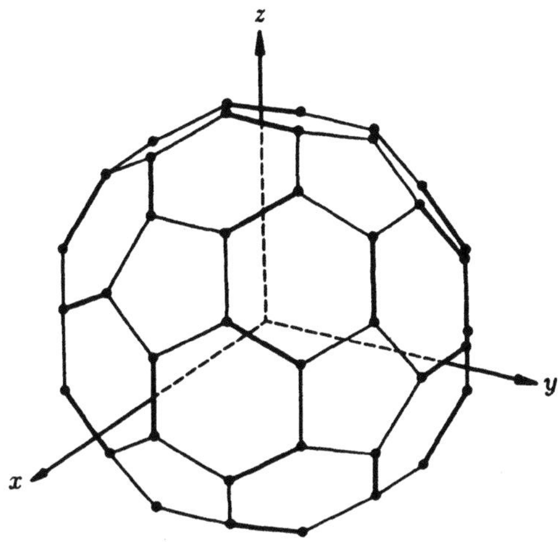

Figure 1. A C_{60} molecule in standard orientation.

site point group, ρ labels representations that occur more than once, and i labels the rows of a given representation. The symmetry adapted functions $S_l^{\tau}(\Omega)$ form a complete set when all l and τ are included. The coefficients $\alpha_l^{m\,\tau}$ form a unitary matrix. They are tabulated by Bradley and Cracknell.[23] Symmetry adapted functions were originally introduced as "kubic harmonics" by Bethe[24] for the description of the crystal field due to valence electrons in solids. The first application to molecular crystals goes back to Devonshire[25] who described the motion of a linear molecule in a field of cubic symmetry.

The situation for nonlinear molecules is somewhat more complex, as three Euler angles $\vec{\omega} \equiv (\alpha, \beta, \gamma)$ are needed to describe the orientation of a nonlinear molecule relative to a fixed frame such as the crystal axes. Rotator functions are then appropriate symmetry adapted linear combinations of the Wigner rotation matrix elements $D_l^{mn}(\vec{\omega})$. These functions are the three-dimensional equivalents of the spherical harmonics, and form a complete set in the space of Euler angles.[23, 26] Rotator functions were introduced for the description of tetrahedral molecules in a cubic crystal (CD_4) by James and Keenan.[27] Their use for the interpretation of experimental data was demonstrated in Ref.[28] The theory of rotator functions was further developed and generalized in Refs.[29, 30] A first application to the disordered phase in C_{60} fullerite was given in Ref.[31] As we shall see, C_{60} is rather particular since the symmetry of the molecule is unusually high (I_h).

We now construct the rotator functions for solid C_{60} in the orientationally disordered phase (space group $Fm\bar{3}m$). We follow closely Ref.[32] We consider a crystal which consists of N rigid molecules with centers rigidly located at fcc (face centered cubic) lattice sites. Molecules may rotate about their centers of mass, but we exclude inter- and intra- molecular vibrations. We first consider the case where the molecule consists of 60 interaction centers located at the sites of the 60 carbon atoms (label a). We start with a molecule in a standard orientation,[5] such that three of its twofold axes coincide with the three cubic axes of the crystal as shown in Fig. 1.

We introduce coefficients

$$c_l^{n1} = \sum_{\nu_a} Y_l^n(\Omega'(\nu_a)), \tag{2}$$

where the sum runs over the 60 atoms and the spherical harmonics Y_l^n are defined according to Ref.[23] The superscript 1 of c_l^{n1} refers to the unit symmetry (I_h) of the molecule. Molecular symmetry implies that the only non-zero coefficients c_l^{n1} are those with l =0, 6, 10, 12, ... and that all coefficients with odd n are identically zero. We now consider molecular symmetry adapted functions

$$S_{l(I)}^1(\Omega) = \sum_{n=-l}^{l} \alpha_l^{n1} Y_l^n(\Omega), \tag{3}$$

which have orthonormalized coefficients

$$\alpha_l^{n1} = c_l^{n1}/g_l^1, \tag{4}$$

where

$$g_l^1 = \sqrt{\sum_n (c_l^{n1})^2}. \tag{5}$$

We regard the latter quantity as a "molecular shape factor" (it is sometimes called "molecular form factor"). A rotation from the standard orientation $\{\Omega'(\nu_a)\}$ to an arbitrary molecular orientation $\{\Omega(\nu_a)\}$ may be described by Euler angles. The spherical harmonics transform according to

$$Y_l^n(\Omega) = R(\vec{\omega}) Y_l^n(\Omega') = \sum_{n'=-l}^{l} Y_l^{n'}(\Omega') D_l^{n'n}(\vec{\omega}), \tag{6}$$

where $D_l^{n'n}(\vec{\omega})$ are the Wigner rotation matrices. We imagine that an orientational fluctuation of type (1) in the crystal is the result of a rotation of the C_{60} molecule away from its standard orientation. We insert Eq. (6) into (1). Since the molecule is taken as rigid, only those fluctuations can occur which conserve full molecular symmetry. Projecting the result of Eq. (6) onto Eq. (3) we obtain the rotator function

$$U_l^{\tau 1}(\vec{\omega}) = \int d\Omega' \, S_l^{\tau}(\Omega) \, S_{l(I)}^{1\,*}(\Omega'). \tag{7}$$

Carrying out the integration over Ω' and observing that $\alpha_l^{n1\,*} = \alpha_l^{-n1}$, we find

$$U_l^{\tau 1}(\vec{\omega}) = \sum_n \sum_m \alpha_l^{n1} D_l^{nm}(\vec{\omega}) \alpha_l^{m\tau}. \tag{8}$$

Here the coefficients α_l^{n1} and $\alpha_l^{m\tau}$ account for molecular and site symmetry, respectively. It is not difficult to generalize the rotator function (8) in two respects. Firstly the description of the molecule by its atomic centers is insufficient to account for the subtleties of interactions between different molecules. It was soon realized, on the basis of structural data in the ordered phase,[5] that double bonds which fuse two hexagons of a C_{60} molecule act as repulsive centers between molecules.[33] This fact is plausible in view of the electronic π-bond character of the double bonds. There is also some evidence that centers on single bonds should be taken into account.[34] It can be shown that the rotator functions for different types of interaction centers (say b for double bonds and s for single bonds) are proportional to $U_l^{\tau 1}$ for a-centers, expression (8), with a proportionality factor ξ_l^A, $A = a, b$ or s where $\xi_l^A = -1$ for $A = b$ and l =6, 12 and $\xi_l^A = +1$ otherwise.[35] As a consequence the same set of rotator functions (8) can still be used while the prefactors ξ_l^A are explicitly written down. A second generalization concerns the deformability of the molecule. While Eqs. (3) and (7) refer to the unit representation of I_h, we can take into account molecular deformations by considering the non unit representations of the icosahedral group.[36] These representations also will become relevant if we describe the occupation of the LUMO state of the neutral molecule by an extra electron as a consequence of charge transfer. This problem will be treated in Section 6.

3. MOLECULAR INTERACTIONS IN SOLID C_{60}

Since so far there is no complete determination of the intermolecular potential between rotating molecules (for recent work, see Refs.[37, 47]), we have to resort to phenomenological concepts. The potential is written as a sum of pair potentials between interaction centers ν_A and $\nu_{A'}$ on different molecules at lattice sites $\vec{n}$ and $\vec{n}'$:

$$V = \frac{1}{2} \sum_{\vec{n}\vec{n}'} \sum_{\nu_A \nu_{A'}} V(\vec{n}, \nu_A;\ \vec{n}', \nu_{A'}). \tag{9}$$

Here the index A specifies the type of interaction center: $A = a$, b, or s, where a stands for 60 atomic centers, b for a set of three centers along each of the 30 double bonds and s for single centers on each of the 60 single bonds.[38] The interaction potential between centers includes repulsive Born-Mayer (BM) and attractive Van der Waals (W) terms:

$$V^{BM}_{AA'}(r) = C_1^{AA'} \exp(-C_2^{AA'} r) \tag{10}$$

and

$$V^{W}_{AA'}(r) = B_{AA'} r^{-6}, \tag{11}$$

where $r = r(\vec{n}, \nu_A;\ \vec{n}', \nu_{A'})$ is the distance between centers. Following Ref.[31] we expand the right-hand side of (9) in terms of rotator functions, obtaining

$$V = \frac{1}{2} \sum_{\vec{n}\vec{n}'} \sum_{ll'} \sum_{\tau\tau'} J^{\tau\tau'}_{ll'}(\vec{n} - \vec{n}')\, U^{\tau 1}_{l}(\vec{n})\, U^{\tau' 1}_{l'}(\vec{n}'). \tag{12}$$

The argument $\vec{n}$ in $U^{\tau 1}_l(\vec{n})$ represents the Euler angles $\vec{\omega}(\vec{n})$ which specify the orientation of the molecule at site $\vec{n}$. The index $\tau \equiv (\Gamma, \rho, i)$ accounts for the irreducible representations of the cubic group O_h. We limit our discussion to multipoles with $l \leq 12$. The matrix elements in (12) are given by

$$J^{\tau\tau'}_{ll'}(\vec{n} - \vec{n}') = \sum_{AA'} g^A_l g^{A'}_{l'}\ v^{\tau\tau'}_{ll'}(\vec{n}A;\ \vec{n}'A'). \tag{13}$$

Here g^A_l is the molecular shape factor for interaction centers of type A. The expansion coefficients $v^{\tau\tau'}_{ll'}$ are given[38] by convolution integrals of the potentials $V(\vec{n}, \nu_A;\ \vec{n}', \nu_{A'})$ with the site symmetry adapted surface harmonics $S^{\tau}_l(\Omega_\nu)$ and $S^{\tau'}_{l'}(\Omega_{\nu'})$. All interaction centers are assumed to be distributed on a spherical shell with radius $d = 3.52$ Å. In addition to V^W and V^{BM}, a multipolar Coulomb potential was taken into account,[38] with electric multipoles of C_{60} deduced from first principles calculations.[39] It turns out that the effect of the Coulomb potential is then about 10% of $(V^{BM} + V^W)$.

We separate V, Eq. (12), into two terms: the crystal field potential V_{CF}, and the orientational pair potential V_{RR}. The crystal field potential, which is due to contributions with $l = 0$ but $l' \neq 0$, and vice versa, reads

$$V_{CF} = \sum_{\vec{n}} {\sum_{l}}' \sum_{\tau_{1g}} w^{\tau_{1g}}_l\, U^{\tau_{1g} 1}_l(\vec{n}), \tag{14}$$

where $w_l^{\tau_{1g}}$ are crystal field coefficients.[32] The superscript τ_{1g} signifies the unit representation $\Gamma = A_{1g}$ of O_h, with $\rho = 1$ for $l = 6$ and $l = 10$, $\rho = 1$ and 2 for $l = 12$. Defining Fourier transforms

$$U_l^{\tau 1}(\vec{q}) = \frac{1}{\sqrt{N}} \sum_{\vec{n}} e^{i\vec{q}\cdot\vec{X}(\vec{n})}\, U_l^{\tau 1}(\vec{n}), \tag{15}$$

and

$$J_{ll'}^{\tau\tau'}(\vec{q}) = \sum_{\vec{n}'} e^{i\vec{q}\cdot(\vec{X}(\vec{n})-\vec{X}(\vec{n}'))}\, J_{ll'}^{\tau\tau'}(\vec{n}-\vec{n}'), \tag{16}$$

where $\vec{q}$ is the wave vector, we rewrite the orientational pair potential as

$$V_{RR} = \frac{1}{2} \sum_{\vec{q}} \sum_{ll'}{}' \sum_{\tau\tau'} J_{ll'}^{\tau\tau'}(\vec{q})\, U_l^{\tau 1}(\vec{q})\, U_{l'}^{\tau' 1}(-\vec{q}). \tag{17}$$

Since there are $(2l+1)$ components τ for any given l, $J(\vec{q})$ is a 59×59 matrix. Diagonalizing $J(\vec{q})$ we obtain its eigenvalues $\Lambda_{\alpha\alpha}(\vec{q})$ and its eigenvector components $e_\alpha(\genfrac{}{}{0pt}{}{\tau}{l}|\vec{q})$. Here $\alpha = 1, ..., 59$ labels the eigenvalues and $(\genfrac{}{}{0pt}{}{\tau}{l})$ labels eigenvector components. Since $J(\vec{q})$ is real and symmetric, the eigenvalues and eigenvectors are real. Introducing normal coordinates[40]

$$Q_\alpha(\vec{q}) = \sum_l \sum_\tau e_\alpha(\genfrac{}{}{0pt}{}{\tau}{l}|\vec{q})\, U_l^{\tau 1}(\vec{q}), \tag{18}$$

the potential V_{RR} becomes

$$V_{RR} = \frac{1}{2} \sum_{\vec{q}} \sum_\alpha Q_\alpha(\vec{q})\, \Lambda_{\alpha\alpha}(\vec{q})\, Q_\alpha(-\vec{q}). \tag{19}$$

These normal coordinates generalize those introduced in Ref.[30] where linear combinations of representations belonging to a single manifold have been taken into account.

We recall that for $l = 6$, the irreducible representations of O_h are A_{1g}, A_{2g}, E_g, $2T_{2g}$, T_{1g}; for $l = 10$ they are A_{1g}, A_{2g}, $2E_g$, $3T_{2g}$, $2T_{1g}$; etc. It is useful to partition the matrix $J(\vec{q})$ into submatrices $J_{GG'}(\vec{q})$, where $G \equiv (\Gamma, \rho, l)$. The dimensions of $J_{GG'}$ are equal to the numbers of components of Γ and Γ'. The analysis of the structure of these submatrices helps us to understand the contributions to the orientational pair potential at symmetry points of the Brillouin zone. As an example we consider the case $\Gamma = T_{2g}$ and $\Gamma' = T_{2g}$. Since T_{2g} representations are 3-dimensional, we see that $J_{GG'}(\vec{q})$ is a 3×3 matrix. We find

$$J_{GG'}(\vec{q}) = 4 \begin{bmatrix} \gamma C_{yz} + \alpha(C_{zx} + C_{xy}) & -\beta S_{xy} & -\beta S_{zx} \\ -\beta S_{xy} & \gamma C_{zx} + \alpha(C_{xy} + C_{yz}) & -\beta S_{yz} \\ -\beta S_{zx} & -\beta S_{yz} & \gamma C_{xy} + \alpha(C_{yz} + C_{zx}) \end{bmatrix}. \tag{20}$$

The factors C_{yz}, S_{xy} etc. depend on the wave vector. If we restrict our attention to nearest neighbor interactions on a fcc lattice, we get

$$C_{zx}(\vec{q}) = \cos(q_z \frac{a}{2})\, \cos(q_x \frac{a}{2}), \tag{21}$$

$$S_{zx}(\vec{q}) = \sin(q_z \frac{a}{2})\, \sin(q_x \frac{a}{2}), \tag{22}$$

and similar expressions for the other combinations of indices. Here a is the cubic lattice constant.

We recall that the three basis functions $i = 1,2,3$ of a T_{2g} representation transform as the Cartesian components yz, zx and xy, respectively. At the X point $\vec{q}_x^X = (2\pi/a)(1,0,0)$ the matrix $J_{GG'}(\vec{q}_x^X)$ becomes diagonal with elements $4[\gamma - 2\alpha, -\gamma, -\gamma]$. Thus the second and the third of the T_{2g} functions are equivalent and the corresponding eigenvalue $-\gamma$ is degenerate. The matrices $J_{GG'}(\vec{q})$ with $\vec{q} = \vec{q}_y^X$ and $\vec{q} = \vec{q}_z^X$ are obtained by permutation of elements. With a repulsive potential (10) the coefficients γ obtained for $G = G'$ are positive and the largest positive value corresponds to the representation $\Gamma = T_{2g}$, $\rho = 3$, $l = 10$.[31] The $l = 10$ mode dominates because SAF's belonging to this manifold give a more accurate description of the molecular structure than those with $l = 6$. The corresponding molecular shape factors are $g_{10}^1 = 19.35$ and $g_6^1 = 2.56$. The relative importance of the $l = 10$ term is also demonstrated by quasielastic neutron scattering results.[41] The negative sign of the degenerate eigenvalue $-\gamma$ means that the interaction between the corresponding eigenmodes is attractive at the X point. Condensation of these modes leads to a phase transition[31] $Fm\bar{3}m \rightarrow Pa\bar{3}$. An expression similar to (20) is obtained for the case $\Gamma = \Gamma' = T_{1g}$ at $\vec{q}_x^X$. One also finds[40] that at the X point the degenerate T_{2g} and T_{1g} modes couple. From group theory[42, 43] it follows that the phase transition from the disordered structure $Fm\bar{3}m$ to the ordered structure $Pa\bar{3}$ is driven by the condensation of order parameters with components of T_{2g} and T_{1g} symmetry at the X point of the Brillouin zone. In Ref.[40] the eigenvalues of the matrix $J(\vec{q})$, Eq. (16), have been studied at the X point. One finds that there are 14 degenerate eigenvalues which we label by $\tilde{\alpha} = 1, \ldots, 14$. Notice that 14 is the number of T_{2g} and T_{1g} representations contained in the manifolds with $l = 6$, 10 and 12. For each of the degenerate eigenvalues $\Lambda_{\tilde{\alpha}\tilde{\alpha}}(\vec{q}_x^X)$ one rotates the corresponding eigenvectors so that one of them, denoted $e_{\tilde{\alpha}1}(\vec{q}_x^X)$, only has nonzero components $e_{\tilde{\alpha}1}(^{\Gamma,\rho,i}_{l}|\vec{q}_x^X)$ with $i = 3$ whereas the other one, $e_{\tilde{\alpha}2}(\vec{q}_x^X)$, only has nonzero components $e_{\tilde{\alpha}2}(^{\Gamma,\rho,i}_{l}|\vec{q}_x^X)$ with $i = 2$. The corresponding normal coordinates are called $Q_{\tilde{\alpha}}^{(3)}(\vec{q}_x^X)$ and $Q_{\tilde{\alpha}}^{(2)}(\vec{q}_x^X)$, respectively. These functions form a two dimensional basis (E_g) of the group D_{4h}. Similarly one proceeds at $\vec{q}_y^X$ and $\vec{q}_z^X$. The functions $Q_{\tilde{\alpha}}^{(3)}(\vec{q}_x^X)$, $Q_{\tilde{\alpha}}^{(2)}(\vec{q}_x^X)$, $Q_{\tilde{\alpha}}^{(1)}(\vec{q}_y^X)$, $Q_{\tilde{\alpha}}^{(3)}(\vec{q}_y^X)$, $Q_{\tilde{\alpha}}^{(2)}(\vec{q}_z^X)$ and $Q_{\tilde{\alpha}}^{(1)}(\vec{q}_z^X)$ are the generalized order parameters which form a basis of the six dimensional irreducible representation X_5^+ of the space group[42] $Fm\bar{3}m$. A possible condensation scheme for the transition $Fm\bar{3}m \rightarrow Pa\bar{3}$ is given by

$$Q_{\tilde{\alpha}}^{(3)e}(\vec{q}_x^X) = Q_{\tilde{\alpha}}^{(1)e}(\vec{q}_y^X) = Q_{\tilde{\alpha}}^{(2)e}(\vec{q}_z^X) \equiv \eta_{\tilde{\alpha}}\sqrt{N} \neq 0, \tag{23}$$

$$Q_{\tilde{\alpha}}^{(2)e}(\vec{q}_x^X) = Q_{\tilde{\alpha}}^{(3)e}(\vec{q}_y^X) = Q_{\tilde{\alpha}}^{(1)e}(\vec{q}_z^X) = 0, \tag{24}$$

where the superscript e stands for a thermal expectation value and $\eta_{\tilde{\alpha}}$ is the order parameter amplitude. A condensation scheme corresponds to a domain and there are 8 possible condensation schemes corresponding to the eight possible domains[6] of $Pa\bar{3}$.

Also at the L point of the Brillouin zone, $\vec{q}^L = (2\pi/a)(1/2, 1/2, 1/2)$, several eigenvalues are large and negative. From Eq. (20) we find that for $\Gamma = \Gamma' = T_{2g}$

$$J_{GG'}(\vec{q}^L) = -4 \begin{bmatrix} 0 & \beta & \beta \\ \beta & 0 & \beta \\ \beta & \beta & 0 \end{bmatrix}, \tag{25}$$

and a similar expression is obtained for the case $\Gamma = \Gamma' = T_{1g}$. In the next section we will see that the large attractive interaction at the X point and the L point of the Brillouin zone leads to large intensity peaks in the corresponding diffuse scattering spectra in the disordered phase.

4. PHASE TRANSITION AND DIFFUSE SCATTERING

In order to study the phase transition $Fm\bar{3}m \rightarrow Pa\bar{3}$ it is necessary to investigate the Landau free energy F. In terms of normal coordinates

$$F = \frac{1}{2} \sum_{\vec{q}} \sum_{\alpha\beta} Q^e_\alpha(\vec{q}) \, [\chi^{-1}_{QQ}(\vec{q})]_{\alpha\beta} \, Q^e_\beta(-\vec{q}) + H(Q^e). \tag{26}$$

Here

$$\chi_{QQ}(\vec{q}) = [T \, y^{-1}(\vec{q}) + \Lambda(\vec{q})]^{-1} \tag{27}$$

is the static susceptibility, $Q^e_\alpha(\vec{q})$ is the instantaneous expectation value of $Q_\alpha(\vec{q})$ and $H(Q^e)$ stands for higher order terms. The matrix $y(\vec{q})$ depends on temperature T and is given by

$$y_{\alpha\beta}(\vec{q}) = \sum_{ll'} \sum_{\tau\tau'} e_\alpha(^{\tau}_{l}|\vec{q}) \, [UU]^{\tau\tau'}_{ll'} \, e_\beta(^{\tau'}_{l'}| - \vec{q}). \tag{28}$$

where

$$[UU]^{\tau\tau'}_{ll'} = [\langle U^{\tau 1}_l(\vec{n}) \, U^{\tau' 1}_{l'}(\vec{n})\rangle_{CF} - \langle U^{\tau 1}_l(\vec{n})\rangle_{CF} \langle U^{\tau' 1}_{l'}(\vec{n})\rangle_{CF}] \, \delta_{\Gamma\Gamma'}\delta_{ii'}. \tag{29}$$

Here $\langle \; \rangle_{CF}$ is a single particle expectation value which is calculated using the crystal field V_{CF}. In a nonzero crystal field the matrix $y(\vec{q})$ is non diagonal. If the phase transition were of second order, the transition temperature T_c would be the most positive solution for T of the transcendental equation

$$det[T \, y^{-1}(\vec{q}^{\,X}_x) + \Lambda(\vec{q}^{\,X}_x)] = 0. \tag{30}$$

The susceptibility (27), expressed in the normal coordinate representation, is related to the susceptibility

$$\chi_{UU}(\vec{q}) = \{T \, [UU]^{-1} + J(\vec{q})\}^{-1}, \tag{31}$$

in the rotator functions representation, by a unitary transformation of type (18).

Originally a microscopic theory of the phase transition $Fm\bar{3}m \rightarrow Pa\bar{3}$ was formulated by retaining a single representation in the orientational pair potential V_{RR}, namely $T^{(3)}_{2g}$ belonging to the manifold[31, 38] l=10. The potential presented in Ref.[38] gave a satisfactory description of the crystal field[38, 44] in agreement with X-ray single crystal[45] and high resolution neutron powder diffraction results,[46] the values of T_c and of the corresponding first order transition temperature T_1 being 178 K and 202 K, respectively whereas the experimental value is ≈260 K. It was shown earlier[35] that coupling among representations enhances T_c, but the numerical estimates of this effect were made with inadequate knowledge of the intermolecular potential. More recently, stimulated by the diffuse scattering work of Launois et al.,[9] J. R. D. Copley and K. H. Michel carried out calculations where all representations up to the manifold l = 12 were taken into account. With the model of molecular interactions described in Ref.[38] it was found that the transition temperature is greatly increased, to $T_c \approx 500$ K. The same conclusion was reached in Ref.[47]; calculating diffuse scattering spectra[47, 48] for various potential models,[33, 38, 43, 49] the authors of Ref.[47] found that the potential in Ref.[38] leads to the closest agreement with experiment, but the temperature scale deduced from this potential was at variance with experiment. Notice that a simple scaling of the potential[38] is not sufficient since it would spoil the agreement with experimental results[45, 46, 50] for

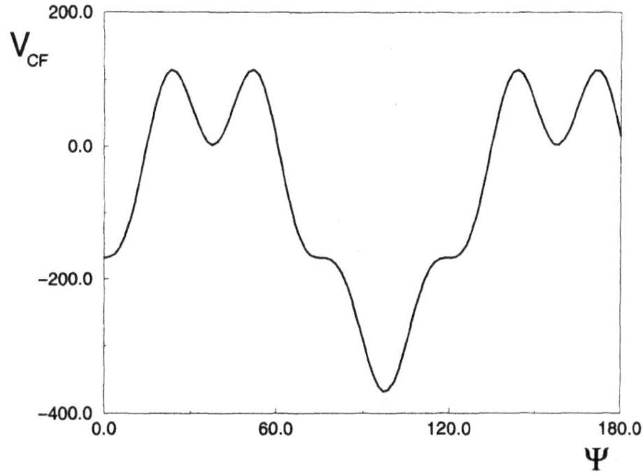

Figure 2. The crystal field $V_{CF}(\Psi)$, in degrees K, for the disordered phase of C_{60}.

the crystal field. In revising the intermolecular potential, the authors of work[40] have respected two imperatives. Firstly, in order to reduce the transition temperature (for the description which takes into account the coupling of order parameter components in comparison with the single component $T_{2g}^{(3)}$, l=10) it is necessary to reduce the strength of the repulsive interaction between double bond centers. Secondly, in order to retain an acceptable description of the crystal field in the disordered phase, it is necessary to increase the van der Waals interaction.[40] In Fig. 2 we show the angular dependence of the crystal field potential $V_{CF}(\Psi)$ when the C_{60} molecule is rotated counterclockwise by an angle Ψ about a [111] axis away from the standard orientation (convention of Ref[5]). Minima in the crystal field potential reflect maxima in the orientational density distribution function[44, 32] which is probed by diffraction experiments. The plot of Fig. 2 is in a satisfactory agreement with a determination of V_{CF} from recent neutron diffraction data.[50] We recall that the absolute minimum at $\Psi = 98°$ was first deduced from diffraction data at low T.[5, 51] It corresponds to the setting angle at low T on the four sublattices in the fully ordered $Pa\bar{3}$ structure. Subsequently it was found that a significant fraction of the molecules occupy the 38° position.[52] The occurrence of these two orientations at low T has led to the suggestion of an orientational glass.[52] It was first shown by theory that the 98° and 38° orientations have their precursors in the high temperature phase.[44]

We now discuss the transition temperature. Considering the 14 representations T_{2g} and T_{1g} belonging to the manifolds l = 6, 10, and 12 and taking a = 14.15 Å, one finds from expression (30) with the potential parameters in Ref[40] that the virtual second order transition temperature T_c = 163 K. If we take into account the contraction of the lattice (to a = 14.11 Å) at the phase transition,[52, 53] we find T_c = 186 K. In order to obtain the first order transition temperature T_1, one has to include explicitly terms of third and fourth order in the Landau free energy. Although this is feasible[38] for the case of order parameter components belonging to a single irreducible representation, say $T_{2g}^{(3)}$, l =10, the procedure becomes technically very cumbersome for the case of a large number of coupled irreducible representations. It is then preferable, and also more rigorous, to solve the coupled molecular field equations directly by numerical methods.[54] In addition to the primary order parameters $\eta_{\tilde{\alpha}}$ which enter the condensation scheme

(23), there are secondary order parameters of A_{1g} and A_{2g} symmetry which condense at the Γ point $\vec{q}=0$. Within the manifolds $l=6$, 10 and 12, there are 14 primary order parameters $\eta_{\tilde{\alpha}}$ corresponding to the number of irreducible representations T_{2g} and T_{1g}; in addition there are 4 secondary order parameters of symmetry A_{1g}, denoted by $\eta_{\alpha}^{A_{1g}}$, and 3 secondary order parameters $\eta_{\beta}^{A_{2g}}$. We can write the molecular field equations for these order parameters. Due to the crystal field V_{CF}, all order parameters are coupled so that there are 21 coupled functional equations (for $l \leq 12$):

$$\eta_{\tilde{\alpha}} = f_{\tilde{\alpha}}[\eta_{\tilde{\alpha}'}, \eta_{\alpha'}^{A_{1g}}, \eta_{\beta'}^{A_{2g}}], \tag{32}$$

$$\eta_{\alpha}^{A_{1g}} = f_{\alpha}^{A_{1g}}[\eta_{\tilde{\alpha}'}, \eta_{\alpha'}^{A_{1g}}, \eta_{\beta'}^{A_{2g}}], \tag{33}$$

$$\eta_{\beta}^{A_{2g}} = f_{\beta}^{A_{2g}}[\eta_{\tilde{\alpha}'}, \eta_{\alpha'}^{A_{1g}}, \eta_{\beta'}^{A_{2g}}], \tag{34}$$

where

$$[\eta_{\tilde{\alpha}'}, \eta_{\alpha'}^{A_{1g}}, \eta_{\beta'}^{A_{2g}}] \equiv [\eta_{\tilde{\alpha}'}, \tilde{\alpha}'=1-14;\ \eta_{\alpha'}^{A_{1g}}, \alpha'=1-4;\ \eta_{\beta'}^{A_{2g}}, \beta'=1-3]. \tag{35}$$

We have numerically solved these coupled molecular field equations as a function of temperature. The results of such a calculation are shown in Fig. 3.

Notice that the components $\eta_{\alpha'}^{A_{1g}}$ are already different from zero in the high temperature phase. At the phase transition these components become renormalized by the condensation of the primary order parameters. The order parameters can be used to calculate Bragg reflections in the ordered and disordered phases. The theoretical results can then be compared with diffraction experiments.[53,55 - 56] Such a comparison provides a test of the intermolecular potential. We are currently working on this problem.

Diffuse scattering experiments in the disordered phase allow a detailed investigation of the static collective orientational susceptibility and constitute a test of the intermolecular potential. We have calculated the diffuse scattering cross section[9,32] in the disordered phase as follows:

$$\left.\frac{d\sigma}{d\Omega_{\vec{Q}}}\right|_d = 16\pi^2 N f^2(Q)\{\sum_{l,l'\leq 12}\sum_{\tau\tau'} R_l^{\tau}(\vec{Q}) R_{l'}^{\tau'*}(\vec{Q}) T[\chi_{UU}(\vec{q})]_{ll'}^{\tau\tau'} + \sum_{l>12} F_l(\vec{Q})\}. \tag{36}$$

Here $\hbar\vec{Q}$ is the momentum transfer in the scattering process and $f(Q)$ is the X-ray or neutron atomic scattering factor. The static susceptibility $\chi_{UU}(\vec{q})$ is given by expression (31). We recall that $\vec{Q}=\vec{G}+\vec{q}$, where $\vec{G}$ is a vector of the reciprocal lattice and $\vec{q}$ belongs to the first Brillouin zone. The prefactor $R_l^{\tau}(\vec{Q})$ reads

$$R_l^{\tau}(\vec{Q}) = i^l\, g_l^1\, j_l(Qd)\, S_l^{\tau}(\Omega_{\vec{Q}}). \tag{37}$$

Here g_l^1 is the molecular shape factor for atomic centers, j_l is a Bessel function, S_l^{τ} are again site SAF's and $\Omega_{\vec{Q}}$ specifies the direction of $\vec{Q}$. The first term within braces on the right hand side of (36) takes account of intermolecular interactions; all representations up to and including the manifold $l=12$ are included. This term is the same as the expression given in Ref.[48] The second term within braces, with

$$F_l(\vec{Q}) = \frac{[g_l^1\, j_l(Q_d)]^2}{2l+1}\sum_{\tau}[S_l^{\tau}(\Omega_{\vec{Q}})], \tag{38}$$

represents contributions for $l>12$ which are treated in the free molecule approximation (setting $J(\vec{q})=0$ and $V_{CF}=0$). Intermolecular interactions become very small for large l whereas intramolecular scattering, whose form is determined by the shape and size

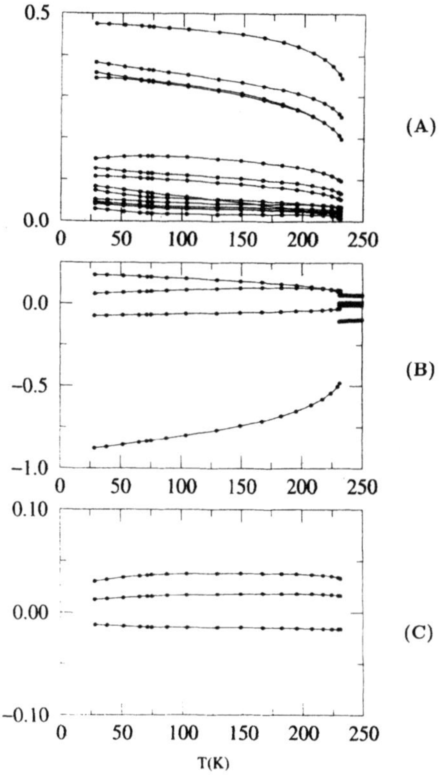

Figure 3. Coupled order parameter amplitudes for C_{60}: (a) primary $(T_{2g} + T_{1g})$ order parameters $\eta_{\tilde{\alpha}}$, $\tilde{\alpha} = 1 - 14$; (b) secondary parameters $\eta_{\alpha}^{A_{1g}}$, $\alpha = 1 - 4$; (c) secondary parameters $\eta_{\beta}^{A_{2g}}$, $\beta = 1 - 3$.

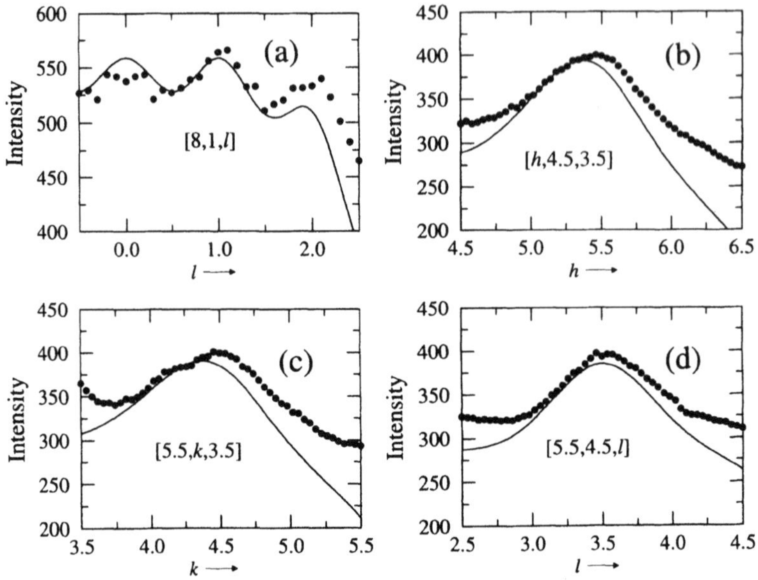

Figure 4. Diffuse scattering intensities: (a) for a scan through X points from $\vec{Q} = (2\pi/a)(8, 1. -0.5)$ to $\vec{Q} = (2\pi/a)(8, 1, 2.5)$; (b), (c) and (d) for orthogonal scans through the L point at $\vec{Q} = (2\pi/a)(5.5, 4.5, 3.5)$. The points are the experimental data of Launois et al.[9], Fig.4, corrected (P.Launois, private communications) for the X-ray polarization factor $(1 + \cos^2 2\Theta)/2$ where 2Θ is the scattering angle. The lines were calculated using Eq. (36). The calculations for the L point scans were divided by 1.7 to account for the transmission of the beryllium window (as described in the caption to Fig.4 of Ref.[9]). The calculations were also multiplied by a single scale factor.

of the C_{60} molecule, remain very important; indeed the largest contribution to this scattering is associated with $l = 18$.[41]

Starting from a revised intermolecular potential whose details are given in Ref.[40], we have calculated the diffuse X-ray scattering cross section for wave vectors along various paths in $\vec{Q}$-space using Eq. (36). In Fig. 4a we show scans through a series of X points and in Figs. 4b, 4c and 4d we show scans through the L point $(2\pi/a)(5.5, 4.4, 3.5)$ (T=300 K, disordered phase). Our results are in a good agreement with single crystal X-ray diffuse scattering measurements, Figs. 4a–4d of Ref.[9]

In order to provide further comparison with experiment we have calculated the diffuse X-ray scattering for wave vectors $\vec{Q}$ along a longitudinal path through the X point $(2\pi/a)(2, 3, 6)$. Our theoretical results at T=300 K are shown in Fig. 5 together with unpublished single crystal X-ray diffraction results due to Wochner et al. (see also Fig. 4 in Ref.[12]).

Although the overall agreement with experiment is satisfactory, it is not perfect at low Q where the experimental spectra show more detailed structure. We attribute this to the fact that our present parametrization of the potential, while giving the right value of the first order transition temperature T_1, probably underestimates the virtual second order transition temperature T_c. Finally we have calculated at $T = 300$ K the diffuse neutron scattering intensity along the line $\vec{Q} = (2\pi/a)(\xi, \xi, 0.35)$ and our results (Fig. 6) are in a good agreement with the neutron data[11].

The results of Figs. 5 and 6 show the importance of the free molecule term in Eq. (36) for large values of the momentum transfer $\hbar\vec{Q}$.

The maxima of the diffuse scattering intensity in certain regions of the Brillouin

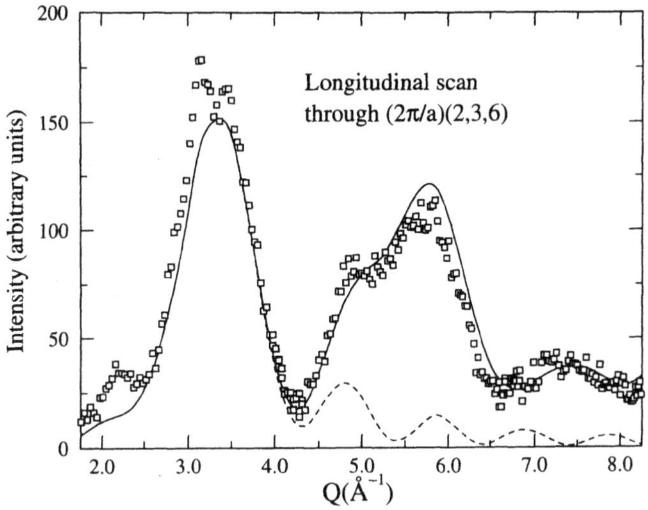

Figure 5. Diffuse X-ray scattering intensity for a longitudinal scan through the X point at $\vec{Q} = (2\pi/a)(2,3,6)$ (which is at Q=3.1 Å^{-1}). The symbols show the room temperature experimental measurements of Wochner et al. with the Compton scattering subtracted (see Fig. 4 in Ref.[12]); since the normalization of the Compton scattering is not exactly known the subtraction procedure introduces a small amount of additional uncertainty in the experimental results (P.Wochner, private communication). The dashed line shows a calculation ignoring the second term within the braces in Eq. (36).

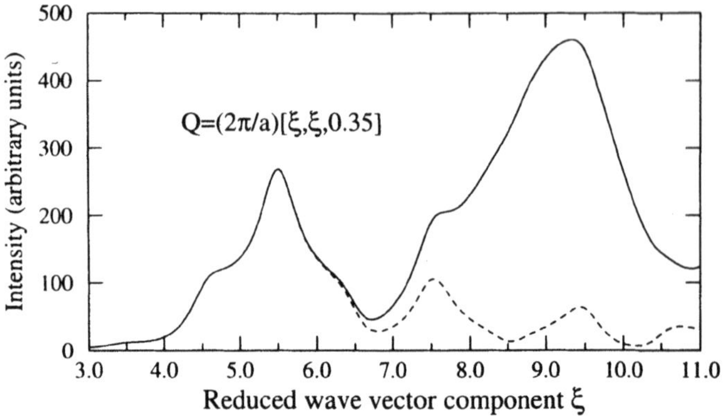

Figure 6. Calculated diffuse neutron scattering intensity along $\vec{Q}=(2\pi/a)(\xi,\ \xi,\ 0.35)$. The solid line represents a full calculation using Eq. (36). The dashed line shows a calculation ignoring the second term within the braces in Eq. (36).

zone reflect correspondingly large values of the collective susceptibility $\chi_{UU}(\vec{q})$. Since $\chi_{UU}(\vec{q})$ and $\chi_{QQ}(\vec{q})$ have the same eigenvalue spectrum, diffuse scattering maxima correspond to large negative eigenvalues of the matrix $[y(\vec{q})\,\Lambda(\vec{q})]$. We have studied the eigenvalues $\Lambda_{\alpha\alpha}(\vec{q})$ and eigenvectors $\vec{e_\alpha}(\vec{q})$ throughout the Brillouin zone and we find that the largest negative eigenvalue, which is –2410 K, occurs at the X point and is due to a normal mode with E_g and A_{2g} components. This mode, which we call the leading X_2^+-mode, corresponds to a basis function of the irreducible representation X_2^+ of $Fm\bar{3}m$, and its condensation would lead to a $P4_2/mnm$ structure.[42, 43] The second largest negative eigenvalue, –2382 K, is doubly degenerate and belongs to normal modes with T_{2g} and T_{1g} components. These modes, which are called leading X_5^+-modes, are basis functions of the irreducible representation X_5^+ of $Fm\bar{3}m$, and are therefore order parameters of the $Pa\bar{3}$ structure.

While the eigenvalue of the X_2^+ mode is larger in absolute value than that of the leading X_5^+ modes, the opposite holds for the corresponding elements of the single particle matrix, Eq. (28). At $T = 300\ K$ we obtain values of 3.97×10^{-2} for the X_2^+ case and 4.77×10^{-2} for the X_5^+ case, and with decreasing of temperature the X_2^+ expectation value decreases while the X_5^+ expectation value increases. Consequently the contribution of the X_5^+ modes to $[y(\vec{q})\,\Lambda(\vec{q})]$ prevails. We recall that the single particle thermal averages are calculated using the crystal field potential V_{CF}.

We have also investigated the interactions which lead to maxima in the diffuse scattering at the L point (see Fig. 4). From Eq. (25) we see that components of T_{2g} and T_{1g} symmetry contribute to fluctuations at the L point. Accidentally the contribution of the mode $T_{2g}^{(3)}$, $l = 10$ is very weak at the L point but is dominant at the X point. On the other hand some of the other modes of T_{2g} and T_{1g} symmetry are large at the L point. The single crystal X-ray diffuse scattering experiments[9] most clearly illustrate the importance of couplings among the order parameter components.

5. CHARGE TRANSFER AND INTERACTION POTENTIAL IN MC_{60}

In the disordered phase the alkali metal fullerides MC_{60} (M=Rb,Cs,K)[13-15,18-22, 58, 59] adopt the rocksalt structure, typical of ionic crystals (Fig. 7).

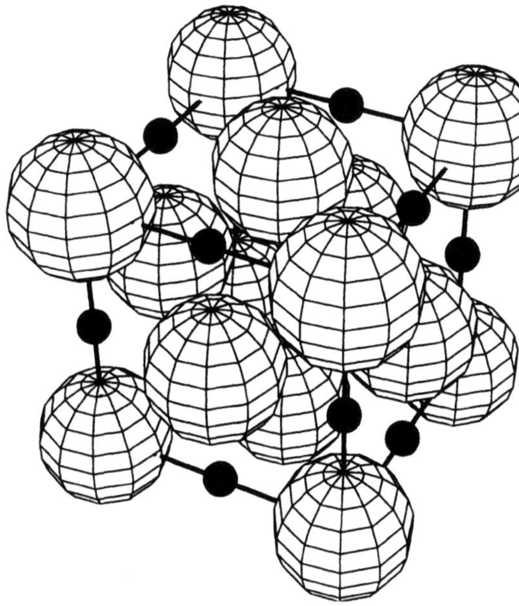

Figure 7. The rocksalt structure of the disordered cubic phase of MC_{60}, space group $Fm\bar{3}m$.

At $T \approx 400$ K, the cubic lattice constant a is typically 14.04–14.09 Å, considerably smaller than the value of 14.15 Å for the disordered fcc phase of pristine C_{60}. Quasi-elastic neutron scattering data[60] at 400 K may be described using a rotational diffusion constant D_r=2.4×10^{10} s^{-1} which implies even faster molecular reorientations than those for pristine C_{60} at 300 K.[61] The charge transfer (of one electron) from the alkali atom to the C_{60} molecule is nearly complete,[62] changing the electronic structure of the molecule. In the following paragraphs we study the change in the intermolecular potential due to charge transfer.

We consider a crystal of N rigid C_{60}^- ions whose centers of mass are located on a face centered cubic (fcc) lattice. (We follow closely Ref.[63]) Each C_{60}^- ion is surrounded by six nearest neighbor M^+ ions in octahedral positions. The intermolecular potential is written as

$$V = V^0 + V^{el}. \tag{39}$$

Here V^0 stands for the intermolecular potential of the MC_{60} crystal where the electronic structure of the C_{60} molecule is assumed unchanged, whereas V^{el} accounts for changes in the intermolecular potential induced by charge transfer. The form of V^0 can be inferred from previous work[31, 44] on pristine C_{60}. Each molecule is modeled by a rigid cluster of interaction centers I distributed on a spherical shell. The intermolecular potential V^0 is written as a sum of pair potentials between interaction centers ν_I and $\nu_{I'}$ on different molecules at lattice sites $\vec{n}$ and $\vec{n}'$, Eq. (9). The pair potential depends on the instantaneous orientations of the molecules. Performing an expansion in terms of rotator functions, we separate V^0 into two terms[31, 44]

$$V^0 = V^0_{RR} + V^0_{CF}, \tag{40}$$

Table 1. Interaction potential parameters between C, D, S centers of interaction. The multiplicity for 3 centers on double bonds D is included. Scaling factors are $a_C = 0.82$, $a_D = 0.47$, a_S=0.197.

	units	C, C	C, D	C, S	D, D	D, S	S, S
$C^{(1)}_{\nu\nu'}$	6.742×10^7 K	$a_C \times a_C$	$a_C \times a_D$	$a_C \times a_S$	$a_D \times a_D$	$a_D \times a_S$	$a_S \times a_S$
$C^{(2)}_{\nu\nu'}$	Å^{-1}	3.6	3.4	3.5	3.2	3.3	3.4
$B_{\nu\nu'}$	10^5 K×Å^6	6.872	0	0	0	0	0

Table 2. Crystal field coefficients $w_l^{\tau_{1g}}$, $\tau_{1g} = (A_{1g}, \rho)$ for the disordered phase of pristine C_{60}, and corresponding contributions to crystal field coefficients for the alkali fullerides.

l, δ	6,1	10,1	12,1	12,2
MC_{60}	669	-334	-105	-481
C_{60}	621	-274	-93	-427

where V_{RR} is the orientational pair potential (17), and V_{CF} is the crystal field (14). The superscript τ_{1g} on the crystal field coefficients $w_l^{\tau_{1g}}$ in Eq. (14) signifies the unit representation $\Gamma = A_{1g}$ of O_h, with $\rho = 1$ for $l = 6$ and $l = 10$ and $\rho = 1$ and 2 for $l = 12$. The matrix elements $J_{ll'}^{\tau\tau'}$ in Eq. (12) are calculated by the methods of Ref.[44] We assume that the radius of the C_{60} molecular ion is $d = 3.55$ Å. The potential parameters are specified in Table 1, and in Table 2 we list the crystal field coefficients calculated for two different lattice constants: $a = 14.08$ Å, corresponding to MC_{60} and a=14.16 Å to pristine C_{60}.

We now turn to the electronic structure and recall some results for the neutral C_{60} molecule.[16, 17] To start with we model the molecule as a composite 'spherical' structure with icosahedral symmetry. The carbon cores are located on a sphere, three of the four valence electrons on each atom participating in sp^2 hybridized σ bonds. The remaining 60 electrons form delocalized molecular orbitals (π-orbitals) which are classified according to the irreducible representations of the group I_h. We neglect the radial distribution of the delocalized electronic wave functions and retain only the orientational part. The molecular orbitals are then given by symmetry adapted linear combinations[64] of spherical harmonics Y_l^m. In general, functions with different angular momentum number l can contribute to a given orbital. However, since there is no overlap between spherical harmonics with different l and in general integrals over spherical harmonics decrease rapidly with higher l values, it is a good approximation to retain only the lowest value of l compatible with the symmetry of the given molecular orbital. Indeed Hückel molecular orbital (HMO) calculations[17] on the neutral C_{60} molecule give the eigenfunctions $a_g(l = 0)$, $t_{1u}(l = 1)$, $h_g(l = 2)$, $t_{2u}(l = 3)$, $g_u(l = 3)$, $g_g(l = 4) + h_g(l = 4)$, $h_u(l = 5)$. The corresponding molecular states accommodate 60 electrons, so that the seven shells are completely filled and according to the generalized Unsöld theorem[26] the electron density has unit icosahedral symmetry.

The situation is very different for the case of the alkali fullerides where one extra electron occupies the LUMO (lowest unoccupied molecular orbital, referring to the neutral molecule) state, which is the threefold degenerate $t_{1u}(l = 5)$ level. Extending

Table 3. Coefficients of expansion of t_{1u} functions.

	ψ_1	ψ_2	ψ_3
$n =$	$Y_5^{n,c}$	$Y_5^{n,c}$	$Y_5^{n,s}$
0	-0.13941	-0.22556	0
1	-0.31020	0.19171	-0.70699
2	-0.42589	-0.68911	0
3	0.68279	-0.42199	0.30659
4	0.27491	0.44482	0
5	0.40147	-0.24812	0.63731

the concepts of the previous paragraph we introduce three molecular orbitals belonging to the irreducible representation $t_{1u}(l = 5)$ and define them on the same composite 'spherical' structure. This means that we keep only the orientational part of the orbitals and as a first approximation ignore the radial distribution. One can argue that since there is a triple degeneracy, we should take into account deformations of the molecule induced by the Jahn-Teller effect. However, we will show that the degeneracy is lifted as soon as the molecule is rotated away from its standard orientation in the crystal. Since the C_{60} molecule is very stable and there is no experimental evidence of its deformation in the high temperature disordered phase, we assume in the following analysis that the molecular skeleton formed by the carbon cores keeps its icosahedral symmetry.

If we choose as z axis one of the twelve five-fold axes of the icosahedral group, and as y axis one of the remaining two-fold axes perpendicular to z, the normalized spherical harmonics belonging to the representation $t_{1u}, l = 5$ of the molecular group I_h are[36]

$$\psi_1(\Omega) = 0.84853\ Y_5^0 + 0.52915\ Y_5^{5,c}, \tag{41}$$
$$\psi_2(\Omega) = 0.54772\ Y_5^{1,c} - 0.83666\ Y_5^{4,c}, \tag{42}$$
$$\psi_3(\Omega) = 0.54772\ Y_5^{1,s} + 0.83666\ Y_5^{4,s}. \tag{43}$$

The real spherical harmonics $Y_l^{m,c}$, $Y_l^{m,s}$ are defined with the phase convention of Ref.[23], where $\Omega \equiv (\Theta, \phi)$ denotes polar angles. We rotate the molecule anticlockwise about the y axis by the angle $\beta = \arccos[2/\sqrt{10 + 2\sqrt{5}}]$ in order to transform it to the standard orientation[5, 32] where molecular two-fold axes lie along the Cartesian x, y and z directions. In the following we use this standard orientation. The resulting decomposition of ψ_1, ψ_2, ψ_3 over real spherical harmonics is shown in Table 3.

When C_{60}^- and M^+ ions are assembled in the rocksalt structure of the alkali fullerides, two main additional interactions are introduced in the crystal. First, there is an attractive Coulomb interaction between alkali and C_{60} ions. Second, there is a Coulomb intermolecular interaction between valence electrons sitting on neighboring C_{60} molecules. Since there is only one extra electron on each C_{60} molecule there is no intramolecular Coulomb interaction.

We start with the crystal field due to Coulomb interactions experienced by one (central) C_{60}^- ion when the neighboring alkali ions are represented by point positive unit charges ($+e$) and neighboring C_{60} ions are treated in the spherical approximation, which means that for electrostatic considerations they can be replaced with point negative charges ($-e$). In general the Coulomb interaction in a molecular crystal is reduced by several effects such as static screening, polarization of the molecule and more complicated dynamic phenomena such as retardation of the electron wave function. We assume that the polarization of the delocalized core states of the C_{60}^- molecule leads to a weakening of the Coulomb forces. We introduce a parameter $0 < p < 1$ to describe this reduction quantitatively. The resulting electrostatic potential then reads

$$V_{CF}^{el}(\vec{r}) = p\left[\sum_{k_1=1}^{12} \frac{e^2}{|\vec{r} - \vec{r}_{k_1}|} - \sum_{k_2=1}^{14} \frac{e^2}{|\vec{r} - \vec{r}_{k_2}|}\right], \tag{44}$$

where $\vec{r} = (r, \Theta, \phi)$ refers to the location of the extra electron on the central C_{60}^-. Since we neglect the radial distribution $r = d$, the radius of the C_{60} 'sphere'. In Eq. (44) the sum k_1 runs over 12 nearest C_{60}^- neighbors with center of mass coordinates $\vec{r}_{k_1}$ and the sum k_2 runs over 14 alkali ions with coordinates $\vec{r}_{k_2}$ (6 in octahedral positions and 8 in the second shell). If the central ion is centered at lattice site $\vec{n}$ we denote the location of the corresponding extra electron by $\vec{r}(\vec{n})$ and we use the notation $V_{CF}^{el}(\vec{n})$ for $V_{CF}^{el}(\vec{r}(\vec{n}))$. In the next section we shall see that $\vec{r}(\vec{n})$ and hence the potential, depends on the instantaneous orientation of the molecule at site $\vec{n}$. Extending these concepts we consider the Coulomb interaction

$$V_{RR}^{el}(\vec{n}, \vec{n}') = \frac{pe^2}{|\vec{r}(\vec{n}) - \vec{r}(\vec{n}')|} \tag{45}$$

between electronic distributions at sites $\vec{n}$ and $\vec{n}'$.

6. ELECTRONIC ORIENTATIONAL COORDINATES

In the previous section we saw that electronic charge transfer results in occupation of the threefold degenerate LUMO levels of t_{1u} symmetry. In the presence of the crystal field given by Eq. (44), the degeneracy is lifted. We will now study this process as a function of molecular orientation, treating the molecular rotations classically and the valence electron as a quantum particle. In the orientationally disordered phase, the potential $V_{CF}^{el}(\vec{n})$ has cubic symmetry. Expansion in site symmetry adapted functions (SAF's)[23] $S_l^{\tau_{1g}}(\Omega)$ yields

$$V_{CF}^{el}(\vec{n}) = \sum_l{}' \sum_{\tau_{1g}} v_l^{\tau_{1g},\, el} S_l^{\tau_{1g}}(\Omega(\vec{n})). \tag{46}$$

The superscript $\tau_{1g} \equiv (A_{1g}, \rho)$, with $\rho = 1$ for $l = 4, 6, 8, 10$, and $\rho = 1, 2$ for $l = 12$; A_{1g} is the unit representation of the cubic group O_h. Notice that the allowed values of l are no longer restricted by the icosahedral symmetry of the molecule as was the case for expressions (14) and (17). The expansion coefficients are given by

$$v_l^{\tau_{1g},\, el} = \int d\Omega \; V_{CF}^{el}(\Omega) S_l^{\tau_{1g}}(\Omega), \tag{47}$$

with $d\Omega = \sin\Theta d\Theta d\phi$. The site SAF's are linear combinations of spherical harmonics

$$S_l^{\tau}(\Omega) = \sum_m \alpha_l^{m\tau} Y_l^m(\Omega), \tag{48}$$

with $\tau = (\Gamma, \rho, i)$ (see Sec. 2), with the coefficients $\alpha_l^{m\tau}$ given in Ref.[23] (see also Eq. (8)). So far we have taken the molecular system of axes in coincidence with the crystal (site) fixed axes. An arbitrary molecular orientation (electron orientational coordinate Ω) is obtained from the standard orientation (electron coordinate Ω') by a rotation $R(\vec{\omega})$ with Euler angles $\vec{\omega} \equiv (\alpha, \beta, \gamma)$:

$$Y_l^m(\Omega) = \sum_n Y_l^n(\Omega') D_l^{nm}(\vec{\omega}), \tag{49}$$

Table 4. Shape factors of electronic valence density matrix $g_l^{el}(i,j)$, i,j=1,2,3.

l	$i,j=$ (1,1), (2,2), (3,3)	(1,2), (1,3), (2,3)
0	0.28209	0
2	0.04852	0.04202
4	0.13020	0.11275
6	0.15648	0.02983
8	0.17823	0.15435
10	0.19037	0.15684

where $D_l^{nm}(\vec{\omega})$ are the Wigner functions.[23] Using the t_{1u} functions (41)–(43), we define electronic density coefficients

$$c_l^n(ij) \equiv \langle i|Y_l^n|j\rangle = \int d\Omega' \psi_i(\Omega') Y_l^n(\Omega') \psi_j(\Omega'), \tag{50}$$

where i,j are 1, 2 or 3 and only coefficients with $l = 0,2,4,6,8,10$ are allowed. From Eqs. (48)–(50) we get

$$S_l^\tau(ij;\, \vec{\omega}) \equiv \langle i|S_l^\tau|j\rangle_{\vec{\omega}} = \sum_{mn} c_l^n(ij)\alpha_l^{m\tau} D_l^{nm}(\vec{\omega}). \tag{51}$$

The normalization condition reads

$$\int d\vec{\omega} S_{l'}^{\tau'*}(ij;\, \vec{\omega}) S_l^\tau(ij;\, \vec{\omega}) = \frac{8\pi^2}{2l+1}[g_l^{el}(ij)]^2 \delta_{\tau\tau'}\delta_{ll'}, \tag{52}$$

where

$$g_l^{el}(ij) \equiv \sqrt{\sum_n c_l^{n*}(ij) c_l^n(ij)}, \tag{53}$$

are electronic molecular shape factors. In obtaining (52) we have used the normalization of the Wigner functions and the fact that the transformation (48) is unitary. In Table 4 we give numerical values of the elements $g_l^{el}(ij)$.

Introducing the coefficients

$$\alpha_l^{n,el}(ij) = c_l^n(ij)/g_l^{el}(ij), \tag{54}$$

we define electronic molecular rotator functions

$$U_l^{\tau,el}(ij;\, \vec{\omega}) = \sum_{nm} \alpha_l^{n,el}(ij) D_l^{nm}(\vec{\omega}) \alpha_l^{m\tau}. \tag{55}$$

Expressions (53) and (54) are a generalization of concepts that have been previously used for the description of the molecular structure when the molecule has been modeled by a cluster of interaction centers.[31, 32]

Having introduced the formalism that relates the electronic density due to occupation of the t_{1u} LUMO states with the molecular orientations, we study the effect of the crystal field. Applying perturbation theory for degenerate states, we obtain the secular equation

$$|E^{el}\delta_{ij} - \langle i|V_{CF}^{el}|j\rangle_{\vec{\omega}}| = 0. \tag{56}$$

Taking into account expressions (46), (51), (54) and (55) we have for a molecule at lattice site $\vec{n}$

$$\langle i|V_{CF}^{el}|j\rangle_{\vec{\omega}(\vec{n})} = {\sum_l}' \sum_{\tau_{1g}} w_l^{\tau_{1g},el}(ij) U_l^{\tau_{1g},el}(ij;\vec{n}), \tag{57}$$

Table 5. Electronic crystal field coefficients $w_l^{\tau_{1g},el}$.

	i,j = (1,1), (2,2), (3,3)			(1,2), (1,3) (2,3)		
l	alkali ions	C_{60}	total	alkali ions	C_{60}	total
4	-368.96	-34.28	-403.24	-319.51	-29.69	-349.20
6	-47.71	-13.11	-60.82	-9.09	-2.50	-11.59
8	-23.54	1.17	-22.37	-20.39	1.01	-19.38
10	-3.27	0.00	-3.27	-2.70	0.00	-2.70

where

$$w_l^{\tau_{1g},el}(ij) = g_l^{el}(ij) v_l^{\tau_{1g},el}, \tag{58}$$

and the argument $\vec{n}$ of the rotator function again represents $\vec{\omega}(\vec{n})$. Table 5 lists calculated electronic crystal field coefficients $w_l^{\tau_{1g},el} = w_l^{\tau_{1g},el}(C_{60}^-) + w_l^{\tau_{1g},el}(M^+)$ together with their partial contributions from neighboring C_{60}^- ions and alkali metal ions, using $p = 0.36$. This value of p has been selected because the resulting crystal field implies crystal phases that reasonably approximate experimental findings, as discussed in Section 8. Note that the largest contribution occurs for $l = 4$, which is now an allowed value because of charge transfer. The 3×3 secular equation leads to three eigenvalues $E_\alpha^{el}(\vec{\omega})$ which depend on the instantaneous molecular orientation and for an arbitrary orientation the initial t_{1u} degeneracy will be lifted. A thorough discussion of these effects will be given in Sec. 8.

Finally the electronic density-density interaction, Eq. (45), between two rotating molecules at sites $\vec{n}$, $\vec{n}'$ is

$$\langle ii'|V_{RR}^{el}(\vec{n},\vec{n}')|jj'\rangle_{\vec{\omega}(\vec{n}),\vec{\omega}'(\vec{n}')} = \sum_{ll'}{}' \sum_{\tau\tau'} J_{ll'}^{\tau\tau'}(\vec{n}, ij;\, \vec{n}', i'j')\, U_l^{\tau,el}(ij;\, \vec{n})\, U_{l'}^{\tau',el}(i', j';\, \vec{n}'), \tag{59}$$

where

$$J_{ll'}^{\tau\tau'}(\vec{n}, ij;\, \vec{n}', i'j') = v_{ll'}^{\tau\tau';\,el}(\vec{n},\vec{n}') g_l^{el}(ij) g_{l'}^{el}(i'j') \tag{60}$$

and

$$v_{ll'}^{\tau\tau',\,el}(\vec{n},\vec{n}') = \int d\Omega\, d\Omega'\, S_l^\tau(\Omega)\, S_{l'}^{\tau'}(\Omega')\, V_{RR}^{el}(\Omega,\Omega'). \tag{61}$$

Expressions (57) and (59) represent contributions to the crystal field and to the rotation-rotation interaction respectively, due to a modification of the electronic structure of the molecule by charge transfer. They should be compared with expressions (14) and (17), respectively, for the neutral molecule. We have numerically calculated the elements $J_{ll'}^{\tau\tau'}(\vec{n}, ij;\, \vec{n}', i'j')$, finding that they are negligible in comparison with the elements $J_{ll'}^{\tau\tau'}(\vec{n} - \vec{n}')$ of the orientational pair potential (12) of the neutral molecule. On the other hand we will find (Sec. 8) that modifications to the crystal field as a result of electron charge transfer are relatively important.

7. PHASE TRANSITIONS IN MC_{60}

We now study the interactions that drive the phase transition from the orientationally disordered phase $Fm\bar{3}m$ to the polymer phase[19] $Pmnn$ in MC_{60}. We also compare this phase transition with the $Fm\bar{3}m \to Pa\bar{3}$ transition[4-6] in solid C_{60}. In order to describe collective phenomena we introduce Fourier transforms of the rotator functions and of the orientational interaction, Eqs. (15) and (16). Here $\vec{q}$ is a wave vector and $\vec{X}(\vec{n})$ is the lattice vector at site $\vec{n}$. Since the effect of charge transfer on the orientational pair potential is negligible, we write the potential as in (17). The orientational Landau free energy, analogous to the case of C_{60}-fullerite[31] reads

$$F = \frac{1}{2} \sum_{\vec{q}} \sum_{ll'}{}' \sum_{\tau\tau'} [T\,(x^{(2)})^{-1} + J(\vec{q})]_{ll'}^{\tau\tau'}\, U_l^{\tau 1\,e}(\vec{q})\, U_l^{\tau' 1\,e}(-\vec{q}) + H[U^e], \tag{62}$$

where

$$x^{(2)} \equiv [\langle U_l^{\tau 1}(\vec{n})\, U_{l'}^{\tau' 1}(\vec{n})\rangle_{CF}] \tag{63}$$

is a single particle expectation value to be calculated with the crystal field and $H[U^e]$ refers to higher order terms. The superscript e denotes an instantaneous expectation value. Notice that the effect of charge transfer on the crystal field has to be included. If the transition were of second order, the transition temperature T_c would be given by the largest eigenvalue of $-[x^{(2)}J(\vec{q})]$. We observe that the largest eigenvalue is found for $\vec{q}$ at the X point of the Brillouin zone and τ corresponding to the irreducible representation $\Gamma = T_{2g}$, $\rho = 3$, belonging to the manifold $l = 10$. There are couplings among the T_{2g} irreducible representations at the X point, discussed in Secs. 3 and 4, but in a first approach we will restrict ourselves to the most important irreducible representation. We recall that the three components of T_{2g} symmetry transform under the operation of O_h as the Cartesian tensors yz, zx and xy for $i = 1, 2$ and 3 respectively. We denote the components of the irreducible representation T_{2g}, $\rho = 3$ and $l = 10$, by $U^{(i)}(\vec{q})$. At the X point two of the three eigenvalues are degenerate. For example at $\vec{q}_x^X = (2\pi/a)(1,0,0)$ the components $i = 2$ and $i = 3$ are degenerate and the functions $U^{(2)}(\vec{q}_x^X)$ and $U^{(3)}(\vec{q}_x^X)$ form a two dimensional basis E_g of the little group D_{4h} of $\vec{q}_x^X$. Similarly the pairs of functions $U^{(1)}(\vec{q}_y^X)$, $U^{(3)}(\vec{q}_y^X)$ and $U^{(2)}(\vec{q}_z^X)$, $U^{(1)}(\vec{q}_z^X)$ are degenerate at $\vec{q}_y^X$ and $\vec{q}_z^X$ respectively. The functions $U^{(i)}(\vec{q}_\alpha^X)$, where $i = 2, 3$ for $\alpha = x$ etc., form a six dimensional representation (called[42] X_5^+ or $\hat{\tau}^9$) of the space group $Fm\bar{3}m$. In C_{60}-fullerite the condensation scheme for the transition to the $Pa\bar{3}$ phase reads[31]

$$\begin{aligned} Fm\bar{3}m:\ (\vec{q}^X,\ X_5^+,\ & U^{(3)e}(\vec{q}_x^X) = U^{(1)e}(\vec{q}_y^X) = U^{(2)e}(\vec{q}_z^X) \equiv \eta\sqrt{N} \neq 0; \\ & U^{(2)e}(\vec{q}_x^X) = U^{(3)e}(\vec{q}_y^X) = U^{(1)e}(\vec{q}_z^X) = 0) \to Pa\bar{3}, \end{aligned} \tag{64}$$

and here three arms of the star of $\vec{q}^X$ are involved.

For the alkali fullerides the following condensation scheme to an orthorhombic structure has been proposed[65]

$$\begin{aligned} Fm\bar{3}m:(\vec{q}^X,\ X_5^+,\ & U^{(3)e}(\vec{q}_x^X) = U^{(2)e}(\vec{q}_x^X) = U^{(3)e}(\vec{q}_y^X) = U^{(1)e}(\vec{q}_y^X) = 0; \\ & U^{(1)e}(\vec{q}_z^X) = -U^{(2)e}(\vec{q}_z^X) \equiv \eta\sqrt{N} \neq 0) \to Pmnn. \end{aligned} \tag{65}$$

This involves the freezing out of two rotator functions for the arm $\vec{q}_z^X$ only. The condensation scheme (65) corresponds to one of twelve possible domains. We see that in the transitions to $Pmnn$ and $Pa\bar{3}$, the same irreducible representation of $Fm\bar{3}m$ is involved. Condensation schemes of the form (64) and (65) have previously been established[67] for orientational phase transitions $Fm\bar{3}m \to Pa\bar{3}$ and $Fm\bar{3}m \to Pmnn$ in NaO_2.

The orientational interactions between C_{60} molecules in pristine C_{60} and between molecular ions C_{60}^- in MC_{60} consist essentially of repulsive Born–Mayer (BM) and attractive Van der Waals (W) potentials, Eqs. (10) and (11). However, due to the different cubic lattice constants ($a = 14.08$ Å for RbC_{60} and $a = 14.16$ Å for solid C_{60}), the strengths of the interactions and hence of the matrix elements $J^{3,3}_{10,10}(\vec{q}^X_x) \equiv J^X$ are different. Taking the same potential interaction parameters but different values for a, we find $J^X = -7181$ K for MC_{60} and $J^X = -5187$ K for pristine C_{60}. The quantity $x^{(2)} = \langle (U^{(3)}(\vec{n}))^2 \rangle_{CF}$ has values of 0.0449 and 0.0433 for MC_{60} and C_{60} respectively. Hence the virtual second order transition temperatures are $T_c^{MC_{60}} = 322$ K and $T_c^{C_{60}} = 225$ K. The smaller lattice constant of MC_{60} is due to the fact that in an ionic crystal a decrease of the lattice constant favors a more negative Madelung energy. A small lattice constant in turn leads to a larger value of the repulsive BM interactions which are responsible for the antiferrorotational (X-point) ordering.

We now discuss the symmetry of the order parameter for the $Pmnn$ phase. For a molecule at lattice site $\vec{n}$ the condensation scheme (65) leads to

$$U^{(1)e}(\vec{n}) \equiv \langle U^{(1)}(\vec{n}) \rangle = \eta \, \cos(\vec{q}^X_z \cdot \vec{X}(\vec{n})), \tag{66}$$

$$U^{(2)e}(\vec{n}) \equiv \langle U^{(2)}(\vec{n}) \rangle = -\eta \, \cos(\vec{q}^X_z \cdot \vec{X}(\vec{n})). \tag{67}$$

Angle brackets stand for thermal averages of the MC_{60} crystal in the ordered phase. The lattice vectors $\vec{X}(\vec{n})$ specify the center of mass positions of the molecules. We distinguish two sublattices which are characterized by different molecular orientations; with respect to the cubic system of axes, we label by I the sublattice containing the origin 0 0 0 and by II the sublattice containing 0 1/2 1/2. We then obtain from Eqs. (66) and (67) on sublattice I

$$\langle U^{(1)}(\vec{n}_I) \rangle = \eta, \quad \langle U^{(2)}(\vec{n}_I) \rangle = -\eta, \tag{68}$$

and on sublattice II

$$\langle U^{(1)}(\vec{n}_{II}) \rangle = -\eta, \quad \langle U^{(2)}(\vec{n}_{II}) \rangle = \eta. \tag{69}$$

If we use the orthorhombic system of axes with basis vectors $\vec{a}_o$, $\vec{b}_o$ and $\vec{c}_o$ of Ref.[19], where the center of mass displacements are taken into account, the site 1/2 1/2 1/2 belongs to sublattice II. We notice that all of the C_{60}^- particles in the same basal plane $(\vec{a}_o, \vec{b}_o)$ have the same orientation, but orientations alternate in neighboring basal planes separated by the distance $c_o/2$.

It is instructive to analyze the symmetry reduction $Fm\bar{3}m \rightarrow Pmnn$ by considering the corresponding orientational distribution function. In the disordered phase the thermally averaged orientational density distribution of the carbon nuclei is given by[32]

$$\rho(\Omega) = \frac{60}{4\pi} + \sum_l{}' \sum_{\tau_{1g}} \gamma_l^{\tau_{1g}} S_l^{\tau_{1g}}(\Omega) \tag{70}$$

with

$$\gamma_l^{\tau_{1g}} = g_l^C \langle U_l^{\tau_{1g}}(\vec{\omega}) \rangle. \tag{71}$$

Here $l = 6, 10, ..$, the subscript τ_{1g} refers to the irreducible representations A_{1g} of O_h, the thermal averages are the same on all fcc lattice sites, and Ω is the direction of observation. The function $\rho(\Omega)$ is invariant under the operations of $Fm\bar{3}m$ and can be identified as the Landau density function[66] of the disordered phase. In the ordered phase ($Pmnn$) the symmetry reduction leads to a new contribution[66] $\delta\rho$ to the orientational

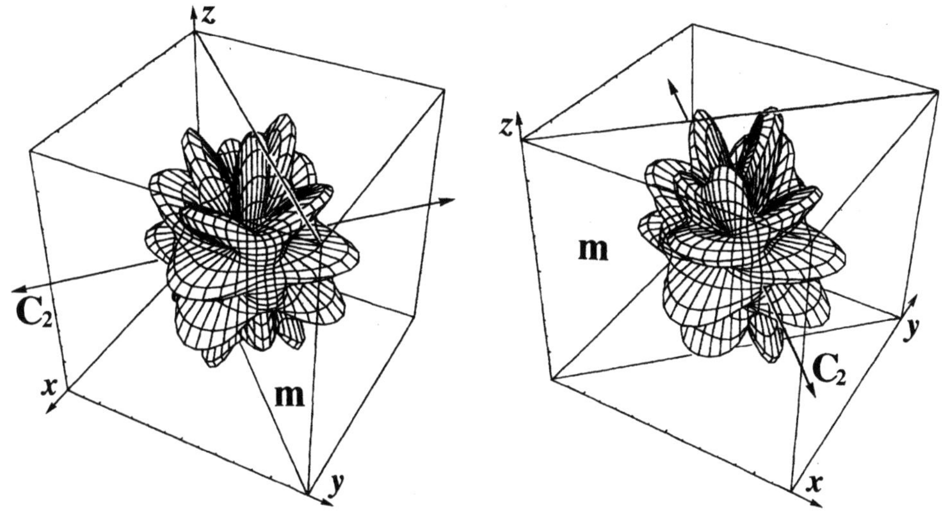

Figure 8. Two views of the function $(S_-(\Omega)+1.5)$, which is closely related to the function $\delta\rho_\sigma(\Omega)$. The $C_{2h}(2/m)$ site symmetry of the function is apparent from these pictures.

distribution, which now depends on the sublattice. We label the sublattices by an index σ =I, II and write

$$\delta\rho_\sigma(\Omega) = g_{10}^C \sum_{\alpha=1}^{2} \eta_\sigma^{(\alpha)} S^{(\alpha)}(\Omega). \tag{72}$$

Here α labels the two components of $\vec{\eta}_I \equiv (\eta, -\eta)$ and $\vec{\eta}_{II} \equiv (-\eta, \eta)$ and $S^{(\alpha)}$ denotes $S_{10}^{T_{2g},\rho=3,1}$ for $\alpha = 1$ and $S_{10}^{T_{2g},\rho=3,2}$ for $\alpha = 2$. We then have

$$\delta\rho_I(\Omega) = g_{10}^C \eta S_-(\Omega) \tag{73}$$

and

$$\delta\rho_{II}(\Omega) = -g_{10}^C \eta S_-(\Omega), \tag{74}$$

where

$$S_-(\Omega) = S_{10}^{T_{2g},3,1} - S_{10}^{T_{2g},3,2}. \tag{75}$$

Since $\delta\rho$ must be invariant under the symmetry operations of the low temperature phase, it should be possible to obtain the symmetry of this phase by studying the symmetry properties of the function S_-. This function plays the role of an order parameter variable[66] and belongs to the unit representation of *Pmnn*.[63] Unlike the situation in C_{60}-fullerite, the condensation scheme (65) comprises only functions of T_{2g} symmetry[63] and excludes those of T_{1g}.

The symmetry properties of the order parameter are visualized in Fig. 8, where the orientational distribution $(S_-(\Omega)+1.5)$ is plotted as a function of Ω. The site symmetry $C_{2h}(2/m)$ of the polymer phase is clearly visible.

8. CRYSTAL FIELD AND MOLECULAR FIELD IN MC_{60}

We shall now investigate the role of the charge transfer induced crystal field which leads to preferential molecular orientations in the disordered phase. We will then consider preferential orientations induced by the molecular field in the orthorhombic phase. We will study the eigenvalues of the secular equation (56) with p=0.36. We start with a molecule in a standard orientation where three of the twofold molecular axes coincide with the cubic crystal axes and the double bond intersected by the x-axis is parallel to the z-axis, as in Fig. 1. The [111] direction of the crystal is a threefold axis of the molecule.[6] Rotating the molecule counterclockwise about this axis, we have calculated eigenvalues as a function of the rotation angle Ψ. The result is shown in Fig. 9A.

The lifting of the degeneracy of the three t_{1u} levels (Fig. 9A) is incomplete. One of the curves remains doubly degenerate because the rotation axis retains symmetry elements (group C_3) shared by the molecule and the crystal, and the group C_3 has two dimensional irreducible representations. We observe that the level splitting between (1) and (2) in Fig. 9A is large. In Fig. 9B we show the crystal field $V^0_{CF}(\Psi)$, Eq. (14), which is experienced by the molecule in the absence of charge transfer; we also plot the total crystal field $V_{CF}(\Psi) = V^0_{CF}(\Psi) + E^{el}_l(\Psi)$. Here $E^{el}_l(\Psi)$ denotes the lower eigenvalue in Fig. 9A for the given value of Ψ. In addition to the local minimum at $\Psi = 0°$, V^0_{CF} has a weak local minimum M_2 at 38° and a global minimum M_1 at 98°. The occurrence of these minima in the crystal field potential of the disordered phase is well known in C_{60}-fullerite,[44, 12]; see also Section 4. The global minimum is a precursor of the 98° preferred orientation in the $Pa\bar{3}$ phase.[5] The local M_2 minimum becomes more pronounced with decreasing of T in the $Pa\bar{3}$ phase and ultimately leads to low temperature trapping of about 15% of the molecules in the 38° orientation. Perfect orientational order of the $Pa\bar{3}$ structure is prevented, and an orientational glass[52] appears in C_{60}-fullerite at low T. On the other hand we see from Fig. 9B that for the case of MC_{60}-fulleride, even in the disordered phase, charge transfer has relatively little effect on the global minimum M_1 but strongly deepens the minimum M_2. The magnitude of this effect clearly depends on the electrostatic potential reduction coefficient p, Eq. (44). Indeed for $p \geq 0.53$, M_2 becomes the global minimum. We conclude that the electronic crystal field contribution, depending on its magnitude, can actually suppress the $Fm\bar{3}m \rightarrow Pa\bar{3}$ transition in MC_{60}-fulleride. On the other hand it has been reported that a $Pa\bar{3}$ structure is reached in RbC_{60} and CsC_{60} on deep quenching.[68, 69]

We now examine the possibility that certain molecular orientations preferentially occur, guided in a large part by crystal structure studies of the $Pmnn$ phase.[19] We start again from the standard orientation, rotate the molecule clockwise by 45° about the z-axis, bringing the molecular y axis into coincidence with the orthorhombic basis vector $\vec{a}_o$. We call this situation the orthorhombic starting orientation. Taking the $\vec{a}_o$ direction (which also corresponds to the cubic [110] direction) as axis of rotation, we study the eigenvalues of the secular equation as a function of the rotation angle Φ. The result is shown in Fig. 10A. We see that the threefold degeneracy is completely lifted and the level splitting is again large.

The crystal field $V^0_{CF}(\Phi)$ and the sum $V_{CF}(\Phi) = V^0_{CF}(\Phi) + E^{el}_l(\Phi)$ are plotted (for p=0.36) in Fig. 10B. The shape of the crystal field V^0 is again close to that reported for C_{60}-fullerite, with degenerate minima L_1 and L_2 at $\Phi = 45°$ and 135° respectively, and degenerate local minima L_0 and L'_0 at 0° and 90° respectively. Charge transfer results in the global and local minima becoming almost equally deep.

In order to study the preferential orientations that lead to the low temperature

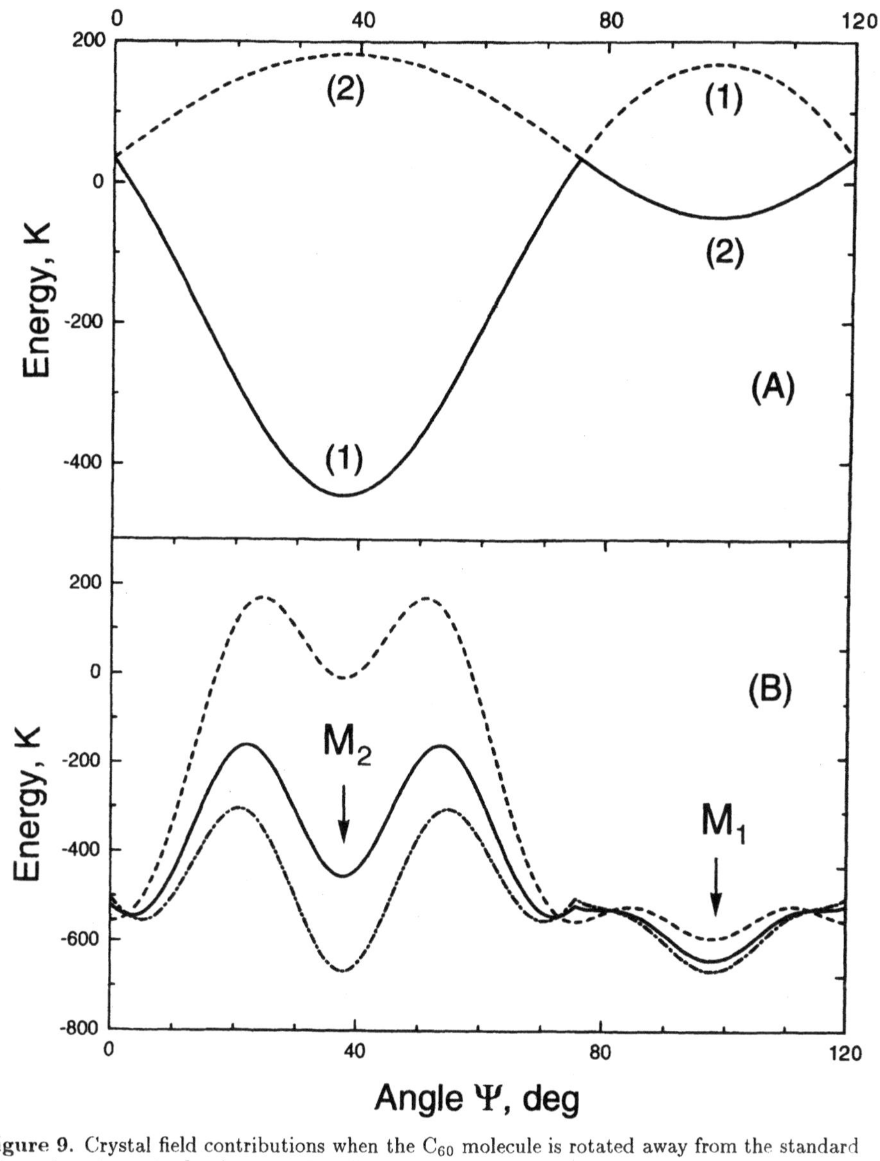

Figure 9. Crystal field contributions when the C_{60} molecule is rotated away from the standard orientation about the [111] axis of the cubic system. Global M_1 and secondary M_2 minima are shown. (A) Valence electron density contribution $V_{CF}^{el}(\Psi)$, with p=0.36. The triply degenerate t_{1u} levels are split into a single level and a doubly degenerate level, labeled (1) and (2) respectively. (B) The dashed line shows the core contribution $V_{CF}^{0}(\Psi)$ (analogous to pure C_{60}). The full line is the total crystal field of MC_{60} with p=0.36. The dash-dotted line corresponds to the total crystal field with p=0.53.

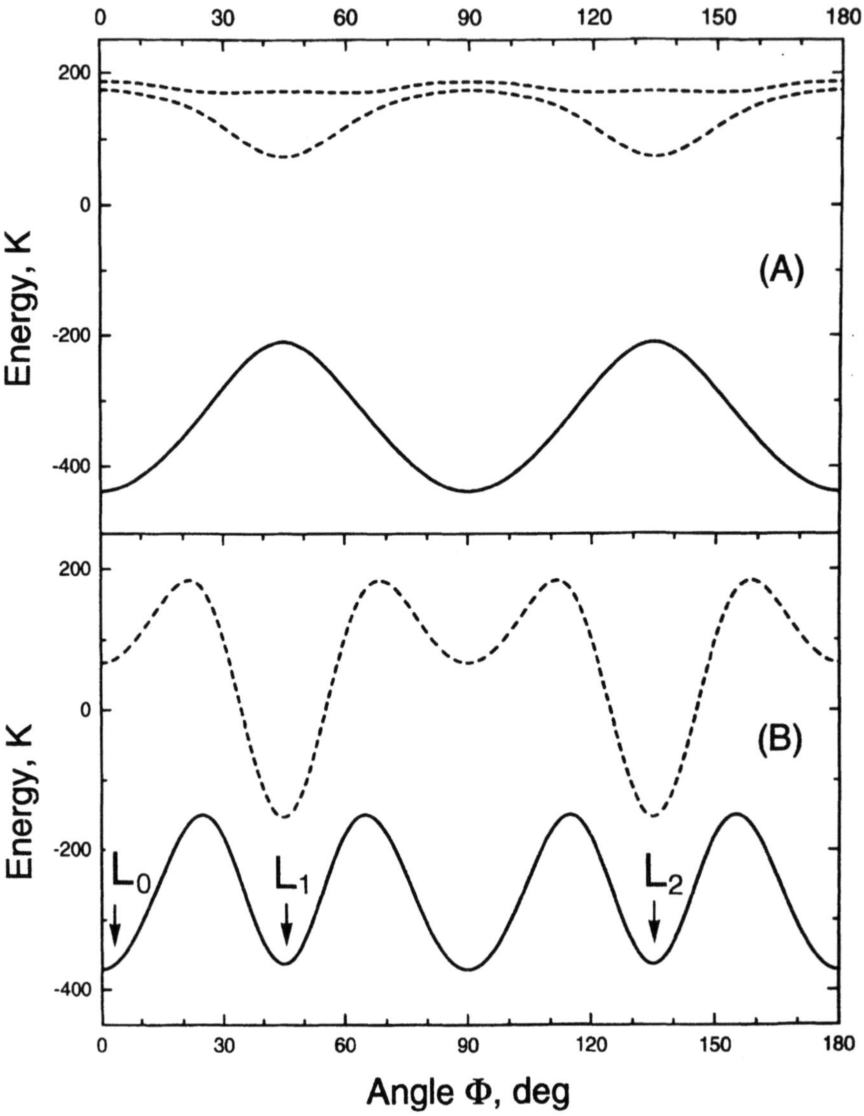

Figure 10. Crystal field contributions when the C_{60} molecule is rotated away from the orthorhombic starting orientation about the [110] axis of the cubic system, p=0.36. Three minima, L_0, L_1, and L_2, are indicated. (A) Valence electron density contribution $V_{CF}^{el}(\Phi)$, showing the splitting of the three t_{1u} levels in the cubic environment. (B) The core contribution $V_{CF}^{0}(\Phi)$ (analogous to pure C_{60}) is shown as the dashed line, and the total crystal field of MC_{60} is shown as the full line.

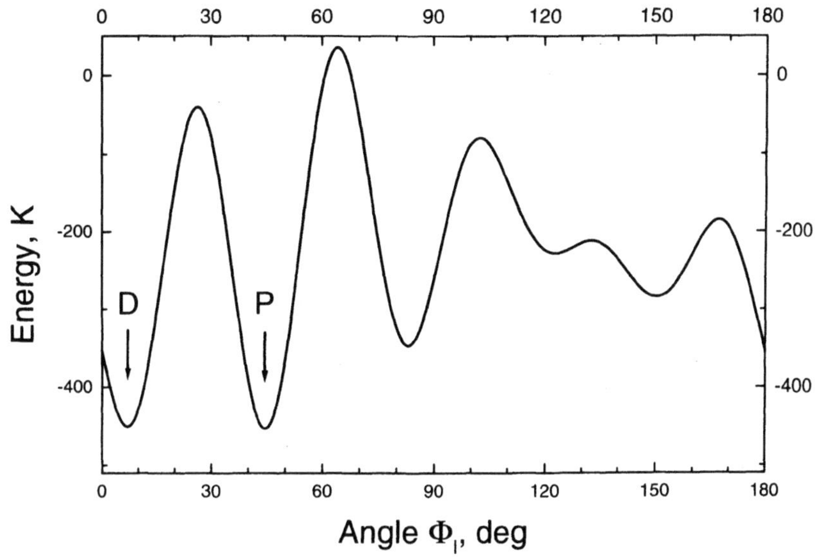

Figure 11. The molecular field $V^{MF}(\Phi_I)$ (Eq. (78)) when the C_{60} molecule is rotated away from the orthorhombic starting orientation about the [110] axis of the cubic system, with p=0.36. Adding the pair potential V_{RR}^{MF} (Eq. (76)) to the crystal field (which is shown in Fig. 10B), we see that minima develop at 45° and $\approx 9°$, corresponding to competitive ortho-1 (polymer) and ortho-2 (intermediate) phases respectively.

orthorhombic phase we introduce the molecular field at a given lattice site $\vec{n}$. Considering the condensation scheme (65) as well as Eqs. (17), (15), (66) and (67), we obtain the orientational pair potential

$$V_{RR}^{MF}(\vec{n}) = J^X \eta\, U_-(\vec{n})\, \cos(\vec{q}_z^{\,X} \cdot \vec{X}(\vec{n})), \tag{76}$$

where (c. f. Eq. (75))

$$U_-(\vec{n}) = U^{(1)}(\vec{n}) - U^{(2)}(\vec{n}). \tag{77}$$

Distinguishing between the two sublattices as in Sec. 7, and adding the total crystal field $V_{CF}(\vec{n})$, we obtain molecular field potentials

$$V^{MF}(\vec{\omega}_I) = J^X \eta\, U_-(\vec{\omega}_I) + V_{CF}(\vec{\omega}_I) \tag{78}$$

and

$$V^{MF}(\vec{\omega}_{II}) = -J^X \eta\, U_-(\vec{\omega}_{II}) + V_{CF}(\vec{\omega}_{II}) \tag{79}$$

at sites on sublattices I and II, respectively. For molecular rotations about the cubic [111] axis, the first term in the right hand side of Eq. (78) (or Eq. (79)) is unaffected so that the molecular field coincides with the crystal field (see Figs. 9A,B). We consider rotations about the $\vec{a}_o$ axis (the cubic [110] axis), starting from the orthorhombic starting orientation. Taking $J^X\,\eta = 350$ K we plot V^{MF}, Eq. (78), as a function of the rotation angle Φ_I in Fig. 11.

In Fig. 11 we observe the development of two competitive orientations with Φ_I at 45° and $\approx 9°$, which we associate with ortho-1 and ortho-2 phases, respectively. These

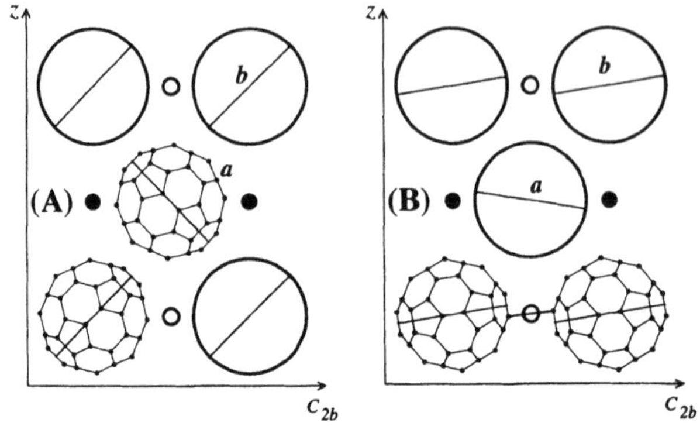

Figure 12. The orthorhombic unit cell of MC_{60}. The central molecule and the two nearest M^+ ions (hollow circles) are in the (100) plane at $x = 1/2$. Surrounding molecules and M^+ ions (full circles) are in the (100) plane at $x = 0$. The x-axis C_{2a} is perpendicular to the plane of the picture. Positions of hexagon-hexagon fusing bonds for C_{60} molecules at the preferred orientations 45° and ≈ 9° are shown. (A) The ortho-1 (polymer) phase is accomplished by strong contraction along the C_{2a} axis (direction of polymerization). (B) The ortho-2 phase is accomplished by contraction along the C_{2b} axis (direction of dimerization). The closest contact between carbon atoms of two neighboring C_{60} molecules is shown.

minima result from overlap of minima of V_{RR}^{MF} with minima at L_1 and L_0 of the crystal field (see Fig. 10B, full curve). It is straightforward to associate the minimum L_1 with the polymer phase (ortho-1). Clearly the main effect of the order parameter terms in the molecular field equations, (78) and (79), is the breaking of the degeneracy of the L_1 and L_2 minima, stabilizing the first one on sublattice I and the second one on sublattice II. Assuming that the molecules occupy the most favorable orientations, corresponding to $\Phi = 45°$ for sublattice I and $\Phi = 135°$ for sublattice II, we obtain two chains of C_{60} molecules related through rotations by 90° about the diagonal axis C_{2a} (equivalently [110]) of the cubic coordinate system (see Fig. 12A). These findings are in full agreement with experiment.[19]

The transition $Fm\bar{3}m \rightarrow Pmnn$ is accompanied by lattice distortion, in particular by a contraction along the $\vec{a}_o$ axis or chain direction. Our calculations of lattice deformations depend on the magnitude of the bare elastic constants but generally indicate a tendency towards contraction along $\vec{a}_o$. Taking into account that the phase transition is accompanied by a contraction along the chains, we observe that polymerization and progressive formation of chemical bonding, although important, are secondary effects. At a later stage quantum chemistry and deformations[19] of C_{60} molecules should be taken into account, but *ab initio* calculations of these effects are beyond the scope of the present paper. However we want to stress that within the present scenario polymerization is a consequence of the cooperative changes (molecular rotations and center of mass displacements) associated with the structural phase transition. This view, in which cooperative orientational ordering and lattice distortions are a prerequisite for subsequent polymer and dimer formation, is supported by calorimetric results. As has been mentioned by Stephens et al,[19] a puzzle in the interpretation of the structure of the polymer phase of MC_{60} is its low stability. The enthalpy of transformation associated with the depolymerization[70] into fullerides with the rocksalt structure is only four times the enthalpy change[71] at the 250 K orientational phase transition in pristine C_{60}. From thermodynamic studies of polymorphism in MC_{60}, the authors of Ref.[70] conclude that "the rotation of the C_{60}^- ions seems to be a precondition of polymerization".

The origin of the ortho-2 minimum, at $\Phi_I \approx 9°$, is more intriguing. In Fig. 12B we see that the rotation by 9° about the C_{2a} axis brings the equatorial carbon atoms of two neighboring molecules to positions where the distance between them along the C_{2b} axis is small, reminiscent of the dimer phase. The closest contact would occur if the C_{60} molecules were rotated by $\approx 12°$ and it is easy to imagine that the formation of a chemical bond between the two C atoms could eventually lock the two molecules into a dimer. We are then led to associate the resulting ortho-2 phase (space group $Pmnn$) with an intermediate phase which is followed by a monoclinic distortion to the dimer phase. (It is worth noting that in early experiments [22] the dimer phase was reported as an ortho-2 phase.) Here again, as in the case of the ortho-1 (polymer) phase, we would like to emphasize that there are two stages in the scenario: first the cooperative effect of preferential orientation ($\Phi_I \approx 9°$), and then the formation of a chemical bond between closest C atoms ($\Phi_I \approx 12°$). Unlike the polymerization in the ortho-1 phase, here the step that leads to a monoclinic structure (space group $P2_1/a$) can itself be considered as a phase transition. On the other hand dimerization is not the only possibility for the ortho-2 phase at low temperature. Recently a phase comprising one-dimensional polymer chains of C_{60}^{3-} ions connected by single carbon-carbon bonds (as in the dimer phase of MC_{60}) has been reported[72] for Na_2RbC_{60}. The crystal structure is monoclinic (space group $P2_1/a$) with only a small distortion from orthorhombic symmetry. The value of the optimized angle Φ_I is $\approx 8°$. In both Na_2RbC_{60} and MC_{60} the order in the $\vec{b}_o\vec{c}_o$ mirror plane ($\vec{a}\vec{c}$ for monoclinic axes) is important while the angle of rotation Φ_I about the $\vec{a}_o$ axis varies depending on the system.

9. CONCLUSIONS

We have given a unified description of the orientational properties of pristine C_{60} in the disordered phase, involving the coupling of all representations belonging to the manifolds l=6, 10 and 12, throughout the Brillouin zone. We have introduced new collective orientational coordinates and have shown that the coupling of orientational fluctuations of T_{2g} and T_{1g} symmetry at the X point of the Brillouin zone leads to a strong enhancement of the transition temperature for the structural change $Fm\bar{3}m \rightarrow Pa\bar{3}$. In order to obtain a transition temperature close to the experimental value $\approx$260 K, we have revised a previous model[38] of the intermolecular potential. In particular the repulsive forces among double bond centers and atomic centers have been reduced while the attractive Van der Waals interactions between atomic centers have been increased. We have studied the eigenvalues and eigenvectors of the orientational pair potential and defined orientational order parameters for the transition $Fm\bar{3}m \rightarrow Pa\bar{3}$ in terms of normal coordinates. We have investigated the stability of the $Pa\bar{3}$ phase in comparison with a tetragonal phase and have found that the crystal field of the disordered phase plays an essential role in favoring a transition to the ordered phase $Pa\bar{3}$. We have calculated the diffuse scattering cross section in the disordered phase, taking into account molecular interactions for the manifolds l=6, 10 and 12 and treating the manifolds with $l > 12$ in the free molecule approximation. Our results are in a good agreement with the results of various diffuse scattering investigations.[9, 11, 12, 48, 47]

With regard to phase changes in MC_{60} fullerides we have presented a microscopic theory which leads to polymerization and dimerization in these compounds. In particular we have studied precursor effects and consequences of the structural phase transition from the orientationally disordered cubic phase $Fm\bar{3}m$ to the orientationally ordered orthorhombic phase $Pmnn$. While in C_{60}-fullerite the orientational phase transition

$Fm\bar{3}m \to Pa\bar{3}$ is driven by repulsive interactions between neutral molecules, additional interactions due to electronic charge transfer have been found to be of great importance in MC_{60}. Charge transfer has several consequences. Ionic forces in M^+-C_{60}^- lead to a decrease in the cubic lattice constant (in comparison with pristine C_{60}) which entails an increase in the repulsive interactions and hence an increase in the transition temperature for orientational ordering. Such a scenario alone would lead to a $Pa\bar{3}$ structure in MC_{60}, although with a higher T_c than for C_{60}. However, charge transfer modifies the electronic structure of the C_{60}^- molecular ion (in comparison with the neutral C_{60} molecule) and opens new channels for molecular ordering. Charge transfer results in partial occupation of the degenerate t_{1u} LUMO levels. We have studied the interaction between t_{1u} orbitals belonging to different molecules, and the interactions of t_{1u} orbitals with surrounding alkali metal ions. Since these interactions depend on molecular orientation, we have extended the formalism of rotator functions to a description of molecular orbitals (Sec. 6). We have applied perturbation theory to the case of degenerate t_{1u} levels and have solved the secular equation for the potential V_{CF}^{el}. The spectrum of eigenvalues depends on the instantaneous orientation of the central molecule, and as a general rule the original t_{1u} degeneracy is lifted. We observe that the largest contributions to V_{CF}^{el} are due to terms belonging to the manifold with angular momentum number $l = 4$. In the case of the neutral C_{60} molecule $l = 4$ terms are forbidden since there is no unit representation of I_h within the manifold $l = 4$. We have investigated the total molecular field potential, which includes the crystal field, in the orthorhombic phase (Sec. 8). For molecular rotations about the orthorhombic axis $\vec{a}_o$, away from the orthorhombic starting orientation, we find minima for the 45° and ≈9° orientations, which we associate with ortho-1 and ortho-2 phases. We have then concluded from structural considerations that the ortho-1 phase favors polymerization (in agreement with experimental results[19]), while the ortho-2 phase favors formation of singly C-C bonded dimers[22, 21] along the $\vec{b}_o$ axis with a monoclinic structure close to the orthorhombic one. In both cases we have proposed a two stage scenario where cooperative structural transformations (orientational ordering and lattice distortions) are a prerequisite for subsequent polymer and dimer formation.

The formalism of Sec. 6, combining the properties of rotator functions with changes of the electronic molecular structure, should be applicable in a broad range of charge transfer systems.

Acknowledgments

This work has been financially supported by Fonds voor Wetenschappelijk Onderzoek, Vlaanderen, and the Onderzoeksfonds, Universiteit Antwerpen, UIA. Part of this work was performed while A.V.N. was a visitor at the University of Sussex at Brighton. He thanks K.Prassides for hospitality during the visit and acknowledges financial support from the Engineering and Physical Sciences Research Council (EPSRC), UK.

REFERENCES

1. N. Parsonage, and L. A. K. Stavely. "Disorder in Crystals," Clarendon, Oxford (1978).
2. W.Krätschmer, L. D. Lamb, K. Fostiropoulos, and D. R. Huffman, *Nature* 347:354 (1990).
3. H. W. Kroto, J. R. Heath, S. C. O'Brien, R. F. Curl, and R. E. Smalley, *Nature* 318:162 (1985).
4. R. Sachidanandam and A. B. Harris, *Phys. Rev. Lett.* 67:1467 (1991); P. A. Heiney, J. E. Fischer, A. R. McGhie, W. J. Romanov, A. M. Denenstein, J. P. McCauley Jr., A. B. Smith III, and D. E. Cox, *Phys. Rev. Lett.* 67:1468 (1991).

5. W. I. F. David, R. M. Ibberson, J. C. Matthewman, K. Prassides, T. J. S. Dennis, J. P. Hare, H. W. Kroto, and D. R. M. Walton, *Nature* 353:147 (1991).
6. A. B. Harris, and R. Sachidanandam, *Phys. Rev.* B 46:4944 (1992).
7. "Physics and Chemistry of the Fullerenes," Edited by K. Prassides, NATO ASI Series C: Mathematical and Physical Sciences - vol.443, Kluwer, Dordrecht (1994).
8. H. Kuzmany, J. Fink, M. Mehring, and S. Roth (eds.) "Fullerenes and Fullerene Nanostructures," World Scientific, Singapore (1996).
9. P. Launois, S. Ravy, and R. Moret, *Phys. Rev.* B 52:5414 (1995).
10. O. Blaschko, R. Glas, Ch. Maier, M. Haluska, and H. Kuzmany, *Phys. Rev.* B 48:14638 (1993).
11. L. Pintschovius, S. L. Chaplot, G. Roth, and G. Heger, *Phys.Rev.Lett.* 75:2843 (1995).
12. J. D. Axe, S. C. Moss, and D. A. Neumann, *in*: "Solid State Physics," vol.48, p.149, H.Ehrenreich, F.Spaepen (eds.), Academic Press and Erratum, New York, P.Wochner, private communication (1994).
13. J. Winter, and H. Kuzmany, *Solid State Commun.* 84:935 (1992).
14. D. M. Poirier, T. R. Ohno, G. H. Kroll, P. J. Benning, F. Stepniak, J. H. Weaver, L. P. F. Chibante, and R. E. Smalley, *Phys. Rev. B* 47:9870 (1993).
15. A. Janossy, O. Chauvet, S. Pekker, J. R. Cooper, and L. Forro, *Phys. Rev. Lett* 71:1091 (1993).
16. D. A. Bochvar, E. G. Gal'perin, *Dokl. Akad. Nauk SSSR Chem.* 209:N.3, 610 (1973).
17. R. C. Haddon, L. E. Brus, and K. Raghavachari, *Chem. Phys. Lett.* 125:459 (1986); M. Braga, S. Larsson, A. Rosen, and A. Volosov, *Astron. Astrophys.* 245:232 (1991).
18. S. Pekker, L. Forro, L. Mihaly, and A. Janossy, *Solid State Commun.* 90:349 (1994).
19. P.W. Stephens, G. Bortel, G. Faigel, M. Tegze, A. Janossy, S. Pekker, G. Oszlanyi, and L. Forro, *Nature* (London) 370:636 (1994).
20. S. Pekker, A. Janossy, L. Mihaly, O. Chauvet, M. Carrard, and L. Forro, *Science* 265:1077 (1994).
21. G. Oszlanyi, G. Bortel, G. Faigel, L. Granasy, G. M. Bendele, P. W. Stephens, and L. Forro, *Phys. Rev.* B 54:11849 (1996).
22. G. Oszlanyi, G. Bortel, G. Faigel, M. Tegze, L. Granasy, S. Pekker, P. W. Stephens, G. Bendele, R. Dinnebier, G. Mihaly, A. Janossy, O. Chauvet, and L. Forro, *Phys. Rev.* B 51:12228 (1995); Q. Zhu, D. E. Cox, and J. E. Fischer, *Phys. Rev.* B 51:3966 (1995).
23. C. J. Bradley, and A. P. Cracknell. "The Mathematical Theory of Symmetry in Solids," Clarendon, Oxford (1972).
24. H. Bethe, *Ann.Physik* 3:133 (1929); F. C. Von der Lage, and H. A. Bethe, *Phys. Rev.* 71:612 (1947).
25. A. F. Devonshire, *Proc. Roy. Soc. London* A153:601 (1936).
26. M. Tinkham. "Group Theory and Quantum Mechanics," McGraw-Hill, New York (1964).
27. H. M. James, and T.A.Keenan, *J. Chem. Phys.* 31:12 (1959).
28. W. Press, and A. Hüller, *Acta Crystallogr.* A29:252 (1973a).
29. M. Yvinec, and R. M. Pick, *J. Phys.* (Paris) 41:1045 (1980).
30. K. H. Michel, and K. Parlinski, *Phys. Rev.* B 31:1823 (1985).
31. K. H. Michel, J. R. D. Copley, and D. A. Neumann, *Phys. Rev. Lett.* 68:2929 (1992); K. H. Michel, *Z. Phys. B Cond. Matter* 88:71 (1992).
32. J. R. D. Copley, and K. H. Michel, *J. Phys. Condens. Matter* 5:4353 (1993).
33. M. Sprik, A. Cheng, and M. L. Klein, *J. Phys. Chem.* 96:2027 (1992).
34. Q. -M. Zhang, J. -Y. Yi, and J. Bernholc, *Phys. Rev. Lett.* 66: 2633 (1991).
35. K. H. Michel, *J.Chem. Phys.* 97:5155 (1992).
36. N. V. Cohan, *Proc. Camb. Phil. Soc. Math. Phys. Sci.* 54:28 (1957).
37. S. Savin, A. B. Harris, and T. Yildirim, *Phys. Rev.* B 55:14182 (1997).
38. D. Lamoen, and K. H. Michel, *J. Chem. Phys.* 101:1435 (1994).
39. T. Yildirim, A. B. Harris, S. C. Erwin, and M. R. Pederson, *Phys. Rev.* B 48:1888 (1993).
40. K. H. Michel, and J. R. D. Copley, *Z. Phys. B Cond. Matter* 103:369 (1997).
41. D. A. Neumann, J. R. D. Copley, R. L. Cappelletti, W. A. Kamitakahara, R. M. Lindstrom, K. M. Creegan, D. M. Cox, W. J. Romanow, N. Coustel, J. P. McCauley Jr., N. C. Maliszewskyj, J. E. Fischer, and A. B. Smith III, *Phys. Rev. Lett.* 67:3808 (1991).
42. H. T. Stokes, and D. M. Hatch. "Isotropy Subgroups of the Crystallographic Space Groups," World Scientific, Singapore (1988).
43. R. Heid, *Phys. Rev.* B 47:15912 (1993).
44. D. Lamoen, and K. H. Michel, *Z. Phys.* B 92:323 (1993).
45. P. C. Chow, X. Jiang, G. Reiter, P. Wochner, S. C. Moss, J. D. Axe, J. C. Hanson, R. K. McMullen, R. L. Meng, and C. W. Chu, *Phys. Rev. Lett.* 69:2943 (1992); H. -B. Bürgi, R. Restori, and D. Schwarzenbach, *Acta Crystallogr.* B 49:832 (1993).

46. W. I. F. David, R. M. Ibberson, and T. Matsuo, *Proc. Roy. Soc. London Ser.* A 442:129 (1993).
47. P. Launois, S. Ravy, and R. Moret, *Phys. Rev.* B 55:2651 (1997).
48. S. Ravy, P. Launois, and R. Moret, *Phys. Rev.* B 53:10532 (1996).
49. J. P. Lu, X. -P. Li, and R. M. Martin, *Phys. Rev. Lett.* 68:1551 (1992).
50. P. Schiebel, K. Wulf, W. Prandl, G. Heger, R. Papoular, and W.Paulus, *Acta Crystallogr.* A 52:176 (1996).
51. H. -B. Bürgi, E. Blanc, D. Schwarzenbach, S. Lin, Y. -J. Lu, M. M. Kappes, and J. A. Ibers, *Angew. Chem. Int. Ed. Engl.* 31:640 (1992).
52. W. I. F. David, R. M. Ibberson, T. J. S. Dennis, J. P. Hare, and K. Prassides, *Europhys. Lett.* 18:219 (1992).
53. P. A. Heiney, G. B. M. Vaughan, J. E. Fischer, N. Coustel, D. E. Cox, J. R. D. Copley, D. A. Neumann, W. A. Kamitakahara, K. M. Creegan, D. M. Cox, J. P. McCauley, Jr., and A. B. Smith III, *Phys. Rev.* B 45:4544 (1992); J. R. D. Copley, W. I. F. David, and D. A. Neumann, *Neutron News* 4:21 (1993).
54. J. R. D. Copley, and K. H. Michel, to be published. A first account has been given by these authors in the Proceedings of the International Conference on Neutron Scattering, Toronto 1997, to be published in Physica B (1998).
55. R. Moret, P. A. Albony, V. Agafonov, R. Ceolin, D. André, A. Dworkin, H. Swarc, C. Fabré, A. Rassat, A. Zahab, and P. Bernier, *J. Phys.*I France 2:511 (1992).
56. R. Moret, P. Launois, and S. Ravy, *Fullerene Science and Technology* 4:1298 (1996).
57. K. Sakaue, N. Toyoda, H. Kasatani, H. Terauchi, T. Arai, Y. Murakami, and H. Suematsu, *J. Phys. Soc. Japan*, 63:1237 (1994).
58. Q. Zhu, O. Zhou, J. E. Fischer, A. R. McGhie, W. J. Romanow, R. M. Strongin, M. A. Cichy, and A. B. Smith III, *Phys. Rev.* B 47:13948 (1993).
59. O. Chauvet, G. Oszlanyi, L. Forro, P. W. Stephens, M. Tegze, G. Faigel, and A. Janossy, *Phys. Rev. Lett.* 72:2721 (1994).
60. B. Renker, H. Schober, and R. Heid, *Appl. Phys.*A 64:271 (1997).
61. B. Renker, F. Gompf, R. Heid, P. Adelmann, A. Heiming, W. Reichardt, G. Roth, H. Schober, and H. Rietschel, *Z. Phys.*B 90:325 (1993).
62. R. Tycko, G. Dabbagh, D. W. Murphy, Q. Zhu, and J. E. Fischer, *Phys. Rev.* B 48:9097 (1993).
63. A. V. Nikolaev, K. Prassides, and K. H. Michel, *J. Chem. Phys.* 108:4912 (1998).
64. P. W. Atkins, and R. S. Friedman. "Molecular Quantum Mechanics," 3rd edition, Oxford University Press, Oxford (1997).
65. V. L. Aksenov, V. S. Shakhmatov, and Y. A. Osipyan, *JETP Lett.* 62:428 (1995); V. L. Aksenov, V. S. Shakhmatov, and Y. A. Osipyan, *JETP Lett.* 64:120 (1996).
66. L. D. Landau, *Phys. Z. Sowjetunion* 11:26545 (1937); L. D. Landau and E. M. Lifshitz. "Statistical Physics," Vol.5 Pergamon, Bristol (1995).
67. P. Zielinski, and K. Parlinski, *J. Phys.* C 17:3301 (1984).
68. M. Kosaka, K. Tanigaki, T. Tanaka, T. Atake, A. Lappas, and K. Prassides, *Phys. Rev.* B 51:12018 (1995).
69. A. Lappas, M. Kosaka, K. Tanigaki, and K. Prassides, *J. Am. Chem. Soc.* 117:7560 (1995).
70. L. Granasy, S. Pekker, and L. Forro, *Phys. Rev.* B 53:5059 (1996).
71. T. Matsuo, H. Suga, W. I. F. David, R. M. Ibberson, P. Bernier, A. Zahab, C. Fabre, A. Rassat, and A. Dworkin, *Solid State Commun.* 83:711 (1992); G. Pitsi, J. Caerels, and J. Thoen, *Phys. Rev.* B 55:915 (1997).
72. K. Prassides, K. Vavekis, K. Kordatos, K. Tanigaki, G. M. Bendele, and P. W. Stephens, *J. Am. Chem. Soc.* 119:834 (1997); G. M. Bendele, P. W. Stephens, K. Prassides, K. Vavekis, K. Kordatos, and K. Tanigaki, *Phys. Rev. Lett.* 80:736 (1998).

DESORPTION INDUCED BY ELECTRONIC TRANSITIONS: APPLICATION TO IONIC COMPOUNDS

Mario Piacentini[1,2] and Nicola Zema[2]

[1] Dipartimento di Energetica, Università di Roma La Sapienza, and Istituto Nazionale di Fisica della Materia, Sezione di Roma 1, Rome, Italy
[2] Istituto di Struttura della Materia, Area di Ricerca del CNR a Tor Vergata, Rome, Italy

1. INTRODUCTION

The ejection of atoms and molecules from the surface of solids subject to the bombardment of energetic particles is a well known phenomenon that is employed routinely in laboratories and industrial plants for cleaning a surface, determining surface composition or ejecting material from a solid for deposition on adjacent targets. This emission, called sputtering, is the result of a direct momentum transfer from the incident particles to the atoms of the solid.

Also low intensity beams of photons and electrons produce the emission of constituent atoms or surface adsorbates (photon– or electron– stimulated desorption, respectively). However, in this case the energy of the incident particles is substantially lower than that necessary to create the elastic knock–on process. Thus, the desorption is triggered by electronic transitions induced by the incident photons or electrons.

Several mechanisms have been proposed for desorption. One model proposed for the desorption of adsorbed species derives from the dissociation theory of gas–phase molecules: desorption is the result of an electronic transition from the bonding ground state to an anti bonding excited state of a surface atom or molecule with final energy above the dissociation limit.[1, 2] Another model has been proposed for several transition metal oxides.[3] For these materials the electron–stimulated desorption yield of positive ions (O^+, OH^+) has a threshold at the highest–lying atomic core levels. The interatomic Auger decay of the excited ion core state removes two valence electrons from the anion, which is transformed into a neutral atom or even in a positive ion. If this ion is on the surface, it will experience a repulsive force from the neighbouring cations and then it will be desorbed.

The desorption of constituent particles from alkali halides does not fall in either scheme. The electronic transition occurring in the bulk material is followed by several relaxation processes that may involve atomic displacements and defect formation. The three step model that is generally adopted for desorption is sketched in Fig. 1:

Correlations, Coherence, and Order. Edited by Shopova and Uzunov
Kluwer Academic/Plenum Publishers, 1999

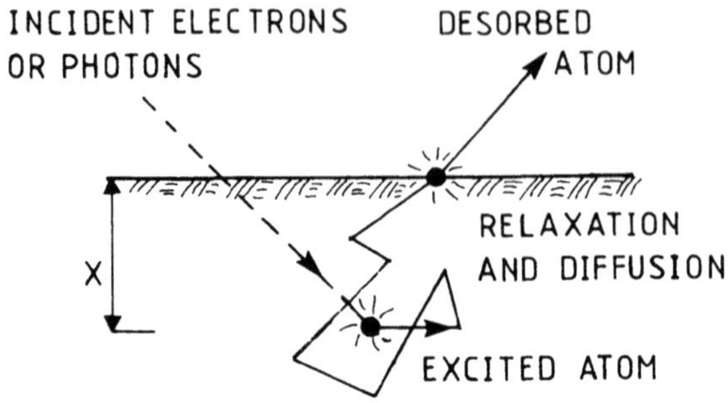

Figure 1. Photon– or electron–stimulated desorption model. An incident particle excites a halogen atom at depth x from the surface. The excited atom performs a random walk and possibly it reaches the surface, from which it can be desorbed directly or induce the emission of alkali atoms.

1. *Electronic transition.* An incident particle, either a photon or an electron, excites an atom inside the crystal.

2. *Defect formation and transport.* Either the excited state migrates directly near the crystal surface, or it decays first into lattice defects, which migrate to the surface.

3. *Desorption.* When the excited state reaches the surface, an atom can be emitted.

In spite of the large amount of photon– and electron– stimulated desorption experiments performed on alkali halides,[4, 5] the entire set of processes sketched in Fig. 1 is not yet completely understood. The main experimental findings reported in the literature are:

i) The emission of both alkali and halogen neutral atoms in the ground state is indeed the most efficient process, being some orders of magnitude larger than the emission of excited atoms. The desorption of excited alkali atoms as well as of alkali ions have been observed too.

ii) The desorption yield is temperature dependent. At room temperature it is very small and it increases fast with the sample temperature.

iii) At room temperature the sample surface becomes rich of alkali metal. This layer evaporates above the boiling point of the alkali metal and the crystal surface becomes stoichiometric.

iv) The desorption yield of both atomic species is a function of the energy of the incoming particles. It shows a threshold at the onset of the absorption coefficient of the alkali halide sample. Further thresholds and structures occur at the energies of multiple excitations and at the energies of core level thresholds.

v) The velocity distribution of the desorbed neutral atoms follows the Maxwell–Boltzmann distribution of a gas of free particles with the sample temperature. In several, but not all, alkali halides the desorption of the halogen neutral atoms shows an additional non–thermal component, that becomes dominant at room temperature.

vi) Alkali and halogen atoms with thermal energy are emitted isotropically with a cosine distribution around the normal to the sample surface, whereas the non–thermal halogen component is emitted along the $\langle 100 \rangle$ direction.

vii) Time dependent measurements show a delay in the desorption process, with at least three different time constants: a very fast one in the nanosecond region, a fast component, of the order of 10^{-3} s, and a slow component of the order of several seconds.

In this review we shall focus our attention to the desorption of neutral atoms from alkali halides following a historical approach. At the end we shall discuss the state of the art of the understanding of desorption induced by electronic transitions in these materials and we shall report also on some not well understood aspects, such as the metalization of the sample surface around room temperature.

2. OPTICAL PROPERTIES

The optical properties of alkali halides have been discussed in several review papers[6, 7] and are well understood. We shall limit ourselves to a very short summary. As an example, we show in Fig. 2a the absorption coefficient spectrum of KI,[8] mea-

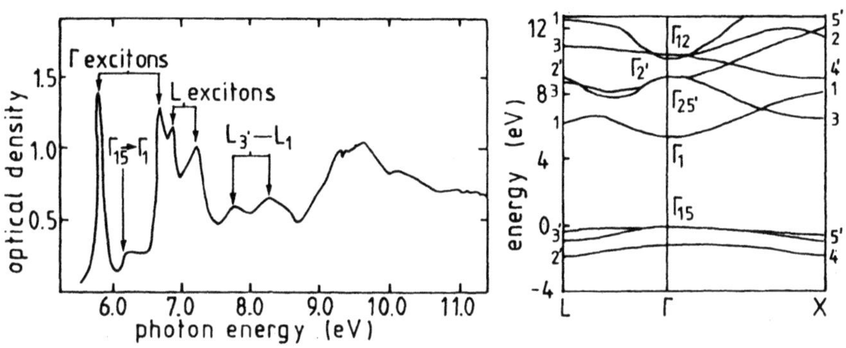

Figure 2. (a) KI absorption spectrum measured at 80 K. The important interband edges and excitons are labelled. (b) Electron energy bands of KI along ΓL and ΓX directions. The single group notation for the energy levels has been used, although spin–orbit interaction is included in the calculation.

sured at 80 K from 5 eV to 11 eV. The spectrum is characterized by several structures assigned to electronic transitions from the valence band to the conduction bands. As a general rule, the valence band of alkali halides has a p–like character, since the wavefunctions of the valence electrons are localized around the halogen ions maintaining the atomic–like character of the halogen p–orbitals. The lowest conduction band has a s–like character, being mainly composed of the alkali ion empty s–orbitals. The second and third conduction bands, for increasing energy, partially derive from the alkali d–orbitals. In Fig. 2b the energy bands of KI are shown.[9] The assignment of several features to transitions occurring at particular high symmetry points of the Brillouin zone are indicated in Fig. 2a. In a crude model, interband transitions correspond to ionizing halogen ions. In the following sections we shall refer to these final states as free electron–hole pairs.

The sharpest peaks of Fig. 2a are assigned to exciton states, i.e. to bound states due to the Coulomb interaction between the excited electron and the valence hole. The lowest energy peak at 5.8 eV is associated with the n=1 $\Gamma_{15}(J = 3/2)$ exciton, formed by a hole at the top of the valence band and an electron at the bottom of the conduction band, both of which lie at the Γ point in k–space. The spin–orbit interaction of the halogen p valence electrons splits the valence band at the Γ point by about 0.9 eV for KI. Thus, the spin–orbit partner $\Gamma_{15}(J = 1/2)$ exciton is degenerate with the interband transitions and is assigned to the 6.7 eV maximum. A second group of exciton states is associated with higher energy interband minima: in the case of the energy bands shown in Fig. 2b this occurs around the L point. Also these exciton states are resonant with the interband transitions continuum and are assigned to the 6.8–7.3 eV structures.

The spectrum of the absorption coefficient of KI is characteristic for alkali halides. For lighter alkali metals (Li, Na) the d–like conduction states have higher energy and the energies of the structures assigned to the X transitions increase. Instead, the spin–orbit interaction of the halogen valence p electrons decreases with the halogen atomic number, so that the spin–orbit doublets are less separated. In KCl the Cl $3p$ spin–orbit splitting of 0.1 eV is of the same order as the thermal broadening of structures at room temperature and it is resolved only at low temperature.[10, 11] In this case both Γ_{15} exciton lines are bound states. At low temperatures the structures of the absorption coefficient spectrum shift to higher energies and become narrower. Higher members of the exciton series can be resolved.[10]

At higher photon energies also the core electrons of both alkali and halogen atoms can be excited, so that the optical spectra present several new features. For example, between 19 eV and 22 eV the reflectivity spectra of the potassium halides show several sharp features, shown in Fig. 3 for KCl and KI,[12] assigned to transitions from the K^+

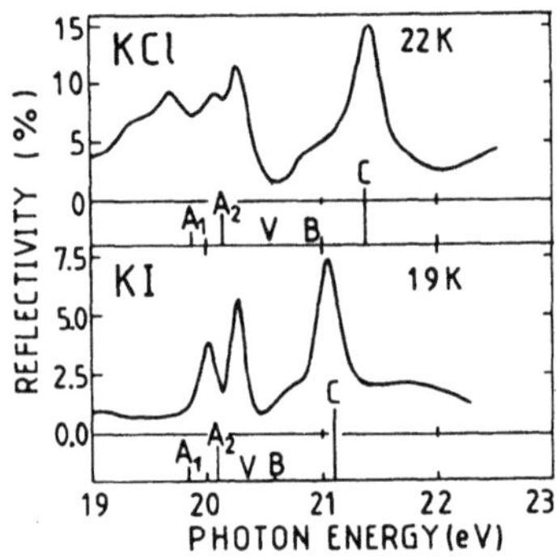

Figure 3. Low temperature reflectivity spectra of KCl and KI single crystals together with the predictions of the ligand field model for the K^+ $3p$ core excitons (vertical bars at the bottom).

$3p$ core level to the K^+ $4s$ and $3d$ levels. These transitions can be considered as core excitons with both the core hole and the conduction electron orbitals well localized on an alkali ion. The energy dependence and the intensity variation of the K^+ $3p$ core excitons with the halide ion have been explained within the ligand field framework with two p–s and three p–$d\gamma$ transitions.[13] These core excitons are degenerate in energy with the ionization continuum of the valence bands.

Fast electrons entering inside a crystal lose their energy by inelastic scattering with the other electrons. At each scattering event any type of the above electronic transitions can be excited.

3. DEFECT FORMATION

Several types of lattice defects can be formed in alkali halides. Here we shall consider only the three most important ones for the purpose of this paper, represented schematically in Fig. 4:

i) the F centre: a halogen ion vacancy, that has trapped an electron;

ii) the H centre: an interstitial halogen atom bound to the neighbouring halogen ions along the $\langle 110 \rangle$ direction;

iii) the V_k centre: a halogen atom bound to a neighbouring halogen ion at a distance shorter than in the unperturbed lattice.

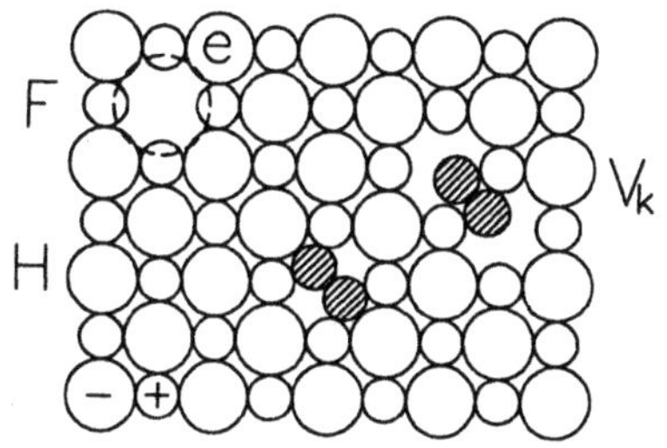

Figure 4. A schematic representation of a F centre, a H centre and a V_k centre.

The defect formation and migration are induced by electron-lattice coupling. In fact, an excited halogen ion has lost one of its valence p electrons, becoming a neutral halogen atom; the inter–ionic forces between it and the neighbouring ions change significantly and the lattice relaxes to a new equilibrium configuration.

In the case of free electron–hole pairs, the conduction electron is mobile and moves freely through the crystal. Instead the hole becomes self-trapped within a picosecond, forming a V_k centre. The halogen atom forms a bond with a neighbouring halogen ion; the two attract each other and their distance is shorter than in the unperturbed lattice. The V_k centre can be considered as a X_2^- (X represents a halogen atom) molecular ion with its axis along the ⟨110⟩ direction.

An electron may be trapped by a V_k centre at an excited state and it can decay to the lowest state of the (V_k + e) system. This system corresponds to a self–trapped exciton state. During this process there is a further lattice relaxation. Several self–trapped exciton states exist, the energy of which depends on the lattice configuration coordinates.

The energy levels of the exciton ground state and the first excited state are shown schematically in Fig. 5 as a function of the lattice distortion.[14] On the left side exciton

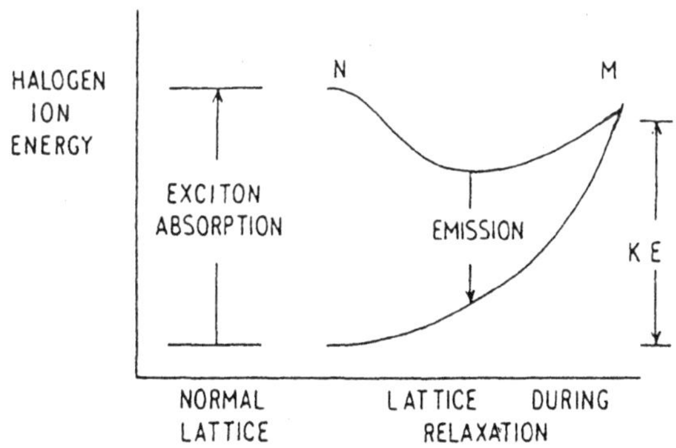

Figure 5. Exciton energy levels for a halogen ion in a normal lattice site (N) and a simplified sketch of the changes in the energy levels which occur during the thermal relaxation along a ⟨110⟩ direction to form a halogen molecular ion (M). Decay is either by a luminescence or by a non–radiative process.

absorption occurs at the normal lattice configuration due to the Frank Condon principle.

Then the exciton relaxes to the minimum of the upper potential curve, where it remains at low temperatures. From this minimum the exciton can decay only radiatively to the ground state potential curve. At higher temperatures the exciton has enough thermal energy to overcome the potential barrier on the right of Fig. 5 and the system goes back to the ground state following the configurational coordinate curves for the excited and the ground states. This will be possible if the self-trapped exciton is in a on-centre configuration with the two bonded ions symmetrically displaced with respect to a lattice site and if the excess energy is shared equally between the two halogen ions. Instead, if the system evolves into an off-center configuration, the energy will be unequally shared and the more energetic halogen ion can be shot away along the ⟨110⟩ axis, initiating a replacement collision sequence, that causes the separation of a F centre and a H centre. This model, proposed originally by Pooley[15] and Hersh,[16] has been refined theoretically in order to explain several subsequent experimental observations on self-trapped exciton luminescence, excited state spectroscopy, pulsed excitation experiments, anticorrelation between luminescence and defect formation yields, and so on.[17]

Several singlet and triplet excited states of the self-trapped exciton exist, the energy of which depends in different ways on atom positions, as shown in the scheme of Fig. 6 calculated for KCl[18]. In Fig. 6, the a_{1g} curve corresponds to the upper curve of Fig. 5, whereas the ground state energy curve is not represented. The b_{3u}, b_{1u} and b_{2u} curves are higher excited levels. The abscissa coordinate R_{FH} is the separation between the F and the H centres. $R_{FH} = 0$ is the on-centre configuration, i.e., the (V_k+e) system. The self-trapped exciton equilibrium configuration, from which the π luminescence is emitted, corresponds to the minimum of the a_{1g} curve. This off-centre configuration closely resembles a nearest-neighbour F-H pair, as it can be better seen in the pictorial lattice representation in the lowest part of Fig. 6. In this case one can imagine that self-trapped excitons and nearest-neighbour F-H pairs transform reversibly from one to the other many times while the processes of radiative decay or conversion to more stable F-H pairs are occurring. A self-trapped exciton that have reached the equilibrium configuration at the off-centre potential minimum, labelled STE in Fig. 6, must overcome a potential barrier for the H centre to pass to a next-nearest-neighbor (NNN) configuration: this can be done only by thermal activation. Thus, at temperatures too low to allow H centre migration, the nascent F centre and the H centre are confined together and the luminescent recombination dominates. The decay of higher excited states of the self-trapped exciton to the a_{1g} state may occur in the on-center configuration, $R_{FH} = 0$ in Fig. 6, which is unstable. The halogen atom experiences a repulsive force towards large values of R_{FH}, gaining approximately 1 eV in KCl. This energy is quickly partitioned among lattice vibrations, but in the same time the H centre can be projected away in a short collision replacement sequence and stabilize even at low temperatures. The difference with respect to thermal activation is the following. In thermal activation the F-H centre separation starts from the bottom of the a_{1g} potential curve with almost zero kinetic energy and the H centre is pushed away from the equilibrium position. In the non-thermal activation the process begins at the top of the a_{1g} curve; the H centre is pushed towards the equilibrium position with a large momentum, so that it overcomes such a position reaching a large F-H separation.

At very low temperatures, in several alkali halides excitons cannot relax in the self-trapped configuration and resonant emission, i.e. emission of luminescent radiation with the same photon energy as that of the exciting photon, has been observed.[19-21] Except for the collision replacement sequence, that can occur at all temperatures, self-trapped excitons as well as lattice defects require a thermal activation to become

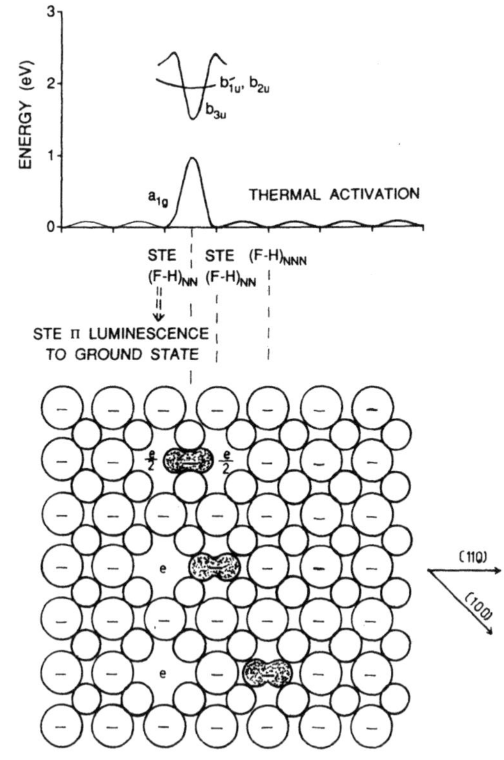

Figure 6. The total energy (sum of the defect electron energy and the change of the lattice energy relative to the perfect lattice) versus the translational coordinate R_{FH}, representing the separation between the F and the H centres, of several exciton levels calculated for KCl. The zero energy is placed at the minimum of the lowest exciton level, which is drown over a broad range of F–H pair separations to exhibit explicitly the symmetry around $R_{FH} = 0$ and the succession of shallow potential minima corresponding to stable, increasing separations of the H centre from the F centre. The lower part of the figure shows pictorial representations of lattice configurations for the "on–centre" and the "off–centre" self trapped exciton and the F–H next–nearest neighbours.

mobile and diffuse through the crystal. Their diffusion has been studied in detail both experimentally and theoretically.

4. FIRST STIMULATED DESORPTION EXPERIMENTS

Since the preliminary studies of defect formation in alkali halides subject to photon or electron beam irradiation, the sputtering of constituent atoms was detected and surface damage and decomposition were observed. Thus, desorption was considered as a consequence of the defect formation.

The first models about photon– and electron–stimulated desorption from alkali halides considered the diffusion of excitons, V_k and H centres from inside the sample, where they have been generated by the incident particles, to the surface. A surface neutral halogen atom is produced; this atom is loosely bound and thus it is easily released by thermal evaporation. Instead, if a focused replacement collision sequence had crossed the surface, an energetic halogen atom was expected to be ejected in a continuation of the $\langle 110 \rangle$ direction, or, possibly, also of the $\langle 211 \rangle$ direction.

Elliott and Townsend[22] and Townsend[23] observed the directional emission of halogen atoms in electron–stimulated desorption experiments on NaCl crystals. They used a silica disc in front of the sample and the directional emission was detected looking at the deposit patterns on the disc. The disc was kept at 77 K to avoid resputtering by back–scattered primary electrons. Under well controlled experimental conditions several spots have been observed in the $\langle 110 \rangle$, $\langle 133 \rangle$ and $\langle 112 \rangle$ directions. Using a different detection system, Townsend[23] have measured the angular distribution of the alkali atoms, obtaining a cosine curve around the $\langle 001 \rangle$ direction. Finally, Townsend[23] found an anti–correlation between the temperature dependence of desorption yield and that of self–trapped exciton luminescence in the temperature range 50÷250 °C. All these observations are consistent with the Pooley[15] and Hersh[16] model for F centre formation as well as that for desorption. Also multiphoton stimulated desorption measurements[24] of halogen atoms induced by a ruby laser beam on KCl, KBr and NaCl show a directional emission along the $\langle 110 \rangle$ and $\langle 211 \rangle$ directions, in agreement with Townsend.[23]

Overeijnder[25] measured the energy distribution of both alkali and halogen atoms using a time of flight technique and compressed powder samples in electron–stimulated desorption experiments. The electrons hit the sample surface at 30° and the particles emitted perpendicularly from the surface were collected. The neutral particles were selected with a quadrupole mass spectrometer. The energy distribution curves of the desorbed alkali atoms could be fit with a Maxwell–Boltzmann distribution function for a free atom gas at the temperature of the sample surface. Also Stoffel[26, 27] obtained a Maxwell–Boltzmann distribution for photon stimulated Na atoms desorbed from Na halides at several temperatures between room temperature and 650 K, measured with the laser induced fluorescence technique. Instead, the energy distribution curves of halogen atoms were composed by two features:[25] a low energy one, following the Maxwell–Boltzmann distribution as for alkali atoms, called the *thermal* component, and a second sharp peak of energetic atoms, called the *non–thermal* component. The non–thermal component was missing for several alkali halides even at room temperature. This has been correlated[25] with the Rabin–Klick criterion for H centre diffusion, which is inhibited when the ratio s/d is smaller than 0.33 at room temperature; s is the interionic space between two halogen ions along a $\langle 110 \rangle$ row, d is the diameter of a halogen atom.

Al Jammal[28] measured the variation of the total desorption yield of electron stimulated pure KI and KI:Tl samples as a function of the electron beam energy. The sputtering rate was measured by recording the time taken for an electron beam to "drill" a hole through a thin sample of alkali halide. They observed that the rate of atom removal per electron was independent of the beam flux, but it was strongly sensitive to the sample temperature. The yield spectra were fit considering the diffusion of excitons for transferring the energy deposited inside the crystal by the primary electron beam to the surface according to the three–step model of Fig. 1. The primary electron beam loses energy inside the sample exciting the crystal electrons. The rate of energy dissipation $H(x, E)$ as a function of the penetration depth x depends on the primary electron beam energy E. The number of excitons produced at depth x is the energy efficiency ϵ for producing excitons multiplied by the energy deposition rate function $H(x, E)$. Supposing that the probability for an exciton subject to random walk to diffuse from depth x and reach the surface, where it will produce a sputtering event, is

$$p_0 e^{-(x^2/L^2)}, \tag{1}$$

where L is a measure of the diffusion range of the exciton, the desorption yield Y has been calculated by Al Jammal[28]:

$$Y(E) = \int_0^\infty \epsilon H(x, E) p_0 e^{-(x^2/L^2)} \mathrm{d}x, \tag{2}$$

obtaining good fits of the experimental curves. The diffusion length values thus determined are 21 nm for pure KI and 13 nm for KI:Tl. The shorter diffusion length in the doped sample could be expected since the impurities act as recombination centres for the excitons, decreasing their lifetime.

5. RECENT DEVELOPMENTS

At the end of the 70's it appeared that the desorption of neutral atoms was a well understood process, since, as discussed above, the experimental observations were in a good agreement with the Pooley[15] and Hersh[16] model of defect formation and diffusion. In the 80's researchers switched their interest to the emission of alkali ions and of excited neutrals, which posed several puzzles, the main of which was whether they are the result of direct emission or if they are generated as a secondary process. The interest in the desorption of both alkali and halogen neutral atoms arose again with a series of papers on electron–stimulated desorption due mainly to Szymonski's group at the Jagellonian University in Krakow, who repeated the angular and energy distribution experiments with improved experimental conditions. In fact, according to Szymonski[4], the collector technique employed by Townsend[23] did not provide any information on the type of atoms deposited on the silica discs. Thus, a thin NaCl layer could have deposited on the disc and the directional spots could have been the result of halogen re-emission due to electrons diffracted by the crystal surface, that left behind Na colloidal particles. At almost the same time our group has started the investigation of near threshold photon–stimulated desorption with the aim of separating the effects of the different excited states on desorption. Also a theoretical effort for the study of defect formation and evolution in the first atomic layers has occurred.

Szymonski[29-36] studied the angular distribution and the energy distribution of halogen and alkali atoms desorbed at several temperatures from KBr and KCl crystals. The measurements were performed on single crystal samples cleaved in air and then heated to high temperature for several hours in the ultra high vacuum chamber in order to obtain clean surfaces. The atoms were detected with a quadrupole mass analyser placed far from the samples for time of flight measurements. The sample and the electron gun could be rotated together in order to change the observation angle while keeping the angle of incidence of the electron beam fixed.

As an example, in Fig. 7 the time of flight distributions of Br atoms leaving a (100) surface of KBr at temperatures between 95 °C and 270 °C along the $\langle 100 \rangle$ direction are shown[34]. The non–thermal component clearly appears as the sharp peak at the lowest time–of–flight values, indicating a high kinetic energy, whereas the thermal component corresponds to the broader structure generated by the slower atoms. These curves are in very good agreement with the measurements by Overeijnder.[25]

A detailed analysis of the experimental curves has shown that the non–thermal component dominates the low temperature spectrum and *decreases* exponentially when increasing the temperature with a "negative" activation energy of 0.09±0.02 eV in KBr, whereas the thermal one increases with temperature[34]. In KBr above 160 °C the increase is exponential with 0.19±0.04 eV activation energy for the Br thermal desorption. Below 160 °C the thermal component deviates from the exponential law, probably

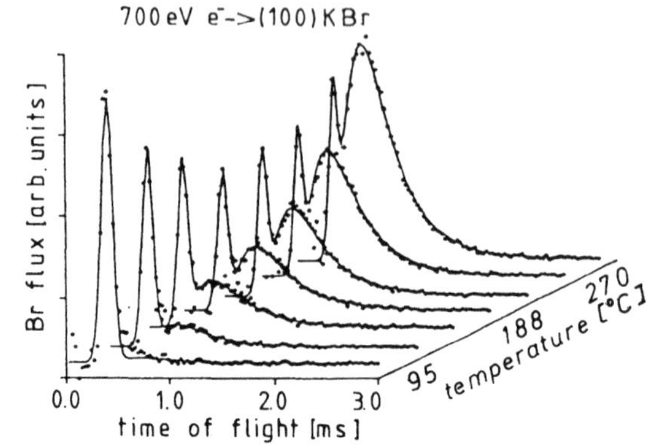

Figure 7. Time of flight distributions of Br atoms leaving a (100) surface of KBr along the ⟨100⟩ direction at several temperatures. The distributions have been normalized to reproduce the temperature dependence of the total intensity.

because of metalization of the sample surface. The ratio between the integrated intensity of the non-thermal emission and the thermal one at the same temperature changes with the material investigated.[5, 25] The non-thermal component is missing for several alkali halides even at room temperature. Szymonski[5] have related this effect to the extra empty space between two neighbouring halogen ions, which is very small (below 0.5 nm) in those alkali halides that do not show the non-thermal emission.

The desorption yield of K atoms and of Br (or Cl) atoms for both components follow characteristic patterns as the direction of collection is changed from the normal to the surface. The experimental angular distribution of K atoms leaving a (110) surface of KBr, shown in Fig. 8,[33] and the (100) surfaces of KBr and KCl have been fit by a cosine distribution, which is characteristic of a thermally induced desorption. Fig. 8 also shows the angular distribution of the total emission of Br atoms leaving the (100) crystal surface in a (010) plane (middle curve) and the (110) surface (top curve)[33]. The thermal emission of halogen atoms follows the cosine law represented by the dashed curves, indicating isotropic emission. The solid lines have been obtained by adding a Gaussian distribution centred at the ⟨100⟩ axis for emission from the (100) surface and two Gaussians centred at +45° and −45° with respect to the (110) surface normal in the second case. From these observations Szymonski[32] have concluded that the emission of alkali atoms as well as that of thermal halogen atoms is isotropically distributed with respect to the surface normal; instead the emission of non–thermal halogen atoms is strongly directed along the ⟨100⟩ axis of the crystal. Clearly, these findings are in contrast with the older experiments and with the model for the non–thermal component based on the replacement collision sequence that intersects the crystal surface, at least for two reasons:

i) the emission of the non–thermal halogen atoms occurs preferentially along the ⟨100⟩ direction;

ii) the non–thermal peak is sharp in energy, whereas it should be broader towards lower energies according to the old model. If the non–thermal atoms have to migrate from inside the crystal, they should loose energy at each step of their migration. Thus, the lack of broadening suggests that the emission process begins at the surface.

Szymonski[32] have concluded that a new model is necessary. They have proposed

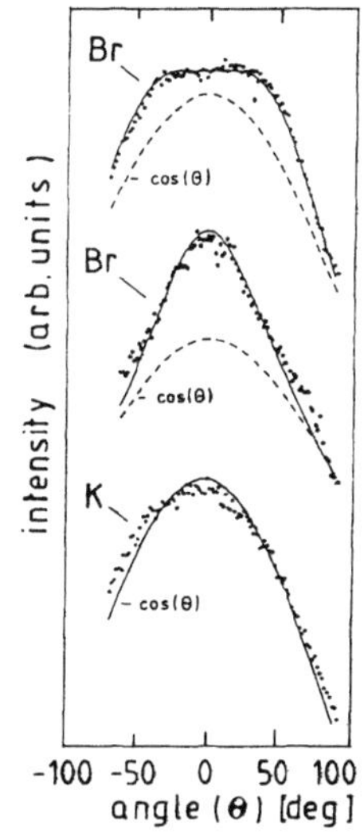

Figure 8. The angular distribution of K atoms leaving a (110) surface of KBr kept at 95 °C (bottom curve). The angular distribution of Br atoms leaving a (100) surface of KBr kept at 140 °C (middle curve). The angular distribution of Br atoms leaving a (100) surface of KBr kept at 140 °C (top curve). The full lines are best–fit curves to the experimental points (dots), as discussed in the text.

that energetic holes "hot holes" created inside the crystal can migrate at considerable distances, thus providing a necessary, efficient transport mechanism from the bulk to the surface. A surface halogen ion, onto which a "hot hole" is trapped, would experience a repulsive force oriented preferentially along the ⟨100⟩ direction.

Postawa[5, 36] have measured the desorption yields of both the thermal and the non–thermal components of bromine atoms desorbed from KBr. The non–thermal desorption component increases steeply with the electron beam energy and then decays slowly. The experimental curve has been fit resolving the diffusion equation for holes, using the model function proposed by Al Jammal[28] for the depth distribution of the deposited energy by the primary electron beam. The best fit has been obtained with 10 nm diffusion length for the holes. The desorption yield of the thermal component, instead, have different behaviours with the electron beam energy, depending on the sample and on the sample temperature, and it is affected by the excess of alkali atoms at the surface.[4, 30]

Photon–stimulated desorption measurements performed with energetic photons in the regions of core transitions gave contrasting results. Stoffel[26] measured the photon–stimulated desorption yield of ground state Na atoms emitted from the sodium halide samples in the 40÷160 eV photon energy range. The shapes of the desorption yield curves are very similar to each other for all the Na halides, presenting a general broad bell–shaped behaviour. Several core level thresholds occur in the experimental region investigated, but they have not been associated with pronounced structures in the curves. Thus, these measurements has provided an indirect evidence for the dominant role of a common valence-excitation mechanism in photon–stimulated desorption. Recently also Zema[37] have not observed structures (or just weak dips) at the K $3p$ core thresholds in KI and KBr and at the Rb $4p$ threshold in RbBr. A different experimental result has been obtained by Szymonski[38], who have measured the desorption yield of Na atoms from a NaCl single crystal at 430 K at the Cl K edge, and found a

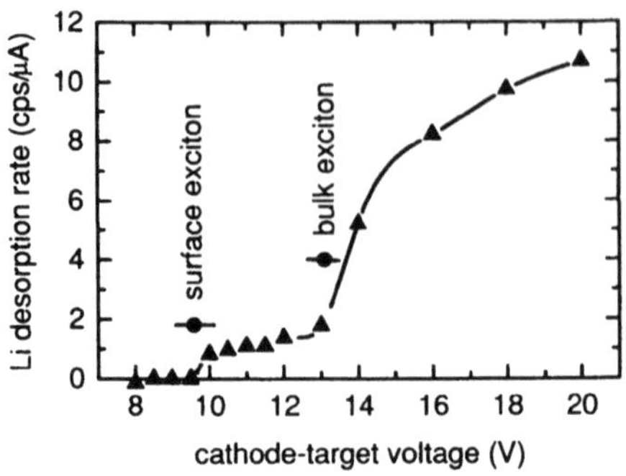

Figure 9. The electron stimulated desorption yield of LiF with the electron energy scanning the fundamental absorption threshold.

good correlation with the photoelectron total yield spectrum. Their conclusion is that the creation of secondary electrons inside the sample leads to the formation of highly mobile holes. The desorption of the Na neutrals is caused by the self-trapping and recombination of the holes with free electrons at the surface.

In the electron-stimulated desorption energetic electrons can excite core levels as well as valence band electrons. Several types of excited states can be created, that go through a cascade of processes before self trapped excitons are formed. It is not yet clear the role played by all these intermediate steps in defect formation and, consequently, in the desorption process of neutral atoms. Thus it is very important to study the behaviour of the desorption yield in the energy interval where a direct exciton formation takes place and slightly above the fundamental absorption threshold, where free electron-hole pairs are created. In both cases a direct information on the evolution of the fundamental electronic excitations involved in the desorption of neutral atoms can be achieved. For this reason, photon stimulated desorption becomes mandatory.

Wurz[39] have measured the electron stimulated desorption of neutral ground state lithium atoms from the (100) surface of single crystal LiF samples held at a temperature of 400 °C irradiated with 5÷20 eV electrons, shown in Fig. 9.[39] The desorption yield curve clearly presents two thresholds, a first one at 9.6 ± 0.6 eV and a second one at 13.1 ± 0.4 eV, that have been assigned to the onset for surface exciton and bulk exciton creation, respectively.

Brinciotti[40, 41] have investigated the photon-stimulated emission of both potassium and iodine atoms from KI single crystals cleaved in ultra high vacuum in the fundamental absorption region from 5 eV to 10 eV, using a deuterium lamp and synchrotron radiation as photon sources. The desorption yield of both atoms follows the same spectral distribution. The correlation with the absorption spectra is very weak: a structure at threshold has been assigned to the $\Gamma_{15}(J = 3/2)$ exciton absorption peak. Above threshold the emission yield has a weak, broad maximum, shifting to lower photon energies for increasing temperatures, and then increased exponentially up to about 10 eV.

The intensities of the spectra as a function of the temperature increased smoothly for both potassium and iodine atoms, except for potassium around 395 K. This spectrum have shown a higher intensity than expected. According to Brinciotti,[40, 41] below this temperature the crystal surface becomes rich of potassium atoms, that begin to evaporate at about 355 K. Thus, the higher intensity of the 395 K spectrum has been associated with the evaporation of the excess potassium ions. Above 400 K the surface becomes stoichiometric and the temperature dependence of the spectra becomes smooth again.

The measurements by Brinciotti[40, 41] have been extended recently with measurements of the desorption yield of Br, I and Rb atoms using synchrotron radiation from 5 eV to 25 eV on KI, KBr and RbI single crystals by Zema,[37] presented for KI in Fig. 10.[37] The spectra show a rather smooth shape, that does not resemble the structured shape

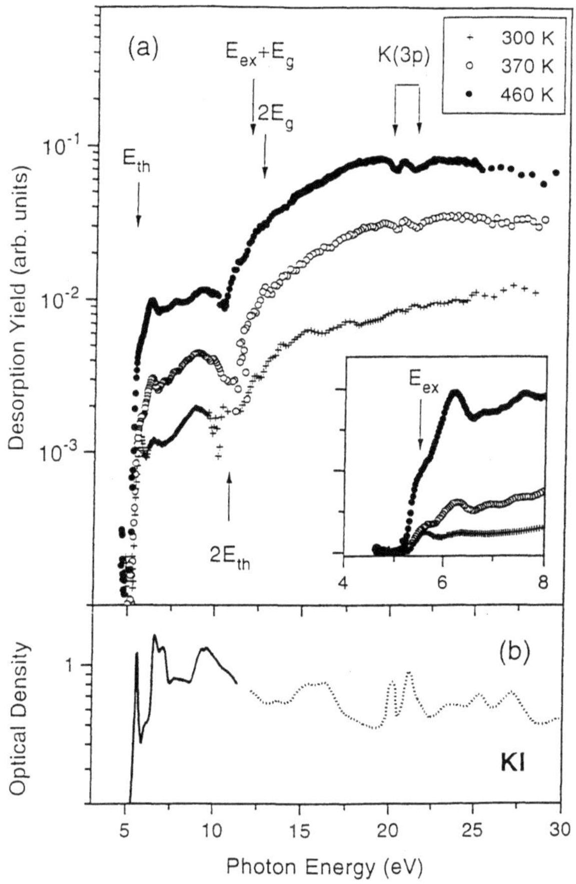

Figure 10. (a) Photon–stimulated desorption yield spectra as a function of photon energy of neutral iodine emission from the KI surface measured at different sample temperatures. The arrows mark the threshold energies for electron-electron scattering as well as the K^+ $3p$ core excitons. The inset shows on a linear scale the desorption yield in the threshold region. (b) Room temperature absorption spectrum of KI.

of the absorption coefficient spectrum. A weak shoulder resolved on the threshold has been assigned to the $\Gamma_{15}(J = 3/2)$ exciton. Several other thresholds have been found in correspondence of multiple excitation energies, i.e. at photon energies such that the excited electrons can produce other electron–hole pairs or other excitons by an inelastic scattering. This multiplication has been observed as an enhancement in the yield of the intrinsic luminescence[43, 44] and as a reduction of the photoemitted electrons in total yield spectra.[45, 46]

For understanding the little correlation with the optical spectra, Piacentini and Zema[47] have extended the expression derived by Al Jammal[28] for electron–stimulated desorption to photon–stimulated desorption. The rate of energy deposition function $H(x, E)$ for electrons in Eq. (2) has been replaced by the number $\mathrm{d}N(x)$ of excited states created at depth x from the surface per unit time, which equals the attenuation of the intensity of the photon beam reaching the depth x in traversing a thickness $\mathrm{d}x$ of the sample:

$$\mathrm{d}N(x) = \frac{1}{\hbar\omega}\alpha I_0(1-R)e^{-\alpha x}\mathrm{d}x, \tag{3}$$

where α is the sample absorption coefficient, R its reflectivity, I_0 the intensity of the incident radiation beam and $\hbar\omega$ the photon energy. The desorption yield $Y(\hbar\omega)$, i. e. the number of atoms desorbed from the surface per incident photon and per unit time, is obtained by integrating over the total sample thickness $\mathrm{d}N(x)$ times the diffusion probability, given by Eq. (1), everything divided by the number $(I_0/\hbar\omega)$ of incident photons:

$$Y(\hbar\omega) \propto \alpha(1-R)\int_0^\infty e^{-\alpha x}e^{-(x^2/L^2)}\mathrm{d}x. \tag{4}$$

If the diffusion length is much shorter than the radiation penetration depth ($L \ll 1/\alpha$), then the desorption yield will be proportional to the absorption coefficient. Instead, when the mean free path is much longer than the radiation penetration depth ($L \gg 1/\alpha$), saturation is obtained: $Y = C(1-R)$, where C is a constant that can be used to normalize the calculated yield spectra.

The comparison between the yield spectra calculated according to Eq. (4), using the values of the room temperature absorption coefficient spectrum of KI,[8] and the experimental ones shown in Fig. 10,[37] indicates that the photon–stimulated desorption model based on a single diffusion length is not adequate. Thus, Piacentini and Zema[47] have implemented the model described above assigning different values for the diffusion length of the $\Gamma_{15}(J=3/2)$ exciton, of the resonant excitons and of the free electron–holes. In the latter case they allow the diffusion length to increase up to a limiting value with photon energy. The result of this calculation is compared with the measured iodine desorption yield from KI for two temperatures in Fig. 11. The values of the diffusion lengths used in this calculation are compared with the diffusion lengths derived by other research groups in Table 1.

With the aim of relating specific optical transitions with either the thermal or the non–thermal emission of halogen atoms, Szymonski [48] have measured the angular distribution of Br atoms desorbed from RbBr at several photon energies in the 5÷25 eV range using synchrotron radiation, finding only the thermal emission behaviour. In these measurements the incident beam direction and the mass spectrometer direction were fixed forming an angle of 45° and the angle was changed rotating the sample. Electron stimulated desorption performed with the same experimental arrangement, instead, showed also the directional emission of the non–thermal component, as shown in Fig. 12.[48]

In order to investigate better the diffusion of the excited particles inside the alkali halides, Kolodziej[49] measured the time of flight distributions of the emission of Br atoms desorbed from KBr thin films grown epitaxially on (100) InSb surfaces using 2 keV electrons. Both the thermal and the non–thermal components of desorption were strongly dependent on the film thickness. Kolodziej[49] determined the diffusion length of the carriers responsible for the thermal component to increase from 3 nm to 70 nm with the temperature in the range 30÷300 °C. From this variation they deduced

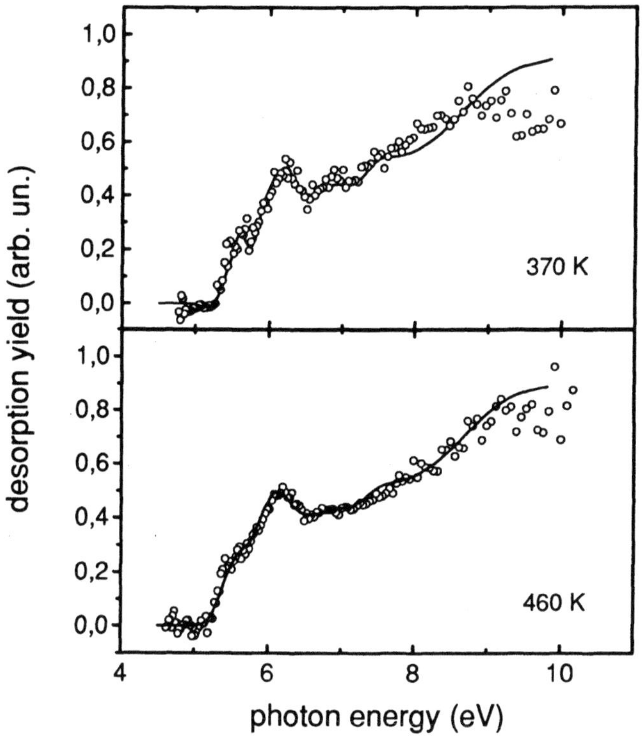

Figure 11. A comparison of the experimental photon–stimulated desorption yield spectra of iodine atoms from KI (empty circles) with the yield spectra calculated using the Piacentini and Zema model with several diffusion lengths.

an activation energy of about 0.27 eV, in agreement with the value of the H centre migration, supporting the H center diffusion as the source for the thermal particles. The diffusion length of the non–thermal component was found to be 14 nm, almost independent of the crystal temperature, suggesting the diffusion of "hot" unrelaxed excitons.

Time dependent measurements give some insight into the dynamics of desorption. Loubriel[50-52] measured the decay of the flux of desorbed ground state Li atoms following the irradiation of LiF crystals by a pulsed 200 eV electron beam, obtaining a decay time of the order of 10^{-3} s. The measurements were performed on samples kept at 628, 699 and 826 K. At such high temperatures the sample surfaces were stoichiometric, whereas they became rich of Li atoms below 588 K.

From this observation, Loubriel[50-52] have assumed that, at high temperature, the time necessary to eject a Li atom after an F centre arrives at the surface is very short and the entire process is governed by the diffusion of F centres to the surface, in analogy to the scheme presented in Fig. 1. Best fit curves have been calculated by solving the diffusion equations for a square wave excitation under the above assumptions. The F centre hopping rate W obtained from the best fit is of the order of 5.4×10^4 s^{-1} at 628 K and increases with temperature to the value of 1.2×10^6 s^{-1} at 826 K.

The desorption of potassium atoms from KI samples illuminated with monochromatic photons of energy between 5.5 eV and 6.5 eV, measured by Brinciotti[41], has shown two time behaviours. When the light was turned off after a steady state situa-

Table 1. Diffusion length of excited particles determined by several groups in electron- or photon-stimulated desorption experiments (ESD and PSD, respectively).

Group	Method	Diffusing particles	Diffusion length (nm)	sample	temperature
Al Jammal[a]	ESD	Excitons	21	KI	—
Postawa[b]	ESD	H centres	5.5	KBr	100 °C
		Focussed replacement sequence	0.63		
Postawa[c]	ESD	Hot holes	10	KBr	
Piacentini[d]	PSD	Exciton $J = 3/2$	5.5	KI	370 K
		Resonant excitons	4.5		
		Free electron–holes	25÷80		
Piacentini[e]	PSD	Exciton $J = 3/2$	7.5	KI	460 K
		Resonant excitons	4.5		
		Free electron–holes	40÷70		
Kolodziej[f]	ESD	non–thermal emission	14	KBr	variable
		thermal emission	3÷70		20÷300 °C

[a]Reference[28]
[b]Reference[30]
[c]Reference[36]
[d]Reference[47]
[e]Present work
[f]Reference[49]

tion, the potassium signal dropped very fast (within the time resolution of 0.5 s of the experimental equipment) to a lower value and then decayed very slowly with a time constant of 17.5 ± 0.6 s. Also the rise time had at first a fast increase, followed by a slow one before reaching the steady state value. The time–resolved spectra, presented in Fig. 13,[41] clearly show that the slow emission is the main desorption channel at low photon energies, in correspondence to the $\Gamma_{15}(J = 3/2)$ exciton transition, while the fast emission becomes dominant at the highest photon energies, in correspondence to the free electron-hole formation. The time dependence of potassium desorption yield, measured also as a function of the sample temperature, shows that the decay of the potassium desorption signal is mainly due to the fast emission at high temperature. Unlike potassium, the desorption of iodine atoms has only the fast component.[41]

Using 36 eV synchrotron radiation photons and the laser induced fluorescence

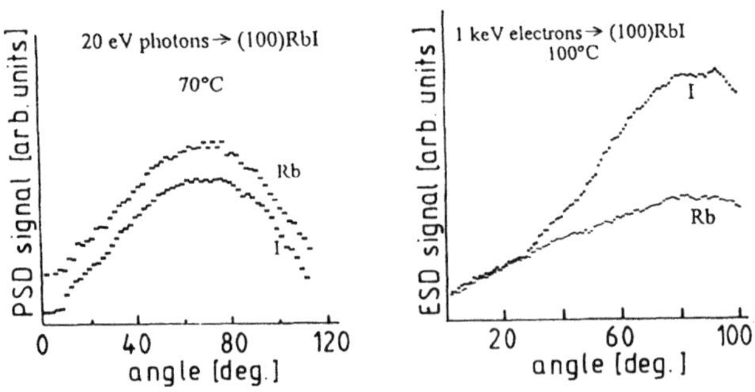

Figure 12. Left panel: the angular characteristics of Rb and I atom emission from (100) surface under synchrotron light irradiation. Right panel: the angular characteristics of Rb and I atom emission from a RbI (100) surface under 1 keV electron irradiation.

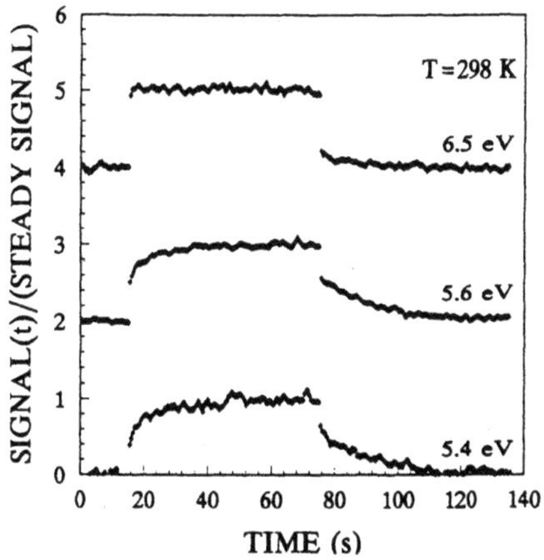

Figure 13. Build-up and decay of the potassium desorption signal under a square wave excitation at several photon energies. Curves are obtained at room temperature and upward displaced.

method to detect ground state potassium desorption, Kamada and Hiroshe[53] have measured the time response of ground state potassium desorption from KCl and KBr at 374 K and 384 K, respectively, obtaining a fast emission in the nanosecond range followed by a slow one (> 180 ns). They have suggested to associate the fast component to the lattice instability induced by the electronic excitation in the surface layer.

6. DISCUSSION

From the above presentation of the experiments it appears that photon– and electron–stimulated desorption are rather complicated processes. The current approach has been to describe them with simple models.

The first desorption experiments have been performed with the aim of probing the models of lattice defect formation in alkali halide samples subject to low–energy electron or photon bombardment. The desorption has been attributed to the possibility that the energy from the incident particles is transferred directly to surface atoms, that become excited to a new electronic state where the attraction to the solid is reduced. Thermal lattice energy is then sufficient to overcome this reduced binding energy and the excited ions can escape by thermal evaporation. Thermal emission can be expected also if the energy is deposited inside the crystal and the excited state diffuses to the surface. On the other hand, if a replacement collision sequence, generated by the F centre production process, intersects the surface, the ejection of non–thermal halogen atoms along preferential directions has been considered possible.

The recent experimental findings by the Szymonski's group[5, 32] have shown that the latter mechanism is very unlikely and that different mechanisms, supported by theoretical calculations, are necessary.

The electronic excitation processes, which correspond to the first step of the three–step model for the stimulated desorption, are rather well understood. Instead, although a large number of experimental work has been performed, many questions arise on the transport and desorption steps, probably for the lack of a sufficient theoretical work. For example, consider the ejection of atoms from the surface:

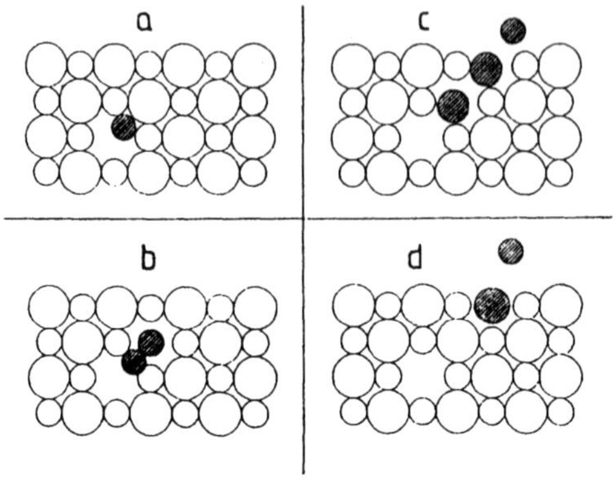

Figure 14. Atomic displacements at several stages of evolution leading from a self-trapped exciton in the third layer to emission of a halogen atom.

i) Can a halogen ion, that has been neutralized by a hole self-trapped on the surface, escape by thermal evaporation? Which is the energy barrier it should overcome?

ii) Can a surface exciton release the halogen atom?

iii) Which is the electronic precursor state for the non-thermal halogen emission and why the non-thermal emission is missing in several alkali halides?

iv) Can surface F centres be responsible for alkali atom ejection?

v) Which is the surface morphology when metalization occurs and how metalization develops?

Only recently detailed theoretical studies of surface defects have been performed for NaCl.[54-56] According to these calculations the emission of both halogen and alkali atoms is due to the lattice relaxation occurring around H centres or around excitons present in the first two atomic layers of the crystal. There is no adiabatic potential barrier for the H centre diffusion from the second layer to the surface top layer. The H centre placed in the third layer posses an extremely low activation energy for diffusion towards the surface. The axis of the H centre approaching to the (100) NaCl surface tilts and tends to be oriented along the ⟨100⟩ direction when arriving in the second layer. On the top layer the H centre decomposes into a halogen ion occupying a lattice point and a halogen atom physically adsorbed above the anion with a very small binding energy, the value of which is in the range 0.03÷0.14 eV, depending on the approximation used in the calculations. These surface halogen atoms can be released by a thermal activation and contribute to the thermal component of desorption, maintaining the surface stoichiometry. The non-thermal halogen desorption component may derive from the exciton formation in the first two-three layers. If an exciton is created in the first layer, the hole will be localized on a surface halogen ion, which experiences a repulsive force in the ⟨100⟩ direction. Therefore the exciton relaxes expelling the Cl atom with a kinetic energy of about 2 eV, and a surface F centre is left behind. If the exciton is generated in the second layer, it can either jump in the first layer with energy gain or it can be first self-trapped and then move to the surface. In both cases the emission of an energetic halogen atom and the formation of a surface F centre occurs. If the exciton is formed in the third layer, it relaxes in the off-centre configuration which undergoes a short replacement sequence sketched in Fig. 14 with the molecule Cl_2^- axis tilting towards the ⟨100⟩ direction at the surface. Also in this case an energetic halogen

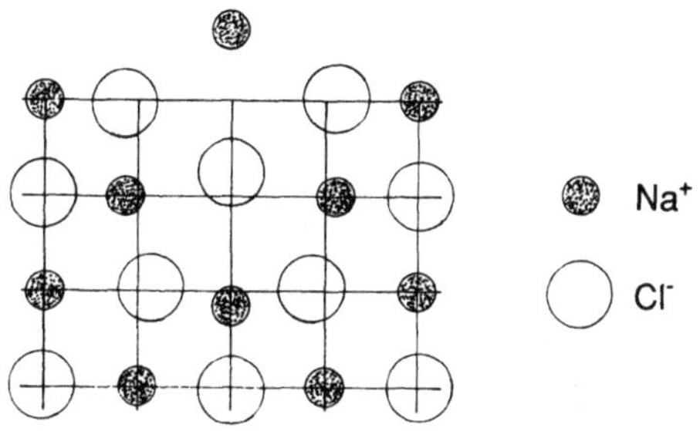

Figure 15. Crystal lattice in the symmetry plane perpendicular to the (100) surface of a NaCl crystal in the relaxed configuration of the triplet self-trapped exciton in the second layer below the surface.

atom can be released along the preferential ⟨100⟩ direction.

F centres are rather stable systems and cannot be considered the source for alkali atom emission, as previously thought, unless they are in an excited state or they form clusters. Instead, excitons in the triplet state, self-trapped in the second layer, can relax in the asymmetric configuration presented in Fig. 15, with the electronic cloud extending on the surface alkali ion above the excited halogen atom. Due to the lattice relaxation, this alkali atom is pushed outwards perpendicular to the surface and it can be ejected by thermal activation.[56]

The calculations reported above explain in part the last step of the desorption process and answer some of the questions formulated above. In particular, they justify the directional properties of desorption, studied in detail by Szymonski.[5, 32] However the non-thermal emission, that is predicted for NaCl,[54] as a matter of fact has not been detected in NaCl,[5] possibly for reasons associated with the transport of the excited state, or because the non-thermal Cl atoms may have a kinetic energy above the experimental set-up capabilities.

The second aspect that needs further analysis is the transport of the excited states to the surface. The models for the atom emission discussed above require the presence of the exited atoms in the first three layers below the surface. How these excited atoms are generated and arrive in these layers is a subject that has not yet been sufficiently analyzed. The diffusion of excitons, of H centres, of hot holes, etc. have been considered as a single transport mechanism from the bulk to the surface for the electron-stimulated desorption of ground state neutral halogen atoms. But in electron-stimulated desorption energetic electrons have been employed: each primary electron performs several collisions with the crystal electrons before thermalizing, and a wealth of excited states are created inside the crystal. The secondary electrons, if sufficiently energetic, may produce additional excited states. Thus, a very complicated situation is generated inside the crystal. Reducing the transport to the surface to a single mechanism is clearly an over-simplification. Also energetic photons, in the soft X-ray and X-ray regions, used for photon-stimulated desorption experiments,[26, 38] induce electronic transitions from core levels, producing a cascade of secondary processes. Again several types of excited states can be created.

The electron-stimulated desorption experiment performed on LiF by Wurtz[39] using low energy electrons has revealed that also surface excitons are responsible for the emission of neutral Li atoms. Thus, assuming that the same could occur also for the halogen atoms, this experiment gives direct evidence that in electron- and photon-stimulated desorption a fraction of the incident beam excites directly the first two—three atomic layers from where a direct emission occurs without the necessity of transport. On the other hand, Piacentini and Zema,[47] continuing their studies on the near-threshold photon-stimulated desorption, have found recently that different excited states have different diffusion lengths that change with the sample temperature and, in the case of free electron-hole pairs also with the photon energy. However, their model has not taken into account the energy released directly in the surface layers, neither the thermal component of halogen emission has been separated from the non-thermal component. Different diffusion lengths for the thermal and the non-thermal components instead have been derived recently by Kolodziej.[49]

Time-dependent measurements have shown that a fraction of the desorbed alkali atoms are emitted within nanoseconds from the excitation,[53] indicating a surface process, some are emitted within milliseconds, interpreted with the diffusion of F centres inside the sample,[50-52] and others are emitted within seconds[41]. However, these results are not homogeneous with each other, since the measurements have been performed at different temperatures. The slow component disappears as the sample temperature increases above the boiling point of the alkali metal,[41] indicating that the slow emission process could be related with the surface metalization.

Surface metalization and morphology are aspects connected with the stimulated desorption process which have been not yet completely resolved. The evidence of a strong surface modification induced by electron bombardment has been reported since the first papers of Elliott and Townsend[22, 23] and the contribution at low temperature of the surface metalization due to non-stoichiometric damage has been pointed out by Cota Araiza and Powell[57]. Since then other authors have undertaken surface damage characterization of alkali halide surfaces by using different techniques, such as SEM/SAM microscopy,[58, 59] AFM imaging techniques[60-62] and He-atom scattering[63]. The origin of the surface damage has been associated with the creation of F and H centres induced by electronic excitation and to their diffusion from the bulk to the surface. At all temperatures where a surface metalization is observed, bulk F centres are stable and have a very small mobility in the ground state. Thus surface F centres can only be generated by the emission of a halogen atom from the surface following the excitonic mechanism previously discussed. The surface metalization through the formation of alkali islands, due to the clustering of surface F centres, indicates a strong mobility of F centres on the surface, because the probability to create F centres close to each other is very low with the electron and photon densities used in the experiments The understanding of surface metalization needs further experimental and theoretical efforts, which will bring new informations on the first stages of the kinetics of alkali atom clustering, and on the different behaviour in the formation of a metal overlayer when selected electronic excited states, such as direct excitons, are created in the crystal.

There are several other aspects associated with photon– or electron– stimulated desorption that we have not addressed in this review, such as the study of the relative intensities of the desorbed atoms, which is rather difficult because of the calibration of the detector sensitivity to the different atomic species. We also did not try to relate the desorption with the other decay mechanisms of the excited states in alkali halides, such as luminescence. In addition to the desorption of neutral, ground state atoms, also the emission of atoms in their excited states, as well as ions and more complex systems have been observed, that require a detailed review and further studies. In conclusion, desorption induced by electronic transitions in alkali halides is indeed a very interesting field of investigation, strongly associated with other aspects of the physics of these compounds, the study of which require refined techniques both experimental and theoretical.

Acknowledgments

We are strongly indebted to Mr. M. Brolatti for preparing the drawings of the manuscript.

REFERENCES

1. D. Menzel, and R. Gomer, *J. Chem. Phys.* 41:3311 (1964).
2. P. A. Redhead, *Can. J. Phys.* 42:886 (1964).
3. M. L. Knotek, and P. J. Feibelman, *Phys. Rev. Letters* 40:964 (1978).
4. M. Szymonski, *K. Dan. Vidensk. Selsk. Mat. Fys. Medd.* 43:495 (1993), and references therein.
5. M. Szymonski, J. Kolodziej, Z. Postawa, P. Czuba, and P. Piatowski, *Progress in Surface Science* 48:83 (1995).
6. J. C. Phillips. "Solid State Physics", Vol. 18: pag. 55, Academic Press (1966).
7. G. Stephan, and S. Robin, "Optical properties of ionic insulators", *in*: "Some Aspects of V.U.V. Radiation Physics", ed. by N. Damoy, Pergamon Press (1974).
8. J. E. Eby, K. J. Teegarden, and D. S. Dutton, *Phys. Rev.* 116:1099 (1959).
9. A. B. Kunz, *J. Phys. Chem. Solids* 31:265 (1970).
10. T. Tomiki, *J. Phys. Soc. Jpn.* 22:463 (1967).
11. K. Teegarden, and G. Baldini, *Phys. Rev.* 155:896 (1967).
12. G. Sprussel, M. Skibowski, and V. Saile, *Solid State Commun.* 32:1091 (1979).
13. C. Satoko, *Solid State Commun.* 13:1851 (1973).
14. P. D. Townsend, *Nucl. Instr. Methods* 198:9 (1982).
15. D. Pooley, *Proc. Phys. Soc.* 87:145 (1966); *ibid.* 87:257 (1966).
16. H. N. Hersh, *Phys. Rev.* 148:928 (1966).
17. N. Itoh, *Advances in Physics* 31:491 (1982). This is a thorough review paper of both experimental and theoretical works on exciton absorption, luminescence, defect formation. Many sinificant theoretical papers have appeared more recently.
18. R. T. Williams, K. S. Song, W. L. Faust, and C. H. Leung, *Phys. Rev.* B 33:7232 (1986).
19. Ch. Lushchick, I. Kuusmann, P. Liblik, N. Lushchick, T. Soovik, and A. Ratas. "International Conference on Luminescence", Tokyo (1975).
20. H. Nishimura, and M. Tomura, *J. Phys. Soc. Jpn.* 39:390 (1975).
21. H. Nishimura, C. Ohhigashi, Y. Tanaka, and M. Tomura, *J. Phys. Soc. Jpn.* 43:157 (1977).
22. D. J. Elliott, and P. D. Townsend, *Phil. Mag.* 23:249 (1971).
23. P. D. Townsend, R. Browning, D. J. Garlant, J. C. Kelly, A. Mahjoobi, A. J. Michael, and M. Saidoh, *Radiat. Eff.* 30:55 (1976).
24. A. Schmid, P. B. Braunlich, and P. Rol, Phys. Rev. Letters 35:1382 (1975).
25. H. Overeijnder, M. Szymonski, A. Haring, and A. E. De Vries, *Radiat. Eff.* 36:63 (1978).
26. N. G. Stoffel, R. Riedel, E. Colavita, G. Margaritondo, R. F. Haglund Jr., E. Taglauer, and N. H. Tolk, *Phys. Rev.* B 32:6805 (1985).
27. R. F. Haglund Jr, R. G. Albridge, D. W. Cherry R. K. Cole, M. H. Mendenhall, D. Niles, G. Margaritondo, N. G. Stoffel, and E. Taglauer, *Nucl. Instr. Methods* B 13:525 (1986).
28. Y. Al Jammal, D. Pooley, and P. D. Townsend, *J. Phys. C: Solid State Phys.* 6:247 (1973).
29. Z. Postawa, and M. Szymonski, *Phys. Rev.* B 39:12950 (1989).
30. Z. Postawa, P. Czuba, A. Poradzisz, and M. Szymonski, *Rad. Eff. Defects Sol.* 109:189 (1989).
31. M. Szymonski, *Nucl. Instr. Methods* B 46:427 (1990).
32. M. Szymonski, J. Kolodziej, P. Czuba, P. Piatowski, A. Poradzisz, N. H. Tolk, and J. Fine, *Phys. Rev. Letters* 67:1906 (1991).
33. M. Szymonski, J. Kolodziej, P. Czuba, P. Piatowski, A. Poradzisz, and N. H. Tolk, *Nucl. Instr. Methods* B 58:485 (1991).
34. J. Kolodziej, P. Czuba, P. Piatowski,, A. Poradzisz, Z. Postawa, M. Szymonski, and J. Fine, *Nucl. Instr. Methods* B 65:507 (1992).
35. M. Szymonski, A. Poradzisz, P. Czuba, J. Kolodziej, P. Piatowski, J. Fine, L. Tanovic, and N. Tanovic, *Surf. Sci.* 260:295 (1992).
36. Z. Postawa, J. Kolodziej, P. Czuba, P. Piatowski, A. Poradzisz, M. Szymonski, and J. Fine. "Springer Series in Surface Science", Vol. 31:pag. 299, Springer Verlag (1993).

37. N. Zema, M. Piacentini, P. Czuba, J. Kolodziej, P. Piatowski, Z. Postawa, and M. Szymonski, *Phys. Rev. B* 55:5448 (1997).
38. M. Szymonski, T. Tyliszczak, P. Aebi, and A. P. Hitchcock, *Surf. Sci.* 271:287 (1992).
39. P. Wurz, J. Sarnthein, W. Husinsky, G. Betz, P. Nordlander, and Y. Wang, *Phys. Rev.* B 43:6729 (1991).
40. A. Brinciotti, N. Zema, and M. Piacentini, *Radiat. Eff. Defects Solids* 119–121:559 (1991).
41. A. Brinciotti, M. Piacentini, and N. Zema, *Radiat. Eff. Defects Solids* 128:89 (1994).
42. A. Ejiri, *Phys. Rev.* B 36:4946 (1987).
43. M. Yanaghira, Y. Kondo, and H. Kanzaki, *J. Phys. Soc. Jpn.* 52:4397 (1983).
44. J. H. Beaumont, A. J. Bourdillon, and M. N. Kabler, *J. Phys.* C 9:2961 (1976).
45. D. Blechschmidt, M. Skibowswki, and W. Steinman, *Opt. Commun.* 1:275 (1970).
46. H. Sugawara, and T. Sasaki, *J. Phys. Soc. Jpn.* 46:132 (1979).
47. M. Piacentini, and N. Zema, *Il Nuovo Cimento*, in press (1998).
48. M. Szymonski, J. Kolodziej, P. Czuba, P. Korecki, P. Piatowski, Z. Postawa, M. Piacentini, and N. Zema, *Surf. Sci.* 363:229 (1996).
49. J. Kolodziej, P. Piatkowski, and M. Szymonski, *Surf. Sci.* 390:152 (1997).
50. G. M. Loubriel, T. A. Green, P. M. Richards, R. G. Albridge, D. W. Cherry, R. K. Cole, R. F. Haglund, Jr., L. T. Hudson, M. H. Mendenhall, D. M. Newns, P. M. Savundararaj, K. J. Snowdon, and N. H. Tolk, *Phys. Rev. Letters* 57:1781 (1986).
51. T. A. Green, G. M. Loubriel, P. M. Richards N. H. Tolk, and R. F. Haglund Jr., *Phys. Rev.* B 35:781 (1987).
52. G. M. Loubriel, T. A. Green, N. H. Tolk, and R. F. Haglund, *J. Vac. Sci. Technol.* B 5:1514 (1987).
53. M. Kamada, and S. Hirose, *Surf. Sci.* 390:194 (1997).
54. V. E. Puchin, A. L. Schluger, and N. Itoh, *Phys. Rev.* B 47:10760 (1993).
55. V. E. Puchin, A. L. Schluger, Y. Nakai, and N. Itoh, *Phys. Rev.* B 49:11364 (1994).
56. V. E. Puchin, A. L. Schluger, and N. Itoh, *J. Phys.: Condens. Matter* 7:L147 (1995).
57. L. S. Cota Araiza, and B. D. Powell, *Surf. Sci.* 51:504 (1975).
58. Q. Dou, D. W. Lynch, and A. Bevolo, *Surf. Sci.* 219:L623 (1989).
59. E. Paparazzo, and N. Zema, *Surf. Sci.* 372:L301 (1997).
60. R. M. Wilson, W. E. Pendleton, and R.T.Williams, *Radiat. Eff. Defects Solids* 128:79 (1994).
61. R. M. Wilson, and R. T. Williams, *Nucl. Instr. Methods* B 101:122 (1995).
62. K. Miura, and K. Maeda, *Nucl. Instr. Methods* B 116:486 (1996).
63. H. Hoche, J. P. Tonnies, and R. Vollmer, *Phys. Rev. Letters* 71:1208 (1993).

STRONG ELECTRON CORRELATION EFFECTS IN COPPER OXIDES

Nikolai M. Plakida

Joint Institute for Nuclear Research,
141980 Dubna, Russia

In developing the theory of the high-temperature superconductivity it is necessary to solve the two most important problems which are definitely interrelated: what is the nature of the normal state for electrons in oxide compounds and what is the mechanism of formation of the superconducting phase? It is generally accepted that electron correlations constitute a key issue in the explanation of many "unconventional" physical properties of copper oxides. After a brief discussion of experimental results which prove this statement we shall introduce theoretical models (phenomenological and microscopic ones) to treat strong electron correlations in CuO_2 planes. By employing the microscopic models (the $p-d$-like Hubbard model and the one-band $t-J$ model) we present the results of calculations for the optical conductivity, the dynamic spin susceptibility, the electronic spectrum and the superconducting pairing of quasiparticles. It is shown that dynamic spin fluctuations, resulting from strong electron correlations in the models, heavily renormalize the quasiparticle spectrum which can explain many unconventional properties of copper oxides in the normal state and also naturally ensure the d-wave superconducting pairing observed experimentally.

1. INTRODUCTION

While in conventional superconductors the picture of the Fermi liquid with a properly determined spectrum of quasiparticles (QP) near the Fermi surface is sufficiently well established, in copper oxides we have many experimental evidences for an anomalous behavior of the low-energy excitation spectra for both spin and charge fluctuations which points to the important role of strong electron correlations in these compounds; see, for example, Kampf (1994), and Brenig (1995).

Therefore, the Bardeen–Cooper–Schrieffer (BCS) theory of the electron-phonon pairing which works perfectly for the system of weakly bounded QP in conventional metals, can be questioned for the system of electrons with strong Coulomb correlations in copper oxides. In spite of an unprecedented scientific activity for more than 10 years we are still far from the solution of these problems and there is no consensus on the

Correlations, Coherence, and Order. Edited by Shopova and Uzunov
Kluwer Academic/Plenum Publishers, 1999

theoretical explanation of the unusual normal and superconducting behavior of high temperature superconductors; for a review of the experiment and the theory see, for example, Plakida (1995).

However, recent experimental evidences in favor of d-wave superconducting pairing in high-T_c cuprates (Van Harlingen, 1995; Kirtley et al., 1995; Tsuei et al. 1996) strongly support the spin-exchange pairing mechanism for the high-temperature superconductivity; see, for example, Schrieffer (1995), and Scalapino (1995). Earlier this mechanism has been proposed by several groups on the basis of some phenomenological models; see, for example, Bickers et al., (1989), Schrieffer et al., (1989), Schrieffer (1991), Pines (1990), and Moria et al. (1990).

In the present lecture we would like to discuss recent results which have been obtained in studies of strong electron correlations on the basis of microscopic models by applying the Green function methods.

1.1 Phase Diagram and AFM Spin Fluctuations

One of the universal features of copper-oxide compounds is the antiferromagnetic (AFM) ordering of copper spins in the CuO_2 planes. In stoichiometric compounds, the copper ions are in the state Cu^{2+} and have one hole with a spin $S = 1/2$ in the $3d$ shell. A strong superexchange interaction (via oxygen ions) between hole spins at copper sites gives rise to a three-dimensional long-range AFM order with relatively high Néel temperatures $T_N = 300 \div 500$ K. However, according to the standard band theory a crystal with one electron (hole) per lattice site should have a half-filled band and should be a good metal. An insulating AFM state in stoichiometric copper-oxide compounds is a very strong evidence for the importance of electron correlations in these materials. A strong on-site, the Hubbard type, Coulomb repulsion for Cu 3d electrons splits the half-filled band and results in an insulating state. By changing the number of charge carriers away from the half filling one can get a narrow band metal with low carrier concentrations which is really observed in copper oxides. It should be also pointed out that the scenario of transition to the AFM insulating state due to the spin density wave instability in copper oxides contradicts to the observation of large cooper site magnetic moment which does not depend on the charge carriers concentration.

Although the long-range three-dimensional order disappears in the metallic and superconducting phases, strong dynamic spin fluctuations with wide spectrum of excitations are observed even at temperatures above 100 K. This fact has led to a number of hypotheses on the possible electron pairing in copper-oxide compounds via magnetic degrees of freedom. The study of AFM properties of high-temperature superconductors is thus important for checking the hypotheses of magnetic mechanisms of superconductivity. The interaction of copper spins in a plane is of two-dimensional nature. Their small value, $S = 1/2$, is the reason behind the important quantum fluctuations. In this respect, besides exploring the interplay of AFM ordering and superconductivity, the study of quantum two-dimensional antiferromagnets is of great interest by itself.

The first indications for existence of the AFM order in copper-oxide compounds have been obtained on the basis of macroscopic measurements of the susceptibility. However, a detailed study of both the magnetic structure and spin correlations in the metallic phase has become possible only with the aid of neutron scattering. Since these experiments require large single crystals, most of the results have been obtained for the compounds $La_{2-x}M_xCuO_4$ (LMCO) and $YBa_2Cu_3O_{6+x}$ (YBCO) which can be synthesized as large crystals of a high quality.

Below the spin correlations and magnetic excitations observed by neutron scattering are briefly discussed. The spin dynamics in the superconducting phase is also responsible for a variation in the spin-lattice relaxation rates of nuclear spins. The main results obtained by the NMR method are also presented.

Spin Dynamics in $La_{2-x}M_xCuO_4$. The stoichiometric La_2CuO_4 is an antiferromagnet with the Néel temperature $T_N \simeq 300$ K. In Fig. 1.1 one can see a phase diagram of $La_{2-x}Sr_xCuO_4$ (LSCO). It shows that the AFM state occurs in the orthorhombic phase (a space group Cmca) with lattice constants $a \simeq c \simeq a_t\sqrt{2}$ and $b \simeq c_t$, where $a_t \simeq 3.78$ Å and $c_t \simeq 13.2$ Å are tetragonal lattice constants.

The spins $S = 1/2$ of Cu^{2+} ions are directed along the orthorhombic axis c, and the AFM modulation is along the a axis with wave vector $\mathbf{Q}_{AF} = (1, 0, 0)$. The value of the magnetic moment at a copper site is $\mu \simeq 0.5\mu_B$. For the copper ion Cu^{2+} with spin $S = 1/2$, the magnetic moment should be equal to $\mu = gS\mu_B = 1.14\mu_B$. The smaller observed value of the magnetic moment may be due to quantum spin fluctuations and influence of the covalent Cu-O bond. In the case of a two-dimensional Heisenberg magnet with spin $S = 1/2$, the spin fluctuations reduce the magnetic moment to 0.62 of its static value, i.e., to $0.68\mu_B$ for Cu^{2+}.

The temperature of the AFM phase transition in LMCO turns out to be very sensitive to the concentration of divalent impurities M = (Ba, Sr) which replace the trivalent La ions and the concentration of oxygen vacancies. The phase diagram in Fig. 1.1 shows that, even at concentration $x = 0.02$, the long-range AFM order already disappears. Only a spin-glass phase, i.e., the phase of frozen spins at copper sites, remains in the region of low temperatures. This phase has been observed most distinctly in μSR experiments. They show that the long-range AFM order disappears at $x \simeq 0.02$, while the static local magnetic field remains up to concentrations $x \simeq 0.08$ including the superconducting phase.

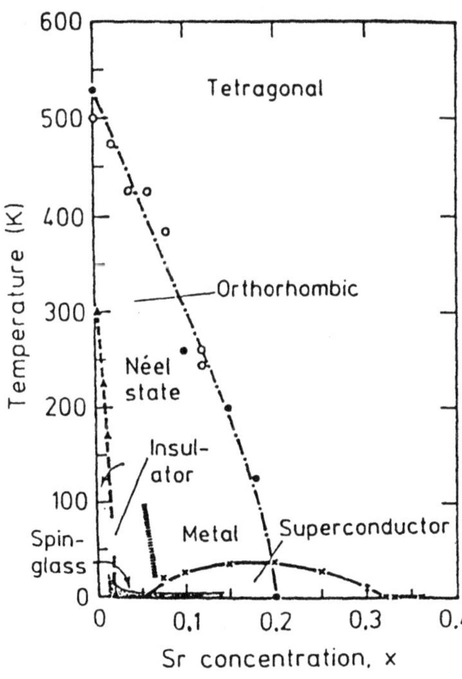

Fig. 1.1 The temperature-concentration phase diagram for $La_{2-x}Sr_xCuO_4$ (Birgeneau and Shirane, 1989).

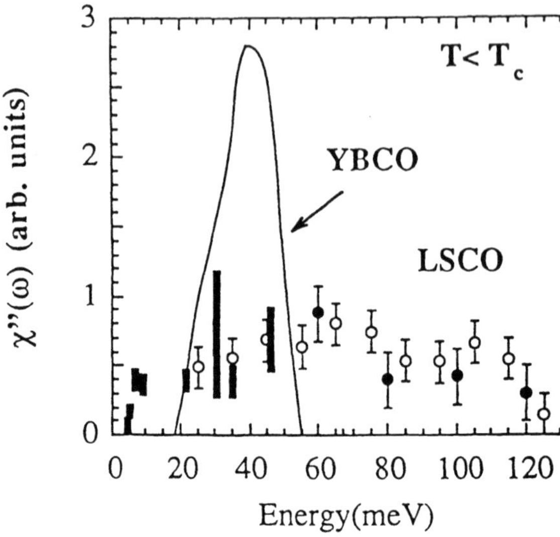

Fig. 1.2 Spin-fluctuation spectra in LSCO and YBCO below T_c (Yamada et al., 1995).

At the same time, experimental studies by means of quasi-elastic and inelastic neutron scatterings in LMCO show the existence of strong AFM correlations of short-range order of copper spins in the region $T > T_N(x)$. In these experiments, a spin pair correlation function has been measured. The Fourier transform of the pair spin correlation function

$$S^{\alpha\alpha}(\mathbf{Q},\omega) = \frac{1}{2\pi N}\int_{-\infty}^{\infty} dt\, e^{i\omega t}\langle S^{\alpha}(-\mathbf{Q},0)S^{\alpha}(\mathbf{Q},t)\rangle \qquad (1.1)$$

determines fluctuations of the spin density

$$S^{\alpha}(\mathbf{Q},t) = \sum_{\mathbf{n}} e^{i\mathbf{Q}\mathbf{n}} S^{\alpha}_{n}(t), \qquad (1.2)$$

where the summation is performed over all lattice sites $\mathbf{n}$. When integrated over all scattering energies (at fixed Q), the function (1.1) determines the static spin correlation function

$$S^{\alpha\alpha}(\mathbf{Q}) = \int d\omega S^{\alpha\alpha}(\mathbf{Q},\omega) = \frac{1}{N}\sum_{\mathbf{n},\mathbf{m}} e^{i\mathbf{Q}(\mathbf{n}-\mathbf{m})}\langle S^{\alpha}_{n} S^{\alpha}_{m}\rangle. \qquad (1.3)$$

Detailed studies of the neutron diffuse magnetic scattering have been carried out by Birgeneau et al. (1989). A measurement of the function $S^{\alpha\alpha}(\mathbf{q}_{\|})$ (1.3) for scattering wave vector in the plane, $\mathbf{q}_{\|}$, has shown that, in La_2CuO_4, the scattering from spin fluctuations is observed far above the Néel temperature. The correlation length $\xi_{\|}$ for magnetic spin correlations in the plane varies from 40 Å at 500 K to 400 Å as the temperature tends to T_N.

Experiments with a three-axes spectrometer where the energy of scattered neutrons is also measured, have enabled the dynamics of spin fluctuations to be investigated. Unlike the low-energy dynamics of spin fluctuations in the vicinity of phase transition in normal three-dimensional magnets, spin excitations in La_2CuO_4 have a high energy at temperatures well above T_N. The dispersion of excitations below T_N is of a two-dimensional character, $\omega(q) = vq_{\|}$, i.e., it does not depend on the component of the wave vector perpendicular to the CuO_2 plane. The velocity of spin excitations turns out to be extremely high, of order of 0.85 eV·Å. This value agrees with the data obtained for spin-wave excitations from the evaluation of the in-plane exchange integral $J_{nn} \simeq 0.132$ eV.

We see that the $S = 1/2$ Heisenberg antiferromagnet La_2CuO_4 reveals unusual properties also in the paramagnetic region of temperatures. They are manifested in the existence of two-dimensional spin correlations at distances of order 200 Å and relatively high-energy spin excitations. Such state was called a quantum spin liquid (QSL) (Birgeneau et al., 1989). The term liquid reflects the fact that the structural factor, i.e., the quantity $S^{\alpha\alpha}(\mathbf{Q})$, is of purely dynamic nature.

Experiments performed on a two-axes spectrometer (Birgeneau et al., 1989) have shown that the correlation length of spin fluctuations drastically falls as the hole concentration increases. The good agreement obtained between the correlation length ξ in these experiments and the average distance between holes $\propto 3.8/\sqrt{x}$ (Å) in the plane shows that holes do destroy the magnetic state in the system of spins at Cu^{2+} ions. The measurement of peak intensities of spin correlations $S^{\alpha\alpha}(\mathbf{Q})$ integrated over the transfer momentum has shown that these quantities which determine the local magnetic moment at Cu^{2+} ion are independent of the dopant concentration. This leads to an important general conclusion that, in doped compounds, holes only affect the correlation of spins at copper ions while they do not change the value of cooper atomic magnetic moment due to its local nature. This observation strongly supports the Mott-Hubbard origin of the insulating state in copper oxides.

The inelastic neutron scattering experiments reveal also an incommensurate antiferromagnetic structure in doped crystals of $La_{2-x}Sr_xCuO_4$. The maxima of inelastic AF wave vectors are given by $Q^*_{AF} = (1, q_y), (q_x, 1)$ where $q_{x,y} = (1 \pm \delta)$, with δ varying from 0 to 0.24 for $0.05 < x < 0.15$ (Mason et al., 1993; Matsuda et al., 1994). In the superconducting state a spin gap has been observed for LSCO at Sr concentration $x = 0.15$ ($T_c = 37$ K) for the incommensurate AFM wave vector Q^*_{AF} (Yamada et al., 1995). The intensity of magnetic scattering drops down for transfer energy $\hbar\omega \lesssim 3.5$ meV at $T < T_c$ and disappears for $T < 15$ K.

The spectrum of spin excitations in the high energy region in the superconducting state has been measured at pulsed neutron sources (Yamada et al., 1995; Hayden et al., 1996). Fig. 1.2 shows the spectrum of spin fluctuation in LSCO at $x = 0.15$ for $T < T_c$ integrated over wave vector q around $Q = (\pi, \pi)$. For a comparison the spectrum of spin fluctuations for YBCO with $T_c \simeq 90$ K is also shown. While LSCO has a very broad spectrum up to the energies of order of 300 meV, the YBCO spectrum has a characteristic peak in the energy region of 20÷60 meV.

Spin Dynamics in $YBa_2Cu_3O_{6+x}$. In the insulating phase of $YBa_2Cu_3O_{6+x}$ (YBCO), i.e., at $x < 0.4$, one observes an AFM ordering of magnetic moments at Cu2* sites in CuO_2 planes. It is quite similar to the antiferromagnetic transition in La_2CuO_4. The antiferromagnetic ordering in the tetragonal phase of YBCO is described by the wave vector $\mathbf{Q}_{AF} = (1/2, 1/2, l)$, which corresponds to a magnetic unit cell with parameters $(a\sqrt{2}, a\sqrt{2}, c)$, where $a \simeq 3.8$ Å and $c \simeq 11.6$ Å are the lattice constants

*Editor's remark: The notation Cu2 is explained by the author in Section 1.2, below Eq. (1.9)

of the tetragonal unit cell. The value of ordered magnetic moment on Cu2 ions at low temperatures in the region $x < 0.2$ is $\mu = 0.64\mu_B$. This value corresponds to the magnetic moment in La_2CuO_4. Due to quantum fluctuations, the value of the magnetic moment in YBCO is also smaller than its static value.

First reliable indications of the antiferromagnetic phase transition have been obtained with the aid of magnetic neutron scattering (Rossat–Mignod et al., 1988). Further investigations of large single crystals have yielded a rather detailed information of both static and dynamic characteristics of the spin subsystem and antiferromagnetic correlations; see, e.g. Tranquada et al. (1988), Shirane et al. (1992), Sato et al. (1993), and Regnault et al. (1994).

Depending on the concentration of oxygen $0 < x < 1$ in $YBa_2Cu_3O_{6+x}$ the AFM spin fluctuation spectrum can be divided into three regions: the spin wave region of the pure antiferromagnetic state at $x < 0.2$, the region of low hole concentration $0.2 < x < 0.4$ and the region of the metallic phase $x > 0.4$, including the superconducting phase with $T_c \simeq 40 \div 90$ K.

The spectrum of spin waves in the pure antiferromagnetic phase can be described with the aid of model Hamiltonian for $S = 1/2$ copper spins in the CuO_2 plane (Rossat-Mignod et al., 1988):

$$H = \sum_{ij}\{J_{ij}^{||} S_i^z S_j^z + J_{ij}^{\perp}(S_i^x S_j^x + S_i^y S_j^y)\}. \tag{1.4}$$

The difference $\Delta J = J^{\perp} - J^{||}$ determines the anisotropy of coupling for the in-plane $J^{\perp}$ and out-of-plane $J^{||}$ spin components which is rather small: $\Delta J = 10^{-4} J$. The coupling constant via oxygen ions for two neighboring spins at Cu2 ions in the plane is $2J \simeq 0.15$ eV, the exchange interaction of spins in the bilayer is $2J_b \simeq 10^{-2} J$ and the exchange interaction of spins at Cu2 ions via Cu1 - O1 chains is $2J' \simeq 10^{-5} J$. The values of these parameters show that the spin dynamics in YBCO should be determined by the Heisenberg model for spins $S = 1/2$ with a very small anisotropy and strong two-dimensional fluctuations, as in LSCO.

Like in LSCO, the appearance of holes in the CuO_2 layers drastically changes the spin dynamics. The correlation length ξ goes down, the damping of spin waves in the plane increases and their velocity decreases. For example, at $x = 0.37$, we have $\xi = 7.5a$ ($n_h = 0.018$), $v = 0.45$ eV· Å. The increase of the hole concentration in the CuO_2 layer and their delocalization at the transition to metallic phase drastically decrease the magnetic correlation length to $\xi \simeq 2.2a$ ($x = 0.45$) which weakly depends on temperature. However, like in LSCO compounds, intensive dynamic antiferromagnetic fluctuations are maintained for wave vectors near rods of the two-dimensional antiferromagnetic diffusion scattering at $Q = (1/2, 1/2, l)$.

The spectrum of spin fluctuations is distributed over a wide energy range from zero up to $40 \div 50$ meV (see Fig. 1.2). The maximum energy $\hbar\omega_{\max}$ increases with the hole concentration. In the region of low energies, $E < E_G$, the intensity of fluctuations falls sharply as temperature decreases. This can be described as the appearance of a gap E_G in the spectrum of spin fluctuations at some temperature T_s close or above the temperature T_c of superconducting phase transition. The gap energy E_G increases with hole concentration and attains the value of $E_G \simeq 28$ meV at $x = 0.92$ ($T_c \simeq 91$ K).

It is interesting to point out that the suppression of the spin-fluctuation spectrum intensity measured by the imaginary part of the dynamic spin susceptibility, $\text{Im}\,\chi(\mathbf{Q}, \omega)$, starts from temperatures T_s higher than $T_c(x)$: for example, $T_s = 75$ K for $T_c = 47$ K while $T_s = 120$ K for $T_c = 91$ K. Below T_s a pseudogap appears in the spin-fluctuation spectrum which is supported by the data on the rate of the spin-lattice relaxation in

NMR experiments.

Spin Dynamics Studied by NMR Methods. The nuclear magnetic resonance (NMR) and the nuclear quadrupole resonance (NQR) studies have played an important role in understanding of nature of low-energy spin and electron excitations in copper-oxide superconductors. These methods yield both static (the Knight shift) and dynamic (the nuclear spin-lattice relaxation rates) characteristics for a given ion and its nearest neighbors. The most complete data have been obtained by NMR and NQR for the nuclei ^{63}Cu, ^{65}Cu, ^{17}O, and ^{89}Y in YBCO compounds. By studying samples with a various oxygen content, one can investigate the variation in electron and spin characteristics of these compounds in the transition from antiferromagnetic insulating phase to the metallic state and superconducting phase; see, e.g, Slichter (1994), and Alloul (1991).

NMR studies show that a picture of single one-component spin liquid made of strongly bound singlet states of two holes at copper and neighboring oxygen sites looks most plausible. The data on the rate of the spin-lattice relaxation obtained in these experiments agree with the density of spin fluctuations measured by neutron experiments.

The width of the NMR line is determined by the rate of longitudinal relaxation $1/T_1$ of magnetization $M_0(t) \propto \exp(-t/T_1)$. In metals the relaxation rate is mainly determined by the interaction of nuclear spins with conduction electrons and can be written as

$$\frac{1}{T_1} = \frac{kT}{\hbar^2\omega_0} \frac{1}{(\hbar\gamma)^2 N} \sum_q A_\perp^2 \mathrm{Im}\chi^{+-}(\mathbf{q},\omega_0). \tag{1.5}$$

Here $\hbar\omega_0 = \hbar\gamma_n H_0(1+K)$ is the Zeeman splitting energy in the dc external magnetic field H_0 which determines the frequency of NMR in transverse (rf) field. K is the total shift of NMR frequency due to the interaction of nuclear moment with electrons, γ is the gyromagnetic ratio for electron $\hbar\gamma = 2\mu_B = e\hbar/mc$. $A_\perp$ is the component transverse with respect to $M_0 \propto H_0$ of hyperfine interaction tensor. $\chi^{+-}(q,\omega_0)$ is the dynamic spin susceptibility for a circularly polarized rf field.

Since in NMR experiments the inequality $\hbar\omega_0 \ll kT$ usually holds, the integration in (1.5) can be performed in the limit $\hbar\omega_0 \to 0$. For conduction electrons with Fermi distribution $f(\epsilon_p) = (\exp(\epsilon_p - \mu)/kT + 1)^{-1}$ we can write

$$\frac{1}{T_1} = \frac{\pi kT A_\perp^2}{\hbar} \int_{-\infty}^{\infty} d\epsilon N^2(\epsilon)(-\frac{df}{d\epsilon}) = \frac{\pi kT A_\perp^2}{\hbar} N^2(0), \tag{1.6}$$

where $N(0)$ is the density of electron states per atom and per spin direction at the Fermi energy. In conventional superconductors with a s-wave superconducting gap the Hebel–Slichter peak appears in the spin-lattice relaxation since the density of states $N(0)$ below T_c in (1.6) becomes divergent. However, when $T \to 0$ the density of states $N(0)$ exponentially decays and the spin-lattice relaxation (1.6) also tends to zero exponentially.

The value of the Knight shift K_α^S is determined by the static electron paramagnetic susceptibility $\chi_0^{\alpha\alpha}$:

$$K_\alpha^S = A_{\alpha\alpha}\frac{\chi_0^{\alpha\alpha}}{\hbar^2\gamma\gamma_n} = A_{\alpha\alpha}\left(\frac{\gamma}{\gamma_n}\right)\frac{1}{2}N(0), \tag{1.7}$$

where γ_n is the nuclear gyromagnetic ratio. If we express the constant of hyperfine interaction in terms of Knight shift $K_\perp^S$, we shall get the Korringa relation

$$\frac{1}{T_1 T(K_\perp^S)^2} = \frac{4\pi k}{\hbar}\left(\frac{\gamma_n}{\gamma}\right)^2 \Lambda. \tag{1.8}$$

The coefficient Λ is introduced to take into account corrections due to interactions in the electron gas. For non-interacting electrons, we have $\Lambda = 1$. Apart from universal constants, the Korringa relation contains only quantities which can be directly measured in experiments such as the Knight shift K^S and the spin-lattice relaxation time T_1. It is therefore convenient for processing of experimental data. A deviation from the Korringa law is usually related to additional relaxation mechanisms of nuclear spins.

Concerning the Knight shift K^S measurement in copper oxides it has been observed that it sharply drops down below T_c for optimally doped samples which proves the singlet pairing. However, the power-law decay of K^S as $T \to 0$, both at Cu and O sites, indicates an unconventional pairing with an energy gap having nodes at some k-points of Fermi surface that supports a d-wave pairing symmetry. For the underdoped samples the power-law temperature dependence of K^S below the characteristic temperature $T_s > T_c$ proves a formation of (pseudo)gap in spin-fluctuation spectrum. A remarkable result is the universal scaling of different Knight shifts by one static spin susceptibility which proves that there are no independent spin susceptibility for copper ($3d$ holes) and oxygen ($2p$ holes) sites (Monien et al., 1991). Therefore, the system of holes in the CuO_2 planes should be described by one-component spin quantum liquid.

The measurement of the nuclear relaxation (1.5) in copper oxides shows an anomalous behavior of $1/T_1$ at copper sites in the CuO_2 plane which strongly deviates from the Korringa law, Eq. (1.8). An attempt to compare the relaxation rate and the Knight shift for ^{63}Cu nuclei on the basis of the Korringa law for optimally doped sample yields $\Lambda = 11$ at $T = 100$ K. However, for the relaxation rate of the oxygen nuclei the Korringa law (1.8) holds with constant $\Lambda = 1.4$. These results indicate the existence of an additional mechanism of relaxation at Cu2 nuclei compared to O and Y nuclei. The relaxation at Cu2 nuclei has a more complicated temperature dependence due to the additional contribution of AFM spin fluctuations.

For the undoped samples the investigations have shown that the temperature dependence of the spin-lattice relaxation rate strongly depends on doping. The simple Korringa law (1.8) does not hold for underdoped samples due to the rather complicated temperature dependence of susceptibility. For oxygen nuclei in CuO_2 planes and Y nuclei, however, one can obtain the relation $T_1 \; T \; K^S = \text{const}$. This relation maintains the model of one-component spin liquid.

Concerning the spin-lattice relaxation at $T < T_c$, we should note two very important facts. The first is the absence of a Hebel–Slichter peak at $T \leq T_c$ for ^{17}O nuclei in YBCO. The second is a power-law decrease in the relaxation rate at $T < T_c$ instead of exponential one observed in conventional superconductors. This anomalous behavior of the relaxation rate supports the $d-$wave nature of the superconducting pairing.

To conclude this Section, we stress the importance of AFM dynamic spin fluctuations in copper oxides which are observed even in the metallic phase at optimal doping. The energy of spin fluctuations may be quite high, of order of exchange energy $J \simeq 0.15$ eV. The strong interaction of charge carriers, doped holes, with AFM spin fluctuations results in the strong renormalization of the quasiparticle hole spectrum and may mediate a superconducting pairing with high T_c. This scenario we shall discuss in Sec. 4.

1.2 Electronic Structure and Charge Fluctuations

The superconducting properties of metals are essentially defined by their electronic properties in the normal state. Therefore, studies of latter are important in clarifying the mechanism of superconductivity in new oxide superconductors. With respect to their electronic structure, these compounds may be related to the class of ionic semiconductors, in which the metallic conductivity appears with changes of stoichiometry. Their electronic structure is defined by a rather complicated interaction of localized and band electronic states, which is sensitive to the short-range order in atomic positions.

For understanding of the electronic structure of oxide superconductors experimental studies are essential: X-ray and electron spectroscopies are useful in the high energy range, and optical methods are important in the domain of low-energy electronic excitations. Studies of transport properties – electric resistivity, Hall effect, heat conductivity, etc. – provide an additional information of the electronic structure of these compounds in the metallic phase. For a review of experimental investigations; see, e.g., Plakida (1995).

To discuss the electronic structure of copper-oxide compounds one can choose as a basis the ionic model, in which the state of each atom is described by a certain degree of oxidation, or formal valence z. In this case, the actual charge of ions such as Y^{2+}, La^{3+}, Ba^{2+}, Sr^{2+}, rare-earth ions RE=Nd^{3+}, Eu^{3+}, and others, really proves close to their formal valence in $RBa_2Cu_3O_{7-y}$ (RBCO, R=Y, RE), $La_{2-x}M_xCuO_4$ (LMCO), and in Bi-, Tl-, and Hg-based compounds. At the same time, due to the strong hybridization of copper $3d$ states and oxygen $2p$ states, the charge of ions Cu^{2+} and O^{2-} may differ significantly from their formal valence, and depend on the degree of doping and the oxygen content. As a result, mixed $3d - 2p$ states appear, in which copper ions with formal valences z =+2, +3 can be represented in the form

$$\begin{aligned} Cu^{2+} &\to \alpha|3d^9 2p^6\rangle + \beta|3d^{10}2p^5\rangle, \\ Cu^{3+} &\to \alpha_1|3d^8 2p^6\rangle + \beta_1|3d^9 2p^5\rangle. \end{aligned} \tag{1.9}$$

The coefficients β and β_1 determine the degree of positive charge (hole) transfer from the $3d$ shell of copper to the filled $2p$ shell of oxygen ion O^{2-}. When $La_{2-x}M_xCuO_{4-y}$ is doped with divalent ions of M=Ba^{2+}, Sr^{2+}, Ca^{2+}, the formal valence of copper becomes $z = 2+x-2y$. In $RBa_2Cu_3O_{7-y}$ compounds there are two nonequivalent positions of copper: in-plane Cu2 and in-chain Cu1. Under increase of oxygen content, the formal valence of in-chain copper, Cu1, varies smoothly from Cu^{1+} ($y = 1$) to Cu^{3+} ($y = 0$), and for in-plane copper, Cu2, the valence remains constant, Cu^{2+}, until a transition to metallic phase at $y \leq 0.6$ when a charge transfer from chains to planes with a formation of states Cu^{3+} (1.9) appears. An essential role in this phenomenon is played by the short-range order in oxygen positions in chains Cu1–O1.

Let us discuss the role of crystal field splitting. For a copper atom in spherically symmetric field, $3d$ levels are degenerate in energy. The value of this energy for copper is the minimum in the series of $3d$ elements, $\epsilon^0(\mathrm{Cu}\ 3d) = -20.14$ eV, and it proves lower than the atomic level of oxygen $\epsilon^0(\mathrm{O}2p) = -14.1$ eV. In the crystal field of cubic symmetry O_h, for proper octahedron CuO_6 the five $3d$ levels split into a doublet $e_g = \{d(x^2-y^2), d(3z^2-r^2)\}$ and a triplet $t_{2g} = \{d(xy), d(xz), d(yz)\}$. The value of this splitting is $1 \div 2$ eV. Upon a decrease of the symmetry to tetragonal D_{4h}, a further splitting of $3d$ levels into states $b_{1g}\{d(x^2-y^2)\}$, $a_{1g}\{(3z^2-r^2)\}$, $b_{2g}\{d(xy)\}$ and $e_g = \{d(xz), d(yz)\}$ occurs. Degenerate atomic $2p$ oxygen levels $p(x)$, $p(y)$ and $p(z)$ split in the crystal field D_{2h} site symmetry into 3 levels: $(p\pi_\parallel)$, $(p\pi_\perp)$ and $(p\sigma)$. The π-type states correspond either to in-plane orbitals $p(x)$ or $p(y)$ $(\pi_\parallel)$, or to out-of-plane states $p(z)$ $(\pi_\perp)$ which are directed perpendicular to bond Cu–O. The σ-type states are formed by in-plane orbitals $p(x)$ or $p(y)$, which are directed along Cu–O bonds. These oxygen σ-type orbitals feel the strongest covalent bonding with copper orbitals $d(x^2-y^2)$, which gives rise to broad bonding (σ) and antibonding (σ^*) bands of hybridized $pd\sigma$-states.

It follows from this picture of the electronic structure formation that such material should be a metal with a half-filled antibonding $pd\sigma$ band. We arrive at this conclusion both for electronic structure of La_2CuO_4 and $YBa_2Cu_3O_6$. However, these conclusions contradict to experiments which demonstrate that stoichiometric compounds LCO and YBCO are insulators with a moderately wide gap of $1 \div 2$ eV. This discrepancy is related to the fact that, in the scheme described, one neglects the Coulomb single-site repulsion of $3d$ electrons. The typical value of the Coulomb correlation energy is $U_d \simeq 8 \div 10$ eV, which is much more than the typical width of $pd\sigma$-band $W \simeq 2$ eV leading to a splitting of this band into one- and two-hole subbands (Cu^{2+} and Cu^{3+} in notation of (1.9)). The possibility of correlation splitting has been first noted by Mott (1974) and investigated in the framework of simple model by Hubbard (1963) and Anderson (1959). In these models it has been assumed that $U_d < \Delta$, where Δ is the energy of the anion-cation charge transfer (in our case, $\Delta_{pd} = E_p - E_d$). In this Mott-Hubbard picture and in the Anderson theory of superexchange, anion states can be neglected if one considers only the narrow d-band due to direct d-d exchange. However, the opposite situation when $U > \Delta$ is also possible, as has been first examined by Sawatzky and Allen (1984). Here the insulator correlation gap is defined not by U_d but rather by the energy of charge transfer Δ. It is this case which is realized in cuprates where $\Delta_{pd} \simeq 3 \div 4$ eV$< U_d \simeq 8 \div 10$ eV, so one can call them charge transfer insulators.

After taking into account the Coulomb correlation energy, the picture of electronic structure of copper-oxide compounds can be described as follows. In the insulating phase there is a filled (O2p, Cu3d) band, primarily of O2p-type, which is the valence band, and empty (Cu 3d, O2p) conduction band, primarily of $3d$ type; they are separated by charge transfer gap $E_g \leq \Delta_{pd}$. The conduction band, which is frequently said to be the upper Hubbard band, is separated by a large Coulomb correlation energy U_d from the two-hole (Cu^{3+}) lower Hubbard band.

A large number of experiments, including angular resolved spectroscopy, optical conductivity, electronic Raman scattering have proven the important role of strong electron correlations in copper-oxide materials described above; see, e.g., Brenig (1995), and Battlog (1997).

1.3 Phenomenological Approach

In describing the normal state of copper-oxide compounds, it is very important to study spectrum of antiferromagnetic spin fluctuations which are observed in many experiments as discussed in Sec. 1.1. Due to the strong coupling between spins and charge carriers they strongly renormalize the QP spectrum and can mediate a superconducting pairing of effective QP. In this connection, let us consider at first a phenomenological approach to the theoretical investigations of spin fluctuation spectrum. In Sec. 3.2 we shall present microscopic calculations of the spin susceptibility for $t-J$ model.

In order to describe the contribution of antiferromagnetic spin fluctuations to the spin-lattice relaxation rate in NMR experiments, Millis et al. (1990) have proposed a phenomenological model in which the dynamic spin susceptibility consists of two terms

$$\chi(\mathbf{q},\omega) = \chi_{QP}(\mathbf{q},\omega) + \chi_{AF}(\mathbf{q},\omega) \,. \tag{1.10}$$

The QP term

$$\chi_{QP}(\mathbf{q},\omega) = \bar{\chi}_0 \frac{1}{1 - i\omega\pi/\hbar\Gamma} \tag{1.11}$$

is determined by two parameters, namely, the static susceptibility $\bar{\chi}_0$ and the specific electronic energy $\hbar\Gamma/\pi$ which is of order of Fermi energy E_F. A term due to antiferromagnetic fluctuations

$$\chi_{AF}(\mathbf{q},\omega) = \chi_Q \frac{1}{1+\xi^2(\mathbf{Q}-\mathbf{q})^2 - i\omega/\omega_{sf}} \tag{1.12}$$

is described by the static susceptibility χ_Q for wave vector $\mathbf{Q}$ and the typical energy of antiferromagnetic fluctuation $\hbar\omega_{sf}$. According to Millis et al. (1990) these parameters are connected to QP parameters by the relations

$$\chi_Q = \bar{\chi}_0(\xi/\xi_0)^2, \quad \omega_{sf} = (\hbar\Gamma/\pi)(\xi_0/\xi)^z,$$

where $\xi(T)$ is the correlation length of antiferromagnetic spin fluctuations and $z = 1$ or 2 (Barzykin and Pines, 1995). It is assumed that $(\xi/\xi_0)^2 \gg 1$, and therefore $\chi_Q \gg \bar{\chi}_0$ and $\Gamma \gg \omega_{sf}$. Within the framework of this model a number of anomalous physical properties in the normal state of cuprates have been explained (Pines 1994, 1997).

The effect of Fermi-surface geometry on the spin dynamics in YBCO and LSCO compounds in the framework of slave-boson technique for the $p-d$ model are considered by (Qimiao Si et al., 1993; Zha et al., 1993; Liu et al., 1995). However, a recently observed phenomenon – the pseudogap formation in the electronic spectrum for underdoped cuprates in the normal phase, is still unexplained; see Batlogg (1997).

As has been pointed out above, antiferromagnetic (AF) spin fluctuations are vitally important in explaining many of the anomalous properties of high-temperature superconductors in the normal phase. It is thus natural to suppose that these boson-like excitations are also responsible for superconducting pairing in copper-oxide compounds. Earlier the magnetic mechanism of pairing has been proposed as an explanation for the superconductivity in systems with heavy fermions (Scalapino et al., 1986; Miyake et al., 1986) where d-wave pairing is observed. This has been then investigated with regard to high-temperature superconductors (Bickers et al., 1987). It has been shown that, because of the exchange of AF paramagnon, an attraction appears in d-channel and acts most effectively near the AF instability (Scalapino, 1995).

The retarded interaction between the quasiparticles on the two-dimensional square lattice under exchange of AF paramagnon was most extensively studied by Monthoux et al. (1991, 1992), Monthoux and Pines (1993, 1994). They have considered the model interaction between QP mediated by spin fluctuation exchange

$$V_{mag}^{eff}(\mathbf{q},\omega) = g^2\chi(\mathbf{q},\omega)\sigma_1\sigma_2, \tag{1.13}$$

where the dynamic spin susceptibility is described by the phenomenological model (1.10). The authors solved numerically the Eliashberg equations in $(\mathbf{q},\omega)$ space and observed a superconductivity in the d-wave channel. The results of their calculations can be represented in form of a BCS type formula:

$$T_c = \alpha(\Gamma(T_c)/\pi^2)\exp[-1/\lambda(T_c)], \tag{1.14}$$

where $\Gamma(T_c)/\pi^2$ is the characteristic spin fluctuation energy, $\lambda(T_c) = \eta g_{eff}^2(T_c)\chi_0(T_c)N(0)$ is the effective coupling constant ($\eta \sim 1$ and $\alpha \sim 1$ are constants). The formula (1.14) gives T_c which is experimentally observed for LSCO and YBCO compounds even for sufficiently weak coupling, $\lambda \simeq 0.33 \div 0.48$ due to the large value of the characteristic electron Fermi energy, $\Gamma(T_c) \simeq 0.4$ eV.

Studies of the temperature dependence of gap $\Delta(T)$ show that near T_c the gap grows rapidly when the temperature decreases and reaches maximum value $2\Delta(0)/kT_c = 6 \div 8$ which agrees with the experiment. In these calculations, high T_c and d-wave pairing are conditioned by the quasi-two-dimensional character of electron spectrum and by strongly anisotropic interaction due to AF spin fluctuations. The strong enhancement of the spin susceptibility (1.10) near $\mathbf{Q} = (\pi, \pi)$ is capable of bringing about high T_c in spite of strong pair-braiking effects due to spin scattering.

Above the paramagnetic ground state has been considered. For the antiferromagnetic ground state, transverse fluctuations of AF order parameter ensure that the effective attraction is more favorable for the manifestation of magnetic pairing mechanism. In this connection, we shall consider the concept of the spin bag proposed by Schrieffer et al. (1989). Although in copper-oxide compounds the superconductivity arises for hole concentration $n_h \geq 0.05$, where the long-range AF order is already absent, due to the small superconducting correlation length ξ_0, we can use the theory in the region $\xi_N > \xi_0$ (where ξ_N is correlation length of AF spin fluctuations).

The theory of Schrieffer et al. (1989) is based on the assumption of local depression by a hole of AF gap in the electron spectrum. As a result a magnetic polaron, spin bag which moves together with the cloud of spin deformation, appears. In this case, two holes, i.e., polarons, attract each other due to the overlap of their regions of deformation of AF order. A more detailed analysis shows that the longitudinal spin fluctuations lead to singlet pairing with the maximum contribution coming from the *d*-wave channel. The transverse fluctuations here suppress the effective attraction.

In the paramagnetic phase, the contribution of spin fluctuations near the AF wave vector, $\mathbf{q} = \mathbf{Q}$, leads to the appearance of a pseudo-gap in the electron spectrum near half-filling (Kampf and Schrieffer, 1990). The additional exchange of AF spin fluctuations for two quasi-particles, spin bags, decreases their energy similar to the case of long-range AF order, and results in their mutual attraction. In this case, the effective interaction potential $V_{\mathbf{k}\,\mathbf{k}'}$ is attractive in the range of small momentum transfer $\mathbf{q} = \mathbf{k} - \mathbf{k}{,}'$ and repulsive for large $\mathbf{q} \simeq \mathbf{Q}$. Therefore, in this case, the symmetry permits a superconducting order parameter, $\Delta(\mathbf{k})$, and the value of T_c strongly depends on the form of Fermi surface for quasiparticles.

Thus, the pairing mechanism due to two-magnon scattering processes proposed by Kampf and Schrieffer (1990) is of universal character but requires complicated self-consistent calculations of the quasi-particle spectrum and spin fluctuations in order to yield quantitative results. These calculations have been performed under the assumption of weak or intermediate couplings and within the framework of one-band Hubbard model (for $U \leq 4t$). A generalization of the theory in case of strong coupling and for multiband *p-d* model is important in order to compare its conclusions with experiments in copper-oxide superconductors.

At the end of this section we shall discuss recent studies of strong coupling Eliashberg equations for the Hubbard model in the weak coupling limit (Pao and Bickers, 1994; 1995; Monthoux and Scalapino, 1994; Lenck et al., 1994; Dahm and Tewordt, 1995).

In the framework of conserving fluctuation exchange approximation (FLEX), a self-consistent system of equations for the diagonal and off-diagonal one-electron Green functions can be written as (Monthoux et al., 1994)

$$G(p, \omega_n) = \frac{i\omega_n Z(p, \omega_n) + X(p, \omega_n)}{[i\omega_n Z(p, \omega_n)]^2 - X^2(p, \omega_n) - \phi^2(p, \omega_n)}, \tag{1.15}$$

$$F(p, \omega_n) = \frac{\phi(p, \omega_n)}{[i\omega_n Z(p, \omega_n)]^2 - X^2(p, \omega_n) - \phi^2(p, \omega_n)}, \tag{1.16}$$

where $X(p, \omega_n) = \epsilon_p - \mu + \xi(p, \omega_n)$. The renormalization parameter $Z(p, \omega_n)$, the energy shift $\xi(p, \omega_n)$, and the gap parameter $\phi(p, \omega_n)$ obey standard Eliashberg equations with interactions mediated by spin and charge fluctuations:

$$V_s = \frac{3}{2}U^2 \frac{\chi_0^s}{1 - U\chi_0^s} - \frac{1}{2}U^2 \chi_0^s, \tag{1.17}$$

$$V_c = \frac{1}{2}U^2 \frac{\chi_0^c}{1 + U\chi_0^c} - \frac{1}{2}U^2 \chi_0^c. \tag{1.18}$$

The irreducible spin, χ_0^s, and charge, χ_0^c, susceptibilities are calculated self-consistently by using Green functions (1.15), (1.16):

$$\chi_0^{s,c}(q, \omega_n) = -T \sum_{k,m} [G(k + q, \omega_n + \omega_m)G(k, \omega_m)$$

$$\pm F(k + q, \omega_n + \omega_m)F(k, \omega_m)], \tag{1.19}$$

where $\pm$ stands for $\chi_0^s(\chi_0^c)$. Numerical solutions of the self-consistent set of equations for intermediate values of Coulomb repulsion, $U \leq 4t$, for full momentum and frequency dependent renormalization parameter $Z(p, \omega)$, energy shift $\xi(p, \omega)$, and the gap function $\Delta(p, \omega) = \phi(p, \omega)/Z(p, \omega)$ have been obtained. The gap function has been found to have a $d_{(x^2-y^2)}$ symmetry. In the vicinity of antiferromagnetic instability near half filling the superconducting temperature T_c reaches values of order $0.02t \simeq 60$K. Quasiparticle and spin fluctuation spectra in the normal and superconducting states, calculated also on the real frequency axis by Dahm and Tewordt (1995), reveal a strong dependence of spectra on temperature and strong feedback effects for effective interactions arising from the spin and charge susceptibilities as the superconducting gap opens. These direct numerical solutions of the strong coupling Eliashberg equations for the Hubbard model in the weak coupling limit unambiguously confirm the possibility of d-wave pairing mediated by the spin fluctuation exchange. However, the superconductivity exists only in a narrow range of U, very close to the antiferromagnetic instability in this model. Higher order corrections to effective interactions (1.17), (1.18) may be also important.

Thus, the electron exchange interaction through AF spin fluctuations can lead to superconducting pairing. The magnetic pairing mechanism is most effectively manifested in the d-channel where the strong local Coulomb repulsion is suppressed. In spite of pair-breaking effects due to spin scattering which considerably decreases T_c the strong enhancement of spin susceptibility near the AF instability results in d-wave pairing.

An exceptionally strong suppression of T_c in all cuprates by nonmagnetic Zn impurities which also destroy the local magnetic order and in this way suppress effective spin-fluctuation pairing provides a "smoking gun" for magnetic mechanism (Pines, 1994).

2. MICROSCOPIC MODELS

Anderson (1987) was the first who has stressed the importance of strong electron correlations in copper oxides and has proposed to take them into account in the framework of one-band Hubbard model

$$H = -t \sum_{\langle ij \rangle \sigma} (a_{i\sigma}^+ a_{j\sigma} + H.c.) + U \sum_i n_{i\uparrow} n_{i\downarrow}, \tag{2.1}$$

where t is an effective transfer integral for nearest neighbor sites, $\langle ij \rangle$, and U is the Coulomb single-site energy (Hubbard, 1963). He also considered the so-called $t - J$

model. It results from the Hubbard model (2.1) in the strong coupling limit, $U \gg t$, when only singly occupied sites are taken into account since a doubly occupied site costs a large additional energy U.

An important contribution to understanding of the low-energy electronic spectrum of copper-oxides was done by Zhang and Rice (1988) who pointed out the remarkable role of the singlet formation for doped oxygen holes due to strong Coulomb correlations. Starting from the $p-d$ model they derived an effective one-band $t-J$ model for singlet states. The appearance of singlet quasiparticle states inside $p-d$ gap was proven by different methods based on exact diagonalization, cluster calculations, projection technique and other calculations. It should be noted that the commonly used local density approximation cannot describe the singlet band formation due to insufficient treatment of electronic correlations which questions the results of these calculations. In this section at first we consider the two-band effective $p-d$ model and then the $t-J$ model and its r spin - polaron representation.

2.1 Two-band $p-d$ Model

Two most important features of the electronic structure of copper-oxide compounds - strong hybridization of O2p and Cu3d states in the $pd\sigma$-band and strong Coulomb correlations of Cu 3d states - may be accounted for by constructing an effective Hamiltonian for CuO_2 planes, which are the most important structural element in all copper-oxide compounds. To define the parameters of this Hamiltonian, one can use the density functional method and some of its modifications; see, e.g., (Pickett, 1989) as well as the method of cluster calculations; see, e.g., (Sawatzky, 1990).

In the course of constructing an effective Hamiltonian, one has to restrict the number of electron states considered, accounting for the rest of them by a corresponding renormalization of the model parameters. Therefore, parameters of the effective Hamiltonian depend on the number of states selected, and they should be considered only in conjunction with a given set of electronic states. In particular, for the basis of one-electron states in the tight-binding approximation they use atomic wave functions of 3d and 2p states, which do not form a basis of orthogonal states in the crystal. Despite the several advantages of the tight-binding method with a non-orthogonal basis, it cannot be utilized to write down the effective Hamiltonian in the secondary quantization representation which requires an orthogonalized basis of one-electron states. The standard approach to construct such functions in the form of Wannier functions (linear combinations of Bloch functions) is ineffective. Because of their essential delocalization, the Wannier functions inadequately describe 3d—states, which are strongly localized at lattice sites.

In this context, it seems more convenient to use the method of atomic wave functions sequential orthogonalization developed by N. N. Bogolubov in 1946; see Bogolubov (1970) applied to the construction of the polar theory of metals. In this method, the wave function $\Psi_{i\lambda}$ for a state λ at the i—th site of a lattice is written as a linear combination of atomic function $\chi_{i\lambda}$ and functions at nearest sites j of the lattice, i.e.,

$$\Psi_{i\lambda} = \chi_{i\lambda} + \sum_{j\beta} S_{ij}^{\lambda\beta} \chi_{j\beta},$$

where coefficients $S_{ij}^{\lambda\beta}$ are chosen so as to fit the orthogonality condition $\langle \Psi_{i\lambda} \mid \Psi_{j\beta} \rangle = \delta_{ij}\delta_{\lambda\beta}$. This method proves effective for small overlaps of functions at neighboring sites, $\mid \langle \chi_{i\lambda} \mid \chi_{j\lambda} \rangle \mid \ll 1$. With the aid of this method, an effective Hamiltonian can be constructed which accounts for 3d (x^2-y^2) states on Cu sites and $2p_\sigma(x,y)$ states on O sites in CuO_2 planes. In a more general approach, one can account for the other states: $3d(3z^2-r^2)$, $2p_\pi(x,y)$, $2p_\pi(z)$ as well. However, a "first principles" calculation of parameters of the effective Hamiltonian is a hard problem, and therefore they are usually fixed with the aid of indirect calculations and a comparison with experimental data.

In theoretical models of copper-oxide compounds the following three-band effective Hamiltonian (Emery, 1987; Varma et al., 1987) is most frequently used:

$$H=\sum_{i\sigma}\epsilon_i n_{i\sigma}+\sum_{ij\sigma}t_{ij}a^+_{i\sigma}a_{j\sigma}+\frac{1}{2}\sum_{i\sigma}U_i n_{i\sigma}n_{i-\sigma}+\frac{1}{2}\sum_{i\neq j\sigma\sigma'}V_{ij}n_{i\sigma}n_{j\sigma'}. \tag{2.2}$$

Here $n_{i\sigma}=a^+_{i\sigma}a_{i\sigma}$ and $a^+_{i\sigma}(a_{i\sigma})$ are the creation (annihilation) operators for holes of spin σ at site i of the square lattice CuO_2, $\epsilon_i=(\epsilon_p,\epsilon_d)$ are energies of $O2p_\sigma(x,y)$-states and Cu $3d(x^2-y^2)$-states, respectively, $t_{ij}=(t_{pd},t_{pp})$ are transfer integrals for $p-d$ and $p-p$ states at nearest Cu-O and O-O sites, respectively; $U_i=(U_d,U_p)$ are single-site Coulomb correlation energies for $3d$ and $2p$ states, and $V_{ij}=(V_{pd},V_{pp})$ are intersite Coulomb interactions. They choose the state $\mid 3d^{10}2p^6>$ with filled $3d$ and $2p$ shells, which contains no holes, to be the vacuum state in (2.2). The Hamiltonian (2.2) can be rewritten in terms of electronic variables by introducing the number operator for electrons $N_{i\sigma}=(1-n_{i\sigma})$.

In the limit of strong correlations at copper sites, $U_d\to\infty$, and by taking into account only the most important terms in (2.2) we can write

$$H=\epsilon_d\sum_{i,\sigma}\tilde{d}^+_{i\sigma}\tilde{d}_{i\sigma}+\epsilon_p\sum_{m,\sigma}p^+_{m\sigma}p_{m\sigma}+t\sum_{i,m,\sigma}S_{im}\,(\tilde{d}^+_{i\sigma}p_{m\sigma}+h.c.), \tag{2.3}$$

where $\tilde{d}^+_{i\sigma}=d^+_{i\sigma}(1-n_{i\bar{\sigma}})$ denotes the creation of a hole at copper site i provided there is no other hole with spin $\bar{\sigma}=-\sigma$, $p^+_{m\sigma}$ creates a hole at oxygen site m. The hopping $p-d$ integral in the CuO_2 unit cell $t_{im}=tS_{im}$, where $S_{im}=\pm 1$, depends on the position of site m in the unit cell i in agreement with Zhang and Rice (1988). The hopping $p-d$ integral t and the difference between hole energy levels for oxygen and copper, $\Delta=\epsilon_p-\epsilon_d$, are the only two parameters in model (2.3).

The Hamiltonian (2.3) can be further reduced to an effective two-band Hubbard model for one-hole d-like states and two-hole singlet states to describe the low-energy electronic spectrum of CuO_2 plane in cuprates. To derive the singlet band proposed by Zhang and Rice (1988) it is reasonable to apply the following perturbation procedure (Lovtsov et al., 1991; Hayn et al., 1993). Introducing the symmetric combination of oxygen operators $p^{(s)}_{i\sigma}$ in the unit cell i according to Zhang and Rice (1988) we can define orthogonal Wannier states $c_{i\sigma}$ by the equation

$$p^{(s)}_{i\sigma}=\frac{1}{2}\sum_m S_{im}p_{m\sigma}=\sum_j \nu_{ij}\;c_{j\sigma}. \tag{2.4}$$

The overlapping parameters

$$\nu_{jl}=\frac{1}{N}\sum_k\sqrt{1-\frac{1}{2}(\cos k_x+\cos k_y)}\;e^{ik(j-l)} \tag{2.5}$$

decrease rapidly, but nonexponentially with distance $(j-l)$: $\nu_0=\nu_{jj}\simeq 0.96$, $\nu_1=\nu_{j\;j\pm a_{x/y}}\simeq -0.14$, and $\nu_2=\nu_{j\;j\pm a_x\pm a_y}\simeq -0.02$. Using orthogonal Wannier states $c_{i\sigma}$ in (2.4) we can write the Hamiltonian in the form

$$H=\sum_{i\sigma}\{\epsilon_d\;\tilde{d}^+_{i\sigma}\tilde{d}_{i\sigma}+\epsilon_p\;c^+_{i\sigma}c_{i\sigma}+V_0\;(\tilde{d}^+_{i\sigma}c_{i\sigma}+h.c.)\}$$

$$+\sum_{i\neq j\sigma} V_{ij}\ \{\tilde{d}^{+}_{i\sigma} c_{j\sigma} + h.c.\}, \tag{2.6}$$

where $V_{ij} = 2t\ \nu_{ij}$ and $V_0 = 2t\ \nu_0$. Since $|V_{ij}| \ll V_0$ one can consider the last inter-cell term in (2.6) as a small perturbation to the intra-cell part given by the first term in (2.6). This approach has been called a cell-perturbation method (Feiner et al., 1996; Raimondi et al., 1996).

According to this approach, the first intra-cell part can be diagonalized within one unit cell. That gives for the lowest one-hole d-type state

$$|D_\sigma\rangle = \cos\theta_1\ d^{+}_{\sigma}|0\rangle - \sin\theta_1\ c^{+}_{\sigma}|0\rangle. \tag{2.7}$$

The two-hole state with the lowest energy is the singlet state

$$|\psi\rangle = \cos\theta_2\ \frac{1}{\sqrt{2}}\ (d^{+}_{\uparrow}c^{+}_{\downarrow} - d^{+}_{\downarrow}c^{+}_{\uparrow})\ |0\rangle - \sin\theta_2\ c^{+}_{\uparrow}c^{+}_{\downarrow}\ |0\rangle, \tag{2.8}$$

where the vacuum state $|0\rangle$ has no holes and $\tan 2\theta_1 = 2V_0/\Delta$, $\tan 2\theta_2 = 2\sqrt{2}V_0/\Delta$. The corresponding one-hole E_D and two-hole energies E_ψ are given by

$$E_D = \frac{1}{2}(\epsilon_d + \epsilon_p) - \frac{1}{2}\sqrt{\Delta^2 + 4V_0^2}\,, \tag{2.9}$$

$$E_\psi = \frac{1}{2}(\epsilon_d + 3\epsilon_p) - \frac{1}{2}\sqrt{\Delta^2 + 8V_0^2}\,. \tag{2.10}$$

Another one-hole p-type state (with sign + in (2.9)) has a higher energy than the d-type state (2.7) and can be neglected in the subspace of one-hole states. The singlet states (2.8) are the lowest among two-hole states and have to be filled first with doping. At small doping we can also neglect triplet states with energy $E_\tau = (\epsilon_d + \epsilon_p)$ since mixing between singlet and triplet bands is rather small (Hayn et al., 1993).

By introducing Hubbard operators in the subspace of one-hole states $|D_\sigma\rangle$ (2.7) and singlet states $|\psi\rangle$ (2.8), i.e.,

$$X_i^{\sigma\sigma} = |D_{i\sigma}\rangle\langle D_{i\sigma}|, \quad X_i^{\sigma 0} = |D_{i\sigma}\rangle\langle 0|, \tag{2.11}$$

$$X_i^{22} = |\psi_i\rangle\langle\psi_i|\,, \quad X_i^{20} = |\psi_i\rangle\langle 0|\,, \quad X_i^{2\sigma} = |\psi_i\rangle\langle D_{i\sigma}|, \tag{2.12}$$

we can write the intra-cell part of the effective Hamiltonian in diagonal form

$$H_0 = E_D \sum_{i\sigma} X_i^{\sigma\sigma} + E_\psi \sum_i X_i^{22}. \tag{2.13}$$

By projecting original p- and d-operators in the inter–cell part of the Hamiltonian (2.6) onto the subspace of one– and two–hole states (2.7), (2.8)

$$c^{+}_{\sigma} = \sigma A_c X^{2\bar{\sigma}} - \sin\theta_1 X^{\sigma 0} \qquad \tilde{d}^{+}_{\sigma} = \sigma A_d X^{2\bar{\sigma}} + \cos\theta_1 X^{\sigma 0}, \tag{2.14}$$

where $\sigma = \pm 1$, we can write the inter–cell term in (2.6) in the form

$$H_{int} = \sum_{i\neq j\sigma} \{t^{\psi}_{ij} X_i^{2\sigma} X_j^{\sigma 2} + t^{D}_{ij} X_i^{\sigma 0} X_j^{0\sigma} + \sigma t^{\psi D}_{ij} (X_j^{2\bar{\sigma}} X_j^{0\sigma} + X_i^{\sigma 0} X_j^{\bar{\sigma} 2})\}. \tag{2.15}$$

Effective hopping parameters are given by (Hayn et al., 1993)

$$\begin{aligned} t^{\psi}_{ij} &= V_{ij} K_{\psi\psi} & K_{\psi\psi} &= 2A_d A_c \\ t^{D}_{ij} &= V_{ij} K_{DD} & K_{DD} &= -2\sin\theta_1\cos\theta_1 \\ t^{\psi D}_{ij} &= V_{ij} K_{\psi D} & K_{\psi D} &= A_c\cos\theta_1 - A_d\sin\theta_1 \end{aligned} \tag{2.16}$$

with coefficients

$$A_d = -\frac{1}{\sqrt{2}} \sin\theta_1 \cos\theta_2 ,$$

$$A_c = \sin\theta_1 \sin\theta_2 + \frac{1}{\sqrt{2}} \cos\theta_1 \cos\theta_2 . \tag{2.17}$$

Therefore, the total Hamiltonian of the two-band model for d-like holes and singlets takes the form

$$H = H_0 + H_{int} - \mu N , \tag{2.18}$$

where we have introduced the chemical potential μ and the number operator

$$N = \sum_i N_i = \sum_i (2X_i^{22} + \sum_\sigma X_i^{\sigma\sigma}). \tag{2.19}$$

It is easy to prove that the number operator (2.19) acting in the subspace of one- and two-hole states (2.7) and (2.8) satisfies the necessary condition $[N, H_0 + H_{int}] = 0$. This condition is not satisfied for the number operator of original p- and d-holes in (2.3) written in terms of Hubbard operators given by (2.14) since higher energy one- and two-hole states have been ignored in the model Hamiltonian (2.18).

To prove the importance of the hybridization term in (2.15) between D-holes and singlets we shall estimate hopping parameters (2.16) for the case of strong intra-cell coupling: $2t = \Delta = \epsilon_p - \epsilon_d$. The direct calculation in (2.16) gives

$$K_{\psi\psi} \simeq -0.477, \quad K_{DD} \simeq -0.887, \quad K_{\psi D} \simeq 0.834 . \tag{2.20}$$

This estimation shows that ψ-D hybridization is rather strong being much larger than the singlet-triplet coupling $K_{\psi\tau}$ considered by Hayn et al. (1993): $K_{\psi D} \gg | K_{\psi\tau} | \simeq 0.08$. In the limit of small p-d hybridization, $t/\Delta \to 0$, while all coefficients $K_{\psi\psi}$, K_{DD} and $K_{\psi\tau}$ tend to zero, $K_{\psi D}$ has a finite value, $K_{\psi D} \to 1/\sqrt{2}$. Therefore, in this limit the effective hopping parameter $t_{ij}^{\psi D}$ vanishes linearly with $V_{ij} \propto t$ while all others, $t_{ij}^{\psi}, t_{ij}^{D}$, are proportional to (t^2/Δ). As a result, the inter-band ψ-D hybridization gives a rather strong renormalization of the singlet band dispersion being of the same order, (t^2/Δ), as original one t_{ij}^{ψ}.

In comparison with the the two-band Hubbard-like model (2.18) original $p-d$ model (2.3) takes into account the formation of a new singlet band for doped p-holes due to strong Coulomb correlations at copper sites. The appearance of the singlet band due to many-body correlations has been proven by different methods (see Feiner et al., 1996; Raimondi et al., 1996, and references therein) but it cannot be obtained in the framework of standard band-structure calculations based on local-density approximation. In general, the two-band $(p-d)$ model (2.18) can be considered as a standard Hubbard model with one-hole and two-hole (lower and upper) subbands but with asymmetric hopping parameters (2.16) and with the single-site correlation energy U being substituted by the charge-transfer energy $\Delta = \epsilon_p - \epsilon_d$. Therefore we can apply to this model well-developed methods in the theory of the standard Hubbard model.

2.2 The $t-J$ Model

To obtain the $t-J$ model from the Hubbard model (2.1) we consider the reduced Hilbert space by introducing projector operators P and $Q = 1 - P$ which project states they act on onto the space of configurations with singly and doubly occupied sites, respectively. By applying the operator perturbation approach (Bogolubov, 1970; for details, see, Fulde, 1991) we can write the reduced Hamiltonian in form

$$\tilde{H} = PHP - \frac{1}{U} PHQHP. \tag{2.21}$$

For the Hubbard model (2.1) we get (Chao et al., 1977; Gros et al., 1987)

$$\tilde{H} = H_{t-J} = -t \sum_{\langle ij\rangle,\sigma} (\tilde{a}^{+}_{i\sigma}\tilde{a}_{j\sigma} + H.c.) + J \sum_{\langle ij\rangle} (\mathbf{S}_i\mathbf{S}_j - \frac{1}{4} n_i n_j)$$

$$-\frac{t^2}{U} \sum_{\langle ijk\rangle,\sigma} (\tilde{a}^{+}_{k\sigma} n_{j-\sigma} \tilde{a}_{i\sigma} - \tilde{a}^{+}_{k\sigma} a^{+}_{j-\sigma} a_{j\sigma} \tilde{a}_{i-\sigma} + H.c.), \tag{2.22}$$

where the first term describes electron hopping with the energy t for nearest neighbors. Electron operators $\tilde{a}^{+}_{i\sigma} = a^{+}_{i\sigma}(1 - n_{i-\sigma})$ act in the space without double occupancy and $n_i = n_{i\uparrow} + n_{i\downarrow}$ is the number operator for electrons. The second term describes the spin–1/2 Heisenberg antiferromagnet (AFM) with an exchange energy J for nearest neighbors which is equal to $J = 4t^2/U$ for the Hubbard model (2.1). The third three-site term for i, k sites, which are the nearest neighbors to j, describes indirect hopping processes between next-nearest neighbor sites i and k. Since the third term is only of order t/U when compared with the first hopping term, it is usually discarded and will be neglected further. The same reduction procedure can be applied to the asymmetric Hubbard model (2.18). As a result we obtain the $t - J$ Hamiltonian in the form

$$H_{t-J} = -\sum_{i\neq j,\sigma} t_{ij} \tilde{a}^{+}_{i\sigma}\tilde{a}_{j\sigma} + \frac{1}{2}\sum_{i\neq j} J_{ij}(\mathbf{S}_i\mathbf{S}_j - \frac{1}{4} n_i n_j) \, . \tag{2.23}$$

Here hopping parameters t_{ij} for the singly occupied subband are equal to $t_{ij} = t^{D}_{ij}$ and for the doubly occupied (singlet) subband are equal to t^{ψ}_{ij}. The exchange interaction is given by the hybridization term, $J_{ij} = 2(t^{D\psi}_{ij})^2/\Delta$ and can be considered as an independent parameter. A proper reduction scheme of the full 3-band model (2.2) to the effective $t - J$ model (2.22) has been given recently by Yushankhai et al. (1997). In the $t - J$ model (2.22) (or (2.23)) two main features of the doped hole motion in copper-oxides are properly taken into account: the constraint on excluding a double occupancy for holes on lattice sites due to strong electron correlations and the interaction of holes with AFM spin fluctuations which result in a strong renormalization of the quasiparticle spectrum; for a review, see, Izyumov (1997).

To take into account on a rigorous basis the exclusion of doubly occupied states in the electronic hopping we employ the Hubbard operator (HO) technique (Hubbard, 1965). The HO's are defined as

$$X_i^{\alpha\beta} = |i,\alpha\rangle\langle i,\beta| \tag{2.24}$$

for three possible states at the lattice site i, where

$$|i,\alpha\rangle = |i,0\rangle, \quad |i,\sigma\rangle \, , \tag{2.25}$$

for an empty site and for a singly occupied site by an electron with spin $\sigma/2$ ($\sigma = \pm 1$), respectively. In the $t - J$ model only singly occupied sites are retained and the completeness relation for HO's reads as

$$X_i^{00} + \sum_{\sigma} X_i^{\sigma\sigma} = 1 \, . \tag{2.26}$$

The spin and density operators in Eq. (2.23) are expressed by HO's as

$$S_i^{\sigma} = X_i^{\sigma\bar{\sigma}}, \quad S_i^z = \frac{1}{2}\sum_{\sigma} \sigma X_i^{\sigma\sigma}, \quad n_i = \sum_{\sigma} X_i^{\sigma\sigma}, \tag{2.27}$$

where $\bar{\sigma} = -\sigma$. HO's obey the following multiplication rules

$$X_i^{\alpha\beta} X_i^{\gamma\delta} = \delta_{\beta\gamma} X_i^{\alpha\delta} \tag{2.28}$$

and commutation relations

$$\left[X_i^{\alpha\beta}, X_j^{\gamma\delta}\right]_{\pm} = \delta_{ij}\left(\delta_{\beta\gamma} X_i^{\alpha\delta} \pm \delta_{\delta\alpha} X_i^{\gamma\beta}\right). \tag{2.29}$$

In Eq. (2.29) the upper sign stands for the case when both HO's are Fermi-like (as, e. g., $X_i^{0\sigma}$). The spin and density operators (2.27) are Bose-like and for them the lower sign in Eq. (2.29) should be taken.

By using the Hubbard operator representation (2.24) for $\tilde{a}_{i\sigma}^{+} = X_i^{\sigma 0}$ and $\tilde{a}_{j\sigma} = X_j^{0\sigma}$ and (2.27) for spin and number operators we can write the Hamiltonian of the $t-J$ model (2.23) as

$$H_{t-J} = -\sum_{i\neq j,\sigma} t_{ij} X_i^{\sigma 0} X_j^{0\sigma} - \mu \sum_{i\sigma} X_i^{\sigma\sigma} + \frac{1}{4}\sum_{i\neq j,\sigma} J_{ij}\left(X_i^{\sigma\bar{\sigma}} X_j^{\bar{\sigma}\sigma} - X_i^{\sigma\sigma} X_j^{\bar{\sigma}\bar{\sigma}}\right), \tag{2.30}$$

where we have introduced the chemical potential μ which can be calculated from the equation for the average number of electrons

$$n = \sum_{i,\sigma} \langle X_i^{\sigma\sigma} \rangle . \tag{2.31}$$

Unconventional commutation relations (2.29) for HO's demand to employ a special, very complicated diagram technique for the treatment of the $t-J$ model; see Izyumov et al. (1993). To overcome this problem we will use in next Sections much more simple equation of motion method for two-time Green functions (Zubarev, 1960) in terms of HO's (2.24).

2.3 Spin Polaron Model

For a low concentration of holes when the long range AFM order is preserved or at least strong AFM correlations for nearest-neighbors still govern the hole motion, the $t-J$ model (2.23) can be reduced to a more simple spin polaron model as has been proposed by Schmitt-Rink et al. (1988), and Kane et al. (1989). Studies of this model (Schmitt-Rink et al., 1988; Kane et al., 1989; Trugman, 1990; Martinez and Horsch, 1991; Liu and Manouskas, 1992; Plakida et al., 1994; Scherman and Schrieber, 1993, 1994; and others) predict that the doped hole dressed by strong antiferromagnetic spin fluctuations can propagate coherently as a quasiparticle – a spin polaron, with a weight $Z_k \simeq J/t$. Besides a narrow quasiparticle band of order J there is a broad incoherent band at higher energies. It is quite natural to suggest that the same spin fluctuations could mediate superconducting pairing of spin polarons.

To take into account explicitly the two-sublattice structure for Heisenberg AFM we introduce two sublattices with spin up ($i \in\uparrow$) and spin down ($i \in\downarrow$). We define the hole spinless fermion operators for two sublattices by the equation

$$\tilde{a}_{i\uparrow} = h_i^{+},\ \tilde{a}_{i\downarrow} = h_i^{+} S_i^{+}\ (i \in\uparrow); \quad \tilde{a}_{i\downarrow} = f_i^{+},\ \tilde{a}_{i\uparrow} = f_i^{+} S_i^{-}\ (i \in\downarrow), \tag{2.32}$$

where S_i^+, S_i^- are spin operators for the corresponding sublattices. In the linear spin-wave approximation (LSWA) the exchange part of the Hamiltonian (2.23) can be written as (Liu and Manouskas, 1992)

$$H_J = \sum_q \omega_q(\alpha_q^+ \alpha_q + \beta_q^+ \beta_q) + E_0^J \ , \tag{2.33}$$

where $\alpha_q^+(\alpha_q)$ and $\beta_q^+(\beta_q)$ are magnon creation (annihilation) operators coupled to spin lowering operators on two sublattices in LSWA: $S_i^+ \simeq a_i$, $(i \in\uparrow)$, $S_i^+ \simeq b_i^+$, $(i \in\downarrow)$ by the Bogolubov canonical transformation in k-space:

$$a_{\mathbf{k}} = v_k \alpha_{\mathbf{k}} + u_k \beta_{-\mathbf{k}}^+, \quad b_{\mathbf{k}} = v_k \beta_{\mathbf{k}} + u_k \alpha_{-\mathbf{k}}^+, \tag{2.34}$$

$$u_k = \left(\frac{1+\nu_k}{2\nu_k}\right)^{1/2}, \quad v_k = -\mathrm{sign}(\gamma_k)\left(\frac{1-\nu_k}{2\nu_k}\right)^{1/2} \tag{2.35}$$

with $\nu_k = \sqrt{1-\gamma_k^2}$, $\gamma_k = (\cos ak_x + \cos ak_y)/2$. The spin-wave energy is given by $\omega_k = SzJ(1-\delta)^2\nu_k$ with $\delta = 1-n$ being the hole concentration and $z = 4$ being the number of nearest neighbors (for a two-dimensional square lattice). In the derivation of exchange part of the Hamiltonian (2.33) the contact interaction between holes has been taken into account only in the mean field approximation that results in a renormalization of the magnon energy proportional to the factor $(1-\delta)^2$.

By employing the two-sublattice representation (2.32) for holes and LSWA we get the following expression for the hopping part of the Hamiltonian (2.23):

$$H_t \simeq \sum_{kq}(h_k^+ f_{k-q}[g(k,q)\alpha_q + g(q-k,q)\beta_{-q}^+] + H.c.)$$
$$+\sum_k(\epsilon_k - \mu)(h_k^+ h_k + f_k^+ f_k) \ , \tag{2.36}$$

where

$$g(k,q) = \frac{zt}{\sqrt{N/2}}(u_q\gamma_{k-q} + v_q\gamma_k) \ , \tag{2.37}$$

and the next nearest neighbor hopping energy is $\epsilon_k = 4t' \cos ak_x \cos ak_y$. The summation over wave-vectors in (2.33), (2.36) is restricted to $N/2$ points in the AF Brillouin zone. The chemical potential μ should be calculated self-consistently as a function of hole concentration δ and temperature T from the equation

$$\delta - \langle h_i^+ h_i\rangle + \langle f_i^+ f_i\rangle \ . \tag{2.38}$$

In Sec. 4 we shall discuss the quasiparticle hole spectrum and superconducting pairing in the framework of spin-polaron model (2.33), (2.36).

3. SPIN AND CHARGE FLUCTUATIONS

As has been discussed in Sec. 1, magnetic and charge responses in copper oxides show an anomalous behavior which is believed to be due to strong electron correlations. These anomalous properties are: a complicated temperature dependence of the spin susceptibility, spin (pseudo)gap in the dynamic spin susceptibility, linear temperature dependence of the resistivity and anomalous frequency dependence of the optical conductivity, the so-called midinfrared absorption, which are considered as manifestations of strong electron correlations. To prove this conjecture one has to calculate spin and charge susceptibilities for models with strong electron correlations. At first we shall consider the optical conductivity (and resistivity) for the $t-J$ model (2.23) and the effective two-band $p-d$ model (2.10).

3.1 Optical Conductivity in the $p-d$ and $t-J$ Models

Several rigorous results for the optical conductivity has been obtained for one-dimensional Hubbard and $t-J$ models; for references, see Dagotto (1994). By using the Bethe ansatz exact solution the charge stiffness, or the Drude weight D, has been calculated near the Mott–Hubbard metal-insulator transition. However, to obtain the frequency dependence of the optical conductivity some approximations or numerical calculations for finite chains have been used.

There are few analytical calculations of the optical conductivity in two- and three-dimensional microscopic models with strong electron correlations. Among them are the early theoretical studies of the optical conductivity in the framework of Kubo (1957) linear response theory done by a high-temperature expansion (Ohata, 1970) and by an equation of motion method for the Green functions in Hubbard I (K. Kubo, 1971) and Hubbard III approximations (Ohata, 1971; Sadakata et al., 1973).

Later the $t-J$ model has been considered by Rice and Zhang (1989) in the limit of small J when the hole motion has a diffusive character described by an incoherent spectrum for single particle excitations. They obtained a non-Drude conductivity, $\sigma(\omega) \propto 1/\omega$, in the high-frequency limit, $\omega \gg J$. A detailed discussion of optical and photoemission sum rules for one- and two-dimensional Hubbard models has been given by Eskes et al. (1994) by compairing strong coupling perturbation theory in powers of t/U with numerical calculations.

Recently the optical conductivity has been investigated within the dynamic mean-field approach which becomes exact in the limit of infinite dimensions (Rozenberg et al., 1995; Georges et al., 1996). The symmetric Hubbard model at half-filling has been considered near the metal-insulator transition and a semiquantitative agreement with experiments is observed. However, for the realistic two- or three-dimensional Hubbard model nonlocal ($\mathbf{q}$ - dependent) corrections to transport vertices and self-energy could be important.

Most extensively the optical conductivity for two-dimensional models of CuO_2 plane has been studied by numerical methods based on an exact diagonalization of small clusters within the framework of Hubbard or $t-J$ models; for a review, see, Dagotto (1994) and recent papers by Tohyama et al. (1995), Eder et al. (1995), Jaklič et al. (1995). For example, Jaklič et al. (1995) have observed a universal behavior, $\sigma(\omega) \propto (1-\exp(-\omega/T)/\omega$, for the $t-J$ model in the high temperature limit, $T > 0.1t \simeq 500$ K.

However, numerical investigations for small clusters have poor frequency resolution and pronounced finite-size effects in the low temperature and low frequency regions and cannot be quantitatively compared with experimental results. Therefore, analytical self-consistent investigations of one-particle and optical spectral functions in the strong coupling limit for realistic two-dimensional Hubbard and $t-J$ models are required.

In the present section we shall consider the optical conductivity for the effective two-band $p-d$ model (Plakida, 1996) and the $t-J$ model (Plakida, 1997a) by applying the memory function method in terms of Hubbard operators.

Memory Function Method. In the linear response theory of Kubo (1957) the frequency dependent conductivity is defined by the current-current correlation function

$$\sigma_{xx}(\omega) = \frac{1}{V} \int_0^\infty dt e^{\imath\omega t} (J_x(t), J_x) \,, \tag{3.1}$$

where V is the volume of the system, $\Im\omega > 0$, and

$$(A(t), B) = \int_0^\beta d\lambda \langle A(t - \imath\lambda)B\rangle \qquad (3.2)$$

is the Kubo–Mori scalar product for operators in the Heisenberg representation, $A(t) = \exp(\imath Ht)A\exp(-\imath Ht)$, $\langle AB\rangle$ denotes an equilibrium statistical averaging for a system with Hamiltonian H, $\beta = 1/T$ (here $\hbar = k_B = 1$).

The real absorptive part of the conductivity (3.1) can be written in the form

$$\Re\sigma_{xx}(\omega) = \frac{1 - \exp(-\beta\omega)}{V\omega}\Re\int_0^\infty dt e^{\imath\omega t}\langle J_x(t)J_x\rangle \,. \qquad (3.3)$$

This fluctuation-dissipation equation is often used in numerical calculations. It is also convenient for estimations of the high-frequency conductivity. However, to obtain an interpolation formula for the dynamic conductivity which will be valid from the hydrodynamic to the optical frequency regions, it is much more convenient to employ the memory function method as has been discussed by Götze and Wölfle (1971, 1972). Later in a vast literature it has been proven that the Götze–Wölfle formulation is very efficient in calculations of the dynamic conductivity. Below we shortly formulate the memory function approach in a slightly different form to overcome the problem of perturbation calculation of the memory function.

To calculate the conductivity (3.1) we will apply the equation of motion method for retarded two-time Green functions (GF) (Zubarev, 1960; Tserkovnikov, 1982) to the scalar product (3.2),

$$((A|B))_\omega = -\imath\int_0^\infty dt e^{\imath\omega t}(A(t)), B), \qquad (3.4)$$

and to commutator GF

$$\langle\langle A|B\rangle\rangle_\omega = -\imath\int_0^\infty dt e^{\imath\omega t}\langle[A(t), B]\rangle \,. \qquad (3.5)$$

Here $\Im\omega > 0$ and operators have zero average values: $\langle A\rangle = \langle B\rangle = 0$. The conventional dynamic susceptibility is given by

$$\chi_{AB}(\omega) = -\langle\langle A|B\rangle\rangle_\omega \,. \qquad (3.6)$$

GF's (3.3), and (3.5) are coupled by the equation

$$\omega((A|B))_\omega = \langle\langle A|B\rangle\rangle_\omega - \langle\langle A|B\rangle\rangle_{\omega=0} \,. \qquad (3.7)$$

We have also following useful relations:

$$((\imath\dot{A}|B))_\omega = ((A| - \imath\dot{B}))_\omega = \langle\langle A|B\rangle\rangle_\omega \,, \qquad (3.8)$$

$$(\imath\dot{A}, B) = (A, -\imath\dot{B}) = \langle[A, B]\rangle \,, \qquad (3.9)$$

where $\imath\dot{A} = \imath dA/dt = [A, H]$.

By using the above given definitions and writing the current operator as time derivative of the polarization operator, $J_x = \dot{P}_x$, we obtain the following equivalent representations for the optical conductivity (3.1):

$$\sigma(\omega) = \frac{\imath}{V}((J|J))_\omega = \frac{1}{V}\langle\langle P|J\rangle\rangle_\omega = \frac{\imath}{V\omega}[\chi_{JJ}(0) - \chi_{JJ}(\omega)] \,, \qquad (3.10)$$

where we have omitted the indices of operators J_x, P_x. By employing the standard dispersion relation for GF (3.5) or susceptibility (3.6) we readily get the sum rule for the real, or absorptive part, of the conductivity (3.10):

$$\int_0^\infty d\omega \Re\sigma_{xx}(\omega) = \frac{1}{V}\int_0^\infty d\omega \frac{\Im\chi_{JJ}(\omega)}{\omega} = \frac{\pi}{2V}\Re\chi_{JJ}(0) = \frac{\imath\pi}{2V}\langle[J_x, P_x]\rangle \ . \quad (3.11)$$

Sum rule (3.11) has been extensively used by many authors to discuss the metal-insulator transitions in the Hubbard model since the right hand side of (3.11) can be calculated from static correlation functions.

To calculate the current-current correlation function for the conductivity (3.10) it is convenient to employ the memory function approach of Mori (1965) in form slightly different from that one used by Götze and Wölfle (1972). We define the memory function $M_{JJ}(\omega) \equiv M(\omega)$ by the equation

$$\Phi_{JJ}(\omega) = ((J|J))_\omega = \frac{\chi_0}{\omega + M(\omega)} \ , \quad (3.12)$$

where $\chi_0 = \chi_{JJ}^0$ and

$$M(\omega \pm \imath\delta) = M'(\omega) \pm \imath M''(\omega).$$

Here $M'(\omega) = \Re M(\omega)$ and $M''(\omega) = \Im M(\omega)$ are real functions.

We calculate the memory function by using the GF equation of motion:

$$\Phi_{JJ}(t - t') = \langle\langle J(t); J(t')\rangle\rangle \ . \quad (3.13)$$

By differentiating it in respect to times t and t' we readily get an equation for its Fourier transform (3.4):

$$\Phi(\omega) = \Phi_0(\omega) + \Phi_0(\omega)M_0(\omega)\Phi_0(\omega) \ , \quad (3.14)$$

where

$$\Phi_0(\omega) = \frac{\chi_0}{\omega} \ , \quad (3.15)$$

and the "scattering matrix"

$$M_0(\omega) = -(1/\chi_0)((F_x|F_x))_\omega(1/\chi_0) \quad (3.16)$$

is given by the correlation function for forces

$$F_x = \imath\dot{J}_x = [J_x, H] \ . \quad (3.17)$$

We have also used the relation of orthogonality for current and force:

$$(F_x, J_x) = (\imath\dot{J}_x, J_x) = \langle[J_x, J_x]\rangle = 0.$$

From Eqs. (3.12) and (3.14) we obtain the following relation for the memory function $M(\omega)$ and $M_0(\omega)$ (3.16):

$$M_0(\omega) = -[M(\omega)/\chi_0] - [M(\omega)/\chi_0]\Phi_0(\omega)M_0(\omega) \ . \quad (3.18)$$

A formal solution of this equation by iteration shows that the memory function is just the irreducible part of scattering matrix (3.16) which has no parts connected by single zero order GF $\Phi_0(\omega)$:

$$M(\omega) = ((F_x|F_x))_\omega^{(irred)}(1/\chi_0) \ . \quad (3.19)$$

In the original formulation of memory function approach for calculation of the dynamic conductivity Götze and Wölfle (1972) have calculated the memory function in Eq. (3.18) in a perturbative way. The "exact" memory function representation in terms of irreducible part of the force-force relaxation function, Eq. (3.19), seems to be more convenient. Though, the exact meaning of irreducibility is really given by Mori (1965) in his definition of the memory function in terms of operators with projected time evolution; see also, Tserkovnikov (1982).

Now we can write the frequency dependent conductivity (3.10) by using the representation for GF (3.12) in form of generalized Drude law

$$\sigma(\omega) = \frac{\chi_0}{V} \; \frac{m}{\tilde{m}(\omega)} \; \frac{1}{\tilde{\Gamma}(\omega) - i\omega}, \tag{3.20}$$

where the effective optical mass and the relaxation rate are given by

$$\frac{\tilde{m}(\omega)}{m} = 1 + \lambda(\omega), \quad \tilde{\Gamma}(\omega) = \frac{\Gamma(\omega)}{1 + \lambda(\omega)}, \tag{3.21}$$

with

$$\lambda(\omega) = \frac{M'(\omega)}{\omega}, \quad \Gamma(\omega) = M''(\omega). \tag{3.22}$$

Real and imaginary parts of the memory function are coupled by the dispersion relation

$$M'(\omega) = \frac{1}{\pi} P \int_{-\infty}^{\infty} dz \; \frac{M''(z)}{z - \omega}. \tag{3.23}$$

It is also convenient, by using the spectral representation for GF, to write the relaxation rate given by Eq. (3.22) in terms of conventional time-dependent force-force correlation function

$$\Gamma(\omega) = \frac{1 - \exp(\beta\omega)}{2\chi_0 \omega} \int_{-\infty}^{\infty} dt e^{i\omega t} \langle F_x F_x(t) \rangle \,. \tag{3.24}$$

Here

$$\chi_0 = (J_x, J_x) = i\langle [J_x, P_x] \rangle \tag{3.25}$$

is the static susceptibility.

Relaxation Rates. Let us consider first the results for the $t - J$ model by employing the Hubbard operator representation (2.24). We start with the definition of polarization operator for the $t - J$ model (2.30):

$$P_x = e \sum_i R_i^x N_i = e \sum_i R_i^x \sum_\sigma X_i^{\sigma\sigma} \,, \tag{3.26}$$

where R_i^α are electron coordinates with a charge e on a square lattice. From this definition the following expression for the current operator results

$$J_x = -i[P_x, H] = ie \sum_{i \neq j\sigma} (R_i^x - R_j^x) \, t_{ij} \, X_i^{\sigma 0} X_j^{0\sigma} \,. \tag{3.27}$$

By introducing a **q**-representation for Hubbard operators and hopping integral

$$X_{\mathbf{q}}^{\alpha\beta} = \frac{1}{\sqrt{N}} \sum_i X_i^{\alpha\beta} e^{-i\mathbf{q}\mathbf{R}_i} \,, \quad t(\mathbf{q}) = \sum_{i \neq 0} t_{i0} e^{-i\mathbf{q}\mathbf{R}_i} \quad , \tag{3.28}$$

the current operator (3.27) can be written as

$$J_x = e \sum_{q\sigma} v_x(q)\, X_q^{\sigma 0} X_q^{0\sigma} \,, \tag{3.29}$$

where $v_x(q) = -\partial t(\mathbf{q})/\partial q_x$ is the electron velocity.

Now we calculate the force (3.17) for the current (3.27). It can be written as a sum of two terms:

$$F_x = F_x^t + F_x^J = [J_x, H_t] + [J_x, H_J] \,. \tag{3.30}$$

The first term comes from the kinematic interaction and has the form

$$F_x^t = -\imath e \sum_{i \neq j \neq l} \sum_{\sigma\sigma'} (R_i^x - R_j^x)(t_{ij}\, t_{jl}\, X_i^{\sigma 0} X_l^{0\sigma'} B^t_{j\sigma\sigma'} + \text{H. c.}) \,. \tag{3.31}$$

The second term is proportional to the exchange interaction. It reads

$$F_x^J = \imath e \sum_{i \neq j \neq l} \sum_{\sigma\sigma'} (R_i^x - R_j^x)(t_{ij}\, J_{jl}\, X_i^{\sigma 0} X_j^{0\sigma'} B^J_{l\sigma\sigma'} + \text{H. c.}) \,. \tag{3.32}$$

Here Bose-like operators have been introduced:

$$B^t_{j\sigma\sigma'} = (X_j^{00} + X_j^{\sigma\sigma})\delta_{\sigma'\sigma} + X_j^{\bar\sigma\sigma}\delta_{\sigma'\bar\sigma} \,,$$

$$B^J_{l\sigma\sigma'} = -X_l^{\bar\sigma\bar\sigma}\delta_{\sigma'\sigma} + X_l^{\bar\sigma\sigma}\delta_{\sigma'\bar\sigma} = B^t_{l\sigma\sigma'} - \delta_{\sigma'\sigma} \,. \tag{3.33}$$

In the second line of this equation we have used the completeness relation, Eq. (2.26), for Hubbard operators. Bose-like operators describe electron scattering on spin and charge fluctuations caused by nonfermionic commutation relations (the kinematic interaction) for Hubbard operators and exchange spin-spin interaction. This can be demonstrated explicitly by using the following representation:

$$B^t_{j\sigma\sigma'} = (1 - \frac{1}{2}N_j + \sigma S_j^z)\delta_{\sigma'\sigma} + S_j^{\bar\sigma}\delta_{\sigma'\bar\sigma} \,, \tag{3.34}$$

where $S_j^z = \pm 1/2$ and $S_j^{\bar\sigma} = S_j^{\mp}$ for $\sigma = \pm 1$. It follows from Eq. (3.34) that operators (3.33) can be written in terms of number N_j and spin S_j^α operators.

Now we can write the total force (3.30) in $\mathbf{q}$-representation as

$$F_x = -\frac{e}{\sqrt{N}} \sum_{k,q} \sum_{\sigma\sigma'} v_x(k)\, [t(k-q) - J(q)](X_k^{\sigma 0} X_{k-q}^{0\sigma'} B_{q\sigma\sigma'} + \text{H. c.}) \,, \tag{3.35}$$

where we have neglected the $\mathbf{q} = \mathbf{0}$ term in functions (3.33) and introduced only one function $B_{q\sigma\sigma'} = B^t_{q\sigma\sigma'} = B^J_{q\sigma\sigma'}$ in $\mathbf{q}$–representation:

$$B_{q\sigma\sigma'} = \frac{1}{\sqrt{N}} \sum_i B_{i\sigma\sigma'} e^{-\imath \mathbf{q}\mathbf{R}_i} \,. \tag{3.36}$$

So we can calculate the relaxation rate (3.24). To calculate many-particle time-dependent correlation functions in the right-hand side of Eq. (3.24) we apply the mode-coupling approximation in terms of independent propagation of electron-hole and charge-spin fluctuations. This approximation is essentially equivalent to the self-consistent Born approximation, in which vertex corrections are neglected. The proposed approximation is defined by following decoupling of time-dependent correlation functions:

$$\langle X_k^{\sigma 0} X_{k-q}^{0\sigma'} B_{q\sigma\sigma'} | X_{k'-q'}^{s'0}(t) X_{k'}^{0s}(t) (B_{q'ss'}(t))^\dagger \rangle$$

$$\simeq \delta_{k,k'}\delta_{q,q'}\delta_{s,\sigma}\delta_{s',\sigma'}\langle X_k^{\sigma 0} X_k^{0\sigma}(t)\rangle\langle X_{k-q}^{0\sigma'} X_{k-q}^{\sigma' 0}(t)\rangle\langle B_{q\sigma\sigma'}(B_{q\sigma\sigma'}(t))^{\dagger}\rangle \,. \quad (3.37)$$

There are four correlation functions of this type in (3.24). However, by using the symmetry relations for correlation functions in terms of Bose-like operators (3.33) we can write the final result for the relaxation rate in a compact form:

$$\Gamma(\omega) = \frac{\exp(\beta\omega) - 1}{\chi_0 \omega} \frac{2e^2}{N} \sum_{k,q} g_x^2(k, k-q) \int_{-\infty}^{\infty}\int_{-\infty}^{\infty} d\omega_1 d\omega_2 n(\omega_1)[1 - n(\omega_2)]$$

$$\times N(\omega - \omega_1 + \omega_2)\chi''_{cs}(q, \omega - \omega_1 + \omega_2)A(k, \omega_1)A(k - q, \omega_2) \,, \quad (3.38)$$

where $n(\omega) = (\exp\beta\omega + 1)^{-1}$, $N(\omega) = (\exp\beta\omega - 1)^{-1}$ and the momentum dependent vertex is given by

$$g_x(k, k-q) = v_x(k-q)\, t(k) - v_x(k)\, t(k-q) + J(q)\, [v_x(k) - v_x(k-q)] \,. \quad (3.39)$$

The charge-spin susceptibility $\chi''_{cs}(q,\omega) = \Im\chi_{cs}(q,\omega)$ is defined by the Fourier component of the retarded Green function

$$\chi_{cs}(q,\omega) = -\langle\langle \rho_{cs}(q) | \rho_{cs}(-q)\rangle\rangle_{\omega} = -\frac{1}{4}\langle\langle N_q | N_{-q}\rangle\rangle_{\omega} - \sum_{\alpha}\langle\langle S_q^{\alpha} | S_{-q}^{\alpha}\rangle\rangle_{\omega} \,. \quad (3.40)$$

We have introduced also the spectral function which does not depend on spin σ in the paramagnetic state:

$$A(q,\omega) = -\frac{1}{\pi}\Im\langle\langle X_q^{0\sigma} \mid X_q^{\sigma 0}\rangle\rangle_{\omega + i\delta} \,. \quad (3.41)$$

It defines the spectrum of electronic excitations by one-electron (fully "dressed") Green function for the $t - J$ model.

So we can compare the relaxation rate (3.38) for the $t - J$ model with the relaxation rate for the two-band Hubbard model (2.13), (2.15). The polarization operator now has contributions from two subbands:

$$P_x = e\sum_i R_i^x N_i = e\sum_i R_i^x (\sum_{\sigma} X_i^{\sigma\sigma} + 2X_i^{22}) \,. \quad (3.42)$$

From this definition the following expression can be obtained for the current operator in **q**-representation:

$$J_x = e\sum_{q\sigma}\{v_x^{11}(q)X_q^{\sigma 0}X_q^{0\sigma} + v_x^{22}(q)X_q^{2\sigma}X_q^{\sigma 2} + \sigma v_x^{12}(q)(X_q^{2\bar\sigma}X_q^{0\sigma} + X_q^{\sigma 0}X_q^{\bar\sigma 2})\} \,, \quad (3.43)$$

where $v_x^{\alpha\beta}(q) = -\partial t^{\alpha\beta}(\mathbf{q})/\partial q_x$ are electron velocities. We see that there are three contributions to the current: from each of the two subbands and from the interband transitions. The corresponding forces can be written as a sum of two terms:

$$F_x = F_x^0 + F_x^{int} = [J_x, H_0] + [J_x, H_t] \,. \quad (3.44)$$

The first term comes from on-site energy (2.13). It has a contribution only from the interband transitions and is proportional to interband energy U = (Δ in the $p - d$ model):

$$F_x^0 = -eU\sum_{q\sigma}\sigma v_x^{12}(q)(X_q^{2\bar\sigma}X_q^{0\sigma} - X_q^{\sigma 0}X_q^{\bar\sigma 2}) \,.$$

There is no such term in the one-band $t - J$ model. The second term, being proportional to the square of hopping integrals as in (3.31) for the $t - J$ model, has contributions both from electron hopping in each of two subbands and from interband transitions. It has a more complicated form than (3.31) and we shall not write it here explicitly; see Plakida (1996).

As a result, for the relaxation rate we have also two contributions. For the interband relaxation rate we obtain

$$\Gamma_0(\omega) = \frac{2\pi e^2 U^2}{\chi_0 \omega} \int_{-\infty}^{\infty} dz \sum_q (v_x^{12}(q))^2$$

$$\times \{[n(z-\omega)-n(z)]A_1(q,z-\omega)A_2(q,z)-[n(z+\omega)-n(z)]A_1(q,z+\omega)A_2(q,z)\} . \quad (3.45)$$

Here the spectral functions

$$A_1(q,\omega) = -(1/\pi)\Im\langle\langle X_q^{0\sigma} \mid X_q^{\sigma 0}\rangle\rangle_{\omega+\imath\delta} ; \quad A_2(q,\omega) = -(1/\pi)\Im\langle\langle X_q^{\sigma 2} \mid X_q^{2\sigma}\rangle\rangle_{\omega+\imath\delta}$$

define the spectra of electronic excitations by full one-electron GF for lower (upper) Hubbard subbands. Since the energy difference between the subbands is of order U, the interband relaxation rate is nonzero only for this region of high energy around $|\omega| \simeq U$. There is no interband absorption in the one-band $t-J$ model.

The intraband relaxation rate, being calculated in the same mode coupling approximation as (3.37), has a contribution from sixteen correlation functions of the type given by Eq. (3.37) from different subbands. However, by using the symmetry relations for correlations functions in terms of Bose-like operators $B^t_{j\sigma\sigma'}$ (3.33) we can write the final result for the intraband relaxation rate in a compact form:

$$\Gamma_{int}(\omega) = \frac{\exp(\beta\omega)-1}{\chi_0\omega} \int_{-\infty}^{\infty} d\omega_1 \int_{-\infty}^{\infty} d\omega_2 n(\omega_1)[1-n(\omega_2)]N(\omega-\omega_1+\omega_2)$$

$$\times \frac{2e^2}{N}\sum_{k,q} \quad g_x^2(k,k-q) \quad \chi''_{cs}(q,\omega-\omega_1+\omega_2)$$

$$\times \{(t_{11}^2-t_{12}^2)^2 A_1(k,\omega_1)A_1(k-q,\omega_2) + (t_{22}^2-t_{12}^2)^2 A_2(k,\omega_1)A_2(k-q,\omega_2)\} . \quad (3.46)$$

The momentum dependent vertex is given by

$$g_x(k,k-q) = u_x(k)\gamma(k-q) - u_x(k-q)\gamma(k) . \quad (3.47)$$

Here we have used the representation for the hopping energy $t^{\alpha\beta}(\mathbf{q}) = t_{\alpha\beta}\gamma(\mathbf{q})$ and introduced a dimensionless electron velocity: $u_x(q) = -\partial\gamma(\mathbf{q})/\partial q_x$

We see that in the two-band Hubbard model the kinematic interaction gives contribution to the relaxation rate (3.46) proportional to $[(t_{\alpha\alpha})^2 - (t_{12})^2]^2$ which cancels out for the conventional Hubbard model (2.1), ($t_{\alpha\beta} = t$). However, in the $t-J$ model we observe a finite kinematic contribution. It is defined by the first part of the vertex (3.39), $g_t(k,k-q) = v_x(k)\, t(k-q) - v_x(k-q)\, t(k) \propto t^4$. In the $t-J$ model we have additional spin-exchange scattering, $g_J(k,k-q) = J(q)\, [v_x(k) - v_x(k-q)] \propto t^2 J^2 \propto t^6$ since $J \propto t_{12}^2/\Delta$. So we observe a nonequivalence of the two-band Hubbard model in the strong correlation limit to the $t-J$ model. We can also suggest that higher order in t^2/Δ contributions for conventional Hubbard model could give a final Drude relaxation rate depending on the spin-exchange scattering.

To conclude this Section we shall calculate the static current-current susceptibility (3.25) for the $t-J$ model that defines the sum rule for the conductivity, Eq. (3.11). By performing the commutation for the polarization operator (3.26) and the current (3.27) we readily get

$$\chi_0 = (J_x, J_x) = \imath\langle[J_x, P_x]\rangle = e^2 \sum_{i\neq j,\sigma} (R_i^x - R_j^x)^2 t_{ij} \langle X_i^{\sigma 0} X_j^{0\sigma}\rangle . \quad (3.48)$$

For the model with only nearest neighbors hopping, $(R_i^x - R_j^x)^2 = a_x^2$, the static susceptibility (3.48) is equal to the average kinetic energy, the hopping term H_t in Eq. (2.23), multiplied by a constant:

$$\chi_0 = -e^2 a_x^2 (1/2)\langle H_t \rangle \, .$$

In **q**-representation the static susceptibility (3.48) reads

$$\chi_0 = -e^2 \sum_{q,\sigma} \frac{\partial^2 t(q)}{\partial q_x^2} \langle X_q^{\sigma 0} X_q^{0\sigma} \rangle = \frac{Ne^2}{m_{eff}} \, , \tag{3.49}$$

where the doping and temperature dependence of the effective mass m_{eff} are defined by the correlation function $\langle X_q^{\sigma 0} X_q^{0\sigma} \rangle$. Latter can be calculated by using the solution for the one-electron Green function of $t - J$ model.

Numerical Estimations. Now we shall estimate the relaxation rate (3.38) and the conductivity (3.20) by adopting some approximations for the one-electron spectral function (3.41) and the charge spin susceptibility (3.40). As has been proven in many calculations with $t - J$ model the spectral function for one-hole excitations can be written as a sum of coherent contribution from the quasiparticle propagation in a narrow band of order $2J$ and incoherent part due to diffusive motion of holes in a broad band of order of $2W \simeq 8t$. So we can use the approximation

$$A(k,\omega) = Z_k \delta(\omega + \mu - \epsilon_k) + A_{inc}(k,\omega) \, , \tag{3.50}$$

where Z_k is the quasiparticle weight for excitations with dispersion ϵ_k. For second, incoherent contribution we can write

$$A_{inc}(k,\omega) \simeq N_{inc} \, \theta(W - |\omega + \mu|) \, . \tag{3.51}$$

Here the incoherent density of states is $N_{inc} \simeq (1 - Z_k)/2W$. The total density of states (3.50) obeys the sum rule

$$\frac{1}{N} \sum_k \int_{-\infty}^{\infty} d\omega A(k,\omega) = \langle X_i^{00} + X_i^{\sigma\sigma} \rangle = 1 - \frac{n}{2} \, , \tag{3.52}$$

where $n = 2\langle X_i^{\sigma\sigma} \rangle$ is the occupation number. In this approximation we get estimations for chemical potential $\mu/W \simeq (1 - 3\delta)/(1 + \delta)$ and static current-current susceptibility (3.49) $\chi_0/N \simeq e^2 a^2 W \delta$ with $\delta = 1 - n$.

By using Eqs. (3.50), (3.51) we obtain the following expressions for the relaxation rate (3.38) depending on coherent contribution

$$\Gamma_{coh}(\omega) = \frac{\exp(\beta\omega) - 1}{\chi_0 \omega} \frac{2e^2}{N} \sum_{k,q} g_x^2(k, k-q) Z_k \, Z_{k-q}$$

$$\times n(\epsilon_k)[1 - n(\epsilon_{k-q})] N(\omega - \epsilon_k + \epsilon_{k-q}) \chi''_{cs}(q, \omega - \epsilon k + \epsilon_{k-q}) \tag{3.53}$$

and incoherent contribution

$$\Gamma_{inc}(\omega) = \frac{\exp(\beta\omega) - 1}{\chi_0 \omega} \frac{2e^2}{N} \sum_{k,q} g_x^2(k, k-q) \int_{-W-\mu}^{W-\mu} \int_{-W-\mu}^{W-\mu} d\omega_1 d\omega_2$$

$$\times N_{inc}^2 n(\omega_1)[1 - n(\omega_2)] N(\omega - \omega_1 + \omega_2) \chi''_{cs}(q, \omega - \omega_1 + \omega_2) \, . \tag{3.54}$$

The coherent part of the relaxation rate (3.53) has a conventional form for the Drude relaxation rate calculated in the Born approximation. However, the relaxation rate and the conductivity (3.20) have quite complicated temperature and doping dependences due to very specific dependence of the one-electron spectra ϵ_k and spin susceptibility on those parameters. The formula (3.53) has been evaluated by Ihle and Plakida (1994) numerically for $p-d$ and Millis-Monien-Pines spin susceptibility models (Millis et al., 1990). It has been shown that the relaxation rate (3.53) has a crossover from T^2 at $T \to 0$ to linear temperature dependence in the static limit, $(\omega \to 0)$, and a crossover from ω^2 at $\omega \to 0$ to linear frequency dependence for low enough temperatures in agreement with experiments for copper oxides. The absolute value of the conductivity is also in a qualitative agreement with experiments.

It is also interesting to consider the incoherent part (3.54). By using the model for the spin susceptibility suggested in the numerical calculations (Jaklič et al., 1995)

$$\chi_s''(q,\omega) \simeq \chi_s(q)\, \chi_s''(\omega) \simeq \chi_s(q) \,\tanh \frac{\omega}{2T}\, \frac{1}{1+(\omega/\omega_s)^2}\,, \tag{3.55}$$

we get after some algebra the following estimation for incoherent relaxation rate:

$$\Gamma_{inc}(\omega) = \omega_s\, \Gamma(\nu,\tau)\, A = \omega_s\, \Gamma(\nu,\tau)\, \frac{2e^2}{\chi_0 N} \sum_{k,q} g_x^2(k,k-q) N_{inc}^2 \chi_s(q)\,, \tag{3.56}$$

where the dimensionless function for frequency $\nu = \omega/2T$ and temperature $\tau = T/\omega_s$ is given by

$$\Gamma(\nu,\tau) = 2\tau\, \gamma(\nu,\tau) = 2\tau \frac{\tanh \nu}{\nu} \int_0^{+\infty} \frac{dx}{(1+4\tau^2 x^2)}\, \frac{1}{\cosh^2 x}\, \frac{\nu \tanh \nu - x \tanh x}{\tanh^2 \nu - \tanh^2 x}\,. \tag{3.57}$$

It is remarkable that for the incoherent spectrum (3.51) $q-$ and frequency dependences are factorized as a product of scaling function $\gamma(\nu,\tau)$ in (3.57) and integral over scattering vectors, the constant A in (3.56). Therefore, the problem of hot spots on the Fermi surface discussed by Hlubina and Rice (1995) is irrelevant to incoherent scattering.

The frequency dependence of the relaxation rate $\Gamma(\nu,\tau)$ is shown in Fig. 3.1 for several temperatures $\tau = T/\omega_s$ where $\omega_s \simeq J \simeq 1500$ K. A linear frequency dependence of relaxation rate $\Gamma(\nu,\tau)$ is observed for $\nu = \omega/2T \geq 1$ at low temperatures, $\tau \leq 0.2$ $(T \leq 300$ K). In the static limit, $\nu = 0$, we have linear temperature dependence even in the low temperature limit, $T \to 0$, since

$$\Gamma(\nu = 0, \tau \to 0) \simeq \tau \int_0^{+\infty} \frac{x\, dx}{\sinh x} = \tau \frac{\pi^2}{2}\,. \tag{3.58}$$

This low temperature linear behavior is due to specific temperature dependence of the model spin susceptibility (3.55).

For the absorptive, real part of the conductivity, neglecting the optical mass renormalization, $\lambda(\omega) \simeq 1$, we obtain from (3.20), (3.56) the following representation for incoherent scattering:

$$\sigma(\nu,\tau) = \frac{\chi_0}{V\omega_s}\, \frac{1}{2\tau}\, \frac{A\gamma(\nu,\tau)}{(\nu^2 + (A\gamma(\nu,\tau))^2)}\,. \tag{3.59}$$

The frequency dependence of the dimensionless conductivity $\sigma(\nu,\tau)/(\chi_0/V\omega_s)$ for $A = 1$ in Eq. (3.56) is shown in Fig. 3.2 for several temperatures. Since $\Gamma(\nu,\tau) \propto \nu$ in the low temperature region the conductivity will be $\sigma(\nu, \tau) \propto 1/\nu$ in a wide frequency range including the midinfrared band. This universal $1/\omega$ behavior has been observed by Rice et al. (1989) in a model with purely incoherent spectrum of holes, also in numerical calculations for small clusters (Jaklič et al., 1995) and in the Hubbard model in the limit of infinite dimension (Rozenberg et al., 1995; Gerges et al., 1996). So we can confirm by our analytical calculations that midinfrared absorption or universal $(1/\omega)$

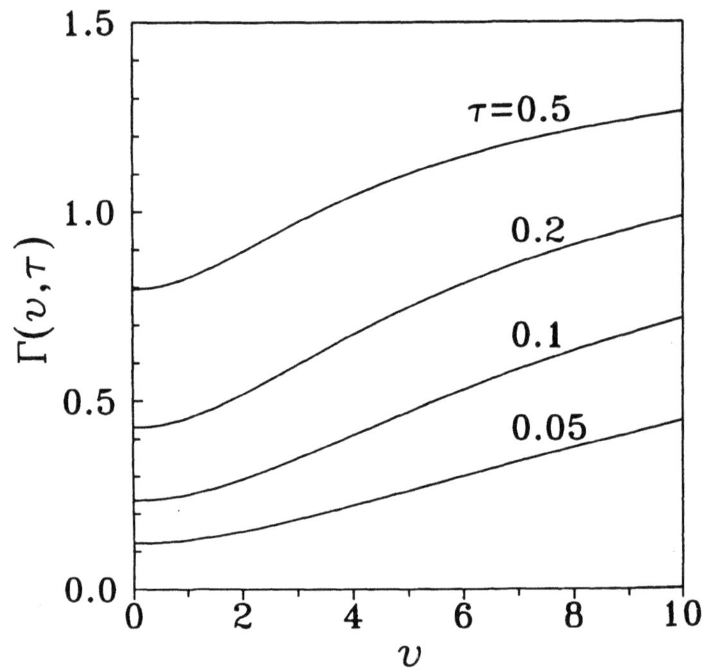

Fig. 3.1. The relaxation rate $\Gamma(\nu, \tau)$ as a function of frequency $\nu = \omega/2T$ and temperature $\tau = T/\omega_s$.

dependence of the optical conductivity can be explained by the diffusive character of hole motion in systems with strong correlations, as in copper oxides.

The results presented in this Section have been obtained for the model incoherent spectrum (3.50) and the model spin-fluctuation susceptibility (3.55). Both the coherent part in (3.50) in form of undamped quasiparticle spectrum and the incoherent part (3.51) in form of frequency independent contribution are really very crude approximations to one-hole ARPES spectra observed in copper oxides. However, since the relaxation rate (3.38) depends only on averaged values of spectral functions we can argue that a qualitative behavior of the relaxation rate should not depend on the details of former. Therefore, the main result of our analytical study – the non-Drude behavior of the relaxation rate caused by the diffusive character of hole motion in systems with strong correlations, seems to be justified.

To calculate the resistivity $\rho(\tau) = 1/\sigma(\tau)$ one should consider the static limit, $(\nu \to 0)$, for the relaxation rate. In the limit of low temperature from (3.58) we obtain a linear temperature dependence for the resistivity, even at $T \to 0$. Therefore the incoherent scattering caused by the diffusive character of hole motion in systems with strong correlations could also be a reason for anomalous linear temperature dependence of the resistivity in copper-oxide materials.

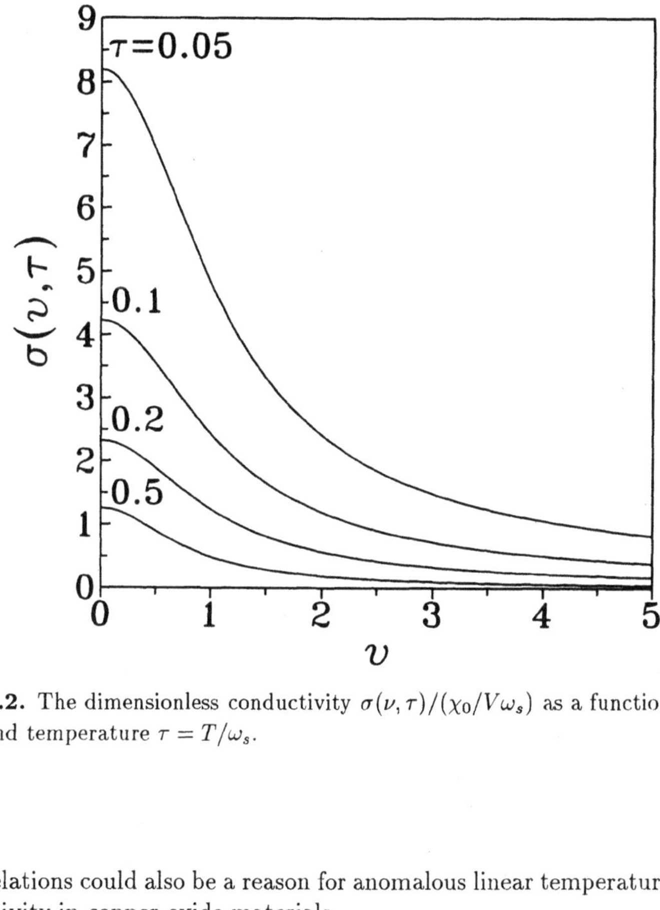

Fig. 3.2. The dimensionless conductivity $\sigma(\nu,\tau)/(\chi_0/V\omega_s)$ as a function of frequency $\nu = \omega/2T$ and temperature $\tau = T/\omega_s$.

We can obtain analogous numerical results for the two-band effective Hubbard model (2.18) since the intraband relaxation rate for this model, Eq. (3.46), has the same analytical form as the $t-J$ model, Eq. (3.38). To improve the above given estimations that enable one to perform a quantitative comparison with experiments a numerical self-consistent solution of the system of equations for the one-electron Green function for doped holes and spin susceptibility should be done within the $t-J$ and two-band Hubbard models. In the next, Sec. 3.2, and Sec. 4 we shall discuss the problem of spin susceptibility and one-hole energy spectrum calculations.

3.2 Spin-Fluctuation Susceptibility in $t-J$ Model.

The various approaches, e.g., slave-boson (Tanamoto et al., 1994; Stemman et al., 1994) or slave-fermion (Hedegard et al., 1991; Kane et al., 1990; Auerbach et al., 1991) methods have been used to study spin dynamics within the $t-J$ model. In the slave-field approaches the local constraint is usually replaced by global one thus restricting the validity of the approach. Whereas in any approximation formulated in terms of HO's the constraint of excluding the double occupancy can be rigorously preserved.

A special diagrammatic technique for HO's has been also developed to calculate the spin susceptibility for the $t-J$ model (Izymov et al., 1990, 1994; Onufrieva et al., 1995). However, it is difficult to estimate the validity of employed approximations since in the summation of diagrams only a special set has been taken into account; e.g., in the generalized random phase approximation (Izyumov et al., 1990) or in the leading order approximation in Larkin equation (Onufrieva et al., 1995).

In the present Section we shall discuss the results of the dynamic spin susceptibility, calculated (Jackeli and Plakida, 1997) for the paramagnetic phase with the help of the $t-J$ model by employing the memory function method considered in the previous Section.

Dynamic Spin Susceptibility The dynamic spin susceptibility is defined as

$$\chi_{\mathbf{q}}(\omega) = -\langle\langle S_{\mathbf{q}}^{+}|S_{\mathbf{q}}^{-}\rangle\rangle_{\omega} = -\sum_{\mathbf{R}_{ij}} e^{-i\mathbf{q}\mathbf{R}_{ij}} \langle\langle S_i^{+}|S_j^{-}\rangle\rangle_{\omega} , \tag{3.60}$$

where $\mathbf{R}_{ij} = \mathbf{R}_i - \mathbf{R}_j$ and $\langle\langle S_{\mathbf{q}}^{+}|S_{\mathbf{q}}^{-}\rangle\rangle_{\omega}$ denotes Fourier transformed two-time retarded commutator GF (3.5).

By using Eq. (3.7) the dynamic spin susceptibility can be written in terms of the relaxation function $\Phi_{\mathbf{q}}(\omega) \equiv ((S_{\mathbf{q}}^{+}|S_{\mathbf{q}}^{-}))_{\omega}$:

$$\chi_{\mathbf{q}}(\omega) = \chi_{\mathbf{q}} - \omega\Phi_{q}(\omega) , \tag{3.61}$$

where $\chi_{\mathbf{q}} \equiv \chi_{\mathbf{q}}(0)$ is the static spin susceptibility. To calculate the spin-spin relaxation function $\Phi_{\mathbf{q}}(\omega)$ it is convenient to employ the Mori memory function approach presented in Sec. 3.1. We define the memory function $M(\mathbf{q},\omega)$ by the equation

$$\Phi_{\mathbf{q}}(\omega) = \frac{\chi_{\mathbf{q}}}{\omega - M(\mathbf{q},\omega)/\chi_{\mathbf{q}}} . \tag{3.62}$$

To calculate the memory function we use the equation of motion for relaxation functions

$$\omega((S_{\mathbf{q}}^{+}|S_{\mathbf{q}}^{-}))_{\omega} = \chi_{\mathbf{q}} + ((i\dot{S}_{\mathbf{q}}^{+}|S_{\mathbf{q}}^{-}))_{\omega} , \tag{3.63}$$

$$\omega((i\dot{S}_{\mathbf{q}}^{+}|S_{\mathbf{q}}^{-}))_{\omega} = ((i\dot{S}_{\mathbf{q}}^{+}| - i\dot{S}_{\mathbf{q}}^{-}))_{\omega} , \tag{3.64}$$

where we have omitted the term $(i\dot{S}_{\mathbf{q}}^{+}, S_{\mathbf{q}}^{-}) = \langle[S_{\mathbf{q}}^{+}, S_{\mathbf{q}}^{-}]\rangle = 2/\sqrt{N}\langle S_{\mathbf{q}}^{z}\rangle\delta_{\mathbf{q}0}$ in Eq. (3.64) which is zero in paramagnetic phase.

The system of equations (3.63), (3.64) can be formally solved, as discussed in Sec. 3.1, by introducing the irreducible part of the scattering matrix which is just the memory function:

$$M(\mathbf{q},\omega) = ((i\dot{S}_{\mathbf{q}}^{+}| - i\dot{S}_{\mathbf{q}}^{-}))_{\omega}^{irr} . \tag{3.65}$$

Therefore, the dynamic spin susceptibility (3.61), by applying the Eq. (3.62), can be written in terms of memory function as

$$\chi_{\mathbf{q}}(\omega) = -\chi_{\mathbf{q}} \frac{M(\mathbf{q},\omega)/\chi_{\mathbf{q}}}{\omega - M(\mathbf{q},\omega)/\chi_{\mathbf{q}}} . \tag{3.66}$$

Mode–Coupling Approximation. First we express the memory function in terms of the irreducible current-current time-dependent correlation function by using the spectral representation for GF

$$M(\mathbf{q},\omega) = \sum_{\mathbf{R}_{ij}} e^{-\mathbf{q}\mathbf{R}_{ij}} \frac{1}{2\pi} \int_{-\infty}^{\infty} d\omega' \frac{e^{\beta\omega'} - 1}{\omega'(\omega - \omega' + i\eta)} \int_{-\infty}^{\infty} dt e^{-i\omega' t} \langle J_i^{\dagger}(t) | J_j \rangle^{irr} , \tag{3.67}$$

where the current operator in site representation is defined as $J_j = i\dot{S}_j^+$.

Current operator can be written as a sum of two terms

$$J_j = J_j^t + J_j^J = [S_j^+, H_t] + [S_j^+, H_J] . \tag{3.68}$$

Here the first term comes from the kinetic energy and is proportional to the hopping integral:

$$J_j^t = -\sum_m t_{jm} (X_j^{+0} X_m^{0-} - X_m^{+0} X_j^{0-}). \tag{3.69}$$

The second term in (3.68) comes from the exchange interaction between localized spins and has the form

$$J_j^J = 2\sum_m J_{jm} (S_j^z S_m^+ - S_m^z S_j^+). \tag{3.70}$$

To calculate the irreducible time-dependent correlation function in the right hand -side of Eq. (3.67) we employ the mode-coupling approximation in terms of independent propagation of dressed particle-hole and spin fluctuations as in the calculation of optical conductivity, Eq. (3.37). This scheme is essentially equivalent to the self-consistent Born approximation, in which the vertex corrections are neglected. The proposed scheme is defined by the following decoupling of time-dependent correlation functions:

$$\langle X_m^{-0}(t) X_i^{0+}(t) X_j^{+0} X_l^{0-} \rangle \simeq \langle X_m^{-0}(t) X_l^{0-} \rangle \langle X_i^{0+}(t) X_j^{+0} \rangle , \tag{3.71}$$

$$\langle S_i^z(t) S_m^-(t) S_j^z S_l^+ \rangle \simeq \langle S_i^z(t) S_j^z \rangle \langle S_m^-(t) S_l^+ \rangle . \tag{3.72}$$

The cross-correlations like $\langle J_i^t | (J^J)_j^+ \rangle$ are ignored within the proposed approximation and they will be omitted.

By using the above defined decoupling scheme (3.71) and (3.72) and the spectral representation for GF, the memory function (3.67) can be written as

$$M(\mathbf{q},\omega) = M_t(\mathbf{q},\omega) + M_J(\mathbf{q},\omega), \tag{3.73}$$

where $M_t(\mathbf{q},\omega)$ is the contribution from the itinerant hole subsystem and it reads as

$$M_t(\mathbf{q},\omega) = \frac{1}{N} \sum_{\mathbf{k}} t_{\mathbf{kq}}^2 \iint_{-\infty}^{\infty} d\omega_1 d\omega' \left[n(\omega_1 - \omega') - n(\omega_1) \right] \frac{A_{\mathbf{k}}(\omega_1) A_{\mathbf{k-q}}(\omega_1 - \omega')}{\omega'(\omega - \omega' + i\eta)} , \tag{3.74}$$

where $n(\omega) = (e^{\beta\omega} + 1)^{-1}$, $t_{\mathbf{kq}} = t_{\mathbf{k}} - t_{\mathbf{k-q}}$ with $t_{\mathbf{k}} = z(t\gamma_{\mathbf{q}} + t'\gamma'_{\mathbf{q}})$, $z = 4$, $\gamma_{\mathbf{q}} = 1/2[\cos q_x + \cos q_y]$ and $\gamma'_{\mathbf{q}} = \cos q_x \cos q_y$ for 2D square lattice (the lattice constant is taken to be unity) and $A_{\mathbf{k}}(\omega) = -(1/\pi)\mathrm{Im}\langle\langle X_{\mathbf{q}}^{\sigma 0} X_{\mathbf{q}}^{0\sigma} \rangle\rangle_\omega$ is a hole particle spectral function which is spin independent in the paramagnetic phase.

The second contribution $M_J(\mathbf{q},\omega)$ in Eq. (3.73) comes from the localized spin subsystem and is given by

$$M_J(\mathbf{q},\omega) = \frac{2}{\pi^2 N} \sum_{\mathbf{k}} J_{\mathbf{kq}}^2 \iint_{-\infty}^{\infty} d\omega_1 d\omega' \left[N(\omega_1 - \omega') - N(\omega_1) \right]$$

$$\times \frac{\mathrm{Im}\chi_{\mathbf{k}}(\omega_1)\mathrm{Im}\chi_{\mathbf{k}-\mathbf{q}}(\omega_1-\omega')}{\omega'(\omega-\omega'+i\eta)} , \qquad (3.75)$$

where $N(\omega) = (e^{\beta\omega}-1)^{-1}$ and $J_{\mathbf{kq}} = J_{\mathbf{k}} - J_{\mathbf{k}-\mathbf{q}}$, and $J_{\mathbf{q}} = zJ\gamma_{\mathbf{q}}$. In obtaining (3.75) the relation $\langle\langle S^z_{\mathbf{q}}|S^z_{\mathbf{q}}\rangle\rangle_\omega = 1/2\langle\langle S^+_{\mathbf{q}}|S^-_{\mathbf{q}}\rangle\rangle_\omega$ which is valid in rotationally invariant system has been used.

The real, $\mathrm{Re}M(\mathbf{q},\omega)$, and imaginary, $\mathrm{Im}M(\mathbf{q},\omega)$, parts of the memory function are odd and even functions of ω, respectively, and they are coupled by the dispersion relation (3.23). Therefore, only the imaginary part of the memory function should be evaluated.

Now we examine asymptotic behavior of the dynamic spin susceptibility. First we consider the hydrodynamic limit $\mathbf{q}\to 0$ and $\omega\to 0$. In this limit, $\mathrm{Re}M(\mathbf{q},\omega)$ being an odd function of ω vanishes while $\mathrm{Im}M(\mathbf{q},\omega)$ remains finite. By using Eqs. (3.73) and (3.75) we can express it as $\mathrm{Im}M(\mathbf{q},\omega) \simeq -Dq^2$, with $D = D_t + D_J$, and

$$D_t = \frac{\pi}{N}\sum_{\mathbf{k}}(\hat{\mathbf{q}}\nabla_{\mathbf{k}}t_{\mathbf{k}})^2 \lim_{\omega\to 0}\lim_{\mathbf{q}\to 0}\int_{-\infty}^{\infty} d\omega_1 |n'(\omega_1)| A_{\mathbf{k}}(\omega_1)A_{\mathbf{k}-\mathbf{q}}(\omega_1-\omega),$$

$$D_J = \frac{2}{\pi N}\sum_{\mathbf{k}}(\hat{\mathbf{q}}\nabla_{\mathbf{k}}J_{\mathbf{k}})^2 \lim_{\omega\to 0}\lim_{\mathbf{q}\to 0}\int_{-\infty}^{\infty} d\omega_1 |N'(\omega_1)| \mathrm{Im}\chi_{\mathbf{k}}(\omega_1)\mathrm{Im}\chi_{\mathbf{k}-\mathbf{q}}(\omega_1-\omega) , \qquad (3.76)$$

where $\hat{\mathbf{q}} = \mathbf{q}/q$, $\nabla_{\mathbf{k}} = dt_{\mathbf{k}}/d\mathbf{k}$, $n'(\omega) = dn(\omega)/d\omega$, and $N'(\omega) = dN(\omega)/d\omega$. Finally in hydrodynamic limit the dynamic spin susceptibility can be expressed in conventional form (see, Forster (1975)) as

$$\chi_{\mathbf{q}}(\omega) = \chi_0 \frac{i\tilde{D}q^2}{\omega + i\tilde{D}q^2} , \qquad (3.77)$$

where $\tilde{D} = D/\chi_0$ is the spin diffusion coefficient and χ_0 is the static uniform susceptibility.

Unlike the hydrodynamic limit, in the high energy limit, $\omega\to\infty$, the dominant contribution to memory functions comes from the real part: $M(\mathbf{q},\omega) \simeq m_{\mathbf{q}}/\omega$, where $m_{\mathbf{q}}$ is the first nonvanishing moment in $1/\omega$ expansion of the memory function defined as

$$m_{\mathbf{q}} = -\frac{1}{\pi}\int_{-\infty}^{\infty} d\omega \mathrm{Im}M(\mathbf{q},\omega) = \langle[i\dot{S}^+_{\mathbf{q}}, S^+_{\mathbf{q}}]\rangle . \qquad (3.78)$$

Thus in the high energy limit the dynamic susceptibility takes the form:

$$\chi_{\mathbf{q}}(\omega) = -\chi_{\mathbf{q}} \frac{\omega^2_{\mathbf{q}}}{\omega^2 - \omega^2_{\mathbf{q}}} , \qquad (3.79)$$

where $\omega^2_{\mathbf{q}} = m_{\mathbf{q}}/\chi_{\mathbf{q}}$. The expressions for $\omega^2_{\mathbf{q}}$ can be derived in the same mode coupling approximation for the static spin susceptibility $\chi_{\mathbf{q}}$; for details, see, Jackeli and Plakida (1997)). In that approximation we reproduce the result for spin susceptibility $\chi_{\mathbf{q}}$ obtained with the help the $t-J$ model by Shimahara and Takada (1992) in the framework of Kondo and Yamaji (1972) theory which is essentially a self-consistent mean field approximation.

To summarize, based on $t-J$ model and memory function approach, we have derived a general representation for the dynamic spin susceptibility (3.66) in terms of memory function (3.67). Our approach is formulated in terms of HO's and, therefore, the constraint of no double occupancy is rigorously preserved. The memory function is calculated by using the equation of motion method for two-time retarded GF within the mode coupling approximation (3.71), (3.72). The two contributions to the memory function are obtained. First one (3.74) comes from the itinerant hole subsystem and is due to the kinematic interaction. Second one (3.75) comes from the localized interacting spin subsytem. In the limit of small concentration of doped holes latter gives the main contribution which describes the spin dynamics characteristic for the Heisenberg model. Whereas in the opposite limit of large hole concentration particle-hole excitations characteristic of the itinerant magnetism give the main contribution to the spin dynamics. We have shown that in the paramagnetic phase there are two regimes in the spin dynamics. In the hydrodynamic limit ($q \to 0$, $\omega \to 0$) the spin susceptibility (3.77) describes diffusion spin dynamics with a diffusion coefficient (3.76), which has essentially two contributions. While in the high-frequency limit ($\omega \to \infty$) spin-wave-like excitations described by Eq. (3.79) are observed. Their dispersion, $\omega_{\mathbf{q}}^2 = m_{\mathbf{q}}/\chi_{\mathbf{q}}$, obtained in the mode coupling approximation for equal time correlation function recovers the earlier result of Shimahara and Takada (1992) in the mean field approximation.

To compare our results with that obtained by diagrammatic methods we would like to point out that our approach, based on a general representation of spin susceptibility (3.66), is equivalent to a summation of infinite diagram series generated by the memory function (3.67). Latter being calculated in the mode coupling approximation can be schematically represented by two loop-diagrams: first one of order t^2 due to the particle-hole loop and second one of order J^2 due to the spin fluctuation loop. In Onufrieva et al. (1995) all contributions in the denominator of the Larkin equation are proportional to J and, therefore, disappear in the limit $J = 0$ ($U \to \infty$). While in our approach the contribution due to the first loop (for which the kinematic interaction is responsible) remains. Whereas, in the generelized random phase approximation of Izyumov et al. (1990) the spin fluctuation contribution given by our second loop is neglected while several other diagrams beyond our simple one-loop diagram due to kinematic interaction are taken into account.

At present time it is difficult to justify any of the discussed schemes, including our mode coupling approximation. To check the validity of our approximation one has to solve numerically self-consistent equations and compare the obtained results with experimental data.

4. QUASIPARTICLES AND SUPERCONDUCTIVITY

In the present Section we shall consider the theory of quasiparticle spectra and superconductivity by applying the equation of motion method for the Green functions in terms of the Hubbard operators for $t - J$ (3.30) and $p - d$ (3.18) models. Results of numerical solution of the self-consistent system of equations for the Green functions will be given for the spin-polaron model (3.36).

It should be pointed out that superconduting pairing due to the kinematic interaction in the Hubbard model in the limit of strong electron correlations ($U \to \infty$) was first obtained by Zaitsev and Ivanov (1987). However, they have considered only the mean field approximation which results in s-wave pairing irrelevant to strongly correlated systems; for a discussion, see Plakida et al. (1989). Later on the mean field approximation has been applied to $t - J$ model within the Green function approach (Plakida et al., 1989; Yushankhai et al., 1991) and d-wave spin fluctuation superconducting pairing is obtained as a result of the exchange interaction J.

Below we present the results of a theory which goes beyond the mean field approximation. In this theory the matrix self-energy operator is calculated in the noncrossing approximation for kinematic and exchange interactions which neglects vertex corrections as in the Migdal-Eliashberg theory. The self-energy operator allowing for finite life-time effects for electrons plays an essential role both in renormalization of the quasiparticle spectrum and superconducting pairing. This has been clearly demonstrated by Plakida et al. (1997) where the $t-J$ model in polaron representation is considered. A self-consistent numerical solution of Eliashberg equations proves that there is a strong renormalization of the quasiparticle hole spectrum due to spin fluctuations and d-wave pairing at finite concentration of doped holes.

The two-time Green function approach in terms of Hubbard operators has been also used to consider superconducting pairing in the $t-J$ model with electron-phonon interaction by Plakida and Hayn (1994) and to discuss the electron and hole spectra in the normal state by Prelovśek (1997). A diagram technique for the Hubbard operators has been also used by Izyumov et al. (1991) to consider spin fluctuations and superconductivity in the $t-J$ model.

4.1 Green Functions for the $t-J$ and $p-d$ Models

To discuss the superconducting pairing within the $t-J$ model (2.30) we consider the matrix Green function

$$\hat{G}_{ij,\sigma}(t-t') = \langle\langle \Psi_{i\sigma}(t) | \Psi^{+}_{j\sigma}(t') \rangle\rangle \tag{4.1}$$

in terms of Nambu operators:

$$\Psi_{i\sigma} = \begin{pmatrix} X_i^{0\sigma} \\ X_i^{\bar{\sigma}0} \end{pmatrix}, \qquad \Psi^{+}_{i\sigma} = \begin{pmatrix} X_i^{\sigma 0} \; X_i^{0\bar{\sigma}} \end{pmatrix}, \tag{4.2}$$

where the Zubarev (1960) notations for the anticommutator Green function (4.1) are used.

By differentiating GF (4.1) over time t we get for the Fourier component the following equation:

$$\omega \hat{G}_{ij\sigma}(\omega) = \delta_{ij}\hat{Q}_{\sigma} + \langle\langle \hat{Z}_{i\sigma} \mid \Psi_{j\sigma} \rangle\rangle_{\omega} , \tag{4.3}$$

where $\hat{Z}_{i\sigma} = [\Psi_{i\sigma}, H]$ and

$$\hat{Q}_{\sigma} = \begin{pmatrix} Q_{\sigma} & 0 \\ 0 & Q_{\bar{\sigma}} \end{pmatrix} \tag{4.4}$$

with matrix elements $Q_{\sigma} = \langle X_i^{00} + X_i^{\sigma\sigma} \rangle$. Here and in what follows we consider a spin-singlet state for which the correlation functions do not depend on spin σ. In that case by using the completeness relation (2.26) we get $Q_{\sigma} = Q = 1 - n/2$, where the average number of electrons is given by the equation

$$n = \langle n_i \rangle = \sum_{\sigma} \langle X_i^{\sigma\sigma} \rangle . \tag{4.5}$$

Now, we project many–particle GF in (4.3) on one–hole GF by introducing the irreducible (irr) part of $\hat{Z}_{i\sigma}$ operator

$$\langle\langle \hat{Z}_{i\sigma} \mid \Psi^{+}_{j\sigma} \rangle\rangle = \sum_{l} \hat{E}_{il\sigma} \langle\langle \Psi_{l\sigma} \mid \Psi^{+}_{j\sigma} \rangle\rangle + \langle\langle \hat{Z}^{(irr)}_{i\sigma} \mid \Psi^{+}_{j\sigma} \rangle\rangle . \tag{4.6}$$

The projection is defined by the condition

$$\langle \{ \hat{Z}^{(irr)}_{i\sigma}, \Psi^{+}_{j\sigma} \} \rangle = 0 , \tag{4.7}$$

that results in an equation for the frequency matrix

$$\hat{E}_{ij\sigma} = \langle\{[\Psi_{i\sigma}, H], \Psi^{+}_{j\sigma}\}\rangle \, \hat{Q}^{-1}_{\sigma} \,. \tag{4.8}$$

Here $\{A, B\}$ and $[A, B]$ are the anticommutator and the commutator for A, B operators, respectively. To calculate the matrix (4.8) we can use the equation of motion for HO's as, e. g.,

$$\left(i\frac{d}{dt} + \mu\right) X_i^{0\sigma} = \sum_l t_{il} B_{i\sigma\sigma'} X_l^{0\sigma'} + \sum_l J_{il}(B_{l\sigma\sigma'} - \delta_{\sigma\sigma'}) X_i^{0\sigma'} \,, \tag{4.9}$$

where we have introduced the operator

$$B_{i\sigma\sigma'} = (X_i^{00} + X_i^{\sigma\sigma})\delta_{\sigma'\sigma} + X_i^{\bar{\sigma}\sigma}\delta_{\sigma'\bar{\sigma}} \,. \tag{4.10}$$

The Bose-like operator (4.10) describes electron scattering on spin and charge fluctuations caused by nonfermionic commutation relations for HO's (the first term in (4.10) – the so-called kinematic interaction) and the exchange spin-spin interaction (the second term in (4.10)). As we have shown in (3.34) it can be written in terms of number and spin operators.

By performing commutations in (4.8) we obtain for normal and anomalous parts of the frequency matrix:

$$E^{11}_{ij\sigma} = \delta_{ij} \sum_l \{t_{il}\langle X_i^{\sigma 0} X_l^{0\sigma}\rangle / Q_\sigma + J_{il}(Q_\sigma - 1 + \chi^{cs}_{il}/Q_\sigma)\}$$

$$-t_{ij}(Q_\sigma + \chi^{cs}_{ij}/Q_\sigma) - J_{ij}\langle X_j^{\sigma 0} X_i^{0\sigma}\rangle / Q_\sigma , \tag{4.11}$$

$$E^{12}_{ij\sigma} = \delta_{ij} \sum_l t_{il}\langle X_i^{0\bar{\sigma}} X_l^{0\sigma} + X_l^{0\bar{\sigma}} X_i^{0\sigma}\rangle / Q_\sigma$$

$$-J_{ij}\langle X_i^{0\bar{\sigma}} X_j^{0\sigma} + X_j^{0\bar{\sigma}} X_i^{0\sigma}\rangle / Q_\sigma \,. \tag{4.12}$$

Here we shall introduce charge- and spin-fluctuation correlation functions

$$\chi^{cs}_{ij} = \frac{1}{4}\langle \delta n_i \delta n_j \rangle + \langle \mathbf{S_i S_j} \rangle \,, \tag{4.13}$$

with $\delta n_i = n_i - \langle n_i \rangle$.

Now we can introduce zero–order GF in the generalized mean–field approximation by neglecting finite lifetime effects described by the operator $\hat{Z}^{irr}_{i\sigma}$ in Eq. (4.6)

$$\hat{G}^0_{ij\sigma}(\omega) = \{\omega\hat{\tau}_0\delta_{ij} - \hat{E}_{ij\sigma}\}^{-1}\hat{Q}_\sigma \,, \tag{4.14}$$

where $\hat{\tau}_0$ is the unity matrix. By writing the equation of motion for the irreducible part of GF in (4.6) with respect to second time t' for right–hand side operator $\Psi^{+}_{j\sigma}(t')$ and performing the same projection procedure as in (4.6), we get

$$\langle\langle \hat{Z}^{(irr)}_{i\sigma} \mid \Psi^{+}_{j\sigma} \rangle\rangle_\omega = \sum_l \langle\langle \hat{Z}^{(irr)}_{i\sigma} \mid (Z^{(irr)}_{l\sigma})^{+} \rangle\rangle_\omega \, \hat{Q}^{-1}_\sigma \, \hat{G}^0_{lj\sigma}(\omega) \,. \tag{4.15}$$

By using (4.3), (4.6) and (4.15) we can obtain the Dyson equation for GF (4.1) in the form

$$\hat{G}_{ij\sigma}(\omega) = \hat{G}^0_{ij\sigma}(\omega) + \sum_{kl} \hat{G}^0_{ik\sigma}(\omega)\, \hat{\Sigma}_{kl\sigma}(\omega)\, \hat{G}_{lj\sigma}(\omega) \,, \tag{4.16}$$

where the self–energy operator $\hat{\Sigma}_{kl\sigma}(\omega)$ is defined by the equation

$$\hat{T}_{ij\sigma}(\omega) = \hat{\Sigma}_{ij\sigma}(\omega) + \sum_{kl} \hat{\Sigma}_{ik\sigma}(\omega)\, \hat{G}^0_{kl\sigma}(\omega)\, \hat{T}_{lj\sigma}(\omega) \,. \tag{4.17}$$

Here the scattering matrix is given by the equation

$$\hat{T}_{ij\sigma}(\omega) = \hat{Q}_\sigma^{-1} \langle\langle \hat{Z}_{i\sigma}^{(irr)} \mid \hat{Z}_{j\sigma}^{(irr)+} \rangle\rangle_\omega \, \hat{Q}_\sigma^{-1} . \quad (4.18)$$

It follows from Eq. (4.17) that the self-energy operator is given by the irreducible part of the scattering matrix (4.18) which has not single zero-order GF (4.14) lines:

$$\hat{\Sigma}_{ij\sigma}(\omega) = \hat{Q}_\sigma^{-1} \tilde{\Sigma}_{ij\sigma}(\omega) = \hat{Q}_\sigma^{-1} \langle\langle \hat{Z}_{i\sigma}^{(irr)} \mid \hat{Z}_{j\sigma}^{(irr)+} \rangle\rangle_\omega^{(irr)} \, \hat{Q}_\sigma^{-1} . \quad (4.19)$$

Eqs. (4.14), (4.16) and (4.19) give an exact representation for one–hole GF (4.1). To calculate it, however, one has to apply some approximations to many–particle GF in the self-energy matrix (4.19) which describes inelastic scattering of electrons on spin and charge fluctuations.

To solve the Dyson equation (4.16) we introduce the **k**-representation for GF:

$$G_\sigma^{\alpha\beta}(k,\omega) = \sum_j G_{oj\sigma}^{\alpha\beta}(\omega) \, e^{-i\mathbf{kj}} . \quad (4.20)$$

For zero-order GF (4.14) we get

$$\hat{G}_\sigma^{(0)}(k,\omega)^{-1} = \{\omega\hat{\tau}_0 - (E_k^\sigma - \tilde{\mu})\hat{\tau}_3 - \Delta_k^\sigma \hat{\tau}_1\}\hat{Q}_\sigma^{-1} , \quad (4.21)$$

where $\hat{\tau}_0$, $\hat{\tau}_1$, $\hat{\tau}_3$ are Pauli matrices. The energy of quasiparticles E_k^σ, the renormalized chemical potential $\tilde{\mu} = \mu - \delta\mu$ and the gap function Δ_k^σ in MFA, Eqs. (4.11), (4.12), are given by

$$E_k^\sigma = -\epsilon(k)Q_\sigma - \epsilon_s(k)/Q_\sigma - \frac{4J}{N}\sum_q \gamma(k-q)N_{q\sigma} , \quad (4.22)$$

where

$$\epsilon(k) = t(k) = 4t\gamma(k) + 4t'\gamma'(k) \; ; \quad \epsilon_s(k) = 4t\gamma(k)\chi_{1s} + 4t'\gamma'(k)\chi_{2s} ,$$

with $\gamma(k) = (1/2)(\cos a_x q_x + \cos a_y q_y), \gamma'(k) = \cos a_x q_x \cos a_y q_y$,

$$\delta\mu = \frac{1}{N}\sum_q \epsilon(q)N_{q\sigma} - 4J(n/2 - \chi_{1s}/Q_\sigma), \quad (4.23)$$

$$\Delta_k^\sigma = \frac{2}{NQ_\sigma}\sum_q J(k-q)\langle X_{-q}^{0\bar{\sigma}} X_q^{0\sigma}\rangle. \quad (4.24)$$

The average number of electrons (4.5) in **k**-representation can be written in the form

$$n = \frac{1}{N}\sum_{k,\sigma}\langle X_k^{\sigma 0} X_k^{0\sigma}\rangle = \frac{1}{N}\sum_{k,\sigma} Q_\sigma N_{k\sigma}, \quad (4.25)$$

which defines the function $N_{q\sigma}$ in Eqs. (4.22), (4.23). In the calculation of normal part of the frequency matrix (4.22) we have neglected the charge fluctuation (the first term in Eq. (4.13)) and have introduced spin correlation functions for nearest (χ_{1s}) and next-nearest (χ_{2s}) neighbor lattice sites

$$\chi_{1s} = \langle \mathbf{S}_i \mathbf{S}_{i+a_1}\rangle \; ; \quad \chi_{2s} = \langle \mathbf{S}_i \mathbf{S}_{i+a_2}\rangle , \quad (4.26)$$

where $a_1 = (\pm a_x, \pm a_y)$ is nearest and $a_2 = \pm(a_x \pm a_y)$ next-nearest neighbor lattice sites. In the gap equation (4.24) we have omitted the **k**-independent part caused by the kinematic interaction (the first term in Eq. (4.12)) since it gives no contribution to d-wave pairing (Plakida et al., 1989).

To calculate the self–energy operator $\hat{\Sigma}(k,\omega)$ we employ the noncrossing approximation (or the self-consistent Born approximation) for the irreducible part of many–particle Green functions in (4.19). In this approximation vertex corrections are neglected as in the Migdal-Eliashberg approximation and it is given by two-time decoupling for correlation functions in (4.19) as, e. g., shown below:

$$\langle X_{j'}^{\sigma'0} B_{j\sigma\sigma'}^{+} X_{i'}^{0\sigma'}(t) B_{i\sigma\sigma'}(t)\rangle \simeq \langle X_{j'}^{\sigma'0} X_{i'}^{0\sigma'}(t)\rangle \langle B_{j\sigma\sigma'}^{+} B_{i\sigma\sigma'}(t)\rangle \ . \tag{4.27}$$

The proposed decoupling does not violate equal time correlations, since in Eq. (4.27) $j \neq j'$ and $i \neq i'$. Using the spectral representation for GF we obtain the following result for the self-energy in noncrossing approximation:

$$\tilde{\Sigma}_{11}^{\sigma}(k,\omega) = -\tilde{\Sigma}_{22}^{\bar{\sigma}}(-k,-\omega)$$

$$= \frac{1}{N}\sum_{q}\iint\limits_{-\infty}^{+\infty} dz d\Omega N(\omega,z,\Omega)\lambda_{11}(q,k-q \mid \Omega) A_{11}^{\sigma}(q,z) \ , \tag{4.28}$$

$$\tilde{\Sigma}_{12}^{\sigma}(k,\omega) = (\tilde{\Sigma}_{21}^{\sigma}(k,\omega))^{*}$$

$$= -\frac{1}{N}\sum_{q}\iint\limits_{-\infty}^{+\infty} dz d\Omega N(\omega,z,\Omega)\lambda_{12}(q,k-q \mid \Omega) A_{12}^{\sigma}(q,z) \ , \tag{4.29}$$

where

$$N(\omega,z,\Omega) = \frac{1}{2}\ \frac{\tanh(z/2T) + \coth(\Omega/2T)}{\omega - z - \Omega} \ . \tag{4.30}$$

Here we shall introduce the spectral density

$$A_{11}^{\sigma}(q,z) = -\frac{1}{Q_{\sigma}\pi}\mathrm{Im}\ \langle\langle X_{q}^{0\sigma} \mid X_{q}^{\sigma 0}\rangle\rangle_{z+i\delta} = A_{22}^{\bar{\sigma}}(q,-z) \ , \tag{4.31}$$

$$A_{12}^{\sigma}(q,z) = -\frac{1}{Q_{\sigma}\pi}\mathrm{Im}\ \langle\langle X_{q}^{0\sigma} \mid X_{-q}^{0\bar{\sigma}}\rangle\rangle_{z+i\delta} = A_{21}^{\sigma}(q,z) \ , \tag{4.32}$$

and electron-electron interaction functions caused by spin-charge fluctuations

$$\lambda_{11}(q,k-q \mid \Omega) = g^{2}(q,k-q) D^{+}(k-q,\Omega), \tag{4.33}$$

$$\lambda_{12}(q,k-q \mid \Omega) = g^{2}(q,k-q) D^{-}(k-q,\Omega), \tag{4.34}$$

where $g(q,k-q) = t(q) - J(k-q)$ and the spectral density for spin-charge fluctuations is defined by commutator Green functions

$$D^{\pm}(q,\Omega) = -\frac{1}{\pi}\mathrm{Im}\ \left\{\frac{1}{4}\langle\langle n_{q} \mid n_{q}^{+}\rangle\rangle_{\Omega+i\delta} \pm \langle\langle \mathbf{S}_{q} \mid \mathbf{S}_{-q}\rangle\rangle_{\Omega+i\delta}\right\}. \tag{4.35}$$

The solution of the Dyson equation (4.16) can be written in Eliashberg notations as

$$\hat{G}^{\sigma}(k,\omega) = Q_{\sigma}\tilde{G}^{\sigma}(k,\omega) = Q_{\sigma}\frac{\omega Z_{k}^{\sigma}(\omega)\hat{\tau}_{0} + (E_{k}^{\sigma} + \xi_{k}^{\sigma}(\omega) - \tilde{\mu})\hat{\tau}_{3} + \Phi_{k}^{\sigma}(\omega)\hat{\tau}_{1}}{(\omega Z_{k}^{\sigma}(\omega))^{2} - (E_{k}^{\sigma} + \xi_{k}^{\sigma}(\omega) - \tilde{\mu})^{2} - \mid \Phi_{k}^{\sigma}(\omega) \mid^{2}} \ , \tag{4.36}$$

where

$$\omega(1 - Z_{k}^{\sigma}(\omega)) = \frac{1}{2}[\tilde{\Sigma}_{11}^{\sigma}(k,\omega) + \tilde{\Sigma}_{22}^{\sigma}(k,\omega)] \ ,$$

$$\xi_k^\sigma(\omega)) = \frac{1}{2}[\tilde{\Sigma}_{11}^\sigma(k,\omega) - \tilde{\Sigma}_{22}^\sigma(k,\omega)] \,, \tag{4.37}$$

$$\Phi_k^\sigma(\omega) = \Delta_k^\sigma + \tilde{\Sigma}_{12}^\sigma(k,\omega) \,.$$

and $\Sigma_{11}^\sigma(k,\omega) = -\Sigma_{22}^{\bar\sigma}(k,-\omega)$.

For the numerical solution of the system of equations (4.28)–(4.37) it is useful to introduce an imaginary frequency representation for the Green function (4.36) with $\omega = i\omega_n = i\pi T(2n+1)$ and spin-charge Green functions (4.35) with $\Omega = i\omega_n = i\pi T 2n$, where $n = 0, \pm 1, \pm 2, \ldots$. By using the representation for function (4.30) in the form

$$N(i\omega_n, z, \Omega) = -T\sum_m \frac{1}{i\omega_m - z} \; \frac{1}{i(\omega_n - \omega_m) - \Omega} \tag{4.38}$$

and after an integration in Eqs. (4.28), (4.29) we get

$$\tilde{\Sigma}_{11}^\sigma(k, i\omega_n) = -\frac{T}{N}\sum_q\sum_m \tilde{G}_{11}^\sigma(q, i\omega_m)\lambda_{11}(q, k-q \mid i\omega_n - i\omega_m) \,, \tag{4.39}$$

$$\tilde{\Sigma}_{12}^\sigma(k, i\omega_n) = \frac{T}{N}\sum_q\sum_m \tilde{G}_{12}^\sigma(q, i\omega_m)\lambda_{12}(q, k-q \mid i\omega_n - i\omega_m). \tag{4.40}$$

The interaction functions are given by

$$\lambda_{11}(q, k-q \mid i\omega_\nu) = g^2(q, k-q)D^+(k-q, i\omega_\nu) \,, \tag{4.41}$$

$$\lambda_{12}(q, k-q \mid i\omega_\nu) = g^2(q, k-q)D^-(k-q, i\omega_\nu) \,. \tag{4.42}$$

Here we have $G_{11}^\sigma(k, i\omega_m) = -G_{22}^{\bar\sigma}(k, -i\omega_m)$ and $G_{12}^\sigma(k, i\omega_m) = G_{21}^\sigma(k, -i\omega_m)$.

The linearized system of Eliashberg equations (4.37) for $T \le T_c$ has the form

$$\tilde{G}_{11}^\sigma(k, i\omega_n) = \frac{1}{i\omega_n - E_k + \tilde{\mu} - \tilde{\Sigma}_{11}^\sigma(k, i\omega_n)} \,, \tag{4.43}$$

$$\Phi^\sigma(k, i\omega_n) = \Delta_k^\sigma + \phi^\sigma(k, i\omega_n) = \frac{T}{N}\sum_q\sum_m \{2J(k-q)$$

$$-\lambda_{12}(q, k-q \mid i\omega_n - i\omega_m)\}\tilde{G}_{11}^\sigma(q, i\omega_m)\tilde{G}_{11}^\sigma(q, -i\omega_m)\Phi^\sigma(q, i\omega_m) \,. \tag{4.44}$$

At first the system of equations for normal GF (4.43), (4.39) should be solved for a given concentration of electrons

$$\frac{n}{1-n/2} = 1 + \frac{2T}{N}\sum_k \sum_{n=-\infty}^{\infty} \tilde{G}_{11}(k, i\omega_n) \,. \tag{4.45}$$

Then eigenvalues and eigenfunctions of the linear equation for the gap function (4.44) should be calculated to obtain the superconducting transition temperature T_c and the (q,ω) dependent gap function.

For numerical calculations one has to introduce a model for charge- and spin-fluctuation functions in (4.35). By taking into account only the spin-fluctuation contribution we can write them in the form

$$D_s^-(q, i\omega_\nu) = -D_s^+(q, i\omega_\nu) = \chi_s(q)\int_0^{+\infty} \frac{2z\,dz}{z^2 + \omega_\nu^2}\chi_s''(z) \,, \tag{4.46}$$

where we have introduced for the spin-fluctuation susceptibility a model representation (Jaklič et al., 1995b):

$$\chi_s''(q,\omega) = -\frac{1}{\pi}\mathrm{Im}\,\langle\langle \mathbf{S}_q \mid \mathbf{S}_{-q}\rangle\rangle_{\omega+i\delta} = \chi_s(q)\,\chi_s''(\omega)$$

$$= \frac{\chi_0}{1+\xi^2(\mathbf{q}-\mathbf{Q}_{AF})^2}\tanh\frac{\omega}{2T}\frac{1}{1+(\omega/\omega_s)^2} \qquad (4.47)$$

with characteristic AFM correlation length ξ and spin-fluctuation energy $\omega_s \simeq J$. To fix the constant χ_0 in (4.47) we can use for the spin-fluctuation susceptibility the normalizing condition

$$\frac{1}{N}\sum_i \langle \mathbf{S}_i\mathbf{S}_i\rangle = \frac{1}{N}\sum_q \chi_s(q)\int_{-\infty}^{+\infty}\frac{dz}{\exp(z/T)-1}\chi_s''(z) = \frac{3}{4}n\,. \qquad (4.48)$$

By using the model (4.47) we get for static spin correlation functions (4.26):

$$\chi_{1s} = \langle \mathbf{S}_i\mathbf{S}_{i+a_1}\rangle = \frac{1}{N}\sum_q \gamma(q)\langle \mathbf{S}_q\mathbf{S}_{-q}\rangle,$$

$$\chi_{2s} = \langle \mathbf{S}_i\mathbf{S}_{i+a_2}\rangle = \frac{1}{N}\sum_q \gamma'(q)\langle \mathbf{S}_q\mathbf{S}_{-q}\rangle\,, \qquad (4.49)$$

where

$$\langle \mathbf{S}_q\mathbf{S}_{-q}\rangle = \chi_s(q)\int_{-\infty}^{+\infty}\frac{dz}{\exp(z/T)-1}\chi_s''(z) = \chi_s(q)\frac{\pi}{2}\omega_s\,. \qquad (4.50)$$

Therefore, we have obtained a closed system of equations which can be solved numerically. Below we present results of the self-consistent solution of Eqs. (4.39), (4.43) for hole concentration $\delta = 1-n = 0.2$ in Eq. (4.45) and temperature $T = 0.012t$. The parameters of the $t-J$ model (3.30) are: $J = 0.4t$, $t' = 0$. For the spin-susceptibility model (4.47) we take $\omega_s = J$ and $\xi = 1$. In Fig. 4.1 the spectral density $A_{11}^{\sigma}(k,\omega)$ (4.31) is shown for the k-vector along the line $(0,0)\to(\pi,\pi)$ $(0\to 32)$ where the energy ω is measured in units of t.

We see that at $k = (\pi/2,\pi/2)$ we have quite sharp spectral density which proves the existence of well defined quasi-particles. The dispersion relations for the one-hole excitation energy E_k (in units of t) defined by maxima of spectral density (4.31) are shown in Fig. 4.2 along the lines of the two-dimensional square Brillouin zone (BZ): $\Gamma(0,0)\to M(\pi,\pi)\to X(\pi,0)\to\Gamma(0,0)$. We observe a narrow band for quasiparticles, with the largest dispersion from $\Gamma(0,0)$ to (π,π) of order of $1.5t$. Fig. 4.3 shows the density of states which is given by the sum of spectral density over BZ, i.e.,

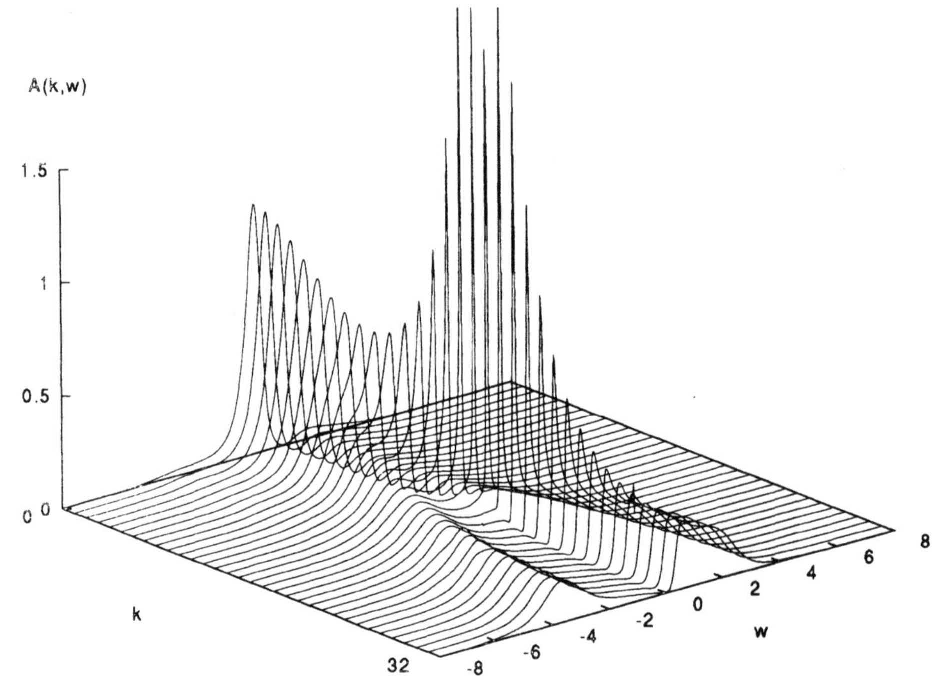

Fig. 4.1 The spectral density $A_{11}^{\sigma}(k,\omega)$ (4.31) for k-vector along the line $(0,0) \to (\pi,\pi)$ $(0 \to 32)$.

$$A_{11}^{\sigma}(z) = \frac{1}{N}\sum_{q} A_{11}^{\sigma}(q,z)\,. \tag{4.51}$$

At the Fermi energy, $\omega = 0$, we have a peak in the density of states (4.51) due to the flat dispersion around the $X(\pi,0)$ point (see Fig. 4.2). These calculations fit quite well the experimentally observed results in ARPES for one-hole excitation spectra in cuprates; see, e.g., Brenig (1995).

Before closing this section we shall briefly discuss the results for the two-band $p-d$ model (2.18) obtained by Plakida et al. (1995). To study the two–band problem we have to introduce the matrix Green function concerning the normal state properties

$$\hat{G}_{ij\sigma}(t-t') = \langle\langle \hat{X}_{i\sigma}(t); \hat{X}_{j\sigma}^{+}(t') \rangle\rangle\,, \tag{4.52}$$

where we have used the two–component operators

$$\hat{X}_{i\sigma} = \begin{pmatrix} X_i^{\sigma 2} \\ X_i^{0\bar{\sigma}} \end{pmatrix} \qquad \hat{X}_{j\sigma}^{+} = \begin{pmatrix} X_j^{2\sigma} & X_j^{\bar{\sigma}0} \end{pmatrix}\,. \tag{4.53}$$

By differentiating GF over time t and t' and using the projection technique described above we get the Dyson equation in a form analogous to (4.16). Plakida et al. (1995) calculated only the zero order GF. The latter was obtained in a form analogous to one band GF (4.14). The two-band spectrum for d-like holes and $p-d$ singlets as well as the density of states were calculated. It is the hybridization between d-like holes and singlets that results in a substantial renormalization of the spectrum. In addition, the dispersion relation depends strongly on antiferromagnetic short-range spin correlations (given by static spin correlation functions, Eqs. (4.26)) in the spin-singlet state. For large spin-correlations at small doping values one finds a next-nearest neighbour dispersion. With doping, which decreases the spin correlations, the dispersion changes to ordinary nearest neighbour one.

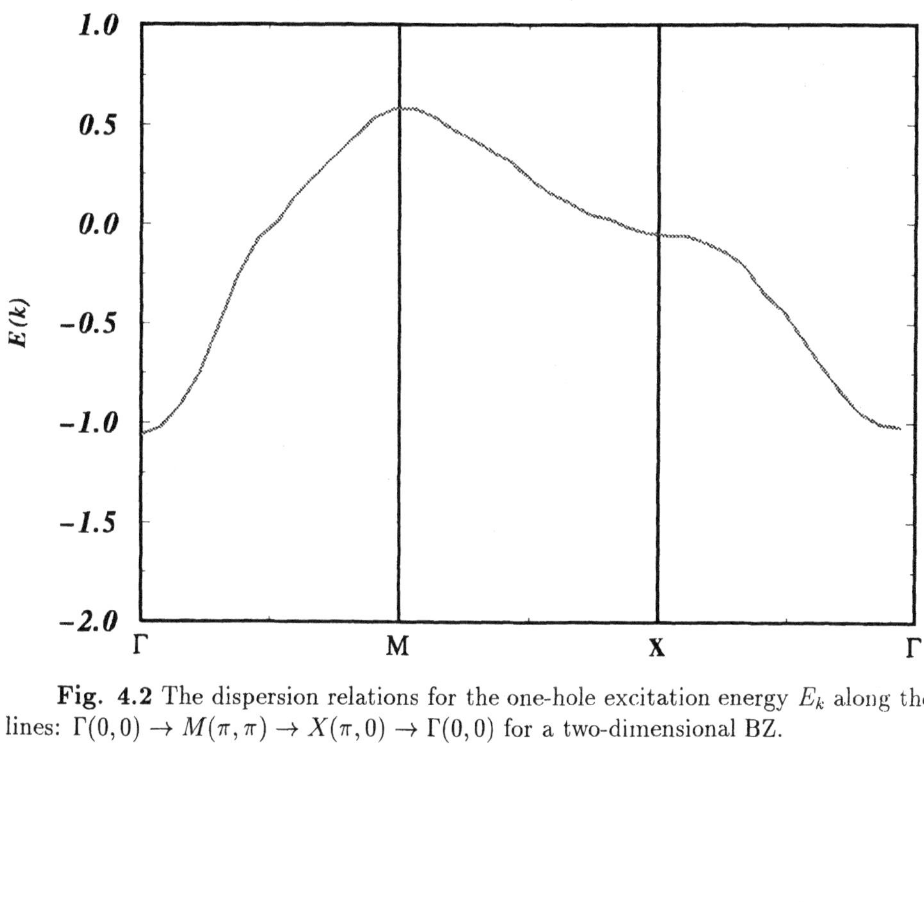

Fig. 4.2 The dispersion relations for the one-hole excitation energy E_k along the lines: $\Gamma(0,0) \rightarrow M(\pi,\pi) \rightarrow X(\pi,0) \rightarrow \Gamma(0,0)$ for a two-dimensional BZ.

However, to consider also superconducting properties for the two-band model we should introduce a 4×4 matrix Green function (Plakida, 1997b)

$$\tilde{G}_{ij\sigma}(t-t') = \langle\langle \hat{X}_{i\sigma}(t); \hat{X}^{+}_{j\sigma}(t') \rangle\rangle \tag{4.54}$$

for four-component operators. For example, H.c. one is given by the row-vector

$$\hat{X}^{+}_{j\sigma} = \left(X_j^{2\sigma} \quad X_j^{\bar{\sigma}0} \quad X_j^{\bar{\sigma}2} \quad X_j^{0\sigma} \right). \tag{4.55}$$

By differentiating GF (4.54) over times t and t' and using the projection technique described above we get the following Dyson equation in (k, ω) space:

$$\tilde{G}_\sigma(k,\omega)^{-1} = \tilde{G}^0_\sigma(k,\omega)^{-1} - \tilde{\Sigma}_\sigma(k,\omega). \tag{4.56}$$

Zero-order matrix GF is given by the generalized mean-field approximation

$$\tilde{G}^0_{ij,\sigma}(\omega) = \{\omega\delta_{i,j}\tilde{\tau}_0 - \tilde{A}_{ij,\sigma}\}^{-1}\tilde{\chi}\ , \tag{4.57}$$

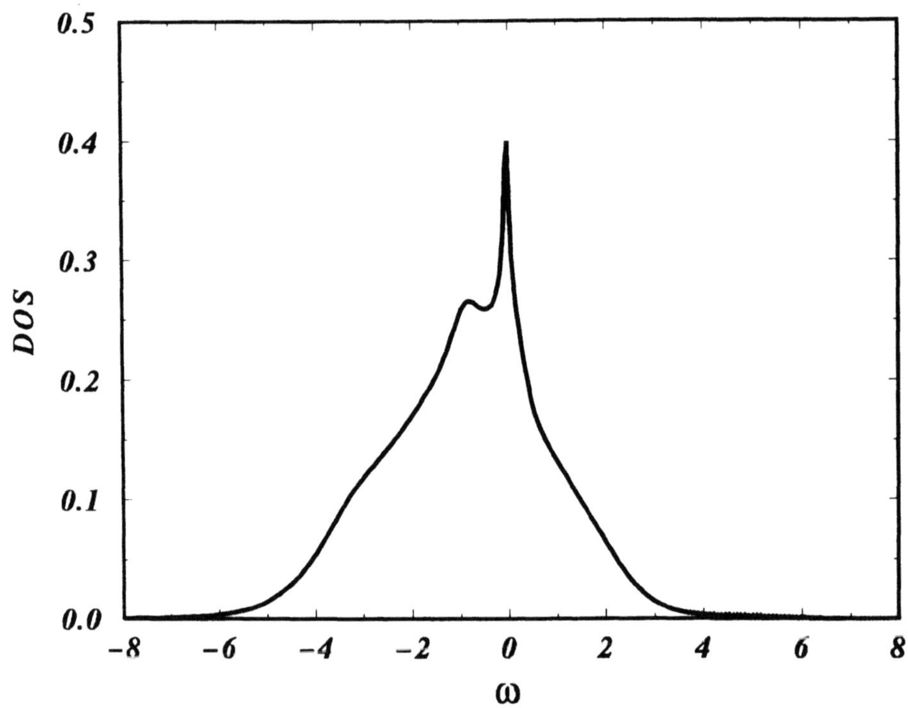

Fig. 4.3 The density of states $A^\sigma_{11}(\omega)$ (4.51)

where

$$\tilde{A}_{ij,\sigma} = \langle\{[\hat{X}_{i\sigma}, H]\ , \hat{X}^+_{j\sigma}\}\rangle\ \tilde{\chi}^{-1} \tag{4.58}$$

is the frequency matrix. The self-energy operator is given by the irreducible part of scattering matrix

$$\tilde{\Sigma}_{ij,\sigma}(\omega) = \tilde{\chi}^{-1}\langle\langle[\hat{X}_{i\sigma}, H]|[H, \hat{X}^+_{j\sigma}]\rangle\rangle^{(irr)}_\omega\tilde{\chi}^{-1}. \tag{4.59}$$

We have introduced unity matrices $\tilde{\tau}_0$ (4×4) and $\hat{\tau}_0$ (2×2) and the matrix $\tilde{\chi} = \hat{\tau}_0 \hat{\chi}$ with

$$\hat{\chi} = \begin{pmatrix} \chi_2 & 0 \\ 0 & \chi_1 \end{pmatrix}, \tag{4.60}$$

where $\chi_2 = \langle X_i^{22} + X_i^{\sigma\sigma} \rangle = 1 - \chi_1$. To solve the Dyson equation (4.56), which can be written in the general form as

$$\tilde{G}_\sigma(k,\omega) = \begin{pmatrix} \hat{G}_\sigma(k,\omega) & \hat{F}_\sigma(k,\omega) \\ (\hat{F}_\sigma(k,\omega))^+ & -\hat{G}_{\bar{\sigma}}(k,-\omega) \end{pmatrix}, \tag{4.61}$$

we have to calculate zero–order GF (4.57) and self-energy matrix (4.59). The anomalous part of zero–order GF in (4.61), $\hat{F}^0_\sigma(k,\omega)$, will vanish if one disregards the mean-field, k-independent gap function (due to the kinematic interaction) which violates the restriction $\langle X_i^{\sigma 2} X_i^{\bar{\sigma}} \rangle = 0$; see Plakida et al. (1989). For the normal part we can use the diagonal approximation

$$\hat{G}^0_\sigma(k,\omega) = \begin{pmatrix} \chi_2/(\omega - \Omega_2(k)) & 0 \\ 0 & \chi_1/(\omega - \Omega_1(k)) \end{pmatrix}, \tag{4.62}$$

where the mean field spectrum for the d-hole like band (singlet band) is given by the dispersion $\Omega_1(k)$ ($\Omega_2(k)$) obtained by Plakida et al. (1994).

To obtain the self-energy matrix (4.59) we have used the non-crossing approximation described above (see Eq. (4.27)). Writing down the self-energy matrix as

$$\tilde{\Sigma}_\sigma(k,\omega) = \tilde{\chi}^{-1} \begin{pmatrix} \hat{M}_\sigma(k,\omega) & \hat{\Phi}_\sigma(k,\omega) \\ \hat{\Phi}^+_\sigma(k,\omega) & -\hat{M}_{\bar{\sigma}}(k,-\omega) \end{pmatrix} \tilde{\chi}^{-1}, \tag{4.63}$$

we get for the normal part of matrix

$$\hat{M}_\sigma(k,\omega) = \frac{t^2}{N} \sum_q \frac{\gamma^2(q)}{\pi} \int_{-\infty}^{\infty} \int_{-\infty}^{\infty} d\nu dz N(\omega,\nu,z) D^+(k-q,z)$$

$$\times \{\hat{P}_2[-\mathrm{Im}G^{22}(q,\nu+i\delta)] + \hat{P}_1[-\mathrm{Im}G^{11}(q,\nu+i\delta)]\}, \tag{4.64}$$

where $\gamma(q) = \sum_j \exp(iqj)\nu_{0j}$; $N(\omega,\nu,z)$ is given by (4.30) and the spin-charge susceptibility $D^+(k-q,z)$ is given by (4.35). The contributions from singlet band and d-hole bands are defined by matrices $\hat{P}_2$ and $\hat{P}_1$, respectively. Their diagonal terms are

$$P_2^{11} = K_{21}^2, \quad P_2^{22} = K_{22}^2, \quad P_1^{11} = K_{11}^2, \quad P_1^{22} = K_{12}^2.$$

Here we have used the notation for hopping integrals (2.16) in the two-band model (2.15) as $t_{ij}^{\alpha\beta} = -K_{\alpha\beta} 2t\nu_{ij}$. The matrix equation (4.64) defines the renormalization of quasiparticle spectra in two bands due to spin and charge fluctuations while the analogous equation for matrix $\hat{\Phi}_\sigma(k,\omega)$ gives the gap equation. In the diagonal approximations for zero-order GF (4.62) and self-energy (4.63), (4.64) the Dyson equation (4.56) can be solved and this enables one to write the equation for superconducting gaps $\phi_\sigma^{\alpha\alpha}(k,\omega) = \Phi_\sigma^{\alpha\alpha}(k,\omega)/\chi_\alpha$ in bands $\alpha = 2,1$ in a closed form:

$$\phi_\sigma^{\alpha\alpha}(k,\omega) = -\frac{t^2}{N} \sum_q \frac{\gamma^2(q)}{\pi} \int_{-\infty}^{\infty} \int_{-\infty}^{\infty} d\nu dz N(\omega,\nu,z) \mathrm{Im}D^-(k-q,z)$$

$$\times \{K_{\alpha\alpha}^2[-\mathrm{Im}\frac{\phi_\sigma^{\alpha\alpha}(q,\nu)}{\Delta_\alpha(q,\nu)}] - \frac{\chi_\beta K_{12}^2}{\chi_\alpha}[-\mathrm{Im}\frac{\phi_\sigma^{\beta\beta}(q,\nu)}{\Delta_\beta(q,\nu)}]\}, \tag{4.65}$$

where $\alpha \neq \beta$ and the denominator

$$\Delta_\alpha(q,\omega) = (\omega - \Omega_\alpha(q) - M_\sigma^{\alpha\alpha}(q,\omega)/\chi_\alpha)(\omega + \Omega_\alpha(q) + M_{\bar{\sigma}}^{\alpha\alpha}(q,-\omega)/\chi_\alpha) - |\phi_\sigma^{\alpha\alpha}|^2 \quad (4.66)$$

gives the spectrum of exitations in the superconducting phase.

To conclude this section we would like to stress that starting from the microscopic $t-J$ or two-band $p-d$ models we have obtained the self-consistent system of equations for Green functions and corresponding self-energies. Both renormalization of quasi-particle spectra given by self-energies and superconducting pairing in gap equations, (4.44) or (4.65) result from spin and charge fluctuations which come from nonfermionic commutation relations for Hubbard operators in the models. Therefore, we have no fitting parameters for electron-spin scattering as in phenomenological approaches.

In the two band model for hole (electron) doped case the chemical potential μ is in the singlet (d-hole) band, $\alpha = 2$ (1), and the main contribution to the integrand in Eq. (4.65) comes from the same band (the first term), while the contribution from the other band is proportional to t/Δ^2. The latter is analogous to the static spin-exchange contribution of order $J \simeq (t/\Delta^2)$ in the one-band $t-J$ model in Eq. (4.44). However, in the two-band model the spin-fluctuation contribution in Eq. (4.65) is given by the frequency dependent susceptibility $D\pm(q,z)$ and interband contributions $\propto K_{12}^2$ cannot be fully allowed in the framework of one-band $t-J$ model. It would be interesting to compare the solution for the gap equations in the $t-J$ model, Eq. (4.44), and in the two-band model, Eq. (4.65). However, this demands quite complicated numerical work. In the next section we shall consider this problem for the gap equation and calculate the superconducting temperature T_c for the more simple spin-polaron model (2.36).

4.2 Superconducting Pairing of Spin Polarons

As we have pointed out in Sec. 2.3 for small concentration of holes when strong AF correlations still play an importatnt role in the hole dynamics, the $t-J$ model can be reduced to a more simple spin polaron model (2.36). It describes motion of spinless fermions on two sublattices with opposite spins. The pairing of two fermions on different sublattices mediated by spin fluctuations could be the possible mechanism of the high-temperature superconductivity in copper oxides.

Recently this problem has been treated in the framework of standard BCS formalism (Dagotto et al., 1995; Belinicher et al., 1995). A simple phenomenological model of quasiparticles with numerically evaluated spectrum and an effective pairing interaction in the atomic limit (Dagotto et al., 1995) or mediated by antiferromagnetic magnon exchange (Belinicher et al., 1995) have been used. By applying the rigid band approximation the high superconducting transition temperature was obtained for d-wave pairing. However, since the pairing spin-fluctuation energy J is of the same order as QP bandwidth, the weak coupling BCS equation is inadequate to treat the problem. Also the rigid band approximation fails to describe the strong doping dependence of QP spectrum. A full self-consistent solution of strong coupling Eliashberg equations both for normal and superconducting states has been given recently by Plakida et al., (1997). Their numerical solution of self-consistent system of equations within hole and magnon Green functions for the spin polaron model demonstrates a possibility of singlet d-wave superconducting pairing. The maximum superconducting temperature T_c of order $0.012t$ has been obtained at optimal hole concentration $\delta \simeq 0.2$. Below we shall discuss their calculations.

To discuss a singlet superconducting pairing within the spin polaron model (2.33), (2.36) we consider the equation of motion method for matrix Green function

$$\hat{G}(k,t-t') = \langle\langle \Psi_k(t) | \Psi_k^+(t') \rangle\rangle\,, \quad (4.67)$$

in terms of Nambu operators

$$\Psi_k = \begin{pmatrix} \tilde{c}_{k\uparrow} \\ \tilde{c}^+_{-k\downarrow} \end{pmatrix} = \begin{pmatrix} h^+_k \\ f_{-k} \end{pmatrix}, \qquad \Psi^+_k = \left(\tilde{c}^+_{k\uparrow}\, \tilde{c}_{-k\downarrow}\right) = \left(h_k\, f^+_{-k}\right) . \tag{4.68}$$

By differentiating the Green function (4.67) with respect to times t and t' we obtain the Dyson equation

$$\hat{G}(k,\omega)^{-1} = \omega\hat{\tau}_0 + (\epsilon_k - \mu)\hat{\tau}_3 - \hat{\Sigma}(k,\omega) , \tag{4.69}$$

where $\hat{\tau}_0$ and $\hat{\tau}_3$ are the standard Pauli matrices. The self-energy operator $\hat{\Sigma}(k.\omega)$ is given by the irreducible part of the many-particle Green function.

The solution of Dyson equation (4.69) can be written in Eliashberg notations as

$$\hat{G}(k,\omega) = \frac{\omega Z_k(\omega)\hat{\tau}_0 + (\chi_k(\omega) - \epsilon_k)\hat{\tau}_3 + \phi_k(\omega)\hat{\tau}_1}{(\omega Z_k(\omega))^2 - (\chi_k(\omega) - \epsilon_k)^2 - \phi_k(\omega)^2} , \tag{4.70}$$

where

$$\omega(1 - Z_k(\omega)) = \frac{1}{2}[\Sigma_{hh}(k,\omega) + \Sigma_{ff}(k,\omega)]$$

$$\chi_k(\omega)) = \frac{1}{2}[\Sigma_{hh}(k,\omega) - \Sigma_{ff}(k,\omega)] ,$$

$$\phi_k(\omega) = \Sigma_{hf}(k,\omega) = (\Sigma_{fh}(k,\omega))^* , \tag{4.71}$$

and $\Sigma_{ff}(k,\omega) = -\Sigma_{hh}(k,-\omega)$.

To obtain self-consistent equations for GF (4.70) we employ the self-consistent Born approximation (SCBA) (or the noncrossing diagram approximation) which has been proven to be quite reasonable in the calculation of one-hole spectrum in the normal state. In SCBA, we get following equations for self-energies of GF (4.70):

$$\Sigma_{hh}(k,i\omega_n) = -T\sum_q\sum_m G_{hh}(q,i\omega_m)\lambda_{11}(k,k-q \mid i\omega_n - i\omega_m) , \tag{4.72}$$

$$\Sigma_{hf}(k,i\omega_n) = -T\sum_q\sum_m G_{hf}(q,i\omega_m)\lambda_{12}(k,k-q \mid i\omega_n - i\omega_m) , \tag{4.73}$$

where Matsubara frequencies are $\omega_n = \pi T(2n+1)$. The interaction functions are

$$\lambda_{11}(k,q \mid i\omega_\nu) = g^2(k,q)D(q,-i\omega_\nu) + g^2(q-k,q)D(-q,i\omega_\nu) , \tag{4.74}$$

$$\lambda_{12}(k,q \mid i\omega_\nu) = g(k,q)g(q-k,q)\{D(q,-i\omega_\nu) + D(-q,i\omega_\nu)\} . \tag{4.75}$$

Here the diagonal magnon GF $D(q,\omega) = \langle\langle \alpha_q \mid \alpha^+_q \rangle\rangle_\omega$ can be written as

$$D(q,\omega) = \frac{\omega + \omega_q + \Pi_{22}(q,\omega)}{[\omega - \omega_q - \Pi_{11}(q,\omega)][\omega + \omega_q + \Pi_{22}(q,\omega)] + \mid \Pi_{12}(q,\omega) \mid^2} . \tag{4.76}$$

Within SCBA the polariazation operators are as follows:

$$\Pi_{11}(q,i\omega_\nu) = T\sum_k\sum_m \{g^2(k,q)G_{hh}(k,i\omega_m)G_{hh}(k-q,i\omega_\nu + i\omega_m)$$

$$-g(k,q)g(q-k,q)G_{hf}(k,i\omega_m)G_{hf}(k-q,i\omega_\nu + i\omega_m)\} , \tag{4.77}$$

$$\Pi_{12}(q,i\omega_\nu) = T\sum_k\sum_m \{g(k,q)g(q-k,q)G_{hh}(k,i\omega_m)G_{hh}(k-q,i\omega_\nu + i\omega_m)$$

$$-g^2(k,q)G_{hf}(k,i\omega_m)G_{hf}(k-q,i\omega_\nu+i\omega_m)\} , \tag{4.78}$$

where $\Pi_{22}(q,i\omega_\nu)=\Pi_{11}(-q,-i\omega_\nu)$.

To calculate the superconducting temperature T_c we can study only the linearized system of Eliashberg equations for normal GF in (4.70)

$$G_{hh}(k,i\omega_n)=\frac{1}{i\omega_n+\epsilon_k-\mu-\Sigma_{hh}(k,i\omega_n)} , \tag{4.79}$$

and for superconducting gap function (4.73)

$$\phi(k,i\omega_n)=\sum_p\sum_m\lambda_{12}(k,k-p\mid i\omega_n-i\omega_m)\phi(p,i\omega_m)$$

$$\times G_{hh}(p,i\omega_m)G_{hh}(-p,-i\omega_m) . \tag{4.80}$$

At first a self-consistent calculation of normal GF (4.79) with a self-energy operator (4.72) has been done for a given concentration of holes (2.38):

$$\delta=\frac{1}{2}+\frac{2T}{N}\sum_k\sum_{n=0}^{\infty}G(k,i\omega_n).$$

Then the gap equation (4.80) has been solved and leading eigenvalues for pairing eigenfuctions $\phi(q,i\omega_n)$ are obtained. The calculations are performed for the hole concentrations in the range $0.02\le\delta\le0.35$ and for parameters of the spin polaron model: $J=0.4$ and $t'=0,\pm0.1$ (all energies here and below are measured in units of t).

The fast Fourier transformation have been used in numerical calculations (Serene and Hess, 1991) for a finite mesh of 64×64 **k**-points in the full Brillouin zone ($0\le k_x,k_y\le1$), in units of $2\pi/a$, and 200-700 points for Matsubara frequencies with a constant cut $\omega_{max}=10t$ in the summation over it. Usually $10-30$ iterations are needed to obtain a solution for the self energy with an accuracy of order 0.001. To calculate the hole spectral function

$$A(k,\omega)=-\frac{1}{\pi}\text{Im}\,\langle\langle h_k\mid h_k^+\rangle\rangle_{\omega+i\epsilon} \tag{4.81}$$

and the density of states (DOS)

$$A(\omega)=\frac{1}{N}\sum_k A(k,\omega) , \tag{4.82}$$

the Pade approximation has been used for analytical continuation from Matsubara points on the imaginary axis.

Calculations of the spin polaron quasiparticle spectrum have been done at finite temperature $T=0.012$ that is slightly higher than the maximal superconducting temperature discussed below. Computations of hole spectral functions $A(k,\omega)$ at different **k**-points show that for small hole concentrations $\delta\le0.10$ there are no much differences for spectral functions calculated with renormalized and unrenormalized magnon energies in the interaction function, Eq. (4.74). In Fig. 4.4 we compare the results of calculations for hole density of states $A(\omega)$ with renormalized (solid line) and unrenormalized (dashed line) magnon spectra for $\delta=0.06$ that demonstrate the small effect of the magnon renormalization.

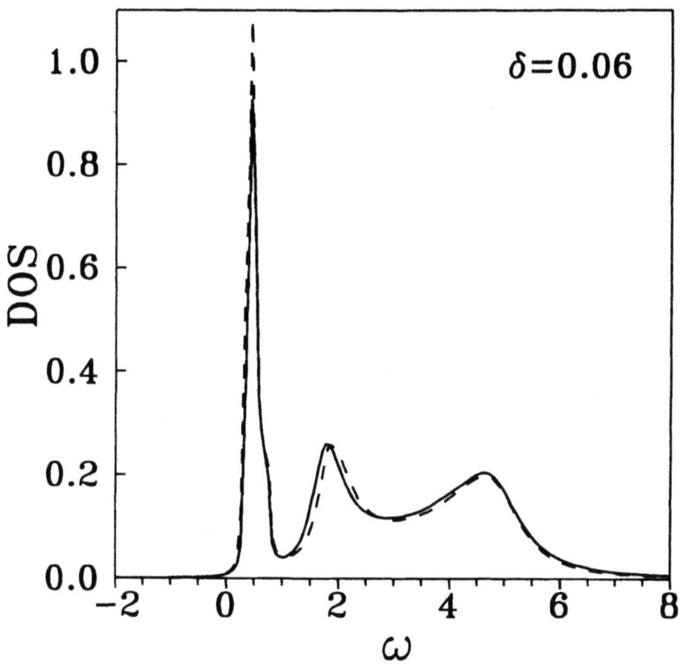

Fig. 4.4 The hole density of states $A(\omega)$ with renormalized (solid line) and unrenormalized (dashed line) magnon spectra for $\delta = 0.06$.

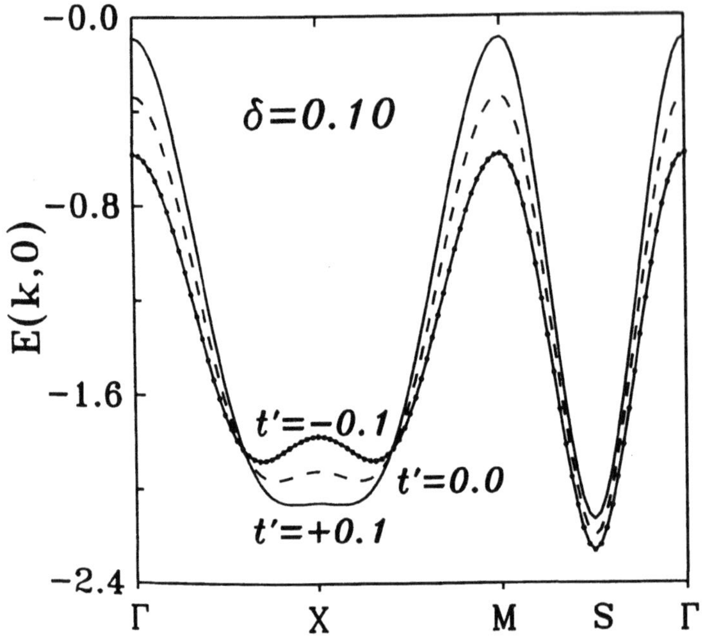

Fig. 4.5 The quasiparticle energy $E(k,0)$ for $t^{'} = -0.1$ (dotted line), $t^{'} = 0$ (dashed line) and $t^{'} = +0.1$ (solid line) at the hole concentration $\delta = 0.1$.

However, at higher hole concentrations a negative contribution to the magnon spectral density appears for $\omega < 0$ due to the excitation of electron-hole pairs. This results in negative values for hole spectral functions in the incoherent part of spectrum. This negative contribution develops at first for long wavelength magnons as has been pointed out by Sherman and Schreiber (1993, 1994). Since the main quasiparticle peak at $\mathbf{k} = (\pi/2, \pi/2)$ does not change much in shape with doping the picture of spin polarons as stable quasiparticle seems to be relevant even for moderate hole concentrations. This robust behavior of spin polarons with doping can be explained by the small size of polarons in comparison with the AFM correlation length due to the large exchange energy. It is interesting to compare the spectral density in Fig. 4.4 for the spin-polaron model and that one for the $t-J$ model (4.51) in paramagnetic phase with only short range AFM correlations, $\xi = 1$, shown in Fig. 4.3. We see that the main peak near Fermi energy $\omega = 0$ in both cases is quite sharp while the incoherent part of the spectrum (for $\omega > 0$ in Fig. 4.4 for holes and for $\omega < 0$ in Fig. 4.3 for electrons) is much broader in the paramagnetic case for $t-J$ model.

The quasiparticle energy defined as $E(k,0) = \epsilon(k) + \mathrm{Re}\ \Sigma(k,0)$ is shown in Fig. 4.5 for $t^{'} = -0.1$ (dotted line), $t^{'} = 0$ (dashed line) and $t^{'} = +0.1$ (solid line) and a hole concentration: $\delta = 0.1$. With doping the hole quasiparticle spectrum does not change much in shape but the bandwidth increases substantially. So, the rigid band approximation adopted by Dagotto et al. (1995), and Belinicher et al. (1995) is inadequate. For $t^{'} = 0$ or $t^{'} = -0.1$ the minimum of dispersion curves is at points $(\pm\pi/2, \pm\pi/2)$ in BZ that results in a 4-pocket like form of the Fermi surface (FS) at low hole concentration. With increasing of hole concentration a transition from 4-pocket like FS to large one occurs quite sharply. However, for $t' = +0.1$ the minimum of the dispersion curves shifts to points of the type $(0, \pm\pi)$ at the BZ boundary. Corresponding FS appears to be large even for small concentration of doped holes. However, the hole spectrum $E(k)$ in Fig. 4.5 is quite different in comparison with the electron spectrum $E(k)$ for the $t - J$ model, Fig. 4.2. The hole spectrum is degenerate at $\Gamma(0,0)$-point and $M(\pi,\pi)$-point since it is calculated for AFM state while the electron spectrum in Fig. 4.2 has a quite large dispersion along the line $(0,0) \to (\pi,\pi)$. The widths of quasiparticle spectrum, nevertheless, in both cases are comparable, of order $1.5t$.

The temperature dependence of momentum distribution for holes in the spin polaron model has been investigated in some details by Plakida et al. (1994) where it is shown that the Fermi surface washes out at some temperature of order of $T_d \simeq 1.5J\,\delta \simeq 0.6t\,\delta$. So, for the low temperatures $T \approx 0.01t$ considered here the Fermi surface does not change much with temperature. It should be also pointed out that the high density of states in present calculations (see Fig. 4.4) results from narrowing of free electron bandwidth due to strong correlations (the spin polaron formation) and has nothing to do with the van Hove singularity at half-filling.

To study the symmetry of superconducting order parameter and to evaluate the superconducting temperature T_c, the linearized Eliashberg equation for pairing energy $\phi(k, i\omega_n)$ (4.80) is considered. Looking for even functions of the wave-vector **k** that are realized in singlet pairing one obtains only a d-type symmetry for the gap function. The **k**-dependence of gap is very close to standard one,

$$\Delta(\mathbf{k}, \omega = 0) \propto (\cos k_x - \cos k_y) \ ,$$

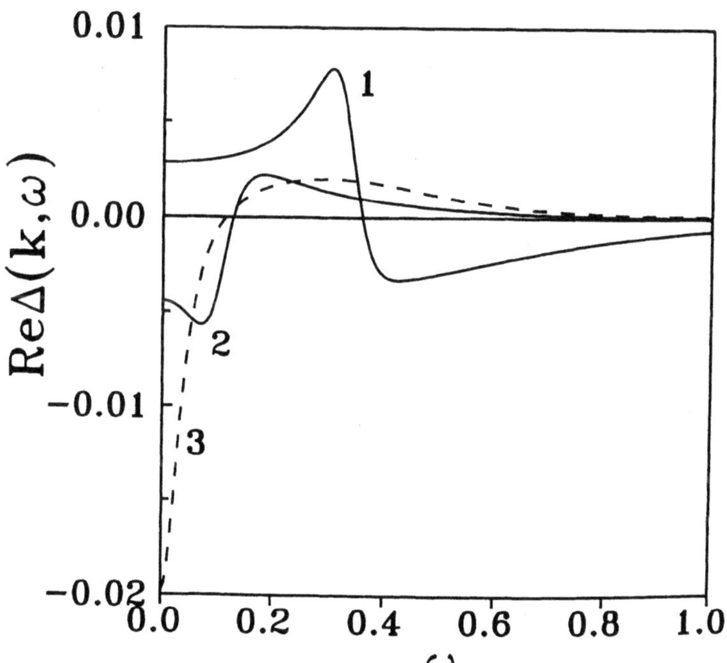

Fig. 4.6 Re $\Delta(\mathbf{k}, \omega)$ for set of (k_x, k_y) points: (1) – inside FS (0, 0.19), (2) – at the AF Brillouin zone boundary (0.31, 0.19), (3) – near FS (dashed line) – (0.38, 0.19).

where $\Delta(\mathbf{k},\omega) = \phi(\mathbf{k},\omega)/Z(\mathbf{k},\omega)$. The frequency dependence of Re $\Delta(\mathbf{k},\omega)$ is shown in Fig. 4.6 and Im $\Delta(\mathbf{k},\omega)$ in Fig. 4.7 for set of (k_x, k_y) points: (1) – inside FS (0, 0.19), at the AF Brillouin zone boundary (2) – (0.31, 0.19), and near FS (dashed line 3) – (0.38, 0.19) .

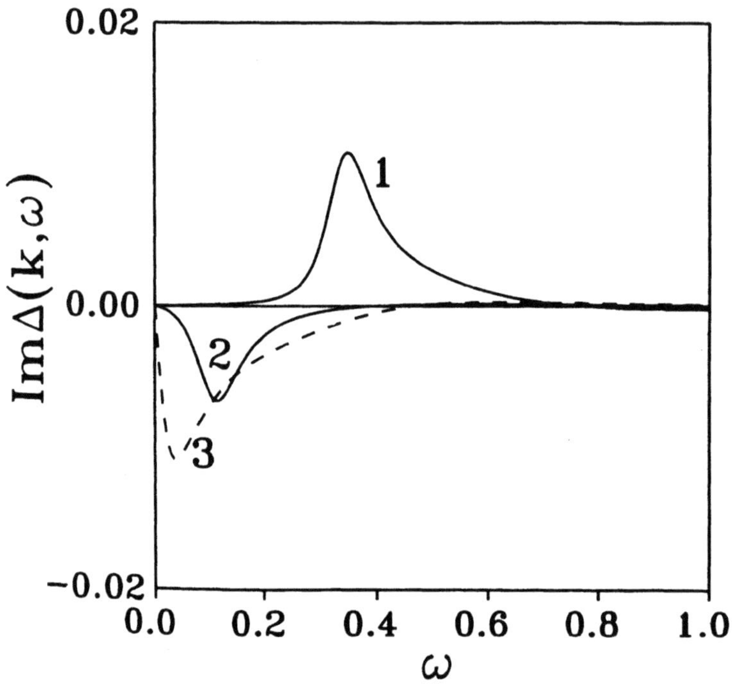

Fig. 4.7 Im $\Delta(\mathbf{k},\omega)$ for same (k_x, k_y) points as in Fig 4.6.

The gap function changes sign after crossing $k_x = k_y = 0.19$ point where it is equal to zero. It is interesting that the characteristic for pairing theory cut off energy of order $J \simeq 0.4$ away from the FS becomes much smaller near FS (see the dashed line 3). Therefore we have really a strong coupling limit for spin polaron pairing where all quasiparticles are paired contrary to weak coupling in conventional superconductors. The large values of Im $\Delta(k,\omega)$ near FS shown in Fig. 4.7 also differ from results for conventional superconductors.

By examining the temperature dependence of the highest eigenvalue in Eq. (4.80) for different hole concentrations we can find the temperature when it passes through unity with decreasing of T. At this temperature the normal state becomes unstable due to singlet pairing of quasiparticles – spin polarons on different sublattices. In Fig. 4.8 the dependence of superconducting temperature on the hole concentration is shown for $t' = +0.1$ (solid line at the right), $t' = 0$ (dashed line) and $t' = -0.1$ (solid line with dots at the left).

The position of the T_c maximum at $\delta \simeq 0.25; 0,20; 0.15$ for $t' = -0.1; 0; +0.1$, respectively, is explained by crossing the maximum of the hole states density by the Fermi level at given hole concentrations. These results are quite different from the monotone increase of T_c obtained within the weak coupling limit from the BCS equation by Belinicher et al. (1995) and the T_c maximum observed near half filling, $\delta = 0$, for small clusters by Ohta et al. (1994).

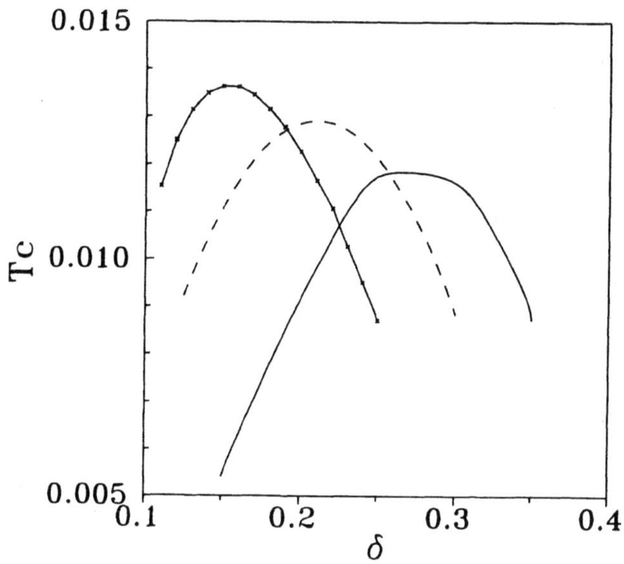

Fig. 4.8 The superconducting temperature T_c against the hole concentration δ for $t' = +0.1; 0; -0.1$ (from right to left).

It has been also observed that T_c increases with J but saturates at $T_c \simeq 0.025$ for $J \simeq 3$. However, a large drop of T_c for $J > 3$ observed in small cluster calculations near the phase separation (Dagotto, 1994) is not obtained. But latter phenomenon is beyond the scope of discussed theoretical approach. We should also mention calculations done by Scherman and Schriber (1995) who have obtained superconducting pairing of spin polarons only by taking into account an additional hole-phonon interaction.

The presented calculations, however are based on two-sublattice representation (2.32) for the spin-polaron model (2.36), which can be proven rigorously for AF background at low hole concentration. However, as we have seen in the previous Section (Fig. 4.1), spin polarons dressed by short range AFM spin fluctuations are the relevant quasiparticles even in the region of moderate hole concentrations in the paramagnetic state. A numerical solution of gap equations for the $t - J$ model, Eq.(4.44), or the two-band model, Eq. (4.65), should support results obtained for the spin polaron model.

5. CONCLUSIONS

Many unconventional properties of copper oxides can be explained by taking into account strong electron correlations which are the origin of the insulating antiferromagnetic ground state for stoichiometric copper-oxide compounds. With doping long range AFM order is destroyed but strong dynamic spin fluctuations persist even for moderate concentration of doped holes. Therefore, the solution of strong electron correlations problem is the key problem now in the theory of high-temperature superconductivity.

In the present lecture we have tried to demonstrate how one can take into account strong electron correlations in studies of different physical properties of copper oxides within microscopic models.

At low concentration of doped holes the relevant quasiparticles are spin polarons which can be described properly in the framework of the spin-polaron model, Sec. 2.3. As shown in Sec. 4.2, they can propagate coherently within narrow band of order of exchange energy. Combining the results obtained from microscopic models: the Hubbard model in the weak coupling limit and the strong coupling spin polaron model, we can argue that spin-exchange pairing could be the true mechanism for high-temperature superconductivity proposed on the basis of phenomenological models.

At moderate hole concentration the strong renormalization of the hole quasiparticle energy spectrum due to spin fluctuations remains important. A rigorous theory in this region can be developed within the Hubbard operator technique in the framework of $t-J$ model (Sec. 2.2) or more realistic two-band $p-d$ model (Sec. 2.3). We have demonstrated this approach in Sec. 3 for calculation of spin and charge susceptibilities and in Sec. 4.1 for calculation of the one-electron spectrum in the strong correlations regime. It has been also shown that the equation of motion method for the Green functions has a great advantage in comparison with a rather complicated diagram technique for Hubbard operator (Izyumov et al., 1993).

At the same time we would like to stress that the employment of the Hubbard operator technique has a twofold advantage. First of all by using equations of motion for the Hubbard operators we automatically (as a result of nonfermionic commutation relations) take into account the scattering of electrons on spin and charge fluctuations due to strong correlations as has been first pointed out by Hubbard (1964). In the Fermi liquid models (see, e.g., Qimiao Si and Levin, 1991; Moria et al., 1991, and Pines, 1994) one has to introduce a phenomenological electron-spin fluctuation scattering mechanism to obtain a nonzero relaxation. To study the physical properties in the auxiliary field representation or in the gauge field technique one has to adopt the spin-charge separation condition which has been rigorously proven only for 1D Hubbard model. By employing the Hubbard operator representation we can also preserve rigorously the restriction for no double occupancy of lower Hubbard subband (or no single occupancy of upper subband) which in auxiliary field and gauge field techniques has to be imposed by the local conservation law of total number of fermions and bosons. The latter can be allowed only approximately as, e.g., in the $1/N$ expansion technique with N being the spin-orbital degeneracy. However, it is difficult to give an unambiguous physical interpretation of obtained results for realistic value $N=2$.

To obtain quantitative results which can be compared with experiments one has to perform very involved self-consistent calculations of the system of equations for one-electron Green function and spin and charge susceptibilities. This problem is under consideration now and we hope to produce physically interesting results which really prove the importance of strong electron correlations in oxide metals.

Acknowledgments

The author thanks Prof. D. I. Uzunov for hospitality during the International Winter Workshop "Cooperative Phenomena in Condensed Matter" (7-15 March, 1998, Pamporovo, Bulgaria). The author acknowledges partial financial supports by the INTAS-RFBR Grant No 95-591, and by NREL, Subcontract AAX-6-16763-01.

REFERENCES

Alloul, H., 1991, *Physica B* 169:51.
Anderson, P. W., 1959, *Phys. Rev.* 115:2; 1987, *Science* 235:1196.

Auerbach, A., and Larson, B. E., 1991, *Phys. Rev. B* 43:7800.
Barzykin, V., and Pines, D., 1995, *Phys. Rev. B* 52:13585.
Bashkaran, G., Zou, Z., Anderson, P. W., 1987, *Solid State Comm.* 63:973.
Batlogg, B., 1997, *Physica C* 282 - 287, XXIY.
Belinicher, V. I., Chernyshov, A, L., Dotsenko, A. V., and Sushov, O. P., 1995, *Phys. Rev. B* 51:6076.
Bickers, N. E., Scalapino, D. J., and Scalettar, R. T., 1987, *Int. J. Mod. Phys. B* 1:687.
Bickers, N. E., Scalapino, D. J., and White, S. R., 1989, *Phys. Rev. Lett.* 62:961.
Bickers, N. E., and White, S. R., 1991, *Phys. Rev. B* 43:8044.
Birgeneau, R. J., and Shirane G., 1989, *in*: "Physical Properties of High Temperature Superconductors," Ed. D. M. Ginsberg, World Scientific, Singapore, V.1. p.151.
Bogolubov, N. N., 1970, "Selected Works," Naukova Dumka, Kiev (in Russian), v.2, p. 390
Brenig, W., 1995, *Phys. Reports* 251:153.
Chao, K. A., Spalek, J., and Oleś, A. M., 1997, *J. Phys. C* 10:L271.
Dagotto, E., 1994, *Rev. Mod. Phys.* 66:763.
Dagotto, E., Nazarenko, A., and Moreo, A., 1995, *Phys. Rev. Lett.* 74:310.
Dahm, T., and Tewordt, L., 1995, *Phys. Rev. Lett.* 74:793.
Ding, H., Yokoya, T., Campuzano, J. C., Takahashi, T., Randiera, M., Norman, N. R., Mochiku, T., Kadowaki, K., and Giapintzakis, J., 1996, *Nature*, 382:51.
Eder, R., Ohta, Y., and Maekawa, S., 1995, *Phys. Rev. Lett.* 74:5124.
Emery, V., 1987, *Phys. Rev. Lett.* 58:2794.
Eskes, H., Oleś, A. M., Meinders, M. B. J., and Stephan, W., 1994, *Phys. Rev. B* 50:17980.
Feiner, L. F., Jefferson, J. H., and Raimondi, R., 1996, *Phys. Rev. B* 53:8751.
Forster, D., 1975, "Hydrodynamic Fluctuations, Broken Symmetry and Correlation Functions," Benjamin, New York.
Fulde, P., 1991, "Electronic Correlations in Molecules and Solids," Springer, Berlin – Heidelberg.
Georges, A., Kotliar, G., Krauth, W., and Rozenberg, M. J., 1996, *Rev. Mod. Phys.* 68:13.
Götze., W., and Wölfle, P., 1971, *J. Low Temp. Phys.* 5:575; 1972, *Phys. Rev. B* 6:1226.
Gros, C., Joint, R., and Rice, T. M., 1987, *Phys. Rev. B* 36:381.
Hayden, S. M., Aeppli, G., Mook, H. A., Perring, T. G., Mason, T. E., Cheong, W., and Fisk, Z., 1996, *Phys. Rev. Lett.* 76:1344.
Hayn, R., Yushankhai, V. Yu., and Lovtsov, S. V., 1993, *Phys. Rev. B* 47:5253.
Hedegard, P., and Pedersen, M. B., 1991, *Phys. Rev. B* 43:11504.
Hlubina, R. A., and Rice, T. M., 1995, *Phys. Rev. B* 51:9523.
Hubbard, J., 1963, *Proc. Roy. Soc. A (London)* 276:238; 1964, *ibid.* 281:401; 1965, *ibid.* 285:542.
Ihle, D., and Plakida, N. M., 1994, *Zeit. Phys. B* 96:159.
Izyumov, Yu. A., 1997, *Soviet Phys. Uspekhi* 167P:465.
Izyumov, Yu. A., and Hedersen, J. A., 1994, *Int. J. Mod. Phys. B* 8:1877.
Izyumov, Yu. A., Katsnelson, M. I., and Skryabin, Yu. N., 1993, "Itinerant Electron Magnetism," Nauka, Moscow.
Izyumov, Yu. A., and Letfulov, B. M., 1990, *J. Phys.: Cond. Matter* 2:8905; 1991, *ibid.* 3:5373.
Jackeli, G., and Plakida, N. M., 1997, *Preprint JINR* E17-97-266, Dubna.
Jaklič, J., and Prelovšek, P., 1995, *Phys. Rev. Lett.* 74:3411; *ibid.* 75:1340.
Kampf, A. P., 1994, *Phys. Reports*, 249:219.
Kampf A. P., and Schrieffer, J. R., 1990, *Phys. Rev. B* 41:6399; *ibid.* 42:7967.
Kane, C. L., Lee, P. A., and Read, N., 1989, *Phys. Rev. B* 39:6880.
Kane, C. L., Lee, P. A., Ng, T. K., Chakraborty, B., and Read, N., 1990, *Phys. Rev. B* 41:2653.
Kirtley, J. R.,Tsuei, C. C., Sun, J., Chi, C. C., Yu-Jahnes, L. S. Gupta, A., Rupp, M., and Ketchan, M. B., 1995, *Nature* 373:225.
Kondo J., and K. Yamaji, K., 1972, *Prog. Theor. Phys.* 47:807.
Kubo, R., 1957, *J. Phys. Soc. Jpn.* 12:570.
Kubo, K., 1971, *J. Phys. Soc. Jpn.* 31:30.
Lenck, S. T., Carbotte, J. P., and Dynes, R. C., 1994, *Phys. Rev. B* 50:10149.
Liu, Z., and Manouskas, E., 1992, *Phys. Rev. B* 45:2425.
Liu, D. Z., Zha, Y., and Levin, K., 1995, *Phys. Rev. Lett.* 75:4130.
Lovtsov, S. V., and Yushankhai, V. Yu., 1991, *Physica C* 179:159.
Martínez, G., and Horsch, P., 1991, *Phys. Rev. B* 44:317.
Mason, T. E., Aeppli, G., Hayden, S. M., Ramirez, A. P., and Mook, H. A., 1993, *Phys. Rev. Lett.* 71:919.
Matsuda, M., Yamada, K., Endok, Y., Thurston, T. R., Shirane, G., Birgeneau, P. J., Kastner, H. A., Tanaka, I., and Hojima, H., 1994, *Phys. Rev. B* 49:6958.

Millis, A., Monien, H., and Pines, D., 1990, *Phys. Rev. B* 42:167.
Miyake, K., Schmitt-Rink, S., and Varma, C. P., 1986, *Phys. Rev. B* 34:6554.
Monien, H., Pines, D., and Takigawa, M., 1991, *Phys. Rev. B* 43:258.
Monthoux, P., Balatsky, A. V., and Pines, D., 1991, *Phys. Rev. Lett.* 67:3448; 1992, *Phys. Rev. B* 46:14803.
Monthoux, P., and Pines, D., 1993, *Phys. Rev. B* 47:6069; 1994, *Phys. Rev. B* 49:4261.
Monthoux, P., and Scalapino, D. J., 1994, *Phys. Rev. Lett.* 72:1874.
Mori, H., 1965, *Prog. Theor. Phys.* 33:423; *ibid.* 34:399.
Moria, T., Takahashi, Y., and Ueda., K., 1990, *J. Phys. Soc. Jpn.* 59:2905.
Moria, T., and Takahashi, Y., 1991, *J. Phys. Soc. Jpn.* 60:776.
Mott, N. F., 1974, "Metal-Insulator Transitions," Taylor and Francis, London.
Ohata, N., and Kubo, R., 1970, *J. Phys. Soc. Jpn.* 28:1402.
Ohata, N., 1971, *J. Phys. Soc. Jpn.* 30:941.
Ohta, Y., Shimozano, T., Eder, R., and Maekawa, S., 1994, *Phys. Rev. Lett.* 73:324.
Onufrieva, F., and Rossat-Mignod, J., 1995, *Phys. Rev. B* 52:7572.
Pao Chien-Hua, and Bickers, N. E., 1994, *Phys. Rev. Lett.* 72:1870; 1995, *Phys. Rev. B* 51:16310.
Pickett, W. E., 1989, *Rev. Mod. Phys.* 61:433.
Pines, D., 1990, *Physica B* 163:78; 1994, *Physica C* 235-240:113; 1997, *Physica C* 282-287:273.
Plakida, N. M., 1995, "High Temperature Superconductivity," Springer, Berlin-Heidelberg.
Plakida, N. M., 1996, *J. Phys. Soc. Jpn.* 65:3964; 1997a, *Zeit. Phys. B* 103:383; 1997b, *Physica C* 282-287:1737.
Plakida, N. M., and Hayn, R., 1994, *Zeit. Phys. B* 93:313.
Plakida, N. M., Hayn, R., and Richard, J-L., 1995, *Phys. Rev. B* 51:16599.
Plakida, N. M., Oudovenko, V. S., Horsch, P., and Liechtenstein A., 1997, *Phys. Rev. B* 57:R11997; 1995, *Preprint JINR*, E17-95-287, Dubna.
Plakida, N. M., Oudovenko, V. S., and Yushankai V. Yu., 1994, *Phys. Rev. B* 50:6431.
Plakida, N. M., Yushankhai, V. Yu., and Stasyuk I. V., 1989, *Physica C* 160:80.
Prelovšek, P., 1997, *Zeit. Phys. B* 103:363.
Qimiao Si, and Levin K., 1991, *Phys. Rev. B* 44:4727.
Qimiao Si, Zha, Y., Levin K., and Lu., J. P., 1993, *Phys. Rev. B* 47:9055.
Raimondi, R., Jefferson, J. H., and Feiner, L. F., 1996, *Phys. Rev. B* 53:8774.
Regnault, L. P., Bourges, P., Burlet, P., Henry, J. Y., Rossat-Mignod, J., Idis, Y., and Vettier, C., 1994, *Physica C* 235-240:59.
Rice, T. M., and Zhang, F. C., 1989, *Phys. Rev. B* 39:1815.
Rossat-Mignod, J., Burlet, P., and Jurgens, M. J., 1988, *Physica C* 152:19.
Rozenberg, M. I., Kotliar, G., Kajueter, H., Thomas, G. A., Rapkine, D. H., Honig, J. M., Metcalf, P., 1995, *Phys. Rev. Lett.* 75:105.
Sadakata, I., and Hanamura, E., 1973, *J. Phys. Soc. Jpn.* 34:882.
Sato, M., Shamoto, S., Kiyokura, T., Kakurai, K., Shirane, G., Sternlieb, B., J., Tranquada, J. M., 1993, *J. Phys. Soc. Jpn.* 62:263.
Sawatzky, G. A., 1990, *in*: "Earlier and Recent Aspects of Superconductivity," ed. by J. G. Bednorz, K. A. Müller, Springer, Berlin – Heidelberg, p.345
Sawatzky, G. A., and Allen, J. W., 1984, *Phys. Rev. Lett.* 53:2239.
Scalapino, D. J., 1994, *Physica C* 235-240:107; 1995, *Phys. Reports* 250:329.
Scalapino, D. J., Loh, E., Jr., and Hirsch, J. E., 1986, *Phys. Rev. B* 34:8190.
Scherman A., and Schrieber M., 1993, *Phys. Rev. B* 48:7492; 1994, *Phys. Rev. B* 50:12887: 1995, *Phys. Rev. B* 52:10621.
Schmitt-Rink, S., Varma, C. M., and Ruckenstein, A. E., 1988, *Phys. Rev. Lett.* 60:2793.
Schrieffer, J. R., 1991, *Physica C* 185-189:17; 1995, *Solid State Commun.* 251:129.
Schrieffer, J. R., Wen X.-G., and Zhang S.-C., 1989, *Phys. Rev. Lett.* 60:944.
Serene, J. W., and Hess, D. W., 1991, *Phys. Rev. B* 44:3391.
Shimahara, H., and S. Takada, S., 1992, *J. Phys. Soc. Jpn.* 61:989.
Slichter, C. P., 1994, *in*: "Strongly Correlated Electronics Materials," Ed. K.Bedell, Addison-Wesley.
Stemman, G., C. Pepin, C., M. Lavagna, M., 1994, *Phys. Rev. B* 50:4075 .
Tanamoto, T., Kuboki, K., and Fukuyama H., 1991, *J. Phys. Soc. Jpn.* 60:3072, 4395.
Tohyama, T., P. Horsch, P., and Maekawa, S., 1995, *Phys. Rev. Lett.* 74:980.
Tranquada, J. M., Cox, D. E., Kunnmann, W., Moudden, H., Shirane, G., Suenaga, M., and Zolliker, P., 1988, *Phys. Rev. Lett.* 60:156.
Tranquada, J. M., Gehring, P. M., Shirane, G., Shamoto, S., and Sato, M., 1992, *Phys. Rev. B* 46:5561.

Trugman, S. A., 1990, *Phys. Rev. B* 41:892.
Tserkovnikov, Yu. A., 1982, *Theor. and Math. Fiz.* 52:147.
Tsuei, C. C., Kirtley, J. R., Rupp, M., Sun, J. Z., Gupta, A., Ketchen, M. B., Wang, C. A., Ren, Z. F., Wang, J. H., and Bhushan, M., 1996, *Science* 271:329.
Van Harlingen, D., 1995, *Rev. Mod. Phys.* 67:515.
Varma, C. M., Schmitt-Rink, S., and Abrahams, E., 1987, *Solid State Commun.* 62:681.
Yamada, K., Endoh, Y., Lee, C., Wakimoto, S., Arai, M., Ubukata, K., Fujita, M., Hosoya, S., and Bennington, S. M., 1995, *J. Phys. Soc. Jpn.* 64:2742.
Yushankhai, V. Yu., Oudovenko, V. S., and Hayn, R., 1997, *Phys. Rev. B* 55:15562.
Yushankhai, V. Yu., Plakida N. M., and Kalinay, P., 1991, *Physica C* 174:401.
Zaitsev, R. O., and Ivanov, V. A., 1987, *Sov. Phys. Solid State* 29:2554; 1989, *Int. J. Mod. Phys. B* 3:1403.
Zha, Y., Levin, K., and Qimiao Si, 1993, *Phys. Rev. B* 47:9124.
Zhang, F. C., and Rice, T. M., 1980, *Phys. Rev. B* 37:3759.
Zubarev, D. N., 1960, *Sov. Phys. Uspekhi* 3:320.

QUANTUM PHASE TRANSITIONS IN 2d QUANTUM LIQUIDS

Adriaan M. J. Schakel
Institut für Theoretische Physik
Freie Universität Berlin
Arnimallee 14, 14195 Berlin

1. PRELUDE

Continuous quantum phase transitions have attracted much attention in this decade both from experimentalists as well as from theorists. (For reviews see Refs.[1-4]). These transitions, taking place at the absolute zero of temperature, are dominated by quantum and not by thermal fluctuations as is the case in classical finite-temperature phase transitions. Whereas time plays no role in a classical phase transition, being an equilibrium phenomenon, it becomes important in quantum phase transitions. The dynamics is characterized by an additional critical exponent, the so-called dynamic exponent, which measures the asymmetry between the time and space dimensions. The natural language to describe these transitions is quantum field theory. In particular, the functional-integral approach, which can also be employed to describe classical phase transitions, turns out to be highly convenient.

The subject is at the border of condensed matter and statistical physics. Typical systems being studied are superfluid and superconducting films, quantum-Hall and related two-dimensional electron systems, as well as quantum spin systems. Despite the diversity in physical content, the quantum critical behavior of these systems shows surprising similarities. It is fair to say that the present theoretical understanding of most of the experimental results is scant.

The purpose of this Lecture is to provide the reader with a framework for studying quantum phase transitions. A central role is played by a repulsively interacting Bose gas at the absolute zero of temperature. The universality class defined by this paradigm is believed to be of relevance to most of the systems studied. Without impurities and a Coulomb interaction, the quantum critical behavior of this system turns out to be surprisingly simple. However, these two ingredients are essential and have to be included. Very general hyperscaling arguments are powerful enough to determine the exact value of the dynamic exponent in the presence of impurities and a Coulomb interaction, but the other critical exponents become highly intractable.

The emphasis in this Lecture will be on effective theories, giving a description of the system under study valid at low energy and small momentum. The rationale for this is the observation that the (quantum) critical behavior of continuous phase transitions is determined by such general features as the dimensionality of space, the

Correlations, Coherence, and Order. Edited by Shopova and Uzunov
Kluwer Academic/Plenum Publishers, 1999

symmetries involved, and the dimensionality of the order parameter. It does not depend on the details of the underlying microscopic theory. In the process of deriving an effective theory starting from some microscopic model, irrelevant degrees of freedom are integrated out and only those relevant for the description of the phase transition are retained. Similarities in the critical behavior in different systems can, accordingly, be more easily understood from the perspective of effective field theories.

The ones discussed in this Lecture are so-called *phase-only* theories. They are the dynamical analogs of the familiar O(2) nonlinear sigma model of classical statistical physics. As in that model, the focus will be on phase fluctuations of the order parameter. The inclusion of fluctuations in the modulus of the order parameter is generally believed not to change the critical behavior. Indeed, there are convincing arguments that both the Landau-Ginzburg model with varying modulus and the $O(n)$ nonlinear sigma model with a fixed modulus belong to the same universality class. For technical reasons a direct comparison is not possible, the Landau-Ginzburg model usually being investigated in an expansion around four dimensions, and the nonlinear sigma model in one around two.

In the case of a repulsively interacting Bose gas at the absolute zero of temperature, the situation is particularly simple as phase fluctuations are the only type of field fluctuations present.

This Lecture covers exclusively lower-dimensional systems. The reason is that in three space dimensions and higher the quantum critical behavior is in general Gaussian and, therefore, not very interesting.

Since time, and how it compares to the space dimensions is an important aspect of quantum phase transitions, Galilei invariance will play an important role in the discussion.

1.1 Notation

We adopt Feynman's notation and denote a spacetime point by $x = x_\mu = (t, \mathbf{x})$, $\mu = 0, 1, \cdots, d$, with d the number of space dimensions, while the energy k_0 and momentum $\mathbf{k}$ of a particle will be denoted by $k = k_\mu = (k_0, \mathbf{k})$. The time derivative $\partial_0 = \partial/\partial t$ and the gradient ∇ are sometimes combined in a single vector $\tilde{\partial}_\mu = (\partial_0, -\nabla)$. The tilde on ∂_μ is to alert the reader for the minus sign appearing in the spatial components of this vector. We define the scalar product $k \cdot x = k_\mu x_\mu = k_0 t - \mathbf{k} \cdot \mathbf{x}$ and use Einstein's summation convention. Because of the minus sign in the definition of the vector $\tilde{\partial}_\mu$ it follows that $\tilde{\partial}_\mu a_\mu = \partial_0 a_0 + \nabla \cdot \mathbf{a}$, with a_μ an arbitrary vector.

Integrals over spacetime are denoted by

$$\int_x = \int_{t,\mathbf{x}} = \int \mathrm{d}t\, \mathrm{d}^d x,$$

while those over energy and momentum by

$$\int_k = \int_{k_0,\mathbf{k}} = \int \frac{\mathrm{d}k_0}{2\pi} \frac{\mathrm{d}^d k}{(2\pi)^d}.$$

When no integration limits are indicated, the integrals are assumed to run over all possible values of the integration variables.

Natural units $\hbar = c = k_\mathrm{B} = 1$ are adopted throughout.

2. FUNCTIONAL INTEGRALS

In this Lecture we shall adopt, unless stated otherwise, the functional-integral approach to quantum field theory. To illustrate the use and the power of functional integrals, let us consider one of the simplest models of *classical* statistical mechanics: the Ising model. It is remarkable that functional integrals cannot only be used to describe quantum systems, governed by quantum fluctuations, but also classical systems, governed by thermal fluctuations.

2.1 Ising Model

The Ising model provides an idealized description of an uniaxial ferromagnet. To be specific, let us assume that the spins of some lattice system can point only along one specific crystallographic axis. The magnetic properties of this system can then be modeled by a lattice with a spin variable $s(\mathbf{x})$ attached to every site $\mathbf{x}$ taking the values $s(\mathbf{x}) = \pm 1$. For definiteness we will assume a d-dimensional cubic lattice. The Hamiltonian is given by

$$H = -\frac{1}{2} \sum_{\mathbf{x},\mathbf{y}} J(\mathbf{x},\mathbf{y})\, s(\mathbf{x})\, s(\mathbf{y}). \tag{1}$$

Here, $\mathbf{x} = a\, x_i\, \mathbf{e}_i$, with a the lattice constant, x_i are integers labeling the sites, and $\mathbf{e}_i$ ($i = 1, \cdots, d$) are unit vectors spanning the lattice. The sums over $\mathbf{x}$ and $\mathbf{y}$ extend over the entire lattice, and $J(\mathbf{x},\mathbf{y})$ is a symmetric matrix representing the interactions between the spins. If the matrix element $J(\mathbf{x},\mathbf{y})$ is positive, the energy will be minimized when the two spins at sites $\mathbf{x}$ and $\mathbf{y}$ are parallel—they are said to have a ferromagnetic coupling. If, on the other hand, the matrix element is negative, anti-parallel spins will be favored—the spins are said to have an antiferromagnetic coupling.

The classical partition function Z of the Ising model reads

$$Z = \sum_{\{s(\mathbf{x})\}} \mathrm{e}^{-\beta H}, \tag{2}$$

with $\beta = 1/T$ the inverse temperature. The sum is over all spin configurations $\{s(\mathbf{x})\}$, of which there are 2^N, with N denoting the number of lattice sites. To evaluate the partition function we linearize the exponent by introducing an auxiliary $\phi(\mathbf{x})$ at each site via the so-called Hubbard-Stratonovich transformation. Such a transformation generalizes the Gaussian integral

$$\exp\left(\tfrac{1}{2}\beta J s^2\right) = \sqrt{\frac{\beta}{2\pi J}} \int_\phi \exp\left(-\tfrac{1}{2}\beta J^{-1}\phi^2 + \beta\phi s\right), \tag{3}$$

where the integration variable ϕ runs from $-\infty$ to ∞. The generalization reads

$$\exp\left[\tfrac{1}{2}\beta \sum_{\mathbf{x},\mathbf{y}} J(\mathbf{x},\mathbf{y})\, s(\mathbf{x})\, s(\mathbf{y})\right] = \tag{4}$$
$$\prod_{\mathbf{x}} \int \mathrm{d}\phi(\mathbf{x}) \exp\left[-\tfrac{1}{2}\beta \sum_{\mathbf{x},\mathbf{y}} J^{-1}(\mathbf{x},\mathbf{y})\, \phi(\mathbf{x})\, \phi(\mathbf{y}) + \beta \sum_{\mathbf{x}} \phi(\mathbf{x}) s(\mathbf{x})\right].$$

Here, $J^{-1}(\mathbf{x},\mathbf{y})$ is the inverse of the matrix $J(\mathbf{x},\mathbf{y})$ and we have ignored—as will be done throughout this Lecture—an irrelevant normalization factor in front of the product at the right-hand side. The equation should not be taken too literally. It will be an identity only if $J(\mathbf{x},\mathbf{y})$ is a symmetric positively definite matrix. This is not true for the Ising model since the diagonal matrix elements $J(\mathbf{x},\mathbf{x})$ are all zero, implying that the sum of the eigenvalues is zero. We will nevertheless use this representation and regard it as a formal one. * The partition function now reads

$$Z = \sum_{\{s(\mathbf{x})\}} \prod_{\mathbf{x}} \int \mathrm{d}\phi(\mathbf{x}) \exp\left[-\tfrac{1}{2}\beta \sum_{\mathbf{x},\mathbf{y}} J^{-1}(\mathbf{x},\mathbf{y})\,\phi(\mathbf{x})\,\phi(\mathbf{y}) + \beta \sum_{\mathbf{x}} \phi(\mathbf{x})\,s(\mathbf{x})\right]. \quad (5)$$

The spins are decoupled in this representation, so that the sum over the spin configurations is easily carried out with the result

$$Z = \prod_{\mathbf{x}} \int \mathrm{d}\phi(\mathbf{x}) \exp\left(-\tfrac{1}{2}\beta \sum_{\mathbf{x},\mathbf{y}} J^{-1}(\mathbf{x},\mathbf{y})\,\phi(\mathbf{x})\,\phi(\mathbf{y}) + \sum_{\mathbf{x}} \ln\{\cosh\left[\beta\phi(\mathbf{x})\right]\}\right), \quad (6)$$

ignoring again an irrelevant constant.

The auxiliary field $\phi(\mathbf{x})$ is not devoid of physical relevance. To see this let us first consider its field equation:

$$\phi(\mathbf{x}) = \sum_{\mathbf{y}} J(\mathbf{x},\mathbf{y})\,s(\mathbf{y}), \quad (7)$$

which follows from (5). This shows that the auxiliary field $\phi(\mathbf{x})$ represents the effect of the other spins at site $\mathbf{x}$. To make this more intuitive let us study the expectation value of the field. For simplicity, we shall take only nearest-neighbor interactions into account by setting

$$J(\mathbf{x},\mathbf{y}) = \begin{cases} J & \text{if site } \mathbf{x} \text{ and } \mathbf{y} \text{ are nearest neighbors} \\ 0 & \text{otherwise,} \end{cases}$$

with J positive, so that we have a ferromagnetic coupling between the spins. The model is now translational invariant and the expectation value $\langle s(\mathbf{x})\rangle$ is independent of $\mathbf{x}$:

$$\langle s(\mathbf{x})\rangle = M. \quad (8)$$

We will refer to M as the magnetization. Upon taking the expectation value of the field equation (7),

$$\langle \phi(\mathbf{x})\rangle = 2dJM, \quad (9)$$

where $2d$ is the number of nearest neighbors, we see that the expectation value of the auxiliary field represents the magnetization.

A useful approximation often studied is the so-called mean-field approximation. It corresponds to approximating the integral over $\phi(\mathbf{x})$ in (6) by the saddle point—the value of the integrand for which the exponent is stationary. This is the case for $\phi(\mathbf{x})$ satisfying the field equation

$$-\sum_{\mathbf{y}} J^{-1}(\mathbf{x},\mathbf{y})\,\phi(\mathbf{y}) + \tanh\left[\beta\phi(\mathbf{x})\right] = 0. \quad (10)$$

We will denote the solution by ϕ_{mf}. In this approximation, the auxiliary field is no longer a fluctuating field taking all possible real values, but a classical one having the value determined by the field equation (10). Being a nonfluctuating field, the expectation value $\langle \phi_{\mathrm{mf}}(\mathbf{x})\rangle = \phi_{\mathrm{mf}}(\mathbf{x})$, and (10) yield a self-consistent equation for the magnetization

$$M = \tanh(2d\beta JM), \quad (11)$$

* *Editor's note:* The approach based on the identity (4) can be justified within the long-length scale approximation (14)–(15); see, D. I. Uzunov, in: "Lectures on Cooperative Phenomena in Condensed Matter", ed. by D. I. Uzunov, Heron, Sofia, 1996; p. 46.

where we have assumed a uniform field solution and invoked Eq. (9). It is easily seen graphically that the equation has a nontrivial solution, i.e., $|M| > 0$, when $2d\beta J > 1$. If, on the other hand, $2d\beta J < 1$ it has only a trivial solution. It follows that

$$\beta_0^{-1} = 2dJ \tag{12}$$

is the critical temperature separating the ordered low-temperature state with a nonzero magnetization from the high-temperature disordered state where the magnetization is zero.

Let us continue by expanding the Hamiltonian in powers of ϕ. To this end we note that the term $\ln[\cosh(\beta\phi)]$ in (6) has the Taylor expansion

$$\ln[\cosh(\beta\phi)] = \frac{1}{2}\beta^2\phi^2 - \frac{1}{12}\beta^4\phi^4 + \cdots \tag{13}$$

Before considering the other term in (6), $\sum_{\mathbf{x},\mathbf{y}} J^{-1}(\mathbf{x},\mathbf{y})\,\phi(\mathbf{x})\,\phi(\mathbf{y})$, let us first study the related object $\sum_{\mathbf{x},\mathbf{y}} J(\mathbf{x},\mathbf{y})\,s(\mathbf{x})\,s(\mathbf{y})$ which shows up in the original Ising Hamiltonian (1). With our choice (8) of the interaction, the Taylor expansion of this object becomes

$$\sum_{\mathbf{x},\mathbf{y}} J(\mathbf{x},\mathbf{y})\,s(\mathbf{x})\,s(\mathbf{y}) = J\sum_{\mathbf{x}} s(\mathbf{x})\,(2d + a^2\nabla^2 + \cdots)\,s(\mathbf{x}), \tag{14}$$

neglecting higher orders in derivatives. From this it follows that

$$\sum_{\mathbf{x},\mathbf{y}} J^{-1}(\mathbf{x},\mathbf{y})\,\phi(\mathbf{x})\,\phi(\mathbf{y}) = J^{-1}\sum_{\mathbf{x}} \phi(\mathbf{x})\left(\frac{1}{2d} - \frac{1}{4d^2}a^2\nabla^2 + \cdots\right)\phi(\mathbf{x}), \tag{15}$$

and the partition function (6) becomes in the small-ϕ approximation

$$Z = \prod_{\mathbf{x}} \int \mathrm{d}\phi(\mathbf{x})\,\mathrm{e}^{-\beta H}, \tag{16}$$

with H the so-called Landau-Ginzburg Hamiltonian

$$H = \sum_{\mathbf{x}} \left[\frac{a^2}{8d^2J}(\nabla\phi)^2 + \frac{1}{2}\left(\frac{1}{2dJ} - \beta\right)\phi^2 + \frac{\beta^3}{12}\phi^4\right]. \tag{17}$$

The model has a classical phase transition when the coefficient of the ϕ^2-term changes sign. This happens when $\beta = 1/2dJ$ in accord with the conclusion obtained by inspecting the self-consistent equation for the magnetization (11).

In the mean-field approximation, the thermal fluctuations around the mean-field configuration are ignored, so that ϕ becomes a nonfluctuating field. The functional integral $\prod_{\mathbf{x}} \int \mathrm{d}\phi(\mathbf{x})$ is approximated by the saddle point.

For future reference we go over to the continuum by letting $a \to 0$. To this end, we replace the discrete sum $\sum_i$ by the integral $a^{-d}\int_{\mathbf{x}}$, and rescale the field $\phi(\mathbf{x})$,

$$\phi(\mathbf{x}) \to \phi'(\mathbf{x}) = \sqrt{\frac{\beta a^{2-d}}{4d^2J}}\,\phi(\mathbf{x}), \tag{18}$$

such that the coefficient of the gradient term in the Hamiltonian takes the canonical form of 1/2. In this way the Hamiltonian becomes

$$\beta H = \int_{\mathbf{x}} \left[\frac{1}{2}(\nabla\phi)^2 + \frac{1}{2}r_0\phi^2 + \frac{1}{4!}\lambda_0\phi^4\right], \tag{19}$$

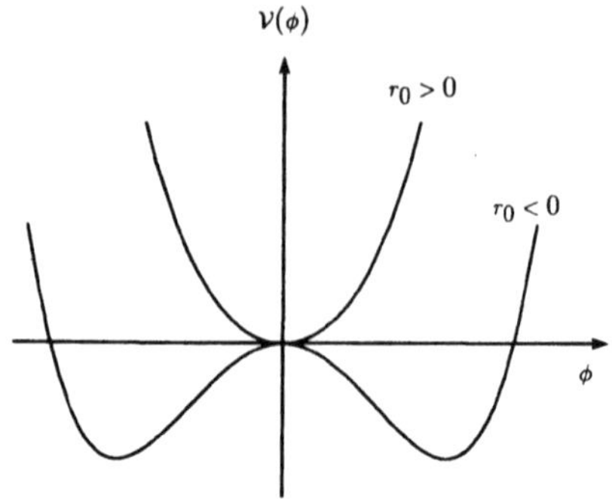

Figure 1. The potential $\mathcal{V}(\phi)$ of the Ising model in the high-temperature ($r_0 > 0$) and low-temperature ($r_0 < 0$) phases.

where we have dropped the prime on the field; the parameter r_0 and the coupling constant λ_0 are given by

$$r_0 = \frac{\beta_0^{-2}}{Ja^2}(\beta_0 - \beta), \quad \lambda_0 = \frac{4\beta^2}{J^2\beta_0^4}a^{d-4}. \tag{20}$$

The partition function now reads

$$Z = \int \mathrm{D}\phi \, \mathrm{e}^{-\beta H}, \tag{21}$$

where the functional integral $\int \mathrm{D}\phi$ denotes the continuum limit of the product of integrals $\prod_{\mathbf{x}} \int \mathrm{d}\phi(\mathbf{x})$. The last two terms in the integrand of (19) constitute the potential

$$\mathcal{V}(\phi) = \frac{1}{2} r_0 \phi^2 + \frac{1}{4!}\lambda_0 \phi^4. \tag{22}$$

In Fig. 1, the potential $\mathcal{V}(\phi)$ is depicted in the high-temperature phase where $r_0 > 0$, and also in the low-temperature phase where $r_0 < 0$. The minimum of the potential in the low-temperature phase is obtained for a value $\phi \neq 0$, whereas in the high-temperature phase the minimum is always at $\phi = 0$.

2.2 Derivative Expansion

We are interested in taking into account field fluctuations around the mean field ϕ_{mf}, which is the solution of the field equation obtained from (19). To this end, we set $\phi = \phi_{\mathrm{mf}} + \tilde{\phi}$, and expand the Hamiltonian around the mean field up to second order in $\tilde{\phi}$:

$$\beta H = \beta H_{\mathrm{mf}} + \frac{1}{2}\int_{\mathbf{x}} \left[(\nabla\tilde{\phi})^2 + (r_0 + \frac{1}{2}\lambda_0\phi_{\mathrm{mf}}^2)\tilde{\phi}^2\right], \tag{23}$$

where H_{mf} denotes the value of the Hamiltonian (19) for $\phi = \phi_{\mathrm{mf}}$. Because of the change of variables, the functional integral $\int \mathrm{D}\phi$ changes to $\int \mathrm{D}\tilde{\phi}$. Since we have neglected higher-order terms, the functional integral is Gaussian and easily carried out. The partition function (21) becomes in this approximation

$$\begin{aligned} Z &= \mathrm{e}^{-\beta H_{\mathrm{mf}}} \int \mathrm{D}\tilde{\phi} \exp\left\{-\frac{1}{2}\int_{\mathbf{x}} \left[(\nabla\tilde{\phi})^2 + (r_0 + \frac{1}{2}\lambda_0\phi_{\mathrm{mf}}^2)\tilde{\phi}^2\right]\right\} \\ &= \mathrm{e}^{-\beta H_{\mathrm{mf}}} \,\mathrm{Det}^{-1/2}(\mathbf{p}^2 + r_0 + \frac{1}{2}\lambda_0\phi_{\mathrm{mf}}^2), \end{aligned} \tag{24}$$

with the derivative $\mathbf{p} = -i\nabla$. The determinant represents the first corrections to the mean-field expression $\exp(-\beta H_{\text{mf}})$ of the partition function due to fluctuations. Using the identity $\text{Det}(A) = \exp[\text{Tr}\ \ln(A)]$, we can collect them in the effective Hamiltonian

$$\beta H_{\text{eff}} = \frac{1}{2}\text{Tr}\ln[\mathbf{p}^2 + r_0 + \frac{1}{2}\lambda_0\phi^2_{\text{mf}}(\mathbf{x})], \tag{25}$$

so that to this order

$$Z = \mathrm{e}^{-\beta(H_{\text{mf}}+H_{\text{eff}})}. \tag{26}$$

As indicated, the mean field $\phi_{\text{mf}}(\mathbf{x})$ may be space dependent.

We next specify the meaning of the trace Tr appearing in (25). Explicitly,

$$\beta H_{\text{eff}} = \frac{1}{2}\int_{\mathbf{x}} \ln\left\{\left[\mathbf{p}^2 + r_0 + \frac{1}{2}\lambda_0\phi^2_{\text{mf}}(\mathbf{x})\right]\delta(\mathbf{x}-\mathbf{y})\big|_{\mathbf{y}=\mathbf{x}}\right\}. \tag{27}$$

The delta function arises because the expression in parentheses at the right-hand side of (24) is obtained as a functional derivative of the Hamiltonian (23),

$$\frac{\delta^2\beta H}{\delta\tilde{\phi}^2(\mathbf{x})} = \left[\mathbf{p}^2 + r_0 + \frac{1}{2}\lambda_0\phi^2_{\text{mf}}(\mathbf{x})\right]\delta(\mathbf{x}-\mathbf{y})\big|_{\mathbf{y}=\mathbf{x}}, \tag{28}$$

which gives a delta function. Since it is the unit operator in function space, the delta function may be taken out of the logarithm and we can write for (27)

$$\begin{aligned}\beta H_{\text{eff}} &= \frac{1}{2}\int_{\mathbf{x}} \ln\left[\mathbf{p}^2 + r_0 + \frac{1}{2}\lambda_0\phi^2_{\text{mf}}(\mathbf{x})\right]\delta(\mathbf{x}-\mathbf{y})\big|_{\mathbf{y}=\mathbf{x}} \\ &= \frac{1}{2}\int_{\mathbf{x}}\int_{\mathbf{k}} \mathrm{e}^{-i\mathbf{k}\cdot\mathbf{x}} \ln\left[\mathbf{p}^2 + r_0 + \frac{1}{2}\lambda_0\phi^2_{\text{mf}}(\mathbf{x})\right]\mathrm{e}^{i\mathbf{k}\cdot\mathbf{x}}.\end{aligned} \tag{29}$$

In the last step, we have used the integral representation of the delta function:

$$\delta(\mathbf{x}) = \int_{\mathbf{k}} \mathrm{e}^{i\mathbf{k}\cdot\mathbf{x}}, \tag{30}$$

have shifted the exponential function $\exp(-i\mathbf{k}\cdot\mathbf{y})$ to the left, which is justified because the derivative $\mathbf{p}$ does not operate on it, and, finally, set $\mathbf{y}$ equal to $\mathbf{x}$. We thus see that the trace Tr in (29) stands for the trace over discrete indices as well as the integration over space and over momentum. The integral $\int_{\mathbf{k}}$ arises because the effective Hamiltonian calculated here is a one-loop result, with $\mathbf{k}$ the loop momentum.

The integrals in (29) cannot in general be evaluated in closed form because the logarithm contains momentum operators and space-dependent functions in a mixed order. To disentangle the integrals resort has to be taken to a derivative expansion[5] in which the logarithm is expanded in a Taylor series. Each term contains powers of the momentum operator $\mathbf{p}$ which acts on *every* space-dependent function to its right. All these operators are shifted to the left by repeatedly applying the identity

$$f(\mathbf{x})\mathbf{p}g(\mathbf{x}) = (\mathbf{p} + i\nabla)f(\mathbf{x})g(\mathbf{x}), \tag{31}$$

where $f(\mathbf{x})$ and $g(\mathbf{x})$ are arbitrary functions and the derivative ∇ acts *only* on the next object to the right. One then integrates by parts, so that all the $\mathbf{p}$'s act to the left where only a factor $\exp(-i\mathbf{k}\cdot\mathbf{x})$ stands. Ignoring total derivatives and taking into account the minus signs that arise when integrating by parts, one sees that all occurrences of $\mathbf{p}$ (an operator) are replaced with $\mathbf{k}$ (an integration variable). The exponential function

$\exp(i\mathbf{k}\cdot\mathbf{x})$ can at this stage be moved to the left where it is annihilated by the function $\exp(-i\mathbf{k}\cdot\mathbf{x})$. The momentum integration can now in principle be carried out and the effective Hamiltonian be cast in the form of an integral over a local density $\mathcal{H}_{\text{eff}}$:

$$H_{\text{eff}} = \int_{\mathbf{x}} \mathcal{H}_{\text{eff}}. \tag{32}$$

This is in a nutshell how the derivative expansion works.

Let us illustrate the method by applying it to (25). When we assume ϕ_{mf} to be a constant field $\bar{\phi}$, the effective Hamiltonian (25) may be evaluated in closed form:

$$\beta\mathcal{V}_{\text{eff}} = \frac{1}{2}\int_{\mathbf{k}} \ln(\mathbf{k}^2 + M^2) = \frac{\Gamma(1-d/2)}{d(4\pi)^{d/2}} M^d, \qquad M = \sqrt{r_0 + \frac{1}{2}\lambda_0\bar{\phi}^2}, \tag{33}$$

where instead of an Hamiltonian we have introduced a potential $\mathcal{V}_{\text{eff}}$ to indicate that we are working with a space-independent field $\bar{\phi}$. To obtain the last equation, we first differentiated $\ln(k^2 + M^2)$ with respect to M^2 and used the dimensional-regularized integral

$$\int_{\mathbf{k}} \frac{1}{(\mathbf{k}^2 + M^2)^\alpha} = \frac{\Gamma(\alpha - d/2)}{(4\pi)^{d/2}\Gamma(\alpha)} \frac{1}{(M^2)^{\alpha - d/2}} \tag{34}$$

to suppress irrelevant ultraviolet divergences, and finally integrated again with respect to M^2. To illustrate the power of dimensional regularization, let us consider the case $d = 3$ in detail. Introducing a momentum cutoff, we find in the large-Λ limit

$$\beta\mathcal{V}_{\text{eff}} = \frac{1}{8\pi^2}\lambda_0\bar{\phi}^2\Lambda - \frac{1}{12\pi}M^3 + \mathcal{O}\left(\frac{1}{\Lambda}\right), \tag{35}$$

where we have ignored irrelevant, $\bar{\phi}$-independent constants proportional to powers of Λ. We see that in (33) only the finite part emerges. That is, all terms that diverge with a strictly positive power of the momentum cutoff are suppressed in dimensional regularization. These contributions, which come from the ultraviolet region, cannot physically be very relevant because the simple Landau-Ginzburg model (19) stops being valid here and new theories are required. It is a virtue of dimensional regularization that these irrelevant divergences are suppressed.

Expanded up to fourth order in $\bar{\phi}$, (33) becomes

$$\beta\mathcal{V}_{\text{eff}} = -\frac{1}{12\pi}r_0^{3/2} - \frac{1}{16\pi}\lambda_0 r_0^{1/2}\bar{\phi}^2 - \frac{1}{128\pi}\frac{\lambda_0^2}{r_0^{1/2}}\bar{\phi}^4 + \cdots, \tag{36}$$

where the first term is an irrelevant $\bar{\phi}$-independent constant. These one-loop contributions, when added to the mean-field potential

$$\beta\mathcal{V}_0 = \frac{1}{2}r_0\bar{\phi}^2 + \frac{1}{4!}\lambda_0\bar{\phi}^4, \tag{37}$$

lead to a renormalization of the bare parameters

$$\lambda = \lambda_0 - \frac{3}{16\pi}\frac{\lambda_0^2}{r_0^{1/2}}, \quad r = r_0 - \frac{1}{8\pi}\lambda_0 r_0^{1/2}. \tag{38}$$

In case ϕ_{mf} is not a constant field, we write the mean field $\phi_{\text{mf}}(\mathbf{x})$, solving the field equation, as $\phi_{\text{mf}}(\mathbf{x}) = \bar{\phi} + \hat{\phi}(\mathbf{x})$, where $\bar{\phi}$ is the constant field introduced above (33), and expand the logarithm in the right-hand side of (25) to second order in $\hat{\phi}$:

$$\beta\hat{H}_{\text{eff}} = \frac{1}{4}\lambda_0\text{Tr}\frac{1}{\mathbf{p}^2 + M^2}(2\bar{\phi}\hat{\phi} + \hat{\phi}^2) - \frac{1}{8}\lambda_0^2\bar{\phi}^2\text{Tr}\frac{1}{\mathbf{p}^2 + M^2}\hat{\phi}\frac{1}{\mathbf{p}^2 + M^2}\hat{\phi}, \tag{39}$$

with

$$\begin{aligned}\hat{H}_{\text{eff}} &= H_{\text{eff}}(\bar{\phi}+\hat{\phi}) - H_{\text{eff}}(\bar{\phi}) \\ &= \int_{\mathbf{x}} \left[\frac{\partial \mathcal{V}_{\text{eff}}}{\partial \bar{\phi}} \hat{\phi} + \frac{1}{2} \frac{\partial^2 \mathcal{V}_{\text{eff}}}{\partial \bar{\phi}^2} \hat{\phi}^2 + \frac{1}{2} \mathcal{Z}(\bar{\phi}) (\nabla \hat{\phi})^2 + \cdots \right]. \end{aligned} \tag{40}$$

Moving the momentum operator $\mathbf{p}$ to the left by using (31), we obtain

$$\beta \hat{H}_{\text{eff}} = \frac{1}{4} \lambda_0 \text{Tr} \frac{1}{\mathbf{p}^2 + M^2} (2\bar{\phi}\hat{\phi} + \hat{\phi}^2) - \frac{1}{8} \lambda_0^2 \bar{\phi}^2 \text{Tr} \frac{1}{\mathbf{p}^2 + M^2} \frac{1}{(\mathbf{p} - i\nabla)^2 + M^2} \hat{\phi}\hat{\phi}, \tag{41}$$

where we recall the definition of the derivative ∇ as operating only on the first object to its right. Using the integral

$$\int_{\mathbf{k}} \frac{1}{\mathbf{k}^2 + M^2} \frac{1}{(\mathbf{k}+\mathbf{q})^2 + M^2} = \frac{1}{4\pi |\mathbf{q}|} \arctan\left(\frac{|\mathbf{q}|}{2M} \right), \tag{42}$$

with $\mathbf{q} = -i\nabla$, we obtain for (41)

$$\beta \hat{H}_{\text{eff}} = -\frac{1}{16\pi} \lambda_0 M (2\bar{\phi}\hat{\phi} + \hat{\phi}^2) - \frac{1}{32\pi} \lambda_0^2 \bar{\phi}^2 \, \hat{\phi} \left[\frac{1}{|\mathbf{q}|} \arctan\left(\frac{|\mathbf{q}|}{2M} \right) \right] \hat{\phi}. \tag{43}$$

We note that only terms with an even number of derivatives appear in the expansion of this expression. The coefficient of the linear term is $\partial \beta \mathcal{V}_{\text{eff}} / \partial \bar{\phi}$, while that of the two quadratic terms independent of $\mathbf{q}$ is $(1/2) \partial^2 \beta \mathcal{V}_{\text{eff}} / \partial \bar{\phi}^2$, as it should be. For $\mathcal{Z}$ we obtain

$$\mathcal{Z}(\bar{\phi}) = \frac{1}{192\pi} \frac{\lambda_0^2 \bar{\phi}^2}{M^3}. \tag{44}$$

Other terms involving higher powers of $\hat{\phi}$, obtained from expanding the logarithm in (25) to higher orders, can be treated in a similar fashion.

3. SUPERFLUIDITY

A central role in this Lecture is played by an interacting Bose gas. In this section we wish to study some of its salient features, notably its ability to become superfluid below a critical temperature. We shall derive the zero-temperature effective theory of the superfluid state, and discuss the effect of the inclusion of impurities and of a $1/|\mathbf{x}|$-Coulomb potential. Finally, vortices both at the absolute zero of temperature and at finite temperature are studied.

3.1 Bogoliubov Theory

The system of an interacting Bose gas is defined by the Lagrangian[6]

$$\mathcal{L} = \phi^* [i\partial_0 - \epsilon(-i\nabla) + \mu_0] \phi - \lambda_0 |\phi|^4, \tag{45}$$

where the complex scalar field ϕ describes the atoms of mass m, $\epsilon(-i\nabla) = -\nabla^2/2m$ is the kinetic energy operator, and μ_0 is the chemical potential. The last term with a positive coupling constant, $\lambda_0 > 0$, represents the repulsive contact interaction. The (zero-temperature) grand-canonical partition function Z is obtained by integrating over

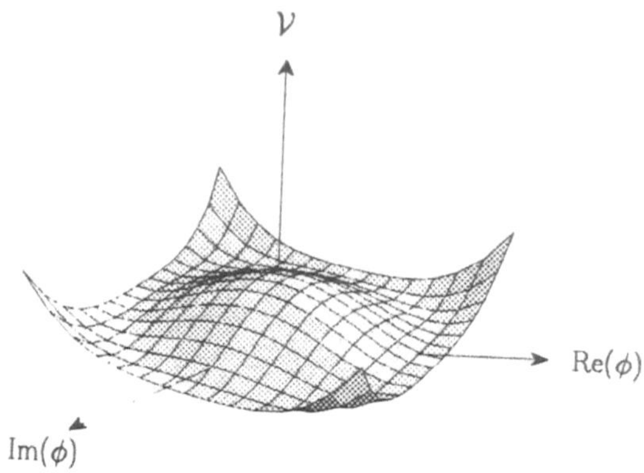

Figure 2. A graphical representation of the potential (48).

all field configurations weighted with an exponential factor determined by the action $S = \int_x \mathcal{L}$:

$$Z = \int \mathrm{D}\phi^* \mathrm{D}\phi \, \mathrm{e}^{iS}. \tag{46}$$

This is the quantum analog of Eq. (21)—the functional-integral representation of a classical partition function.

The theory (45) possesses a global U(1) symmetry under which

$$\phi(x) \to \mathrm{e}^{i\alpha}\phi(x), \tag{47}$$

with α a constant transformation parameter. At zero temperature, this symmetry is spontaneously broken by a nontrivial ground state, and the system is in its superfluid phase. Most of the startling phenomena of a superfluid follow from this symmetry breakdown. The nontrivial groundstate can be easily seen by considering the shape of the potential

$$\mathcal{V} = -\mu_0|\phi|^2 + \lambda_0|\phi|^4, \tag{48}$$

depicted in Fig. 2. It is seen to have a minimum away from the origin $\phi = 0$. To account for this, we shift ϕ by a (complex) constant $\bar{\phi}$ and write

$$\phi(x) = \mathrm{e}^{i\varphi(x)}\,[\bar{\phi} + \tilde{\phi}(x)]. \tag{49}$$

The phase field $\varphi(x)$ represents the Goldstone mode accompanying the spontaneous breakdown of the global U(1) symmetry. At zero temperature and for $\mu_0 > 0$, the constant value

$$|\bar{\phi}|^2 = \frac{1}{2}\frac{\mu_0}{\lambda_0} \tag{50}$$

minimizes the potential energy. It physically represents the number density of particles contained in the condensate for the total particle number density is given by

$$n(x) = |\phi(x)|^2. \tag{51}$$

Because $\bar{\phi}$ is a constant, the condensate is a uniform, zero-momentum state. That is, the particles residing in the ground state are in the $\mathbf{k} = 0$ mode. We will be working in the Bogoliubov approximation which amounts to including only the quadratic terms in $\tilde{\phi}$ and ignoring the higher-order ones. These terms may be cast in the matrix form

$$\mathcal{L}^{(2)} = \frac{1}{2}\tilde{\Phi}^\dagger M_0(p,x)\tilde{\Phi}, \qquad \tilde{\Phi} = \begin{pmatrix} \tilde{\phi} \\ \tilde{\phi}^* \end{pmatrix}, \tag{52}$$

with

$$M_0(p,x) = \tag{53}$$
$$\begin{pmatrix} p_0 - \epsilon(\mathbf{p}) + \mu_0 - U(x) - 4\lambda_0|\bar{\phi}|^2 & -2\lambda_0\bar{\phi}^2 \\ -2\lambda_0\bar{\phi}^{*2} & -p_0 - \epsilon(\mathbf{p}) + \mu_0 - U(x) - 4\lambda_0|\bar{\phi}|^2 \end{pmatrix},$$

where U stands for the combination

$$U(x) = \partial_0\varphi(x) + \frac{1}{2m}[\nabla\varphi(x)]^2. \tag{54}$$

In writing (53) we have omitted a term $\nabla^2\varphi$, containing two derivatives, which is irrelevant in the regime of low momentum, in which we shall be interested. We also have omitted a term of the form $\nabla\varphi \cdot \mathbf{j}$, where $\mathbf{j}$ is the Noether current associated with the global U(1) symmetry,

$$\mathbf{j} = \frac{1}{2im}\phi^* \overleftrightarrow{\nabla} \phi. \tag{55}$$

This term, which after a partial integration becomes $-\varphi\nabla \cdot \mathbf{j}$, is irrelevant too at low energy and small momentum because in a first approximation the particle number density is constant, so that the classical current satisfies the condition

$$\nabla \cdot \mathbf{j} = 0. \tag{56}$$

The spectrum $E(\mathbf{k})$ obtained from the matrix M_0 with the field U set to zero is the famous single-particle Bogoliubov spectrum[7],

$$\begin{aligned} E(\mathbf{k}) &= \sqrt{\epsilon^2(\mathbf{k}) + 2\mu_0\epsilon(\mathbf{k})} \\ &= \sqrt{\epsilon^2(\mathbf{k}) + 4\lambda_0|\bar{\phi}|^2\epsilon(\mathbf{k})}. \end{aligned} \tag{57}$$

The most notable feature of this spectrum is that it is gapless, behaving for small momentum as

$$E(\mathbf{k}) \sim u_0\,|\mathbf{k}|, \tag{58}$$

with $u_0 = \sqrt{\mu_0/m}$ a velocity which is sometimes referred to as the microscopic sound velocity. It has been first shown by Beliaev[8] that the gaplessness of the single-particle spectrum persists at the one-loop order. This has been subsequently proven to hold to all orders in perturbation theory by Hugenholtz and Pines.[9] For large momentum, the Bogoliubov spectrum takes a form

$$E(\mathbf{k}) \sim \epsilon(\mathbf{k}) + 2\lambda_0|\bar{\phi}|^2 \tag{59}$$

typical for a nonrelativistic particle with a mass m moving in a medium. To highlight the condensate we have chosen here the second form in (57) where μ_0 is replaced with $2\lambda_0|\bar{\phi}|^2$.

3.2 Effective Theory

Since gapless modes in general require a justification for their existence, we expect the gaplessness of the single-particle spectrum to be a result of Goldstone's theorem. This is corroborated by the relativistic version of the theory. There, one finds two spectra, one corresponding to a massive Higgs particle which in the nonrelativistic limit becomes too heavy and decouples from the theory, and one corresponding to the Goldstone mode of the spontaneously broken global U(1) symmetry.[10] The latter

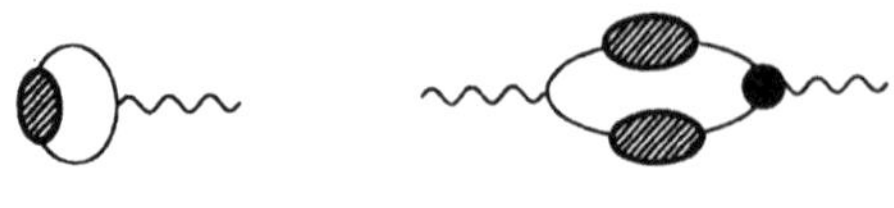

Figure 3. A graphical representation of the effective theory (66). The symbols are explained in the text.

reduces in the nonrelativistic limit to the Bogoliubov spectrum. Also, when the theory is coupled to an electromagnetic field, one finds that the single-particle spectrum acquires an energy gap. This is what one expects to happen with the spectrum of a Goldstone mode when the Higgs mechanism is operating. The equivalence of the single-particle excitation and the collective density fluctuation has been proven to all orders in perturbation by Gavoret and Nozières.[11]

Let us derive[12] the effective theory governing the Goldstone mode at low energy and small momentum by integrating out the fluctuating field $\tilde{\Phi}$. The effective theory is graphically represented by Fig. 3. A line with a shaded bubble inserted stands for i times the *full* Green function G and the black bubble denotes i times the *full* interaction Γ of the $\tilde{\Phi}$-field with the field U which is denoted by a wiggly line. Both G and Γ are 2×2 matrices. The full interaction is obtained from the inverse Green function by a differentiation with respect to the chemical potential,

$$\Gamma = -\frac{\partial G^{-1}}{\partial \mu}. \tag{60}$$

This follows because U, as defined in (54), appears in the theory only in the combination $\mu_0 - U$. To the lowest order, the inverse propagator is given by the matrix M_0 in (53) with $U(x)$ set to zero. It follows that the vertex of the interaction between the $\tilde{\Phi}$ and U-fields is minus the unit matrix. Because in terms of the full Green function G, the particle number density reads

$$\bar{n} = \frac{i}{2} \operatorname{Tr} \int_k G(k), \tag{61}$$

we conclude that the first diagram in Fig. 3 stands for $-\bar{n}U$. The bar over n is to indicate that the particle number density obtained in this way is a constant, representing the density of the uniform system with $U(x)$ set to zero. The second diagram without the wiggly lines denotes i times the (0 0)-component of the *full* polarization tensor, Π_{00}, at zero energy transfer and low momentum $\mathbf{q}$,

$$i \lim_{\mathbf{q} \to 0} \Pi_{00}(0, \mathbf{q}) = -\frac{1}{2} \lim_{\mathbf{q} \to 0} \operatorname{Tr} \int_k G \, \Gamma \, G(k_0, \mathbf{k} + \mathbf{q}). \tag{62}$$

The factor 1/2 is a symmetry factor which arises because the two Bose lines are identical. We proceed by invoking an argument due to Gavoret and Nozières[11] to relate the left-hand side of (62) to the sound velocity. By virtue of the relation (60) between the full Green function G and the full interaction Γ, the (0 0)-component of the polarization tensor can be cast in the form

$$\lim_{\mathbf{q} \to 0} \Pi_{00}(0, \mathbf{q}) = -\frac{i}{2} \lim_{\mathbf{q} \to 0} \operatorname{Tr} \int_k G \, \frac{\partial G^{-1}}{\partial \mu} G(k_0, \mathbf{k} + \mathbf{q})$$

$$= \frac{i}{2}\frac{\partial}{\partial \mu} \lim_{\mathbf{q}\to 0} \mathrm{Tr} \int_k G(k_0, \mathbf{k}+\mathbf{q})$$

$$= \frac{\partial \bar{n}}{\partial \mu} = -\frac{1}{V}\frac{\partial \Omega}{\partial \mu^2}, \tag{63}$$

where Ω is the thermodynamic potential and V the volume of the system. The right-hand side of (63) is $\bar{n}^2\kappa$, with κ the compressibility. Because it is related to the macroscopic sound velocity c via

$$\kappa = \frac{1}{m\bar{n}c^2}, \tag{64}$$

we conclude that the (0 0)-component of the full polarization tensor satisfies the so-called compressibility sum rule of statistical physics[11]

$$\lim_{\mathbf{q}\to 0} \Pi_{00}(0, \mathbf{q}) = \bar{n}^2\kappa = \frac{\bar{n}}{mc^2}. \tag{65}$$

Putting the pieces together, we infer that the diagrams in Fig. 3 stand for the effective theory

$$\mathcal{L}_{\text{eff}} = -\bar{n}\left[\partial_0\varphi + \frac{1}{2m}(\nabla\varphi)^2\right] + \frac{\bar{n}}{2mc^2}\left[\partial_0\varphi + \frac{1}{2m}(\nabla\varphi)^2\right]^2, \tag{66}$$

where we recall that $\bar{n}$ is the particle number density of the fluid at rest. The theory describes a nonrelativistic sound wave, with the dimensionless phase field φ representing the Goldstone mode of the spontaneously broken global U(1) symmetry. It has the gapless dispersion relation $E^2(\mathbf{k}) = c^2\mathbf{k}^2$. The effective theory gives a complete description of the superfluid valid at low energies and small momenta. The same effective theory appears in the context of (neutral) superconductors[13] (see next section) and also in that of classical hydrodynamics.[14]

The chemical potential μ is represented in the effective theory (66) by[15]

$$\mu(x) = -\partial_0\varphi(x), \tag{67}$$

so that

$$\frac{\partial \mathcal{L}_{\text{eff}}}{\partial \mu} = -\frac{\partial \mathcal{L}_{\text{eff}}}{\partial \partial_0 \phi} = n(x), \tag{68}$$

as required. It also follows from this equation that the particle number density $n(x)$ is canonical conjugate to $-\phi(x)$.

The most remarkable aspect of the effective theory (66) is that it is nonlinear. The nonlinearity is necessary to provide a Galilei-invariant description of the gapless mode, as required in a nonrelativistic context. Under a Galilei boost,

$$t \to t' = t, \;\; \mathbf{x} \to \mathbf{x}' = \mathbf{x} - \mathbf{u}t; \quad \partial_0 \to \partial_0' = \partial_0 + \mathbf{u}\cdot\nabla, \;\; \nabla \to \nabla' = \nabla, \tag{69}$$

with $\mathbf{u}$ a constant velocity, the Goldstone field $\varphi(x)$ transforms as

$$\frac{1}{m}\varphi(x) \to \frac{1}{m}\varphi'(x') = \frac{1}{m}\varphi(x) - \mathbf{u}\cdot\mathbf{x} + \frac{1}{2}\mathbf{u}^2 t. \tag{70}$$

As a result, the superfluid velocity $\mathbf{v}_s = \nabla\varphi/m$ and the chemical potential (per unit mass) $\mu/m = -\partial_0\varphi/m$ transform under a Galilei boost in the correct way,

$$\mathbf{v}_s(x) \to \mathbf{v}_s'(x') = \mathbf{v}_s(x) - \mathbf{u}, \quad \mu(x)/m \to \mu'(x')/m = \mu(x)/m - \mathbf{u}\cdot\mathbf{v}_s(x) + \frac{1}{2}\mathbf{u}^2. \tag{71}$$

It is readily checked that the field $U(x)$ defined in (54) and therefore the effective theory (66) is invariant under Galilei boosts.

Since the Goldstone field in (45) is always accompanied by a derivative, we see that the nonlinear terms carry additional factors of $|\mathbf{k}|/mc$, with $|\mathbf{k}|$ the wave number. They can therefore be ignored provided the wave number is smaller than the inverse coherence length $\xi^{-1} = mc$,

$$|\mathbf{k}| < 1/\xi. \tag{72}$$

For example, in the case of ^{4}He the coherence length, or the Compton wavelength, is about 10 nm. In this system, the bound (72), below which the nonlinear terms can be neglected, coincides with the region where the spectrum is linear and the description in terms of solely a sound mode is applicable.

The alert reader might be worrying about the apparent mismatch in the number of degrees of freedom in the normal and the superfluid phase. Whereas the normal phase is described by a complex ϕ-field, the superfluid phase is described by a real scalar field φ. The resolution of this paradox lies in the spectrum of the modes.[16] In the normal phase, the spectrum $E(\mathbf{k}) = \mathbf{k}^2/2m$ is linear in E, so that only positive energies appear in the Fourier decomposition, and one needs—as is well known from standard quantum mechanics—a complex field to describe a single particle. In the superfluid phase, where the spectrum, $E^2(\mathbf{k}) = c^2\mathbf{k}^2$, is quadratic in E, the counting goes differently. The Fourier decomposition now contains positive as well as negative energies and a single real field suffices to describe this mode. In other words, although the number of fields is different, the number of degrees of freedom is the same in both phases.

The particle number density and current that follows from (66) read

$$n(x) = \bar{n} - \frac{\bar{n}}{mc^2}\left\{\partial_0\varphi(x) + \frac{1}{2m}[\nabla\varphi(x)]^2\right\}, \tag{73}$$

$$\mathbf{j}(x) = n(x)\mathbf{v}_s(x). \tag{74}$$

Physically, (73) reflects the Bernoulli principle which states that in regions of rapid flow, the density and therefore the pressure is low.

The diagrams of Fig. 3 can be evaluated in a loop expansion to obtain explicit expressions for the particle number density $\bar{n}$ and the sound velocity c to any given order.[12] In doing so, one encounters—apart from ultraviolet divergences which will be dealt with shortly—also infrared divergences because the Bogoliubov spectrum is gapless. When, however, all one-loop contributions are added together, these divergences are seen to cancel.[12] One finds for $d = 2$ to the one-loop order

$$\bar{n} = \frac{1}{2}\frac{\mu}{\lambda}, \quad c^2 = 2\frac{\lambda\bar{n}}{m}, \tag{75}$$

where μ and λ are the renormalized parameters. Following Ref. 17, we have adopted a dimensional regularization scheme, in which after the integrals over the loop energies have been carried out, the remaining integrals over the loop momenta are analytically continued to arbitrary space dimensions d. As renormalization prescription we have employed the modified minimal subtraction scheme. This leads to the following relation between the bare (λ_0) and renormalized coupling constant [see Eq. (97) below]

$$\frac{1}{\lambda_0} = \frac{1}{\kappa^\epsilon}\left(\frac{1}{\hat{\lambda}} - \frac{m}{\pi\epsilon}\right), \tag{76}$$

where $\epsilon = 2-d$, and κ is an arbitrary renormalization group scale parameter introduced to give the renormalized coupling constant $\hat{\lambda}$ the same engineering dimension as in $d = 2$. The chemical potential is not renormalized to this order.

Incidentally, from the vantage point of renormalization, the mass m is an irrelevant parameter in nonrelativistic theories and can be scaled away (see, e.g., Ref. 17).

The form of the effective theory (66) can also be derived from general symmetry arguments.[18] More specifically. it follows from making the presence of a gapless Goldstone mode compatible with Galilei invariance which demands that the mass current and the momentum density are equal. The latter observation leads to the conclusion that the U(1) Goldstone φ can only appear in the combination (54). To obtain the required linear spectrum for the Goldstone mode it is necessary then to have the form (66). Given the form of the effective theory, the particle density and the sound velocity can then more easily be obtained directly from the thermodynamic potential Ω via

$$\bar{n} = -\frac{1}{V}\frac{\partial \Omega}{\partial \mu}; \qquad \frac{1}{c^2} = -\frac{1}{V}\left(\frac{m}{\bar{n}}\right)\frac{\partial^2 \Omega}{\partial \mu^2}, \tag{77}$$

where V is the volume of the system. In this approach, one only has to calculate the thermodynamic potential which at zero temperature and in the Bogoliubov approximation, in which we are working, is given by the sum $\mathcal{V}$ of the classical potential $\mathcal{V}_0$ and the effective potential $\mathcal{V}_{\rm eff}$ corresponding to the theory (45):

$$\Omega = \int_{\mathbf{x}} (\mathcal{V}_0 + \mathcal{V}_{\rm eff}), \tag{78}$$

where $\mathcal{V}_0$ is given by (48) with ϕ replaced by $\bar{\phi}$. The effective potential for the uniform system is obtained as follows. In the Bogoliubov approximation of ignoring higher than second order in the fields, the integration over $\tilde{\Phi}$ is Gaussian. Carrying out this integral, we obtain for the zero-temperature partition function

$$\begin{aligned} Z &= \mathrm{e}^{-i\int_x \mathcal{V}_0} \int \mathrm{D}\phi^* \mathrm{D}\phi \exp\left(i\int_x \mathcal{L}^{(2)}\right) \\ &= \mathrm{e}^{-i\int_x \mathcal{V}_0}\, \mathrm{Det}^{-1/2}(M_0), \end{aligned} \tag{79}$$

where M_0 stands for the matrix introduced in (53). Setting

$$Z = \exp\left[i\left(-\int_x \mathcal{V}_0 + S_{\rm eff}\right)\right], \tag{80}$$

we conclude from (79) that the effective action in the Bogoliubov approximation is given to the one-loop order by

$$S_{\rm eff} = \frac{i}{2}\mathrm{Tr}\ln[M_0(p,x)], \tag{81}$$

where we again have used the identity $\mathrm{Det}(A) = \exp[\mathrm{Tr}\ln(A)]$. The trace Tr appearing here stands besides for the trace over discrete indices now also for the integral $\int_x$ over spacetime as well as the one $\int_k$ over energy and momentum. The latter integral reflects the fact that the effective action calculated here is a one-loop result with k_μ the loop energy and momentum. To disentangle the integrals one has to carry out similar steps as the ones outlined in Sec. 2.2 and repeatedly apply the identity

$$f(x)p_\mu g(x) = (p_\mu - i\tilde{\partial}_\mu)f(x)g(x), \tag{82}$$

where $f(x)$ and $g(x)$ are arbitrary functions of spacetime and the derivative $\tilde{\partial}_\mu = (\partial_0, -\nabla)$ acts *only* on the next object to the right. The method outlined there can easily be transcribed to the present case where the time dimension is included.

If the field $U(x)$ in M_0 is set to zero, things will simplify because M_0 now will depend on p_μ only. The effective action then becomes $S_{\text{eff}} = -\int_x \mathcal{V}_{\text{eff}}$ with

$$\mathcal{V}_{\text{eff}} = -\frac{i}{2}\text{Tr}\int_k \ln[M_0(k)] \tag{83}$$

the effective potential. The easiest way to evaluate the integral over the loop variable k_μ is to first differentiate the expression with respect to the chemical potential μ_0

$$\frac{\partial}{\partial \mu_0}\text{Tr}\int_k \ln[M_0(k)] = -2\int_k \frac{\epsilon(\mathbf{k})}{k_0^2 - E^2(\mathbf{k}) + i\eta}, \tag{84}$$

with $E(\mathbf{k})$ the Bogoliubov spectrum (57). The integral over k_0 can be carried out with the help of a contour integration, yielding

$$\int_k \frac{\epsilon(\mathbf{k})}{k_0^2 - E^2(\mathbf{k}) + i\eta} = -\frac{i}{2}\int_{\mathbf{k}} \frac{\epsilon(\mathbf{k})}{E(\mathbf{k})}. \tag{85}$$

This in turn is easily integrated with respect to μ_0. Putting the pieces together, we obtain

$$\mathcal{V} = -\frac{\mu_0^2}{4\lambda_0} + \frac{1}{2}\int_{\mathbf{k}} E(\mathbf{k}). \tag{86}$$

The integral over the loop momentum in arbitrary space dimension d yields

$$\mathcal{V} = -\frac{\mu_0^2}{4\lambda_0} - L_d m^{d/2}\mu_0^{d/2+1}, \quad L_d = \frac{\Gamma(1-d/2)\Gamma(d/2+1/2)}{2\pi^{d/2+1/2}\Gamma(d/2+2)}, \tag{87}$$

where we have employed the integral representation of the Gamma function

$$\frac{1}{a^z} = \frac{1}{\Gamma(z)}\int_0^\infty \frac{\mathrm{d}\tau}{\tau}\tau^z \mathrm{e}^{-a\tau} \tag{88}$$

together with dimensional regularization to suppress irrelevant ultraviolet divergences.

For comparison, let us also evaluate the integral in (86) over the loop momentum in three dimensions by introducing a momentum cutoff Λ

$$\begin{aligned}\mathcal{V}_{\text{eff}} &= \frac{1}{2}\int_{\mathbf{k}} E(\mathbf{k}) = \frac{1}{4\pi^2}\int_0^\Lambda k^2 E(|\mathbf{k}|) \\ &= \frac{1}{4\pi^2}\left(\frac{1}{10}\frac{\Lambda^5}{m} + \frac{1}{3}\mu_0\Lambda^3 - m\mu_0^2\Lambda + \frac{32}{15}m^{3/2}\mu_0^{5/2}\right) + \mathcal{O}\left(\frac{1}{\Lambda}\right).\end{aligned} \tag{89}$$

From (87), we obtain by setting $d = 3$ only the finite part, so that all terms diverging with a strictly positive power of the momentum cutoff are suppressed. As we have remarked in Sec. 2.2, these contributions, which come from the ultraviolet region, cannot be physically very relevant because the simple model (45) breaks down here. On account of the uncertainty principle, stating that large momenta correspond to small distances, these terms are always local and can be absorbed by redefining the parameters appearing in the Lagrangian.[19] Since $\mu_0 = 2\lambda_0|\bar{\phi}|^2$, we see that the first diverging term in (89) is an irrelevant constant, while the two remaining diverging terms can be absorbed by introducing the renormalized parameters

$$\mu = \mu_0 - \frac{1}{6\pi^2}\lambda_0\Lambda^3, \tag{90}$$

$$\lambda = \lambda_0 - \frac{1}{\pi^2}m\lambda_0^2\Lambda. \tag{91}$$

Because the diverging terms are — at least to this order — of a form already present in the original Lagrangian, the theory is called "renormalizable". The renormalized parameters are the physical ones that are to be identified with those measured in experiment. In this way, we see that the contributions to the loop integral stemming from the ultraviolet region are of no importance. What remains is the finite part

$$\mathcal{V}_{\text{eff}} = \frac{8}{15\pi^2}m^{3/2}\mu_0^{5/2}, \tag{92}$$

which, as we have seen, is obtained directly without renormalization when using dimensional regularization. In this scheme, divergences proportional to powers of the cutoff never show up. Only logarithmic divergences appear as $1/\epsilon$ poles, where ϵ is the deviation from the upper critical dimension ($d = 2$ in the present case). These logarithmic divergences $\ln(\Lambda/E)$, with E a low-energy scale, are relevant also in the infrared because of the fixed cutoff $\ln(\Lambda/E) \to -\infty$ when E is taken to zero.

In so-called "nonrenormalizable" theories, the ultraviolet-diverging terms are still local but not of a form present in the original Lagrangian. Whereas in former days such theories have been rejected because their supposed lack of predictive power, the modern view is that there are no fundamental theories and that there is no basic difference between renormalizable and nonrenormalizable theories[20]. Even a renormalizable theory like (45) should be extended to include all higher-order terms such as a $|\phi|^6$-term which are allowed by symmetry. These additional terms render the theory "nonrenormalizable". This does not, however, change the predictive power of the theory. The point is that when describing the physics at an energy scale E far below the cutoff, the higher-order terms are suppressed by powers of E/Λ, as follows from the dimensional analysis. Therefore, far below the cutoff, the nonrenormalizable terms are negligible.

That $d = 2$ is the upper critical dimension of the problem at hand can be seen by noting that L_d in (87) diverges when d tends to 2. Special care has to be taken for this case. For $d \neq 2$, one obtains[21] with the help of (77)

$$\bar{n} = \frac{\mu_0}{2\lambda_0}\left[1 + (d+2)L_d m^{d/2}\lambda_0\mu_0^{d/2-1}\right] \tag{93}$$

and

$$c^2 = \frac{\mu_0}{m}\left[1 - (d-2)(d/2+1)L_d m^{d/2}\lambda_0\mu_0^{d/2-1}\right], \tag{94}$$

where to arrive at the last equation an expansion in the coupling constant λ_0 has been made. Up to this point, we have considered the chemical potential to be the independent parameter, thereby assuming the presence of a reservoir that can freely exchange particles with the system under study. The system can thus have any number of particles, only the average number is fixed by external conditions. From the experimental point of view it is, however, often more realistic to consider the particle number fixed. If this is the case, the particle number density $\bar{n}$ should be considered as the independent variable and the chemical potential should be expressed in terms of it. This can be achieved by inverting the relation (93):

$$\mu_0 = 2\lambda_0\bar{n}\left[1 - 2(d-2)(d/2+1)L_d m^{d/2}\lambda_0(2\lambda_0\bar{n})^{d/2-1}\right]. \tag{95}$$

The sound velocity expressed in terms of the particle number density reads

$$c^2 = \frac{2\lambda_0\bar{n}}{m}\left[1 - d(d/2+1)L_d m^{d/2}\lambda_0(2\lambda_0\bar{n})^{d/2-1}\right]. \tag{96}$$

These formulas reproduce the known results[22] $d = 3$ and[23] $d = 1$.

To investigate the case $d = 2$, we expand the potential (87) around $d = 2$:

$$\mathcal{V} = -\frac{\mu_0^2}{4\lambda_0} - \frac{1}{4\pi\epsilon}\frac{m\mu_0^2}{\kappa^\epsilon} + \mathcal{O}(\epsilon^0), \tag{97}$$

with $\epsilon = 2 - d$. This expression is seen to diverge in the limit $d \to 2$. The theory can be rendered finite by introducing a renormalized coupling constant via (76). We also see that the chemical potential is not renormalized to this order. The beta function $\beta(\hat{\lambda})$ follows as[24]

$$\beta(\hat{\lambda}) = \kappa \left.\frac{\partial\hat{\lambda}}{\partial\kappa}\right|_{\lambda_0} = -\epsilon\hat{\lambda} + \frac{m}{\pi}\hat{\lambda}^2. \tag{98}$$

In the upper critical dimension, this yields only one fixed point, viz. the infrared-stable (IR) fixed point $\hat{\lambda}^* = 0$. Below $d = 2$, this point is shifted to $\hat{\lambda}^* = \epsilon\pi/m$. It is now easily checked that Eqs. (93) and (96) also reproduce the two-dimensional results (75).

There is no field renormalization in the one-loop approximation; this is the reason why in (45) we have attributed only the bare parameters μ_0 and λ_0 an index 0, and not ϕ.

We proceed by calculating the fraction of particles residing in the condensate. In deriving the Bogoliubov spectrum (57), we set $|\bar{\phi}|^2 = \mu_0/2\lambda_0$ thereby fixing the number density of particles contained in the condensate,

$$\bar{n}_0 = |\bar{\phi}|^2, \tag{99}$$

in terms of the chemical potential. For our present consideration we have to keep $\bar{\phi}$ as an independent variable. The spectrum of the elementary excitations expressed in terms of $\bar{\phi}$ is

$$E(\mathbf{k}) = \sqrt{[\epsilon(\mathbf{k}) - \mu_0 + 4\lambda_0|\bar{\phi}|^2]^2 - 4\lambda_0^2|\bar{\phi}|^4}. \tag{100}$$

It reduces to the Bogoliubov spectrum when the mean-field value (50) for $\bar{\phi}$ is inserted. The equation (83) for the effective potential is still valid, and so is (78). We thus obtain for the particle number density

$$\bar{n} = |\bar{\phi}|^2 - \frac{1}{2}\frac{\partial}{\partial\mu_0}\int_{\mathbf{k}} E(\mathbf{k})\Bigg|_{|\bar{\phi}|^2=\mu_0/2\lambda_0}, \tag{101}$$

where the mean-field value for $\bar{\phi}$ is to be substituted after the differentiation with respect to the chemical potential has been carried out. We find

$$\bar{n} = |\bar{\phi}|^2 - 2^{d/2-2}\frac{d^2-4}{d-1}L_d m^{d/2}\lambda_0^{d/2}|\bar{\phi}|^d, \tag{102}$$

or for the so-called depletion of the condensate[25]

$$\frac{\bar{n}}{\bar{n}_0} - 1 \approx -2^{d/2-2}\frac{d^2-4}{d-1}L_d m^{d/2}\lambda^{d/2}n^{d/2-1}, \tag{103}$$

where in the last term we have replaced the bare coupling constant with the (one-loop) renormalized one. This is consistent to this order since this term is already an one-loop result. Equation (103) shows that even at zero temperature not all the particles reside in the condensate. Due to the interparticle repulsion, particles are removed from the zero-momentum ground state and put in states of finite momentum. It has been estimated that in bulk superfluid ^{4}He—a strongly interacting system—only about 8% of the particles condense in the zero-momentum state.[26] For $d = 2$, the right-hand side of Eq. (103) reduces to

$$\frac{\bar{n}}{\bar{n}_0} - 1 \approx \frac{m\lambda}{2\pi}, \tag{104}$$

which is seen to be independent of the particle number density.

Despite the fact that not all the particles reside in the condensate, they all participate in the superfluid motion at zero temperature.[27] Apparently, the condensate drags the normal fluid along with it. To show this, let us assume that the entire system moves with a velocity $\mathbf{u}$ relative to the laboratory system. As is known from standard hydrodynamics, the time derivative in the frame following the motion of the fluid is $\partial_0 + \mathbf{u} \cdot \nabla$ [see Eq. (69)]. If we insert this in the Lagrangian (45) of the interacting Bose gas, it becomes

$$\mathcal{L} = \phi^*[i\partial_0 - \epsilon(-i\nabla) + \mu_0 - \mathbf{u} \cdot (-i\nabla)]\phi - \lambda_0|\phi|^4, \tag{105}$$

where the extra term features the total momentum $\int_{\mathbf{x}} \phi^*(-i\nabla)\phi$ of the system. The velocity $-\mathbf{u}$ multiplying this is on the same footing as the chemical potential μ_0 multiplying the particle number $\int_{\mathbf{x}} |\phi|^2$. Whereas μ_0 is associated with particle number conservation, $\mathbf{u}$ is related to the conservation of the momentum.

In the two-fluid picture, the condensate can move with a different velocity $\mathbf{v}_s$ from the rest of the system. To bring this out we introduce new fields, cf. (49)

$$\phi(x) \to \phi'(x) = e^{im\mathbf{v}_s \cdot \mathbf{x}}\phi(x) \tag{106}$$

in terms of which the Lagrangian becomes[28]

$$\mathcal{L} = \phi^*[i\partial_0 - \epsilon(-i\nabla) + \mu_0 - \frac{1}{2}m\mathbf{v}_s \cdot (\mathbf{v}_s - 2\mathbf{u}) - (\mathbf{u} - \mathbf{v}_s) \cdot (-i\nabla)]\phi - \lambda_0|\phi|^4, \tag{107}$$

where we have dropped the primes on ϕ again. Both velocities appear in this expression. Apart from the change $\mathbf{u} \to \mathbf{u} - \mathbf{v}_s$ in the second last term, the field transformation results in a change of the chemical potential

$$\mu_0 \to \mu_{\text{eff}} := \mu_0 - \frac{1}{2}m\mathbf{v}_s \cdot (\mathbf{v}_s - 2\mathbf{u}) \tag{108}$$

where μ_{eff} may be considered as an effective chemical potential.

The equations for the Bogoliubov spectrum and the thermodynamic potential are readily written down for the present case when these two changes are kept in mind. In particular, the effective potential is given by (86) with the replacement Eq. (108). The momentum density, or equivalently, the mass current $\mathbf{g}$ of the system is obtained in this approximation by differentiating the effective potential with respect to $-\mathbf{u}$. We find, using the equation

$$\frac{\partial \mu_{\text{eff}}}{\partial \mathbf{u}} = m\mathbf{v}_s \tag{109}$$

that it is given by

$$\mathbf{g} = \rho_s \mathbf{v}_s, \tag{110}$$

with $\rho_s = m\bar{n}$ the superfluid mass density. This equation, comprising the total particle number density $\bar{n}$, shows that at zero temperature indeed all the particles are involved in the superflow, despite the fact that only a fraction of them resides in the condensate[27].

The superfluid mass density ρ_s, obtained by evaluating the response of the system to an externally imposed velocity field $\mathbf{u}$, should not be confused with the number density $\bar{n}_0$ of particles contained in the condensate introduced in Eq. (99).

Let us close this subsection by pointing out a quick trail to arrive at the effective theory (66) starting from the microscopic model (45). To this end, we set

$$\phi(x) = e^{i\varphi(x)} [\sqrt{\bar{n}} + \tilde{\phi}(x)], \tag{111}$$

and expand the Lagrangian (45) up to quadratic terms in $\tilde{\phi}$. This leads to

$$\mathcal{L}^{(2)} = -\mathcal{V}_0 - \bar{n}U - \sqrt{\bar{n}}U(\tilde{\phi} + \tilde{\phi}^*) - \lambda_0 \bar{n}(\tilde{\phi} + \tilde{\phi}^*)^2, \tag{112}$$

where we have used the mean-field equation $\mu_0 = 2\lambda_0 \bar{n}$. We continue by integrating out the tilde fields—which is a tantamount to substituting the field equation for these fields back into the Lagrangian—to obtain

$$\mathcal{L}_{\text{eff}} = -\bar{n}U(x) + \frac{1}{4}U(x)\frac{1}{\lambda_0}U(x), \tag{113}$$

apart from the irrelevant constant term $\mathcal{V}_0$. This form of the effective theory is equivalent to the one found before in (45). We have cast the last term in a form that can be easily generalized to systems with long-range interactions. A case of particular interest to us is the Coulomb potential

$$V(\mathbf{x}) = \frac{e_0^2}{|\mathbf{x}|}, \tag{114}$$

whose Fourier transform in d space dimensions reads

$$V(\mathbf{k}) = 2^{d-1}\pi^{(d-1)/2}\Gamma\left[\frac{1}{2}(d-1)\right]\frac{e_0^2}{|\mathbf{k}|^{d-1}}. \tag{115}$$

The simple contact interaction $L_\text{i} = -\lambda_0 \int_\mathbf{x} |\phi(x)|^4$ in (45) is now replaced by

$$L_\text{i} = -\frac{1}{2}\int_{\mathbf{x},\mathbf{y}} |\phi(t,\mathbf{x})|^2 V(\mathbf{x}-\mathbf{y})|\phi(t,\mathbf{y})|^2. \tag{116}$$

The rationale for using the three-dimensional Coulomb potential even when considering charges confined to move in a lower dimensional space is that the electromagnetic interaction remains three dimensional. The effective theory (113) now becomes in the Fourier representation

$$\mathcal{L}_{\text{eff}} = -\bar{n}U(k) + \frac{1}{2}U(k_0,\mathbf{k})\frac{1}{V(\mathbf{k})}U(k_0,-\mathbf{k}) \tag{117}$$

and leads to the dispersion relation

$$E^2(\mathbf{k}) = 2^d\pi^{(d-1)/2}\Gamma\left[\frac{1}{2}(d-1)\right]\frac{\bar{n}e_0^2}{m}|\mathbf{k}|^{3-d}. \tag{118}$$

For $d = 3$, this yields the famous plasma mode with an energy gap given by the plasma frequency $\omega_\text{p}^2 = 4\pi\bar{n}e_0^2/m$.

To appreciate under which circumstances the Coulomb interaction becomes important, we note that for electronic systems $1/|\mathbf{x}| \sim k_\text{F}$ for dimensional reasons and the fermion number density $\bar{n} \sim k_\text{F}^d$, where k_F is the Fermi momentum. The ratio of the Coulomb interaction energy ϵ_C to the Fermi energy $\epsilon_\text{F} = k_\text{F}^2/2m$ is therefore proportional to $\bar{n}^{-1/d}$. This means that the lower the electron number density, the more important the Coulomb interaction becomes.

3.3 Quenched Impurities

In most of the quantum systems we will be considering, impurities play an important role. The main effect of impurities is typically to localize states. Localization counteracts the tendency of the system to become superfluid. We shall therefore now include impurities in the interacting Bose gas to see whether this leads to localization and whether the system still has a superfluid phase. It is expected that on increasing the strength of the disorder for a given repulsive interparticle interaction, the superfluid undergoes a zero-temperature phase transition to an insulating phase of localized states. The location and the nature of this transition will be the subject of Sec. 6.

We shall assume that the impurities are fixed and that their distribution is not affected by the host system. This type of impurities is called quenched impurities and is to be distinguished from so-called annealed impurities which change with and depend on the host system. To account for impurities, we add to the theory (45) the term

$$\mathcal{L}_\Delta = \psi(\mathbf{x})\,|\phi(x)|^2, \tag{119}$$

with $\psi(\mathbf{x})$ a random field whose distribution is assumed to be Gaussian[29]

$$P(\psi) = \exp\left[-\frac{1}{\Delta_0}\int_{\mathbf{x}} \psi^2(\mathbf{x})\right], \tag{120}$$

and characterized by the disorder strength Δ_0. The engineering dimension of the random field is the same as that of the chemical potential which is one, $[\psi] = 1$, while that of the parameter Δ_0 is $[\Delta_0] = 2 - d$ so that the exponent in (120) is dimensionless. Since $\psi(\mathbf{x})$ depends only on the d spatial dimensions, the impurities it describes should be considered as grains randomly distributed in space. The quantity

$$Z[\psi] = \int \mathrm{D}\phi^*\mathrm{D}\phi \exp\left(i\int_x \mathcal{L}\right), \tag{121}$$

where $\mathcal{L}$ stands for the Lagrangian (45) with the term (119) added, is the zero-temperature partition function for a given impurity configuration ψ. In the case of quenched impurities, the average of an observable $O(\phi^*, \phi)$ is obtained as follows

$$\langle O(\phi^*, \phi)\rangle = \int \mathrm{D}\psi P(\psi) \langle O(\phi^*, \phi)\rangle_\psi, \tag{122}$$

where $\langle O(\phi^*, \phi)\rangle_\psi$ indicates the grand-canonical average for a given impurity configuration. In other words, first the ensemble average is taken, and only after that the averaging over the random field is carried out.

In terms of the shifted field, the added term reads

$$\mathcal{L}_\Delta = \psi(\mathbf{x})(|\bar\phi|^2 + |\tilde\phi|^2 + \bar\phi\tilde\phi^* + \bar\phi^*\tilde\phi). \tag{123}$$

The first two terms lead to an irrelevant change in the chemical potential, so that we have to consider only the last two terms, which we can cast in the form

$$\mathcal{L}_\Delta = \psi(\mathbf{x})\,\bar\Phi^\dagger\tilde\Phi, \qquad \bar\Phi = \begin{pmatrix} \bar\phi \\ \bar\phi^* \end{pmatrix}. \tag{124}$$

The integral over $\tilde\Phi$ is Gaussian in the Bogoliubov approximation and is easily performed to yield an additional term to the effective action

$$S_\Delta = -\frac{1}{2}\int_{x,y} \psi(\mathbf{x})\bar\Phi^\dagger G_0(x-y)\bar\Phi\psi(\mathbf{y}), \tag{125}$$

where the propagator G_0 is the inverse of the matrix M_0 introduced in (53) with the field $U(x)$ set to zero. Let us first Fourier transform the fields,

$$G_0(x-y) = \int_k e^{-ik\cdot(x-y)}\, G_0(k) \tag{126}$$

$$\psi(\mathbf{x}) = \int_{\mathbf{k}} e^{i\mathbf{k}\cdot\mathbf{x}} \psi(\mathbf{k}). \tag{127}$$

The contribution to the effective action then appears in the form

$$S_\Delta = -\frac{1}{2}\int_{\mathbf{k}} |\psi(\mathbf{k})|^2 \bar{\Phi}^\dagger G(0,\mathbf{k})\bar{\Phi}. \tag{128}$$

Since the random field is Gaussian distributed [see (120)], the average over this field representing quenched impurities yields,

$$\langle|\psi(\mathbf{k})|^2\rangle = \frac{1}{2} V\Delta_0. \tag{129}$$

The remaining integral over the loop momentum in (128) is readily carried out to yield

$$\langle\mathcal{L}_\Delta\rangle = \frac{1}{2}\Gamma(1-d/2)\left(\frac{m}{2\pi}\right)^{d/2} |\bar{\phi}|^2 (6\lambda_0|\bar{\phi}|^2 - \mu_0)^{d/2-1}\Delta_0. \tag{130}$$

This contribution is seen to diverge in the limit $d \to 2$:

$$\langle\mathcal{L}_\Delta\rangle = \frac{1}{4\pi}\frac{m\mu_0}{\lambda_0\kappa^\epsilon}\frac{\Delta_0}{\epsilon}, \tag{131}$$

where we have substituted the mean-field value $\mu_0 = 2\lambda_0|\bar{\phi}|^2$. Recall that κ is an arbitrary scale parameter introduced for dimensional reasons; the engineering dimension of the right-hand side in (131) has the correct value $3-\epsilon$ in this way. The result (131) is a first indication of the importance of impurities in $d=2$, showing that in order to render the random theory finite a modified renormalized coupling constant $\hat{\lambda}$ has to be introduced via, cf. (76),

$$\frac{1}{\lambda_0} = \frac{1}{\kappa^\epsilon}\left[\frac{1}{\hat{\lambda}} - \frac{m}{\pi\epsilon}\left(1 - \frac{\hat{\Delta}}{\mu\hat{\lambda}}\right)\right], \tag{132}$$

which depends on the disorder strength. The renormalized parameter $\hat{\Delta}$ is defined in the same way as $\hat{\lambda}$.

In the previous subsection we have seen that due to the interparticle repulsion, not all the particles reside in the condensate. We expect that the random field causes an additional depletion of the condensate. To obtain this, we differentiate (130) with respect to the chemical potential. This gives[30]

$$\bar{n}_\Delta = \frac{\partial\langle\mathcal{L}_\Delta\rangle}{\partial\mu} = \frac{2^{d/2-5}\Gamma(2-d/2)}{\pi^{d/2}} m^{d/2}\lambda^{d/2-2}\bar{n}_0^{d/2-1}\Delta, \tag{133}$$

where $\bar{n}_0$ denotes the density of particles residing in the condensate. We have here again replaced the bare parameters with the (one-loop) renormalized ones. This is consistent to this order since (133) is already a one-loop result.

The divergence in the limit $\lambda \to 0$ for $d < 4$ signals the collapse of the system when the interparticle repulsion is removed. Note that in $d = 2$, the depletion is independent of the density $\bar{n}_0$:[31]

$$\bar{n}_\Delta = \frac{1}{16\pi} \frac{m}{\lambda} \Delta. \tag{134}$$

The total particle number density $\bar{n}$ is given by

$$\bar{n} = \bar{n}_0 \left(1 + \frac{m\lambda}{2\pi}\right) + \frac{1}{16\pi} \frac{m}{\lambda} \Delta. \tag{135}$$

We next calculate the mass current $\mathbf{g}$ to determine the superfluid mass density, i.e., the mass density flowing with the superfluid velocity $\mathbf{v}_s$. As we have seen in the preceding subsection, in the absence of impurities and at zero temperature all the particles participate in the superflow and move on the average with a velocity $\mathbf{v}_s$. We expect this no longer to hold in the presence of impurities. To determine the change in the superfluid mass density due to impurities, we replace μ_0 with μ_{eff} as defined in (108) and $i\partial_0$ with $i\partial_0 - (\mathbf{u} - \mathbf{v}_s) \cdot (-i\nabla)$ in the contribution (128) to the effective action, and differentiate it with respect to the externally imposed velocity $-\mathbf{u}$. We find to linear order in the difference $\mathbf{u} - \mathbf{v}_s$:

$$\mathbf{g} = \rho_s \mathbf{v}_s + \rho_n \mathbf{u}, \tag{136}$$

with the superfluid and normal mass density[30]

$$\rho_s = m\left(\bar{n} - \frac{4}{d}\bar{n}_\Delta\right), \quad \rho_n = \frac{4}{d} m \bar{n}_\Delta. \tag{137}$$

We see that the normal density is a factor $4/d$ larger than the mass density $m\bar{n}_\Delta$ knocked out of the condensate by the impurities. (For $d = 3$ this gives the factor $4/3$ first found in Ref. 32.) Apparently, part of the zero-momentum states belongs for $d < 4$ not to the condensate, but to the normal fluid. Being trapped by the impurities, this fraction of the zero-momentum states are localized. This shows that the phenomenon of localization can be accounted for in the Bogoliubov theory of superfluidity by including a random field.

3.4 Vortices

We shall now include vortices in the system. A vortex in two space dimensions may be pictured as a point-like object at scales large compared to their core size. It is characterized by the winding number w of the map

$$\varphi(\mathbf{x}) : \mathrm{S}^1_{\mathbf{x}} \to \mathrm{S}^1 \tag{138}$$

of a circle $\mathrm{S}^1_{\mathbf{x}}$ around the vortex into the internal circle S^1 parameterized by the Goldstone field φ. In the microscopic theory (45), the asymptotic solution of a static vortex with a winding number w centered at the origin is well known[33]

$$\phi(\mathbf{x}) = \sqrt{\frac{\mu_0}{2\lambda_0}} \left(1 - \xi_0^2 \frac{w^2}{4\mathbf{x}^2}\right) \mathrm{e}^{iw\theta} + \mathcal{O}\left(\frac{1}{\mathbf{x}^4}\right), \tag{139}$$

where θ is the azimuthal angle and $\xi_0 = 1/\sqrt{m\mu_0} = 1/mc_0$ is the coherence length. The density profile $n(\mathbf{x})$ in the presence of this vortex follows from taking $|\phi(\mathbf{x})|^2$.

To incorporate vortices in the effective theory we employ the powerful principle of defect gauge symmetry developed by Kleinert[34-36]. In this approach, one introduces a so-called vortex gauge field $\varphi_\mu^{\mathrm{P}} = (\varphi_0^{\mathrm{P}}, \boldsymbol{\varphi}^{\mathrm{P}})$ in the effective theory (66) via a minimal coupling to the Goldstone field:

$$\tilde{\partial}_\mu \varphi \to \tilde{\partial}_\mu \varphi + \varphi_\mu^{\mathrm{P}}, \tag{140}$$

with $\tilde{\partial}_\mu = (\partial_0, -\nabla)$. If there are N vortices with winding number w_α $(\alpha = 1, \cdots, N)$ centered at $\mathbf{X}^1(t), \cdots, \mathbf{X}^N(t)$, the plastic field satisfies the relation

$$\nabla \times \boldsymbol{\varphi}^{\mathrm{P}}(x) = -2\pi \sum_\alpha w_\alpha \delta[\mathbf{x} - \mathbf{X}^\alpha(t)], \tag{141}$$

so that we obtain for the superfluid velocity field

$$\nabla \times \mathbf{v}_{\mathrm{s}} = \sum_\alpha \gamma_\alpha \delta[\mathbf{x} - \mathbf{X}^\alpha(t)], \tag{142}$$

as required. Here, $\gamma_\alpha = (2\pi/m) w_\alpha$ is the circulation of the αth vortex which is quantized in units of $2\pi/m$. A summation over the indices labeling the vortices will always be made explicit. The combination $\tilde{\partial}_\mu \varphi + \varphi_\mu^{\mathrm{P}}$ is invariant under the local gauge transformation

$$\varphi(x) \to \varphi(x) + \alpha(x); \quad \varphi_\mu^{\mathrm{P}} \to \varphi_\mu^{\mathrm{P}} - \tilde{\partial}_\mu \alpha(x), \tag{143}$$

with φ_μ^{P} playing the role of a gauge field.

In the gauge $\varphi_0^{\mathrm{P}} = 0$, Eq. (141) can be solved to yield

$$\varphi_i^{\mathrm{P}}(x) = 2\pi \epsilon_{ij} \sum_\alpha w_\alpha \delta_j[x, L_\alpha(t)], \tag{144}$$

where ϵ_{ij} is the antisymmetric Levi-Civita symbol in two dimensions, with $\epsilon_{12} = 1$, and $\boldsymbol{\delta}[x, L_\alpha(t)]$ is a delta function on the line $L_\alpha(t)$ starting at the center $\mathbf{X}^\alpha(t)$ of the αth vortex and running to spatial infinity along an arbitrary path:

$$\delta_i[x, L_\alpha(t)] = \int_{L_\alpha(t)} \mathrm{d}y_i\, \delta(\mathbf{x} - \mathbf{y}). \tag{145}$$

Let us for the moment concentrate on static vortices. The field equation obtained from the effective theory (66) with $\nabla\varphi$ replaced by the covariant derivative $\nabla\varphi - \boldsymbol{\varphi}^{\mathrm{P}}$ and $\partial_0 \varphi$ set to zero simply reads

$$\nabla \cdot \mathbf{v}_{\mathrm{s}} = 0, \quad \text{or} \quad \nabla \cdot (\nabla\varphi - \boldsymbol{\varphi}^{\mathrm{P}}) = 0, \tag{146}$$

when the fourth-order term is neglected. It can be easily solved to yield

$$\varphi(\mathbf{x}) = -\int_{\mathbf{y}} G(\mathbf{x} - \mathbf{y}) \nabla \cdot \boldsymbol{\varphi}^{\mathrm{P}}(\mathbf{y}), \tag{147}$$

where $G(\mathbf{x})$ is the Green function of the Laplace operator

$$G(\mathbf{x}) = \int_{\mathbf{k}} \frac{\mathrm{e}^{i\mathbf{k}\cdot\mathbf{x}}}{\mathbf{k}^2} = -\frac{1}{2\pi} \ln(|\mathbf{x}|). \tag{148}$$

For the velocity field we obtain in this way the well-known expression[37]

$$v_i(\mathbf{x}) = \frac{1}{2\pi} \epsilon_{ij} \sum_{\alpha=1}^{N} \gamma_\alpha \frac{x_j - X_j^\alpha}{|\mathbf{x} - \mathbf{X}^\alpha|^2}, \tag{149}$$

which is valid for $\mathbf{x}$ sufficiently far away from the vortex cores. Let us now specialize to the case of a single static vortex at the origin. On substituting the corresponding solution in (73), we find for the density profile in the presence of a static vortex asymptotically

$$n(\mathbf{x}) = \bar{n}\left(1 - \xi_0^2 \frac{w^2}{2\mathbf{x}^2}\right). \tag{150}$$

This is the same formula as the one obtained from the solution (139) of the microscopic theory. This exemplifies that with the aid of the defect gauge symmetry principle, vortices are correctly accounted for in the effective theory.

Let us proceed to investigate the dynamics of vortices in this formalism and derive the action which governs it. We consider only the first part of the effective theory (66). In ignoring the higher-order terms, we approximate the superfluid by an incompressible fluid for which the particle number density is constant, $n(x) = \bar{n}$, see Eq. (73). We again work in the gauge $\varphi_0^{\mathrm{P}} = 0$ and replace $\nabla\varphi$ by the covariant derivative $\nabla\varphi - \boldsymbol{\varphi}^{\mathrm{P}}$, with the plastic field given by (141). The solution of the resulting field equation for φ is again of the form (147), but now it is time-dependent because the plastic field is. Substituting this in the action $S_{\mathrm{eff}} = \int_x \mathcal{L}_{\mathrm{eff}}$, we find after some straightforward calculus

$$S_{\mathrm{eff}} = m\bar{n}\int_t \left[\frac{1}{2}\sum_\alpha \gamma_\alpha \mathbf{X}^\alpha \times \dot{\mathbf{X}}^\alpha + \frac{1}{2\pi}\sum_{\alpha<\beta} \gamma_\alpha\gamma_\beta \ln(|\mathbf{X}^\alpha - \mathbf{X}^\beta|/a)\right]. \tag{151}$$

The constant a has the dimension of a length and is included in the argument of the logarithm for dimensional reasons. Physically, it represents the core size of a vortex. The first term in (151) leads to a twisted canonical structure which is reminiscent of that found in the so-called Landau problem of a charged particle confined to move in a plane perpendicular to an applied magnetic field H.

To display the canonical structure, let us rewrite the first term of the Lagrangian corresponding to (151) as

$$L_1 = m\bar{n}\sum_\alpha \gamma_\alpha X_1^\alpha \dot{X}_2^\alpha, \tag{152}$$

where we have ignored a total derivative. It follows that the canonical conjugate to the second component X_2^α of the center coordinate $\mathbf{X}^\alpha$ is essentially its first component[38]

$$\frac{\partial L_1}{\partial \dot{X}_2^\alpha} = m\bar{n}\gamma_\alpha X_1^\alpha. \tag{153}$$

It implies that phase space coincides with real space and gives rise to the commutation relation

$$[X_2^\alpha, X_1^\beta] = \frac{i}{w_\alpha}\ell^2\delta^{\alpha\beta}, \tag{154}$$

where

$$\ell = 1/\sqrt{2\pi\bar{n}} \tag{155}$$

is a characteristic length whose definition is such that $2\pi\ell^2$ is the average area occupied by a particle of the superfluid film. The commutation relation leads to an uncertainty in the location of the vortex centers given by

$$\Delta X_1^\alpha \Delta X_2^\alpha \geq \frac{\ell^2}{2|w_\alpha|}, \tag{156}$$

which is inverse proportional to the particle number density.

From elementary quantum mechanics, we know that to each unit cell (of area h) in phase space there corresponds one quantum state. That is, the number of states in an area S of phase space is given by

$$\# \text{ states in } S = \frac{1}{h} \int_S \mathrm{d}p \, \mathrm{d}q, \tag{157}$$

where p and q are a pair of canonically conjugate variables. For the case at hand, this implies that the available number of states in an area S_α of *real* space is

$$\# \text{ states in } S_\alpha = |w_\alpha| \bar{n} S_\alpha, \tag{158}$$

or, equivalently, that the number of states per unit area available to the αth vortex is $|w_\alpha| \bar{n}$.

This phenomenon that phase space coincides with the real space is known to arise also in the Landau problem. There, it leads to the well-known degeneracy $|e_\alpha| H/h$ of each Landau level, where $e_\alpha = v_\alpha e_0$ is the electric charge of the particle, with $e_0 (> 0)$ the unit of charge. In terms of the magnetic flux quantum $\Phi_0 = h/e_0$, the Landau degeneracy can be rewritten as $|v_\alpha| H/\Phi_0 = |v_\alpha| \bar{n}_\otimes$, with $\bar{n}_\otimes$ the flux number density. In other words, whereas the degeneracy in the case of vortices in a superfluid film is given by the particle number density, here it is given by the flux number density. Using this analogy, we see that the characteristic length (155) translates into $\ell_H = 1/\sqrt{2\pi \bar{n}_\otimes}$ which is precisely the magnetic length of the Landau problem.

The first term in the action (151) is also responsible for the so-called geometrical phase[39] acquired by the wavefunction of a vortex when it traverses a closed path. Let us first discuss the case of a charged particle moving adiabatically around a close path Γ_α. Its wavefunction picks up an extra Aharonov-Bohm phase factor given by the Wilson loop:

$$W(\Gamma_\alpha) = \exp[i\gamma(\Gamma_\alpha)] = \exp\left(\frac{ie_\alpha}{\hbar} \oint_{\Gamma_\alpha} \mathrm{d}\mathbf{x} \cdot \mathbf{A} \right) = \exp\left[2\pi i v_\alpha \frac{HS(\Gamma_\alpha)}{\Phi_0} \right], \tag{159}$$

where $\mathbf{A}$ is the vector potential describing the external magnetic field and $HS(\Gamma_\alpha)$ is the magnetic flux through the area $S(\Gamma_\alpha)$ spanned by the loop Γ_α. The geometrical phase $\gamma(\Gamma_\alpha)$ in (159) is seen to be ($2\pi v_\alpha$ times) the number of flux quanta enclosed by the path Γ_α.

On account of the above analogy, it follows that the geometrical phase picked up by the wavefunction of a vortex when it is moved adiabatically around a closed path in the superfluid film is ($2\pi w_\alpha$ times) the number of superfluid particles enclosed by the path.[40]

The second term in the action (151) represents the long-range interaction between two vortices mediated by the exchange of Goldstone quanta. The action yields the well-known equations of motion for point vortices in an incompressible two-dimensional superfluid:[37,41]

$$\dot{X}_i^\beta(t) = \frac{\epsilon_{ij}}{2\pi} \sum_{\alpha \neq \beta} \gamma_\alpha \frac{X_j^\beta(t) - X_j^\alpha(t)}{|\mathbf{X}^\beta(t) - \mathbf{X}^\alpha(t)|^2}. \tag{160}$$

Note that $\dot{X}_i^\beta(t) = v_i \left[\mathbf{X}^\beta(t) \right]$, where $\mathbf{v}(x)$ is the superfluid velocity (149) with the time-dependence of the centers of the vortices included. This nicely illustrates a result due to Helmholtz for ideal fluids, stating that a vortex moves with the fluid, i.e., at the local velocity produced by the other vortices in the system. Experimental support for this conclusion has been reported in Ref. 42.

3.5 Kosterlitz–Thouless Phase Transition

Although we are interested mainly in quantum phase transitions in this Lecture, there is one classical phase transition special to two dimensions which will turn out to be relevant for our discussion later on—the so-called Kosterlitz-Thouless phase transition. It is well known that a superfluid film undergoes such a phase transition at a temperature well below the bulk transition temperature. The superfluid low-temperature state is characterized by tightly bound vortex-antivortex pairs which at the Kosterlitz-Thouless temperature unbind and thereby disorder the superfluid state. The disordered state, at temperatures still below the bulk transition temperature, consists of a plasma of unbound vortices.

Since the phase transition is an equilibrium transition, we can ignore any time dependence. The important fluctuations here, at temperatures below the bulk transition temperature, are phase fluctuations so that we can consider the London limit, where the phase of the $\phi(x)$-field is allowed to vary in spacetime while the modulus is kept fixed. In this limit, we can take as Hamiltonian

$$\mathcal{H} = \frac{1}{2}\rho_s \mathbf{v}_s^2, \tag{161}$$

where ρ_s is the superfluid mass density which we assume to be constant and $\mathbf{v}_s$ is the superfluid velocity

$$\mathbf{v}_s = \frac{1}{m}(\nabla\varphi - \varphi^{\mathrm{P}}), \tag{162}$$

with the vortex gauge field φ^{P} included to account for possible vortices in the system. We shall restrict ourselves to vortices of unit winding number, so that $w_\alpha = \pm 1$ for a vortex and antivortex, respectively.

The canonical partition function describing the equilibrium configuration of N_+ vortices and N_- antivortices in a superfluid film is given by

$$Z_N = \frac{1}{N_+!N_-!}\prod_\alpha \int_{\mathbf{X}^\alpha} \int \mathrm{D}\varphi \exp\left(-\beta \int_{\mathbf{x}} \mathcal{H}\right), \tag{163}$$

with $\mathcal{H}$ the Hamiltonian (161) and $N = N_+ + N_-$ the total number of vortices and antivortices. The factors $N_+!$ and $N_-!$ arise because the vortices and antivortices are indistinguishable, and $\prod_\alpha \int_{\mathbf{X}^\alpha}$ denotes the integration over the positions of the vortices. The functional integral over φ is Gaussian and therefore easily carried out, with the result

$$Z_N = \frac{1}{N_+!N_-!}\prod_\alpha \int_{\mathbf{X}^\alpha} \exp\left[\pi \frac{\beta\rho_s}{m^2}\sum_{\alpha,\beta} w_\alpha w_\beta \ln\left(|\mathbf{X}^\alpha - \mathbf{X}^\beta|/a\right)\right]. \tag{164}$$

Apart from an irrelevant normalization factor, Eq. (164) is the canonical partition function of a two-dimensional Coulomb gas with charges $q_\alpha = q w_\alpha = \pm q$, where

$$q = \sqrt{2\pi\rho_s}/m. \tag{165}$$

Let us rewrite the sum in the exponent appearing in (164) as

$$\begin{aligned}&\sum_{\alpha,\beta} q_\alpha q_\beta \ln\left(|\mathbf{X}^\alpha - \mathbf{X}^\beta|/a\right) \\ &\quad = \sum_{\alpha,\beta} q_\alpha q_\beta \left[\ln\left(|\mathbf{X}^\alpha - \mathbf{X}^\beta|/a\right) - \ln(0)\right] + \ln(0)\left(\sum_\alpha q_\alpha\right)^2,\end{aligned} \tag{166}$$

where we have isolated the self-interaction in the last term in the right-hand side. Since $\ln(0) = -\infty$, the charges must add up to zero so as to obtain a nonzero partition function. From now on we will therefore assume overall charge neutrality, $\sum_\alpha q_\alpha = 0$, so that $N_+ = N_- = N/2$, where N must be an even integer. To regularize the remaining divergence, we replace $\ln(0)$ with an undetermined, negative constant $-c$. The exponent of (164) thus becomes

$$\frac{\beta}{2}\sum_{\alpha,\beta} q_\alpha q_\beta \ln\left(|\mathbf{X}^\alpha - \mathbf{X}^\beta|/a\right) = \frac{\beta}{2}\sum_{\alpha\neq\beta} q_\alpha q_\beta \ln\left(|\mathbf{X}^\alpha - \mathbf{X}^\beta|/a\right) - \beta\epsilon_c N, \tag{167}$$

where $\epsilon_c = cq^2/2$ physically represents the core energy, i.e., the energy required to create a single vortex. In deriving this we have used the identity $\sum_{\alpha\neq\beta} q_\alpha q_\beta = -\sum_\alpha q_\alpha^2 = -Nq^2$ which follows from charge neutrality. Having dealt with the self-interaction, we limit the integrations $\prod_\alpha \int_{\mathbf{X}^\alpha}$ in (164) over the location of the vortices to those regions where they are more than a distance a apart, $|\mathbf{X}^\alpha - \mathbf{X}^\beta| > a$. The grand-canonical partition function of the system can now be cast in the form

$$Z = \sum_{N=0}^{\infty} \frac{z^N}{[(N/2)!]^2} \prod_\alpha \int_{\mathbf{X}^\alpha} \exp\left[\frac{\beta}{2}\sum_{\alpha\neq\beta} q_\alpha q_\beta \ln\left(|\mathbf{X}^\alpha - \mathbf{X}^\beta|/a\right)\right], \tag{168}$$

where $z = \exp(-\beta\epsilon_c)$ is the fugacity. The system is known to undergo a phase transition at the Kosterlitz-Thouless temperature[43,44]

$$T_{\rm KT} = \frac{1}{4}q^2 = \frac{\pi}{2}\frac{\rho_s}{m^2}, \tag{169}$$

triggered by the unbinding of vortex-antivortex pairs. It follows from this equation that the two-dimensional superfluid mass density $\rho_s(T)$, which varies from sample to sample, terminates on a line with a universal slope as the temperature approaches the Kosterlitz-Thouless temperature from below.[45]

3.6 Dual Theory

Let us proceed to represent the partition function (168) by a field theory—a so-called dual theory. The idea behind such a dual transformation is to obtain a formulation, in which the vortices are not described as singular objects as is the case in the original formulation, but by ordinary fields. To derive it we note that $\ln(|\mathbf{x}|)$ is the inverse of the Laplace operator ∇^2,

$$\frac{1}{2\pi}\nabla^2 \ln(|\mathbf{x}|) = \delta(\mathbf{x}). \tag{170}$$

This allows us to represent the exponential function in (168) as a functional integral over an auxiliary field ϕ:

$$\exp\left[\frac{\beta}{2}\sum_{\alpha\neq\beta} q_\alpha q_\beta \ln\left(|\mathbf{X}^\alpha - \mathbf{X}^\beta|/a\right)\right] = \int \mathrm{D}\phi \exp\left\{-\int_{\mathbf{x}}\left[\frac{1}{4\pi\beta}(\nabla\phi)^2 + i\rho_q\phi\right]\right\}, \tag{171}$$

where $\rho_q(\mathbf{x}) = \sum_\alpha q_\alpha \delta(\mathbf{x} - \mathbf{X}^\alpha)$ is the charge density. In this way, the partition function becomes

$$Z = \sum_{N=0}^{\infty} \frac{z^N}{[(N/2)!]^2} \prod_{\alpha=1}^{N} \int_{\mathbf{X}^\alpha} \int \mathrm{D}\phi \exp\left\{-\int_{\mathbf{x}}\left[\frac{1}{4\pi\beta}(\nabla\phi)^2 + i\rho_q\phi\right]\right\}. \tag{172}$$

In a mean-field treatment, the functional integral over the auxiliary field introduced in (171) is approximated by the saddle point determined by the field equation

$$iT\nabla^2\phi = -2\pi\rho_q. \tag{173}$$

When we introduce the scalar variable $\Phi = iT\phi$, this equation becomes formally Gauss law, with Φ the electrostatic scalar potential. The auxiliary field introduced in (171) may therefore be thought of as representing the scalar potential of the equivalent two-dimensional Coulomb gas.[34]

On account of charge neutrality, we have the identity

$$\left[\int_{\mathbf{x}} \left(e^{iq\phi(\mathbf{x})} + e^{-iq\phi(\mathbf{x})}\right)\right]^N = \frac{N!}{[(N/2)!]^2} \prod_{\alpha=1}^{N} \int_{\mathbf{x}^\alpha} e^{-i\sum_\alpha q_\alpha \phi(\mathbf{x}^\alpha)}, \tag{174}$$

where we recall that N is an even number. The factor $N!/[(N/2)!]^2$ is the number of charge-neutral terms contained in the binomial expansion of the left-hand side. The partition function (172) may thus be written as[34]

$$\begin{aligned} Z &= \sum_{N=0}^{\infty} \frac{(2z)^N}{N!} \int \mathrm{D}\phi \exp\left[-\int_{\mathbf{x}} \frac{1}{4\pi\beta}(\nabla\phi)^2\right] \left[\cos\left(\int_{\mathbf{x}} q\phi\right)\right]^N \\ &= \int \mathrm{D}\phi \exp\left\{-\int_{\mathbf{x}} \left[\frac{1}{4\pi\beta}(\nabla\phi)^2 - 2z\cos(q\phi)\right]\right\}, \end{aligned} \tag{175}$$

where in the final form we recognize the sine-Gordon model. This is the dual theory we have been seeking for. Contrary to the original formulation (163), which contains the vortices as singular objects, the dual formulation has no singularities. To see how the vortices and the Kosterlitz–Thouless phase transition are represented in the dual theory we note that the field equation of the auxiliary field now reads

$$iT\nabla^2\phi = 2\pi z q\left(e^{iq\phi} - e^{-iq\phi}\right). \tag{176}$$

A comparison with the previous field equation (173) shows that the right-hand side represents the charge density of the Coulomb gas. In terms of the scalar potential Φ, Eq. (176) becomes the Poisson-Boltzmann equation

$$\nabla^2\Phi = -2\pi q\left(z\, e^{-\beta q\Phi} - z\, e^{\beta q\Phi}\right), \tag{177}$$

describing, at least for temperatures above the Kosterlitz–Thouless temperature, a plasma of positive and negative charges with density $n_\pm$,

$$n_\pm = z\exp(\mp\beta q\Phi), \tag{178}$$

respectively. The fugacity z is the density at zero scalar potential. (It is to be recalled that we suppress factors of a denoting the core size of the vortices.) Equation (177) is a self-consistent equation for the scalar potential Φ giving the spatial distribution of the charges via (178). It follows from this argument that the interaction term $2z\cos(q\phi)$ of the sine-Gordon model represents a plasma of vortices.

The renormalization group analysis of the sine-Gordon model reveals that at the Kosterlitz–Thouless temperature $T_{\mathrm{KT}} = 1/4q^2$ there is a phase transition between a low-temperature phase of tightly bound neutral pairs and a high-temperature plasma phase of unbound vortices.[46] In the low-temperature phase, the (renormalized) fugacity scales to zero in the large-scale limit so that the interaction term, representing the plasma of unbound vortices, is suppressed. The long-distance behavior of the low-temperature phase is therefore well described by the free theory $(\nabla\phi)^2/4\pi\beta$, representing a gapless mode—the so-called Kosterlitz-Thouless mode. This is the superfluid state. The expectation value of a single vortex vanishes because in this gapless state its energy diverges in the infrared.

An important characteristic of a charged plasma is that it has no gapless excitations, the photon being transmuted into a massive plasmon. To see this we assume that $q\Phi \ll T$, so that $\sinh(\beta q\Phi) \approx \beta q\Phi$. In this approximation, the Poisson-Boltzmann equation (177) can be linearized to give

$$(\nabla^2 - m_{\rm D}^2)\Phi = 0, \quad m_{\rm D}^2 = 4\pi\beta z q^2. \tag{179}$$

This shows us that, in contradistinction to the low-temperature phase, in the high-temperature phase, the scalar potential describes a massive mode – the plasmon. In other words, the Kosterlitz–Thouless mode acquires an energy gap $m_{\rm D}$. Since it provides the high-temperature phase with an infrared cutoff, isolated vortices have a finite energy now and accordingly a finite probability to be created. This Debye mechanism of mass generation for the photon should be distinguished from the Higgs mechanism which operates in superconductors (see below) and which also generates a photon mass.

Another property of a charged plasma is that it screens charges. This so-called Debye screening may be illustrated by adding an external charge to the system. The linearized Poisson–Boltzmann Eq. (179) then becomes

$$(\nabla^2 - m_{\rm D}^2)\Phi(\mathbf{x}) = -2\pi q_0\delta(\mathbf{x}), \tag{180}$$

with q_0 the external charge which we have placed at the origin. The solution of this equation is given by $\Phi(\mathbf{x}) = q_0 K_0(m_{\rm D}|\mathbf{x}|)$ with K_0 a modified Bessel function. The mass term in (180) is (2π times) the charge density induced by the external charge, i.e.,

$$\rho_{\rm ind}(\mathbf{x}) = -\frac{1}{2\pi} q_0 m_{\rm D}^2 K_0(m_{\rm D}|\mathbf{x}|). \tag{181}$$

By integrating this density over the entire system, we see that the total induced charge $\int_{\mathbf{x}} \rho_{\rm ind} = -q_0$ completely screens the external charge – at least in the linear approximation we are using here. The inverse of the plasmon mass is the so-called Debye screening length.

To see that the sine-Gordon model gives a dual description of a superfluid film we cast the field equation (173) in the form

$$iT\nabla^2\phi = -mq\nabla \times \mathbf{v}_{\rm s}, \tag{182}$$

where we have employed Eq. (142). On integrating this equation, we obtain up to an irrelevant integration constant

$$iT\partial_i\phi = -q\epsilon_{ij}(\partial_j\varphi - \varphi_j^{\rm P}). \tag{183}$$

This relation, involving the antisymmetric Levi-Civita symbol, is a typical one between dual variables. It also nicely illustrates that although the dual variable ϕ is a regular field, it nevertheless contains the information about the vortices which in the original formulation are described via the singular vortex gauge field $\boldsymbol{\varphi}^{\rm P}$.

Given this observation it is straightforward to calculate the current-current correlation function $\langle g_i(\mathbf{k})g_j(-\mathbf{k})\rangle$, with

$$\mathbf{g} = \rho_{\rm s}\mathbf{v}_{\rm s} \tag{184}$$

the mass current. We find

$$\langle g_i(\mathbf{k})g_j(-\mathbf{k})\rangle = -\frac{\rho_s}{2\pi\beta^2}\epsilon_{ik}\epsilon_{jl}k_k k_l\langle\phi(\mathbf{k})\phi(-\mathbf{k})\rangle, \tag{185}$$

where the average is to be taken with respect to the partition function

$$Z_0 = \int \mathrm{D}\phi \exp\left[-\frac{1}{4\pi\beta}\int_{\mathbf{x}}(\nabla\phi)^2\right], \tag{186}$$

which is obtained from (175) by setting the interaction term to zero. We obtain in this way the standard expression for a superfluid

$$\langle g_i(\mathbf{k})g_j(-\mathbf{k})\rangle = -\frac{\rho_s}{\beta}\frac{1}{\mathbf{k}^2}\left(\delta_{ij}\mathbf{k}^2 - k_i k_j\right). \tag{187}$$

The $1/\mathbf{k}^2$ reflects the gaplessness of the ϕ-field in the low-temperature phase, while the combination $\delta_{ij}\mathbf{k}^2 - k_i k_j$ arises because the current is divergent free, $\nabla \cdot \mathbf{g}(\mathbf{x}) = 0$, or $\mathbf{k} \cdot \mathbf{g}(\mathbf{k}) = 0$.

4. SUPERCONDUCTIVITY

In this section we shall demonstrate a close connection between the Bogoliubov theory of superfluidity discussed in the previous section and the strong-coupling limit of the BCS theory of superconductivity. The phase-only effective theory governing the superconducting state is derived. It is also pointed out that a superconducting film at a finite temperature undergoes a Kosterlitz–Thouless phase transition.

4.1 BCS Theory

Our starting point is the famous microscopic model of Bardeen, Cooper, and Schrieffer (BCS) defined by the Lagrangian[47]

$$\begin{aligned}\mathcal{L} &= \psi_\uparrow^*[i\partial_0 - \xi(-i\nabla)]\psi_\uparrow + \psi_\downarrow^*[i\partial_0 - \xi(-i\nabla)]\psi_\downarrow - \lambda_0\psi_\uparrow^*\psi_\downarrow^*\psi_\downarrow\psi_\uparrow \\ &= \mathcal{L}_0 + \mathcal{L}_\mathrm{i},\end{aligned} \tag{188}$$

where $\mathcal{L}_\mathrm{i} = -\lambda_0\psi_\uparrow^*\psi_\downarrow^*\psi_\downarrow\psi_\uparrow$ is a contact interaction term, representing the effective, phonon mediated, attraction between electrons with a coupling constant $\lambda_0 < 0$, and $\mathcal{L}_0$ is the remainder. In (188), the field $\psi_{\uparrow(\downarrow)}$ is an anticommuting field describing the electrons with a mass m and a spin up (down); $\xi(-i\nabla) = \epsilon(-i\nabla) - \mu_0$, with $\epsilon(-i\nabla) = -\nabla^2/2m$, is the kinetic energy operator with the chemical potential μ_0 subtracted.

The Lagrangian (188) is invariant under global U(1) transformations. Under such a transformation, the electron fields pick up an additional phase factor

$$\psi_\sigma \to \mathrm{e}^{i\alpha}\psi_\sigma \tag{189}$$

with $\sigma = \uparrow, \downarrow$ and α a constant. Notwithstanding its simple form, the microscopic model (188) is a good starting point to describe BCS superconductors. The reason is that the interaction term allows for the formation of Cooper pairs which below a critical temperature condense. This results in a nonzero expectation value of the field Δ describing the Cooper pairs, and a spontaneous breakdown of the global U(1) symmetry. This in turn gives rise to the gapless Anderson–Bogoliubov mode which – after incorporating the electromagnetic field – lies at the root of most startling properties of superconductors.[48]

To obtain the effective theory governing this mode, let us integrate out the fermionic degrees of freedom. To this end, we introduce Nambu's notation and rewrite the Lagrangian (188) in terms of a two-component field

$$\psi = \begin{pmatrix} \psi_\uparrow \\ \psi_\downarrow^* \end{pmatrix}, \qquad \psi^\dagger = (\psi_\uparrow^*, \psi_\downarrow). \tag{190}$$

In this notation, $\mathcal{L}_0$ becomes

$$\mathcal{L}_0 = \psi^\dagger \begin{pmatrix} i\partial_0 - \xi(-i\nabla) & 0 \\ 0 & i\partial_0 + \xi(-i\nabla) \end{pmatrix} \psi, \tag{191}$$

where we have explicitly employed the anticommuting character of the electron fields and neglected terms which are a total derivative. The partition function,

$$Z = \int \mathrm{D}\psi^\dagger \mathrm{D}\psi \exp\left(i \int_x \mathcal{L} \right), \tag{192}$$

must for our purpose be manipulated in a form bilinear in the electron fields. This is achieved by rewriting the quartic interaction term as a functional integral over auxiliary fields Δ and Δ^* (for details see Ref.[49]):

$$\exp\left(-i\lambda_0 \int_x \psi_\uparrow^* \psi_\downarrow^* \psi_\downarrow \psi_\uparrow \right) = \tag{193}$$
$$\int \mathrm{D}\Delta^* \mathrm{D}\Delta \exp\left[-i \int_x \left(\Delta^* \psi_\downarrow \psi_\uparrow + \psi_\uparrow^* \psi_\downarrow^* \Delta - \frac{1}{\lambda_0} \Delta^* \Delta \right) \right],$$

where, as always, an overall normalization factor is omitted. Classically, Δ merely abbreviates the product of two electron fields

$$\Delta = \lambda_0 \psi_\downarrow \psi_\uparrow. \tag{194}$$

It would therefore be more appropriate to give Δ two spin labels $\Delta_{\downarrow\uparrow}$. Since $\psi_\uparrow$ and $\psi_\downarrow$ are anticommuting fields, Δ is antisymmetric in these indices. Physically, it describes the Cooper pairs of the superconducting state.

By employing (193), we can cast the partition function in the desired bilinear form

$$Z = \int \mathrm{D}\psi^\dagger \mathrm{D}\psi \int \mathrm{D}\Delta^* \mathrm{D}\Delta \; \exp\left(\frac{i}{\lambda_0} \int_x \Delta^* \Delta \right) \tag{195}$$
$$\times \exp\left[i \int_x \psi^\dagger \begin{pmatrix} i\partial_0 - \xi(-i\nabla) & -\Delta \\ -\Delta^* & i\partial_0 + \xi(-i\nabla) \end{pmatrix} \psi \right].$$

Changing the order of integration and performing the Gaussian integral over the Grassmann fields, we obtain

$$Z = \int \mathrm{D}\Delta^* \mathrm{D}\Delta \; \exp\left(iS_{\text{eff}}[\Delta^*, \Delta] + \frac{i}{\lambda_0} \int_x \Delta^* \Delta \right), \tag{196}$$

where $S_{\rm eff}$ is the one-loop effective action which, using the identity $\mathrm{Det}(A) = \exp[\mathrm{Tr}\ln(A)]$, can be cast in the form

$$S_{\rm eff}[\Delta^*, \Delta] = -i \, \mathrm{Tr} \ln \begin{pmatrix} p_0 - \xi(\mathbf{p}) & -\Delta \\ -\Delta^* & p_0 + \xi(\mathbf{p}) \end{pmatrix}, \tag{197}$$

where $p_0 = i\partial_0$ and $\xi(\mathbf{p}) = \epsilon(\mathbf{p}) - \mu_0$, with $\epsilon(\mathbf{p}) = \mathbf{p}^2/2m$.

In the mean-field approximation, the functional integral (196) is approximated by the saddle point:

$$Z = \exp\left(iS_{\rm eff}[\Delta^*_{\rm mf}, \Delta_{\rm mf}] + \frac{i}{\lambda_0}\int_x \Delta^*_{\rm mf}\Delta_{\rm mf} \right), \tag{198}$$

where $\Delta_{\rm mf}$ is the solution of the mean-field equation

$$\frac{\delta S_{\rm eff}}{\delta \Delta^*(x)} = -\frac{1}{\lambda_0}\Delta. \tag{199}$$

If we assume the system to be spacetime independent so that $\Delta_{\rm mf}(x) = \bar{\Delta}$, Eq. (199) yields the celebrated BCS gap[47] equation:

$$\begin{aligned} \frac{1}{\lambda_0} &= -i \int_k \frac{1}{k_0^2 - E^2(k) + i\eta} \\ &= -\frac{1}{2}\int_{\mathbf{k}} \frac{1}{E(\mathbf{k})}, \end{aligned} \tag{200}$$

where η is an infinitesimal positive constant that is to be set to zero at the end of the calculation, and

$$E(\mathbf{k}) = \sqrt{\xi^2(\mathbf{k}) + |\bar{\Delta}|^2} \tag{201}$$

is the spectrum of the elementary fermionic excitations. If this equation yields a solution with $\bar{\Delta} \neq 0$, the global U(1) symmetry (189) will be spontaneously broken since

$$\bar{\Delta} \to \mathrm{e}^{2i\alpha}\bar{\Delta} \neq \bar{\Delta} \tag{202}$$

under this transformation. The factor 2 in the exponential function arises because Δ, describing the Cooper pairs, is built from two electron fields. It satisfies the Landau definition of an order parameter as its value is zero in the symmetric, disordered state and nonzero in the state with broken symmetry. It directly measures whether the U(1) symmetry is spontaneously broken.

In the case of a spacetime-independent system, the effective action (197) is readily evaluated. Writing

$$\begin{pmatrix} p_0 - \xi(\mathbf{p}) & -\bar{\Delta} \\ -\bar{\Delta}^* & p_0 + \xi(\mathbf{p}) \end{pmatrix} = \begin{pmatrix} p_0 - \xi(\mathbf{p}) & 0 \\ 0 & p_0 + \xi(\mathbf{p}) \end{pmatrix} - \begin{pmatrix} 0 & \bar{\Delta} \\ \bar{\Delta}^* & 0 \end{pmatrix}, \tag{203}$$

and expanding the second logarithm in a Taylor series, we recognize the form

$$S_{\rm eff}[\bar{\Delta}^*, \bar{\Delta}] = -i \, \mathrm{Tr}\ln \begin{pmatrix} p_0 - \xi(\mathbf{p}) & 0 \\ 0 & p_0 + \xi(\mathbf{p}) \end{pmatrix} - i \, \mathrm{Tr}\ln\left(1 - \frac{|\bar{\Delta}|^2}{p_0^2 - \xi^2(\mathbf{p})}\right), \tag{204}$$

up to an irrelevant constant. The integral over the loop energy k_0 gives for the corresponding effective Lagrangian

$$\mathcal{L}_{\rm eff} = \int_{\mathbf{k}} [E(\mathbf{k}) - \xi(\mathbf{k})]. \tag{205}$$

To this one-loop result we have to add the tree term $|\bar{\Delta}|^2/\lambda_0$. Expanding $E(\mathbf{k})$ in $\bar{\Delta}$, we see that the effective Lagrangian also contains a term quadratic in $\bar{\Delta}$. This term amounts to a renormalization of the coupling constant; we find to this order for the renormalized coupling constant λ:

$$\frac{1}{\lambda} = \frac{1}{\lambda_0} + \frac{1}{2}\int_{\mathbf{k}} \frac{1}{|\xi(\mathbf{k})|}, \tag{206}$$

where it should be remembered that the bare coupling constant λ_0 is negative, so that there is an attractive interaction between the fermions. We shall analyze this equation later on, for the moment it suffices to note that we can distinguish two limits. One, the limit where the bare coupling constant is taken to zero, $\lambda_0 \to 0^-$, which is the famous weak-coupling BCS limit. Second, the limit where the bare coupling is taken to minus infinity $\lambda_0 \to -\infty$. This is the strong-coupling limit, where the attractive interaction is such that the fermions form tightly bound pairs.[50] These composite bosons have a weak repulsive interaction and can undergo Bose–Einstein condensation (see succeeding subsection).

Since there are two unknowns contained in the theory, viz., $\bar{\Delta}$ and μ_0, we need a second equation to determine these variables in the mean-field approximation.[50] To find the second equation we note that the average fermion number N, which is obtained by differentiating the effective action (197) with respect to μ

$$N = \frac{\partial S_{\text{eff}}}{\partial \mu}, \tag{207}$$

is fixed. If the system is spacetime independent, this reduces in the one-loop approximation to

$$\bar{n} = -i \,\text{Tr} \int_k G_0(k)\tau_3, \tag{208}$$

where $\bar{n} = N/V$, with V the volume of the system, is the constant fermion number density, τ_3 is the diagonal Pauli matrix in the Nambu space,

$$\tau_3 = \begin{pmatrix} 1 & 0 \\ 0 & -1 \end{pmatrix}, \tag{209}$$

and $G_0(k)$ is the Feynman propagator,

$$\begin{aligned} G_0(k) &= \begin{pmatrix} k_0 - \xi(\mathbf{k}) & -\bar{\Delta} \\ -\Delta_0^* & k_0 + \xi(\mathbf{k}) \end{pmatrix}^{-1} \\ &= \frac{1}{k_0^2 - E^2(\mathbf{k}) + i\eta} \begin{pmatrix} k_0\, e^{ik_0\eta} + \xi(\mathbf{k}) & \bar{\Delta} \\ \bar{\Delta}^* & k_0\, e^{-ik_0\eta} - \xi(\mathbf{k}) \end{pmatrix}. \end{aligned} \tag{210}$$

Here, η is an infinitesimal positive constant that is to be set to zero at the end of the calculation. The exponential functions in the diagonal elements of the propagator are an additional convergence factor needed in nonrelativistic theories.[51] If the integral over the loop energy k_0 in the particle number equation (208) is carried out, it will take the familiar form

$$\bar{n} = \int_{\mathbf{k}} \left(1 - \frac{\xi(\mathbf{k})}{E(\mathbf{k})}\right). \tag{211}$$

The two equations (200) and (211) determine $\bar{\Delta}$ and μ_0. They are usually evaluated in the weak-coupling BCS limit. However, as has been first pointed out by Leggett[50], they can also be easily solved in the strong-coupling limit (see the succeeding subsection), where the fermions are tightly bound in pairs. More recently, also the crossover between the weak-coupling BCS limit and the strong-coupling composite boson limit has been studied in detail[52-55].

We are now in a position to derive the effective theory governing the gapless Anderson-Bogoliubov mode. To this end, we write the order parameter $\Delta_{\rm mf}$ as

$$\Delta_{\rm mf}(x) = \bar{\Delta}\, e^{2i\varphi(x)}, \tag{212}$$

where $\bar{\Delta}$ is a spacetime-independent solution of the mean-field equation (199) and $\varphi(x)$ represents the Anderson-Bogoliubov mode, i.e., the Goldstone mode of the spontaneously broken U(1) symmetry. This approximation, where the phase of the order parameter is allowed to vary in spacetime while the modulus is kept fixed, is called the London limit. This limit is relevant for our discussion of the zero-temperature superconductor-to-insulator phase transition in Sec. 6 because this transition is driven by phase fluctuations; the modulus of the order parameter remains finite and constant at the transition. The critical behavior can thus be studied with this effective theory formulated solely in terms of the phase field. We proceed by decomposing the Grassmann field as, cf. Ref. 56,

$$\psi_\sigma(x) = e^{i\varphi(x)}\chi_\sigma(x)\,, \tag{213}$$

and substituting the specific form (212) of the order parameter in the partition function (195). Instead of the effective action (197) we now obtain

$$S_{\rm eff} = -i{\rm Tr}\ln\begin{pmatrix} p_0 - \partial_0\varphi - \xi(\mathbf{p}+\nabla\varphi) & -\bar{\Delta} \\ -\Delta_0^* & p_0 + \partial_0\varphi + \xi(\mathbf{p}-\nabla\varphi)\end{pmatrix}, \tag{214}$$

where the derivative $\tilde{\partial}_\mu\varphi$ of the Goldstone field plays the role of an Abelian gauge field. This expression can be handled with the help of the derivative expansion outlined in Sec. 2.2, to yield the phase-only effective theory. We shall not give any details here and merely state the result[13], that the effective theory is again of the form (66).

4.2 Composite Boson Limit

In this subsection we shall investigate the strong-coupling limit of the pairing theory. In this limit, the attractive interaction between the fermions is such that they form tightly bound pairs of mass $2m$. To explicate this limit in an arbitrary dimension d, we swap the bare coupling constant for a more convenient parameter—the binding energy ϵ_a of a fermion pair in vacuum.[57] Both parameters characterize the strength of the contact interaction. To see the connection between the two, let us consider the Schrödinger equation for the problem at hand. In reduced coordinates it reads

$$\left[-\frac{\nabla^2}{m} + \lambda_0\,\delta(\mathbf{x})\right]\psi(\mathbf{x}) = -\epsilon_a, \tag{215}$$

where the reduced mass is $m/2$ and the delta-function potential, with $\lambda_0 < 0$, represents the attractive contact interaction $\mathcal{L}_{\rm i}$ in (188). We stress that this is a two-particle problem in vacuum; it is not the famous Cooper problem of two interacting fermions on top of a filled Fermi sea. The equation is most easily solved by Fourier transforming it. This yields the bound-state equation

$$\psi(\mathbf{k}) = -\frac{\lambda_0}{\mathbf{k}^2/m + \epsilon_a}\psi(0), \tag{216}$$

or

$$-\frac{1}{\lambda_0} = \int_{\mathbf{k}} \frac{1}{\mathbf{k}^2/m + \epsilon_a}. \tag{217}$$

This equation allows us to replace the coupling constant with the binding energy ϵ_a. When substituted in the gap equation (200), the latter becomes

$$\int_{\mathbf{k}} \frac{1}{\mathbf{k}^2/m + \epsilon_a} = \frac{1}{2}\int_{\mathbf{k}} \frac{1}{E(\mathbf{k})}. \tag{218}$$

By inspection, it is easily seen that this equation has a solution given by[50]

$$\bar{\Delta} \to 0, \qquad \mu_0 \to -\frac{1}{2}\epsilon_a, \tag{219}$$

where it should be noted that the chemical potential is negative here. This is the strong-coupling limit. To appreciate the physical significance of the specific value found for the chemical potential in this limit, we note that the spectrum $E_{\mathrm{b}}(\mathbf{q})$ of the two-fermion bound state measured relative to the pair chemical potential $2\mu_0$ reads

$$E_{\mathrm{b}}(\mathbf{q}) = -\epsilon_a + \frac{\mathbf{q}^2}{4m} - 2\mu_0. \tag{220}$$

The negative value for μ_0 found in (219) is precisely the condition for a Bose–Einstein condensation of the composite bosons in the $\mathbf{q} = 0$ state.

To investigate this limit further, we consider the effective action (197) and expand $\Delta(x)$ around a constant value $\bar{\Delta}$ satisfying the gap equation (200),

$$\Delta(x) = \bar{\Delta} + \tilde{\Delta}(x). \tag{221}$$

We obtain in this way,

$$S_{\mathrm{eff}} = i\,\mathrm{Tr}\sum_{l=1}^{\infty}\frac{1}{l}\left[G_0(p)\begin{pmatrix} 0 & \tilde{\Delta} \\ \tilde{\Delta}^* & 0 \end{pmatrix}\right]^l, \tag{222}$$

where G_0 is given in (210). We are interested in terms quadratic in $\tilde{\Delta}$. Employing the derivative expansion outlined in Sec. 2.2, we find

$$\begin{aligned} S^{(2)}_{\mathrm{eff}}(q) = {} & \frac{1}{2} i\,\mathrm{Tr}\,\frac{1}{p_0^2 - E^2(\mathbf{p})}\frac{1}{(p_0+q_0)^2 - E^2(\mathbf{p}-\mathbf{q})} \\ & \times \Big\{\bar{\Delta}^2\tilde{\Delta}^*\tilde{\Delta}^* + [p_0 + \xi(\mathbf{p})][p_0 + q_0 - \xi(\mathbf{p}-\mathbf{q})]\tilde{\Delta}\tilde{\Delta}^* \\ & + \bar{\Delta}^{*2}\tilde{\Delta}\tilde{\Delta} + [p_0 - \xi(\mathbf{p})][p_0 + q_0 + \xi(\mathbf{p}-\mathbf{q})]\tilde{\Delta}^*\tilde{\Delta}\Big\}, \end{aligned} \tag{223}$$

where $q_\mu = i\tilde{\partial}_\mu$. It is to be recalled here that the derivative p_μ operates on everything to its right, while $\tilde{\partial}_\mu$ operates only on the first object to its right. Let us for a moment ignore the derivatives in this expression. After carrying out the integral over the loop energy k_0 and using the gap equation (200), we then obtain

$$\mathcal{L}^{(2)}(0) = -\frac{1}{8}\int_{\mathbf{k}}\frac{1}{E^3(\mathbf{k})}\left(\bar{\Delta}^2\tilde{\Delta}^{*2} + \bar{\Delta}^{*2}\tilde{\Delta}^2 + 2|\bar{\Delta}|^2|\tilde{\Delta}|^2\right). \tag{224}$$

In the composite boson limit $\bar{\Delta} \to 0$, so that the spectrum (201) of the elementary fermionic excitations can be approximated by

$$E(\mathbf{k}) \approx \epsilon(\mathbf{k}) + \frac{1}{2}\epsilon_a. \tag{225}$$

The remaining integrals in (224) then become elementary, i.e.,

$$\int_{\mathbf{k}} \frac{1}{E^3(\mathbf{k})} = \frac{4\Gamma(3-d/2)}{(4\pi)^{d/2}} m^{d/2} \epsilon_a^{d/2-3}. \tag{226}$$

We next consider the terms involving derivatives in (223). Following Ref. 52 we set $\bar{\Delta}$ to zero here. The integral over the loop energy is easily carried out, with the result

$$\begin{aligned}\mathcal{L}^{(2)}(q) = & -\frac{1}{2}\int_{\mathbf{k}} \frac{1}{q_0 - \mathbf{k}^2/m + 2\mu_0 - \mathbf{q}^2/4m} \tilde{\Delta}\tilde{\Delta}^* \\ & -\frac{1}{2}\int_{\mathbf{k}} \frac{1}{-q_0 - \mathbf{k}^2/m + 2\mu_0 - \mathbf{q}^2/4m} \tilde{\Delta}^*\tilde{\Delta}.\end{aligned} \tag{227}$$

The integral over the loop momentum $\mathbf{k}$ gives in the strong-coupling limit using dimensional regularization

$$\int_{\mathbf{k}} \frac{1}{q_0 - \mathbf{k}^2/m - \epsilon_a - \mathbf{q}^2/4m} = -\frac{\Gamma(1-d/2)}{(4\pi)^{d/2}} m^{d/2}(-q_0 + \epsilon_a + \mathbf{q}^2/4m)^{d/2-1}, \tag{228}$$

or expanded in derivatives

$$\begin{aligned}&\int_{\mathbf{k}} \frac{1}{q_0 - \mathbf{k}^2/m - \epsilon_a - \mathbf{q}^2/4m} \\ &= -\frac{\Gamma(1-d/2)}{(4\pi)^{d/2}} m^{d/2}\epsilon_a^{d/2-1} - \frac{\Gamma(2-d/2)}{(4\pi)^{d/2}} m^{d/2}\epsilon_a^{d/2-2}\left(q_0 - \frac{\mathbf{q}^2}{4m}\right) + \cdots\end{aligned} \tag{229}$$

The first term at the right-hand side yields as contribution to the effective theory

$$\mathcal{L}_\lambda^{(2)} = \frac{\Gamma(1-d/2)}{(4\pi)^{d/2}} m^{d/2}\epsilon_a^{d/2-1}|\tilde{\Delta}|^2. \tag{230}$$

To this we have to add the contribution $|\tilde{\Delta}|^2/\lambda_0$ coming from the tree potential, i.e., the last term in the partition function (196). But this combination is no other than the one needed to define the renormalized coupling constant via (206), which in the strong-coupling limit reads explicitly

$$\frac{1}{\lambda} = \frac{1}{\lambda_0} + \frac{\Gamma(1-d/2)}{(4\pi)^{d/2}} m^{d/2}\epsilon_a^{d/2-1}. \tag{231}$$

In other words, the contribution (230) can be combined with the tree contribution to yield the term $|\tilde{\Delta}|^2/\lambda$. Expanding the square root in (228) in powers of the derivative q_μ using the value (219) for the chemical potential, and pasting the pieces together, one obtains[52] for the terms quadratic in $\tilde{\Delta}$

$$\mathcal{L}^{(2)} = \frac{1}{2}\frac{\Gamma(2-d/2)}{(4\pi)^{d/2}} m^{d/2}\epsilon_a^{d/2-2}\,\tilde{\Psi}^\dagger M_0(q)\,\tilde{\Psi}, \quad \tilde{\Psi} = \begin{pmatrix}\tilde{\Delta} \\ \tilde{\Delta}^*\end{pmatrix}, \tag{232}$$

where $M_0(q)$ is the 2×2 matrix,

$$M_0(q) = \begin{pmatrix} q_0 - \mathbf{q}^2/4m - (2-d/2)|\bar{\Delta}|^2/\epsilon_a & -(2-d/2)\bar{\Delta}^2/\epsilon_a \\ -(2-d/2)\bar{\Delta}^{*2}/\epsilon_a & -q_0 - \mathbf{q}^2/4m - (2-d/2)|\bar{\Delta}|^2/\epsilon_a \end{pmatrix}. \tag{233}$$

This Lagrangian is precisely of the form found in (52) describing an interacting Bose gas. On comparing with Eq. (53), we conclude that the composite bosons have — as expected — a mass $m_b = 2m$ twice the fermion mass m, and a small chemical potential

$$\mu_{0,b} = (2 - d/2)\frac{|\bar{\Delta}|^2}{\epsilon_a}. \tag{234}$$

From (233) one easily extracts the Bogoliubov spectrum and the velocity c_0 of the sound mode it describes,

$$c_0^2 = \frac{\mu_{0,b}}{m_b} = (1 - d/4)\frac{|\bar{\Delta}|^2}{m\epsilon_a}. \tag{235}$$

Also the number density $\bar{n}_{0,b}$ of condensed composite bosons,

$$\bar{n}_{0,b} = \frac{\Gamma(2-d/2)}{(4\pi)^{d/2}} m^{d/2} \epsilon_a^{d/2-2} |\bar{\Delta}|^2 \tag{236}$$

as well as the weak repulsive interaction $\lambda_{0,b}$ between the composite bosons,

$$\lambda_{0,b} = (4\pi)^{d/2} \frac{1 - d/4}{\Gamma(2-d/2)} \frac{\epsilon_a^{1-d/2}}{m^{d/2}} \tag{237}$$

follow immediately. We, in this way, have explicitly demonstrated that the BCS theory in the composite boson limit maps onto the Bogoliubov theory.

In concluding this subsection, we remark that in $d = 2$ various integrals we have encountered become elementary for arbitrary values of $\bar{\Delta}$. For example, the gap equation (218) reads explicitly in $d = 2$

$$\epsilon_a = \sqrt{\mu_0^2 + |\bar{\Delta}|^2} - \mu_0, \tag{238}$$

while the particle number equation (211) becomes

$$\bar{n} = \frac{m}{2\pi}\left(\sqrt{\mu_0^2 + |\bar{\Delta}|^2} + \mu_0\right). \tag{239}$$

Since in two dimensions,

$$\bar{n} = \frac{k_F^2}{2\pi} = \frac{m}{\pi}\epsilon_F, \tag{240}$$

with k_F and $\epsilon_F = k_F^2/2m$ the Fermi momentum and energy, the two equations can be combined to yield[57]

$$\frac{\epsilon_a}{\epsilon_F} = 2\frac{\sqrt{\mu_0^2 + |\bar{\Delta}|^2} - \mu_0}{\sqrt{\mu_0^2 + |\bar{\Delta}|^2} + \mu_0}. \tag{241}$$

The composite boson limit we have been discussing in this subsection is easily retrieved from these more general equations. Also note that in this limit, $\bar{n} = 2\bar{n}_{0,b}$, while the renormalization of the coupling constant takes the same form as for an interacting Bose gas

$$\frac{1}{\lambda_0} = \frac{1}{\kappa^\epsilon}\left(\frac{1}{\hat{\lambda}} - \frac{m}{4\pi\epsilon}\right); \tag{242}$$

cf. (76).

4.3 Dual Theory

We now turn to the dual description of a superconducting film at a finite temperature. We thereto minimally couple the model of Sec. 3.5 to a magnetic field described by the magnetic vector potential $\mathbf{A}$. For the time being we ignore vortices by setting the vortex gauge field $\varphi^{\rm P}$ to zero. The partition function of the system then reads

$$Z = \int {\rm D}\varphi \int {\rm D}\mathbf{A}\, \Xi(\mathbf{A}) \exp\left(-\beta \int_{\mathbf{x}} \mathcal{H}\right), \tag{243}$$

where $\Xi(\mathbf{A})$ is a gauge-fixing factor for the gauge field $\mathbf{A}$, and $\mathcal{H}$ is the Hamiltonian

$$\mathcal{H} = \frac{1}{2}\rho_{\rm s}\mathbf{v}_{\rm s}^2 + \frac{1}{2}(\nabla \times \mathbf{A})^2 \tag{244}$$

with

$$\mathbf{v}_{\rm s} = \frac{1}{m}(\nabla\varphi - 2e\mathbf{A}). \tag{245}$$

The double charge $2e$ stands for the charge of the Cooper pairs which are formed at the bulk transition temperature. The functional integral over φ in (243) is easily carried out with the result

$$Z = \int {\rm D}\mathbf{A}\, \Xi(\mathbf{A}) \exp\left\{-\frac{\beta}{2}\int_{\mathbf{x}} \left[(\nabla \times \mathbf{A})^2 + m_A^2 A_i \left(\delta_{ij} - \frac{\partial_i \partial_j}{\nabla^2}\right) A_j\right]\right\}, \tag{246}$$

where the last term, with $m_A^2 = 4e^2\rho_{\rm s}/m^2$, is a gauge-invariant, albeit nonlocal mass term for the gauge field generated by the Higgs mechanism. The number of degrees of freedom does not change in the process. This can be seen by noting that a gapless gauge field in two dimensions represents no physical degrees of freedom. (In Minkowski spacetime, this is easily understood by recognizing that in $1+1$ dimensions there is no transverse direction.) Before the Higgs mechanism took place, the system therefore contains only a single physical degree of freedom described by φ. This equals the number of degrees of freedom contained in (246).

We next introduce an auxiliary field $\tilde{h}$ to linearize the first term in (246),

$$\exp\left[-\frac{\beta}{2}\int_{\mathbf{x}}(\nabla \times \mathbf{A})^2\right] = \int {\rm D}\tilde{h} \exp\left[-\frac{1}{2\beta}\int_{\mathbf{x}} \tilde{h}^2 + i\int_{\mathbf{x}} \tilde{h}(\nabla \times \mathbf{A})\right], \tag{247}$$

and integrate out the gauge-field fluctuations [with a gauge-fixing term $(1/2\alpha)(\nabla \cdot \mathbf{A})^2$]. The result is a manifestly gauge-invariant expression for the partition function in terms of a massive scalar field $\tilde{h}$, representing the single degree of freedom contained in the theory:

$$Z = \int {\rm D}\tilde{h} \exp\left\{-\frac{1}{2\beta}\int_{\mathbf{x}}\left[\frac{1}{m_A^2}(\nabla\tilde{h})^2 + \tilde{h}^2\right]\right\}. \tag{248}$$

To understand the physical significance of this field, we note from (247) that it satisfies the field equation

$$\tilde{h} = i\beta\nabla \times \mathbf{A}. \tag{249}$$

That is, the fluctuating field $\tilde{h}$ represents the local magnetic induction, which is a scalar in two space dimensions. Equation (248) shows that the magnetic field has a finite penetration depth $\lambda_{\rm L} = 1/m_A$. In contrast to the original description where the functional integral runs over the gauge potential, the integration variable in (248) is the physical field.

We next include vortices. The penetration depth λ_L provides the system with an infrared cutoff so that a single magnetic vortex in the charged theory has a finite energy. Vortices can therefore be thermally activated. This is different from the superfluid phase of the neutral model, where the absence of an infrared cutoff permits only tightly bound vortex-antivortex pairs to exist. We expect, accordingly, the superconducting phase to describe a plasma of vortices, each carrying one magnetic flux quantum $\pm\pi/e$. The partition function now reads

$$Z = \sum_{N_+,N_-=0}^{\infty} \frac{z^{N_-+N_+}}{N_+!\,N_-!} \prod_\alpha \int_{\mathbf{x}^\alpha} \int \mathrm{D}\varphi \int \mathrm{D}\mathbf{A}\, \Xi(\mathbf{A}) \exp\left(-\beta \int_{\mathbf{x}} \mathcal{H}\right) \tag{250}$$

where z is the fugacity, i.e., the Boltzmann factor associated with the vortex core energy. The velocity appearing in the Hamiltonian (244) now includes the vortex gauge field

$$\mathbf{v}_s = \frac{1}{m}(\nabla\varphi - 2e\mathbf{A} - \varphi^{\mathrm{P}}). \tag{251}$$

This field can be shifted from the first to the second term in the Hamiltonian (244) by applying the transformation $\mathbf{A} \rightarrow \mathbf{A} - \varphi^{\mathrm{P}}/2e$. This results in the shift

$$\nabla \times \mathbf{A} \rightarrow \nabla \times \mathbf{A} - B^{\mathrm{P}}, \tag{252}$$

with the plastic field

$$B^{\mathrm{P}} = -\Phi_0 \sum_\alpha w_\alpha\, \delta(\mathbf{x} - \mathbf{x}^\alpha) \tag{253}$$

representing the magnetic flux density. Here, $\Phi_0 = \pi/e$ is the elementary flux quantum. Repeating the steps of the previous paragraph we now obtain instead of (248)

$$Z = \sum_{N_\pm=0}^{\infty} \frac{z^{N_++N_-}}{N_+!\,N_-!} \prod_\alpha \int_{\mathbf{x}^\alpha} \int \mathrm{D}\tilde{h} \exp\left\{-\frac{1}{2\beta}\int_{\mathbf{x}} \left[\frac{1}{m_A^2}(\nabla\tilde{h})^2 + \tilde{h}^2\right] + i\int_{\mathbf{x}} B^{\mathrm{P}}\tilde{h}\right\}, \tag{254}$$

where $\tilde{h}$ represents the physical local magnetic induction h:

$$\tilde{h} = i\beta(\nabla \times \mathbf{A} - B^{\mathrm{P}}) = i\beta h. \tag{255}$$

The field equation for $\tilde{h}$ obtained from (254) yields for the magnetic induction:

$$-\nabla^2 h + m_A^2 h = m_A^2 B^{\mathrm{P}}, \tag{256}$$

which is the familiar equation in the presence of magnetic vortices.

The last term in (254) shows that the charge g, with which a magnetic vortex couples to the fluctuating $\tilde{h}$-field, is the product of an elementary flux quantum (contained in the definition of B^{P}) and the inverse penetration depth $m_A = 1/\lambda_L$,

$$g = \Phi_0 m_A. \tag{257}$$

For small fugacities, the summation indices N_+ and N_- can be restricted to the values 0, 1 and we arrive at the partition function of the massive sine-Gordon model[58]

$$Z = \int \mathrm{D}\tilde{h} \exp\left(-\int_{\mathbf{x}} \left\{\frac{1}{2\beta}\left[\frac{1}{m_A^2}(\nabla\tilde{h})^2 + \tilde{h}^2\right] - 2z\cos\left(\Phi_0\tilde{h}\right)\right\}\right). \tag{258}$$

This is the dual formulation of a two-dimensional superconductor. The magnetic vortices of unit winding number $w_\alpha = \pm 1$ have turned the otherwise free theory (248) into an interacting one.

The final form (258) demonstrates the rationales for going over to a dual theory. First, it is a formulation directly in terms of a physical field representing the local magnetic induction. There is no redundancy in this description and therefore no gauge invariance. Second, the magnetic vortices are accounted for in a nonsingular fashion. This is different from the original formulation of the two-dimensional superconductor where the local magnetic induction is the curl of an unphysical gauge potential **A**, and where the magnetic vortices appear as singular objects.

Up to this point we have discussed a genuine two-dimensional superconductor. As a model to describe superconducting films this is, however, not adequate. The reason is that the magnetic interaction between the vortices takes place mostly not through the film but through free space surrounding the film where the photon is gapless. This situation is markedly different from a superfluid film. The interaction between the vortices there is mediated by the Kosterlitz–Thouless mode which is confined to the film. A genuine two-dimensional theory therefore gives a satisfactory description of a superfluid film.

To account for the fact that the magnetic induction is not confined to the film and can roam in outer space, the field equation (256) is modified in the following way[59,60]

$$-\nabla^2 h(\mathbf{x}_\perp, x_3) + \frac{1}{\lambda_\perp}\delta_d(x_3) h(\mathbf{x}_\perp, x_3) = \frac{1}{\lambda_\perp}\delta_d(x_3) B^{\mathrm{P}}(\mathbf{x}). \tag{259}$$

Here, $1/\lambda_\perp = dm_A^2$, with d denoting the thickness of the superconducting film, is an inverse length scale, $\mathbf{x}_\perp$ denotes the coordinates in the plane, h the component of the induction field perpendicular to the film, and $\delta_d(x_3)$ is a smeared delta function of thickness d along the x_3-axis

$$\delta_d(x_3) \begin{cases} = 0 & \text{for} \quad |x_3| > d/2 \\ \neq 0 & \text{for} \quad |x_3| \leq d/2 \end{cases}. \tag{260}$$

The reason for including the smeared delta function at the right-hand side of (259) is that the vortices are confined to the film. The delta function in the second term at the left-hand side is included because this term is generated by screening currents which are also confined to the film.

To be definite, we consider a single magnetic vortex centered at the origin. The induction field found from (259) reads

$$h(\mathbf{x}_\perp, 0) = \frac{\Phi_0}{2\pi} \int_0^\infty \mathrm{d}q \frac{q}{1 + 2\lambda_\perp q} J_0(q|\mathbf{x}_\perp|), \tag{261}$$

with J_0 the zeroth Bessel function of the first kind. At small distances from the vortex core ($\lambda_\perp q \gg 1$)

$$h(\mathbf{x}_\perp, 0) \sim \frac{\Phi_0}{4\pi\lambda_\perp |\mathbf{x}_\perp|}, \tag{262}$$

while far away ($\lambda_\perp q \ll 1$)

$$h(\mathbf{x}_\perp, 0) \sim \frac{\Phi_0 \lambda_\perp}{\pi |\mathbf{x}_\perp|^3}. \tag{263}$$

This last equation shows that the field does not exponentially decay as would be the case in a genuine two-dimensional system. The reason for the long range is that most of the magnetic interaction takes place in the free space outside the film where the photon is gapless. If, as is often the case, the length $\lambda_\perp = 1/dm_A^2$ is much larger than the sample size, it can be effectively set to infinity. In this limit, the effect of the magnetic interaction diminishes, as can be seen from (262), and the vortices behave as in a superfluid film. One therefore expects a superconducting film also to undergo a Kosterlitz–Thouless transition at a temperature $T_{\rm KT}$ characterized by an unbinding of vortex-antivortex pairs. The first experiment to study this possibility has been carried out in Ref. 61. Because the transition temperature $T_{\rm KT}$ is well below the bulk temperature T_c where the Cooper pairs form, the energy gap of the fermions remains finite at the critical point.[62] This prediction has been corroborated by experiments performed by Hebard and Paalanen on superconducting films.[63] For temperatures $T_{\rm KT} \leq T \leq T_c$, there is a plasma of magnetic vortices which disorder the superconducting state. At $T_{\rm KT}$ vortices and antivortices bind into pairs and algebraic long-range order sets in.

5. FRACTIONAL QUANTIZED HALL EFFECT

The nonrelativistic $|\phi|^4$-theory describing an interacting Bose gas is also of importance for the description of the fractional quantized Hall effect (FQHE). As a function of the applied magnetic field, this two-dimensional system undergoes a zero-temperature transition between a so-called quantum Hall liquid, where the Hall conductance is quantized in odd fractions of $e^2/2\pi$, or, reinstalling Planck's constant, e^2/h, and an insulating phase. Here, the nonrelativistic $|\phi|^4$-theory describes – after coupling to a Chern–Simons term – the original electrons bound to an odd number of flux quanta. The Hall liquid is in this picture characterized by a condensate of composite particles.

5.1 Chern–Simons–Ginzburg–Landau Theory

The fractional quantized Hall effect (FQHE) is the hallmark of a new, intrinsically two-dimensional condensed-matter state – the quantum Hall liquid. Many aspects of this state are well understood in the framework of the quantum-mechanical picture developed by Laughlin·64 A considerable effort has nevertheless been invested in formulating an effective field theory which captures the essential low-energy, small-momentum features of the liquid. A similar approach in the context of superconductors has proven most successful. Initially, only the phenomenological model proposed by Ginzburg and Landau[65] in 1950 has been known here. Most of the fundamental properties of the superconducting state such as the superconductivity – the property that has given this condensed-matter state its name, the Meissner effect, the magnetic flux quantization, the Abrikosov flux lattice, and the Josephson effect, can be explained by the model. The microscopic theory was given almost a decade later by Bardeen, Cooper, and Schrieffer.[47] Shortly here after, Gorkov[66] made the connection between the two approaches by deriving the Ginzburg-Landau model from the microscopic BCS theory, thus giving the phenomenological model the status of an effective field theory.

A first step towards an effective field theory of the quantum Hall liquid has been taken by Girvin and MacDonald[67] and has been developed further by Zhang, Hansson and Kivelson[68], who also gave an explicit construction starting from a microscopic Hamiltonian. Their formulation (for a review see Ref. 69) incorporates time dependence which is important for the study of quantum phase transitions.

An essential ingredient for obtaining an effective theory of FQHE has been the identification by Girvin and MacDonald[67] of a bosonic operator ϕ exhibiting (algebraic) off-diagonal long-range order of a type known to exist in two-dimensional bosonic superfluids. They have argued that this field should be viewed as an order parameter in terms of which the effective field theory should be formulated. To account for the incompressibility of the quantum Hall liquid they have suggested to minimally couple ϕ to a so-called statistical gauge field $(a_0, \mathbf{a})$ governed solely by a Chern-Simons term

$$\mathcal{L}_{\rm CS} = \frac{1}{2} e^2 \theta \partial_0 \mathbf{a} \times \mathbf{a} - e^2 \theta a_0 \nabla \times \mathbf{a}, \tag{264}$$

with $\nabla \times \mathbf{a}$ the statistical magnetic field and θ a constant. As we will see below, the gapless Bogoliubov spectrum of the neutral system changes as a result of this coupling into one with an energy gap[68], thus rendering the charged system incompressible.

Because of the absence of a kinetic term (the usual Maxwell term), the statistical gauge field does not represent a physical degree of freedom. In a relativistic setting, a Maxwell term is usually generated by quantum corrections so that the statistical gauge field becomes dynamical at the quantum level. The quantum theory then differs qualitatively from the classical theory. On the other hand, as we shall see below, this need not be the case in a nonrelativistic setting. That is to say, the *Ansatz* of the absence of a Maxwell term here is not necessarily obstructed by quantum corrections.

The effective theory of the quantum Hall liquid is given by the so-called Chern-Simons-Ginzburg-Landau (CSGL) Lagrangian[68]

$$\mathcal{L} = i\phi^* D_0 \phi - \frac{1}{2m} |\mathbf{D}\phi|^2 + \mu_0 |\phi|^2 - \lambda_0 |\phi|^4 + \mathcal{L}_{\rm CS}. \tag{265}$$

The covariant derivatives $D_0 = \partial_0 + ieA_0 + iea_0$ and $\mathbf{D} = \nabla - ie\mathbf{A} - ie\mathbf{a}$ give a minimal coupling to the applied magnetic and electric fields, described by the gauge field $(A_0, \mathbf{A})$ and also to the statistical gauge field. For definiteness we will assume that our two-dimensional sample is perpendicular to the applied magnetic field, defining the z-direction, and we choose the electric field to point in the x-direction. The charged field ϕ represents the Girvin-MacDonald order parameter describing the original electrons bound to an odd number $2l + 1$ of flux quanta. To see that it indeed does, let us consider the field equation for a_0:

$$|\phi|^2 = -e\theta \nabla \times \mathbf{a}. \tag{266}$$

The simplest solution of the CSGL Lagrangian is the uniform mean-field solution

$$|\phi|^2 = \bar{n}, \quad \mathbf{a} = -\mathbf{A}, \quad a_0 = -A_0 = 0, \tag{267}$$

where $\bar{n}$ indicates the constant fermion number density. The statistical gauge field is seen to precisely cancel the applied field. The constraint equation (266) then becomes

$$\bar{n} = e\theta H, \tag{268}$$

with H the applied magnetic field. Now, if we choose $\theta^{-1} = 2\pi(2l + 1)$, it follows on integrating this equation that, as required, with every electron there is associated $2l+1$ flux quanta:

$$N = \frac{1}{2l+1} N_{\otimes}, \tag{269}$$

where $N_{\otimes} = \Phi/\Phi_0$, with $\Phi = \int_{\mathbf{x}} H$ the magnetic flux, indicates the number of flux quanta. Equation (268) implies an odd-denominator filling factor ν_H which is defined by

$$\nu_H = \frac{\bar{n}}{H/\Phi_0} = \frac{1}{2l+1}. \tag{270}$$

The coupling constant $\lambda_0\,(> 0)$ in (265) is the strength of the repulsive contact interaction between the composite particles, and μ_0 is a chemical potential introduced to account for a finite number density of composite particles.

It is well known from anyon physics that the inclusion of the Chern-Simons term changes the statistics of the field ϕ to which the statistical gauge field is coupled[70]. If one composite particle circles another, it will pick up an additional Aharonov–Bohm factor, representing the change in statistics. The binding of an odd number of flux quanta changes the fermionic character of the electrons into a bosonic one for the composite particles, allowing them to Bose condense. The algebraic off-diagonal long-range order of a quantum Hall liquid can in this picture be understood as resulting from this condensation. Conversely, a flux quantum carries $1/(2l+1)$th of an electron's charge[64], and also $1/(2l+1)$th of an electron's statistics.[71]

The defining phenomenological properties of a quantum Hall liquid are easily shown to be described by the CSGL theory.[68,69] From the lowest-order expression for the induced electromagnetic current one finds

$$ej_i = \frac{\delta \mathcal{L}}{\delta A_i} = -\frac{\delta \mathcal{L}_\phi}{\delta a_i} = \frac{\delta \mathcal{L}_{\rm CS}}{\delta a_i} = -e^2\theta\epsilon_{ij}(\partial_0 a_j - \partial_j a_0) = e^2\theta\epsilon_{ij}E_j, \tag{271}$$

with $\mathbf{E}$ the applied electric field and where we have written the Lagrangian (265) as a sum $\mathcal{L} = \mathcal{L}_\phi + \mathcal{L}_{\rm CS}$. It follows that the Hall conductance σ_{xy} is quantized in odd fractions of $e^2/2\pi$, or, reinstalling Planck's constant, e^2/h. This result can also be understand in an intuitive way as follows. Since the composite particles carry a charge e, the applied electric field gives rise to an electric current

$$I = e\frac{\mathrm{d}N}{\mathrm{d}t} \tag{272}$$

in the direction of $\mathbf{E}$, i.e., the x-direction. This is not the end of the story because the composite objects carry in addition to the electric charge also $2l+1$ flux quanta. When the Goldstone field φ encircles $2l+1$ flux quanta, it picks up a factor 2π for each of them

$$\oint_\Gamma \nabla \cdot \varphi = 2\pi(2l+1). \tag{273}$$

Now, consider two points across the sample from each other. Let the phase of these points initially be equal. As a composite particle moves downstream, and crosses the line connecting the two points, the relative phase $\Delta\varphi$ between them changes by $2\pi(2l+1)$. This phase slippage[15] leads to a voltage drop across the sample given by

$$V_{\rm H} = \frac{1}{e}\partial_0 \Delta\varphi = (2l+1)\Phi_0 \frac{\mathrm{d}N}{\mathrm{d}t}, \tag{274}$$

where the first equation can be understood by recalling that due to the minimal coupling $\partial_0\varphi \to \partial_0\varphi + eA_0$. For the Hall resistance we thus obtain the expected value

$$\rho_{xy} = \frac{V_{\rm H}}{I} = (2l+1)\frac{2\pi}{e^2}. \tag{275}$$

If the CSGL theory is to describe an incompressible liquid, the spectrum of the single-particle excitations must have a gap. Without the coupling to the statistical gauge field, the spectrum is given by the gapless Bogoliubov spectrum (57). To obtain the single-particle spectrum of the coupled theory, we integrate out the statistical gauge field. The integration over a_0 has been shown to yield the constraint (266) which in the Coulomb gauge $\nabla \cdot \mathbf{a} = 0$ is solved by

$$a_i = \frac{1}{e\theta}\epsilon_{ij}\frac{\partial_j}{\nabla^2}|\phi|^2. \tag{276}$$

The integration over the remaining components of the statistical gauge field is now simply performed by substituting (276) back into the Lagrangian. The only nonzero contribution arises from the term $-e^2|\phi|^2\mathbf{a}^2/2m$. The spectrum of the charged system acquires as a result an energy gap ω_c

$$E(\mathbf{k}) = \sqrt{\omega_c^2 + \epsilon^2(\mathbf{k}) + 2\mu_0\epsilon(\mathbf{k})}, \tag{277}$$

with $\omega_c = \mu_0/2\theta m\lambda_0$. To lowest order, the gap equals the cyclotron frequency of a free charge e in a magnetic field H

$$\omega_c = \frac{\bar{n}}{\theta m} = \frac{eH}{m}. \tag{278}$$

The presence of this energy gap results in a dissipationless flow with $\sigma_{xx} = 0$.

These facts show that the CSGL theory captures the essentials of a quantum Hall liquid. Given this success, it is tempting to investigate if the theory can also be employed to describe the field-induced Hall-liquid-to-insulator transitions. This will be done in Sec. 6.3. It should however be borne in mind that both the $1/|\mathbf{x}|$-Coulomb potential as well as impurities should be incorporated into the theory in order to obtain a realistic description of FQHE. The repulsive Coulomb potential is believed to play a decisive role in the formation of the composite particles, while the impurities are responsible for the width of the Hall plateaus. As the magnetic field moves away from the magic filling factor, magnetic vortices will materialize in the system to make up the difference between the applied field and the magic field value. In the presence of impurities, these defects get pinned and do not contribute to the resistivities, so that both σ_{xx} and σ_{xy} are unchanged. Only if the difference becomes too large, the system will revert to an other quantum Hall state with a different filling factor.

6. QUANTUM PHASE TRANSITIONS

This section is devoted to continuous phase transitions at the absolute zero of temperature; so-called quantum phase transitions. Unlike in classical phase transitions taking place at a finite temperature and in equilibrium, time plays an important role in quantum phase transitions. Put differently, whereas the critical behavior of classical 2nd-order phase transitions is governed by thermal fluctuations, that of 2nd-order quantum transitions is controlled by quantum fluctuations. These transitions, which have attracted much attention in recent years (for an introductory review, see Ref. 4), are triggered by varying not the temperature, but some other parameter in the system, like the applied magnetic field, the charge carrier density, or the disorder strength. The quantum phase transitions we will be discussing here are all dominated by phase fluctuations.

6.1 Scaling

The natural language to describe quantum phase transitions is quantum field theory. In addition to a diverging correlation length ξ, quantum phase transitions also have a diverging correlation time ξ_t. They indicate, respectively, the distance and time period over which the order parameter characterizing the transition fluctuates coherently. The way the diverging correlation time relates to the diverging correlation length,

$$\xi_t \sim \xi^z, \tag{279}$$

defines the so-called dynamic exponent z. It is a measure for the asymmetry between the time and space directions and tells us how long it takes for information to propagate across a distance ξ. The traditional scaling theory of classical 2nd-order phase transitions, first put forward by Widom[72], is easily extended to include the time dimension[29] because the relation (279) implies the presence of only one independent diverging scale. Let $\delta = K - K_c$, with K the parameter that drives the phase transition, measures the distance from the critical value K_c. A physical observable at the absolute zero of temperature $O(k_0, |\mathbf{k}|, K)$ can in the critical region close to the transition be written as

$$O(k_0, |\mathbf{k}|, K) = \xi^{d_O} \mathcal{O}(k_0 \xi_t, |\mathbf{k}|\xi), \qquad (T = 0), \tag{280}$$

where d_O is the dimension of the observable O. The right-hand side does not depend explicitly on K; only implicitly through ξ and ξ_t. The closer one approaches the critical value K_c, the larger the correlation length and time become.

Since a physical system is always at some finite temperature, we have to investigate how the scaling law (280) changes when the temperature becomes nonzero. The easiest way to include the temperature in a quantum field theory is to go over to imaginary time $\tau = it$, with τ restricted to the interval $0 \leq \tau \leq \beta$. The temporal dimension becomes thus of finite extend. The critical behavior of a phase transition at a finite temperature is still controlled by the quantum critical point provided $\xi_t < \beta$. If this condition is fulfilled, the system will not see the finite extent of the time dimension. This is what makes quantum phase transitions experimentally accessible. Instead of the zero-temperature scaling (280), we now have the finite-size scaling

$$O(k_0, |\mathbf{k}|, K, \beta) = \beta^{d_O/z} \mathcal{O}(k_0 \beta, |\mathbf{k}|\beta^{1/z}, \beta/\xi_t), \qquad (T \neq 0). \tag{281}$$

The distance to the quantum critical point is measured by the ratio $\beta/\xi_t \sim |\delta|^{z\nu}/T$.

6.2 Repulsively Interacting Bosons

The first quantum phase transition we wish to investigate is the superfluid-to-Mott-insulating transition of interacting bosons in the absence of impurities.[73] The transition is described by the nonrelativistic $|\phi|^4$-theory (45), which becomes critical at the absolute zero of temperature at some (positive) value μ_c of the renormalized chemical potential. The Mott insulating phase is destroyed and makes place for the superfluid phase as μ increases. Whereas in the superfluid phase the single-particle (Bogoliubov) spectrum is gapless and the system compressible, the single-particle spectrum of the insulating phase has an energy gap and the compressibility κ vanishes here.

The nature of the insulating phase can be best understood by putting the theory on a lattice. The lattice model is defined by the Hamiltonian

$$H_{\rm H} = -t \sum_j (a_j^\dagger a_{j+1} + \text{h.c.}) + \sum_j (-\mu_{\rm L} \hat{n}_j + U \hat{n}_j^2), \tag{282}$$

where the sum $\sum_j$ is over all lattice sites. The operator $a_j^\dagger$ creates a boson at site j and $\hat{n}_j = a_j^\dagger a_j$ is the particle number operator at that site; t is the hopping parameter, U

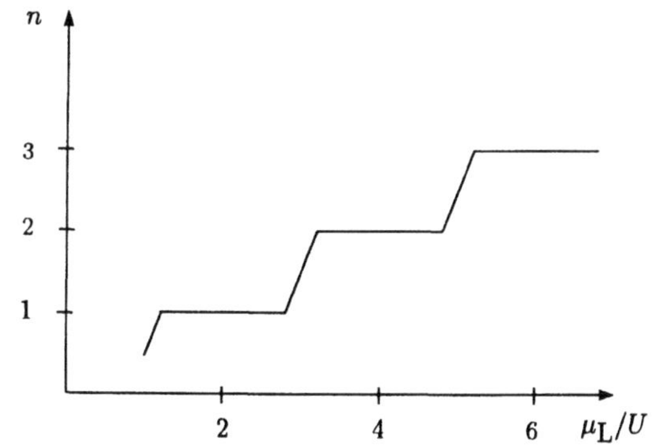

Figure 4. A schematic representation of the average number n of particles per site as a function of the chemical potential μ_L at some finite value of the hopping parameter $t < t_c$.

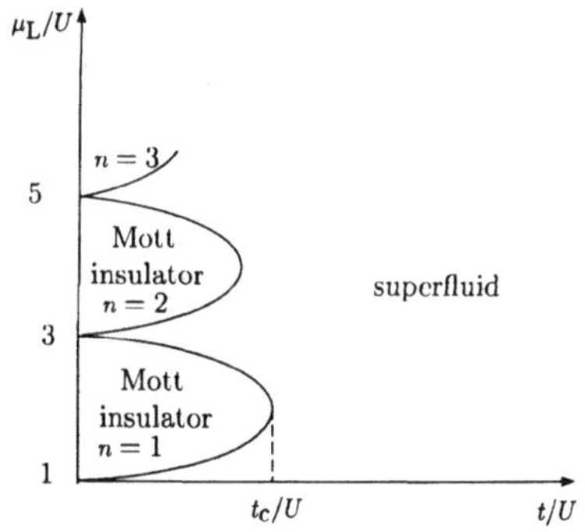

Figure 5. A schematic representation of the phase diagram of the lattice model (282) at the absolute zero of temperature.[73]

the interparticle repulsion, and μ_L is the chemical potential on the lattice. The zero-temperature phase diagram is as follows.[73] In the limit $t/U \to 0$, each site is occupied by an integer number n of bosons which minimizes the on-site energy (see Fig. 4)

$$\epsilon(n) = -\mu_L n + Un^2. \tag{283}$$

It follows that within the interval $2n - 1 < \mu_L/U < 2n + 1$, each site is occupied by exactly n bosons. When the chemical potential is negative, $n = 0$. The intervals become smaller when t/U increases. Within such an interval, where the particles are pinned to the lattice sites, the single-particle spectrum has an energy gap, and the system is in the insulating phase with zero compressibility, $\kappa = n^{-2}\partial n/\partial \mu_L = 0$. Outside these intervals, the particles delocalize and can hop through the lattice. Being at zero temperature, the delocalized bosons condense in a superfluid state. The single-particle spectrum is gapless here and the system compressible ($\kappa \neq 0$).

As t/U increases, the gap in the single-particle spectrum as well as the width of the intervals decrease and eventually vanish at some critical value t_c. For values $t > t_c$ of the hopping parameter, the superfluid phase is the only phase present (see Fig. 5).

The continuum model (45), with $\mu > \mu_c$ describes the condensed delocalized lattice bosons which are present when the density deviates from integer values (see Fig. 4). In the limit $\mu \to \mu_c$ from above, the number of delocalized bosons decreases and eventually becomes zero at the phase boundary $\mu = \mu_c$ between the superfluid and insulating phases.

Various quantum phase transitions belong to the universality class defined by the zero-density transition of repulsively interacting bosons. For example, itinerant quantum antiferromagnets[74-76] as well as lower-dimensional (clean) superconductors belong to this universality class. As we have seen in Sec. 4.2, Cooper pairs become tightly bound composite particles in the strong-coupling limit, which are described by the nonrelativistic $|\phi|^4$-theory with a weak repulsive interaction. For $\mu > \mu_c$, the field ϕ now describes the condensed delocalized Cooper pairs. When the chemical potential decreases, the condensate diminishes, and the system again becomes insulating[62] for $\mu < \mu_c$. By continuity, we expect also the superconductor-to-insulator transition of a (clean) weakly interacting BCS superconductor to be in this universality class. The restriction to lower dimensions is necessary for two different reasons. First, only for $d \leq 2$ the penetration depth is sufficiently large [see, for example, below Eq. (259)], so that it is appropriate to work in the limit $\lambda_L \to \infty$ with no fluctuating gauge field.[77] Second, in lower dimensions, the energy gap which the fermionic excitations face remains finite at the critical point, so that it is appropriate to ignore these degrees of freedom. Moreover, since also the coherence length remains finite at the critical point, the Cooper pairs look like point particles on the scale of the diverging correlation length associated with the phase fluctuations, even in the weak-coupling limit.[62]

In the preceding section, we have argued that the nonrelativistic $|\phi|^4$-theory is also of importance for the description of the fractional quantized Hall effect (FQHE), where it describes—after coupling to the Chern-Simons term—the original electrons bound to an odd number of flux quanta. As a function of the applied magnetic field, this two-dimensional system undergoes a zero-temperature transition between a quantum Hall liquid, where the Hall conductance is quantized in odd fractions of $e^2/2\pi$, and an insulating phase. The Hall liquid corresponds to the phase with $\mu > \mu_c$, while the other phase again describes the insulating phase.

It should be noted however that in most of the applications of the nonrelativistic $|\phi|^4$-theory mentioned here, impurities play an important role; this will be the main subject of Sec. 6.4.

The critical properties of the zero-density transition of the nonrelativistic $|\phi|^4$-theory have been first studied by Uzunov.[24] To facilitate the discussion let us make use of the fact that in nonrelativistic theories the mass is – as far as critical phenomena are concerned – an irrelevant parameter which can be transformed away. This transformation changes, however, the scaling dimensions of the ϕ-field and the coupling constant which is of relevance to the renormalization-group theory. The engineering dimensions become

$$[\mathbf{x}] = -1, \quad [t] = -2, \quad [\mu_0] = 2, \quad [\lambda_0] = 2 - d, \quad [\phi] = \frac{1}{2}d, \tag{284}$$

with d denoting the number of space dimensions. In two space dimensions the coupling constant λ_0 is dimensionless, showing that the $|\phi|^4$-term is a marginal operator, and $d_c = 2$ is the upper critical space dimension. Uzunov has shown that below the upper critical dimension there appears a non-Gaussian infrared-stable (IR) fixed point. He has computed the corresponding critical exponents to all orders in perturbation theory and shown them to have Gaussian values, $\nu = 1/2$, $z = 2$, $\eta = 0$. Here, ν characterizes the divergence of the correlation length, z is the dynamic exponent, and η is the correlation-function exponent which determines the anomalous dimension of the field ϕ. The unexpected conclusion that a non-Gaussian fixed point has nevertheless Gaussian exponents is rooted in the analytic structure of the nonrelativistic propagator at zero

Figure 6. A closed oriented loop.

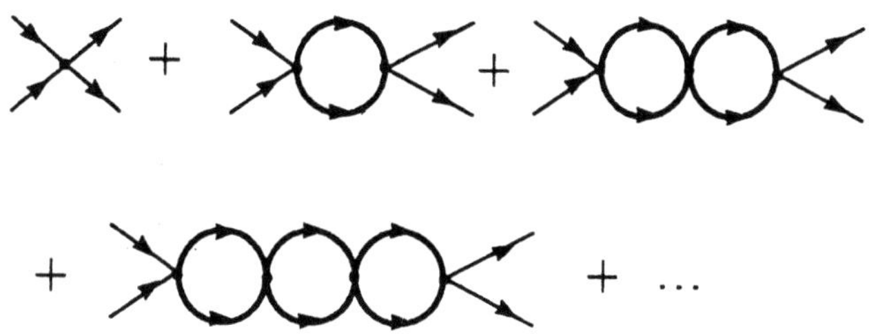

Figure 7. Ring diagrams renormalizing the vertex function of the neutral $|\phi|^4$-theory.

bare chemical potential ($\mu_0 = 0$):

$$\xrightarrow{k_\mu} \quad = G(k) = \frac{i e^{i k_0 \eta}}{k_0 - \frac{1}{2}\mathbf{k}^2 + i\eta}, \tag{285}$$

where, as before, η is a small positive constant that has to be taken to zero after the loop integrations over the energies have been carried out. The factor $\exp(ik_0\eta)$ is an additional convergence factor typical for nonrelativistic theories, which is needed for Feynman diagrams involving only one ϕ-propagator. The rule $k_0 \to k_0 + i\eta$ in (285) expresses the fact that in this nonrelativistic theory particles propagate only forward in time. In diagrams involving loops with more than one propagator, the integrals over the loop energy are convergent and can be evaluated by a contour integration with the contour closed in either the upper or the lower half plane. If a diagram contains a loop which has all its poles in the same half plane, it consequently vanishes. Pictorially, such a loop has all its arrows, representing the Green functions contained in the loop, oriented in a clockwise or anticlockwise direction[78] (see Fig. 6). We will refer to them as closed oriented loops. Owing to this property most diagrams are zero. In particular, all self-energy diagrams vanish. The only surviving ones are the so-called ring diagrams which renormalize the vertex (see Fig. 7). Because this class of diagrams constitute a geometric series, the one-loop result is already exact. The vertex renormalization leads to a non-Gaussian fixed point in $d < 2$, while the vanishing of all the self-energy diagrams asserts that the exponents characterizing the transition are not affected by quantum fluctuations and retain their Gaussian values.[24] These results have been confirmed by numerical simulations[79] $d = 1$ and also by general scaling arguments.[80,73]

To understand the scaling arguments, let us consider the two terms in the effective theory (66) quadratic in the Goldstone field φ with m effectively set to 1:

$$\mathcal{L}^{(2)}_{\text{eff}} = -\frac{1}{2}\rho_{\text{s}}(\nabla\varphi)^2 + \frac{1}{2}\bar{n}^2\kappa(\partial_0\varphi)^2. \tag{286}$$

We have written this in the most general form.[80] The coefficient ρ_s is the superfluid mass density which in the presence of, for example, impurities does not equal $m\bar{n}$—even at the absolute zero of temperature. The other coefficient,

$$\bar{n}^2\kappa = \frac{\partial \bar{n}}{\partial \mu} = \lim_{k\to 0} \Pi_{00}(0,\mathbf{k}), \tag{287}$$

with Π_{00} the (0 0)-component of the full polarization tensor (63), involves the full compressibility and particle number density of the system at rest. This is because the chemical potential is according to (67) represented in the effective theory by $\mu = -\partial_0\varphi$ and

$$\frac{\partial^2 \mathcal{L}_{\text{eff}}}{\partial \mu^2} = \bar{n}^2\kappa. \tag{288}$$

Equation (286) leads to the general expression for the sound velocity

$$c^2 = \frac{\rho_s}{\bar{n}^2\kappa} \tag{289}$$

at the absolute zero of temperature.

Let $\delta \propto \mu - \mu_c$ denote the distance from the phase transition, so that $\xi \sim |\delta|^{-\nu}$. Now, on the one hand, the singular part of the free energy density f_{sing} arises from the low-energy, long-wavelength fluctuations of the Goldstone field. (Here, we have adopted the common practice of using the symbol f for the density Ω/V and of referring to it as the free energy density.) The ensemble averages give

$$\langle(\nabla\varphi)^2\rangle \sim \xi^{-2}, \quad \langle(\partial_0\varphi)^2\rangle \sim \xi_t^{-2} \sim \xi^{-2z}. \tag{290}$$

On the other hand, dimensional analysis shows that the singular part of the free energy density scales near the transition as

$$f_{\text{sing}} \sim \xi^{-(d+z)}. \tag{291}$$

Combining these hyperscaling arguments, we arrive at the following conclusions:

$$\rho_s \sim \xi^{-(d+z-2)}, \quad \bar{n}^2\kappa \sim \xi^{-(d-z)} \sim |\delta|^{(d-z)\nu}. \tag{292}$$

The first conclusion is consistent with the universal jump (169) predicted by Nelson and Kosterlitz[45] which corresponds to taking $z = 0$ and $d = 2$. Since $\xi \sim |\delta|^{-\nu}$, f_{sing} can also be directly differentiated with respect to the chemical potential to yield the singular part of the compressibility

$$\bar{n}^2\kappa_{\text{sing}} \sim |\delta|^{(d+z)\nu-2}. \tag{293}$$

Fisher and Fisher[80] continued to argue that there are two alternatives. Either $\kappa \sim \kappa_{\text{sing}}$, implying $z\nu = 1$; or the full compressibility κ is constant, implying $z = d$. The former is consistent with the Gaussian values $\nu = 1/2$, $z = 2$ found by Uzunov[24] for the pure case in $d < 2$. The latter is believed to apply to repulsively interacting bosons in a random medium. These remarkable simple arguments thus predict the exact value $z = d$ for the dynamic exponent in this case.

For later reference, let us consider the charged case and calculate the conductivity σ. The only relevant term for this purpose is the first one in (286) with $\nabla\varphi$ replaced by $\nabla\varphi - e\mathbf{A}$. We allow the superfluid mass density to vary in space and time. The term in the action quadratic in $\mathbf{A}$ then becomes in the Fourier representation

$$S_\sigma = -\frac{1}{2}e^2 \int_{k_0,\mathbf{k}} \mathbf{A}(-k_0,-\mathbf{k})\rho_s(k_0,\mathbf{k})\mathbf{A}(k_0,\mathbf{k}). \tag{294}$$

The electromagnetic current,

$$\mathbf{j}(k_0, \mathbf{k}) = \frac{\delta S_\sigma}{\delta \mathbf{A}(-k_0, -\mathbf{k})} \tag{295}$$

obtained from this action can be written as

$$\mathbf{j}(k_0, \mathbf{k}) = \sigma(k_0, \mathbf{k}) \mathbf{E}(k_0, \mathbf{k}), \tag{296}$$

with the conductivity

$$\sigma(k) = i e^2 \frac{\rho_s(k)}{k_0} \tag{297}$$

essentially given by the superfluid mass density divided by k_0, where it should be remembered that the mass m is effectively set to 1 here.

The above hyperscaling arguments have been extended by Fisher, Grinstein, and Girvin[77] to include the $1/|\mathbf{x}|$-Coulomb potential. The quadratic terms in the effective theory (117) may be cast in the general form

$$\mathcal{L}_{\text{eff}}^{(2)} = \frac{1}{2} \left(\rho_s \mathbf{k}^2 - \frac{|\mathbf{k}|^{d-1}}{\hat{e}^2} k_0^2 \right) |\varphi(k)|^2, \tag{298}$$

where $\hat{e}$ is the renormalized charge. From (117) we find that to the lowest order:

$$\hat{e}^2 = 2^{d-1} \pi^{(d-1)/2} \Gamma\left[(d-1)/2\right] e_0^2. \tag{299}$$

The renormalized charge is connected to the (0 0)-component of the full polarization tensor (63) via

$$\hat{e}^2 = \lim_{|\mathbf{k}| \to 0} \frac{|\mathbf{k}|^{d-1}}{\Pi_{00}(0, \mathbf{k})}. \tag{300}$$

A simple hyperscaling argument like the ones given above shows that near the transition, the renormalized charge scales as

$$\hat{e}^2 \sim \xi^{1-z}. \tag{301}$$

They then argued that in the presence of random impurities this charge is expected to be finite at the transition so that $z = 1$. This again is an exact result which replaces the value $z = d$ of the neutral system.

We have seen that $d_c = 2$ is the upper critical dimension of the nonrelativistic $|\phi|^4$-theory. Dimensional analysis shows that for an interaction term of the form

$$\mathcal{L}_i = -g_0 |\phi|^{2k} \tag{302}$$

the upper critical dimension is

$$d_c = \frac{2}{k-1}. \tag{303}$$

The two important physical cases are $d_c = 2$, $k = 2$ and $d_c = 1$, $k = 3$, while $d_c \to 0$ when $k \to \infty$. For space dimensions $d > 2$ only the quadratic term, $|\phi|^2$, is relevant so that here the critical behavior is well described by the Gaussian theory.

In the corresponding relativistic theory, the scaling dimensions of t and $\mathbf{x}$ are, of course, equal $[t] = [\mathbf{x}] = -1$ and $[\phi] = (d-1)/2$. This leads to different upper critical (space) dimensions, viz.,

$$d_c = \frac{k+1}{k-1} = \frac{2}{k-1} + 1, \tag{304}$$

Table 1. The upper critical space dimension d_c of a nonrelativistic (NR) and a relativistic (R) quantum theory with a $|\phi|^{2k}$ interaction term.

k	d_c(NR)	d_c(R)
2	2	3
3	1	2
∞	0	1

instead of (303). The two important physical cases are here $d_c = 3$, $k = 2$ and $d_c = 2$, $k = 3$, while $d_c \to 1$ when $k \to \infty$. On comparing with the nonrelativistic results, we see that the nonrelativistic theory has an upper critical space dimension which is one lower than that of the corresponding relativistic theory (see Table 1). Heuristically, this can be understood by noting that in a nonrelativistic context the time dimension counts twice in that it has a scaling dimension twice that of a space dimension [see Eq. (284)], thereby increasing the *effective* spacetime dimensionality by one. From this analysis it follows that for a given number of space dimensions the critical properties of a nonrelativistic theory are unrelated to those of the corresponding relativistic extension.

In closing this subsection we recall that in a one-dimensional relativistic theory, corresponding to the lowest upper critical dimension ($d_c = 1$), a continuous symmetry cannot be spontaneously broken. However, the theory can nevertheless have a phase transition of the Kosterlitz–Thouless type. Given the connection between the relativistic and nonrelativistic theories discussed above, it seems interesting to study the nonrelativistic theory at zero space dimension ($d = 0$) to see if a similar rich phenomenon as in the lower critical dimension of the relativistic theory occurs here. This may be of relevance to so-called quantum dots.

6.3 Quantum Hall Liquid

In this subsection we shall argue that the effective theory of a quantum Hall liquid can be used to describe its liquid-to-insulator transition as the applied magnetic field changes, and study its critical properties.

Experimentally, if the external field is changed so that the filling factor ν_H moves away from an odd-denominator value, the system eventually will become critical and will undergo a transition to an insulating phase. Elsewhere[17], we have argued that this feature is encoded in the CSGL theory. In the spirit of Landau, we took a phenomenological approach towards this field-induced phase transition, and assumed that, when the applied magnetic field H is close to the upper critical field $H^+_{\nu_H}$, at which the quantum Hall liquid with filling factor ν_H is destroyed, the chemical potential of the composite particles depends linearly on H, i.e., $\mu_0 \propto eH^+_{\nu_H} - eH$. This state can, of course, also be destroyed by lowering the applied field. If the system is near the lower critical field $H^-_{\nu_H}$, we assumed that the chemical potential is given instead by $\mu_0 \propto eH - eH^-_{\nu_H}$. This is the basic postulate of our approach.

We modify the CSGL Lagrangian (265) so that it only includes the fluctuating part of the statistical gauge field. That is, we ignore the classical part of a which yields a magnetic field that precisely cancels the externally applied field. We can again transform the mass m of the nonrelativistic $|\phi|^4$-theory away. In addition to the

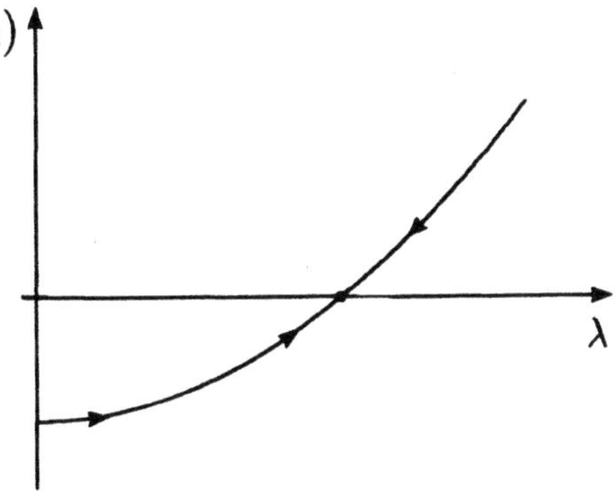

Figure 8. Schematic representation of the beta function (306).

engineering dimensions (284), we have for the Chern-Simons field

$$[ea_i] = 1, \quad [ea_0] = 2, \quad [\theta] = 0. \tag{305}$$

In two space dimensions, the coupling constant λ_0 has been seen to be dimensionless, implying that the $|\phi|^4$-term is a marginal operator. From (305) it follows that also the Chern-Simons term is a marginal operator. Hence, the CSGL theory contains—apart from a random and a Coulomb term—precisely those terms relevant to the description of the liquid-to-insulator transition in a quantized Hall system.

It is well-known,[81,82] that the coefficient of the Chern-Simons term is not renormalized by quantum corrections.

To first order in a loop expansion, the theory in two space dimensions has been known to have an IR fixed point determined by the zero of the beta function,[83]

$$\beta(\lambda) = \frac{1}{\pi}\left(4\lambda^2 - \frac{1}{\theta^2}\right). \tag{306}$$

The calculation of $\beta(\lambda)$ has been extended to fourth order in the loop expansion[17]. This study revealed that the one-loop result (306) is unaffected by these higher-order loops. Presumably, this remains true to all orders in perturbation theory, implying that—just as in the neutral system which corresponds to taking the limit $\theta \to \infty$—the one-loop beta function (306) is exact.

It is schematically represented in Fig. 8, and is seen to yield a nontrivial IR fixed point $\lambda^{*2} = 1/4\theta^2$ determined by the filling factor. More precisely, the strength of the repulsive coupling at the fixed point $\lambda^* = \pi(2l+1)$ increases with the number $2l+1$ of flux quanta bound to the electron. The presence of the fixed point shows that the CSGL theory undergoes a 2nd-order phase transition when the chemical potential of the composite particles tends to a critical value. As in the neutral case, it can be shown that the boson self-energy Σ also vanishes at every loop order in the charged theory, and that the self-coupling parameter λ is the only object that renormalizes. The 2nd-order phase transition described by the nontrivial IR fixed point has consequently again Gaussian exponents[17] $\nu = 1/2$, $z = 2$. It should be noted that only the location of the fixed point depends on θ, the critical exponents—which in contrast to the strength of the coupling at the fixed point are independent of the regularization and renormalization schemes—are universal and independent of the filling factor. This "triviality" is in accord with the experimentally observed universality of FQHE. A dependence of the critical exponents on θ could from the theoretical point of view be hardly made compatible with the hierarchy construction[84] which implies a cascade of phase transitions. From this viewpoint the present results are most satisfying: the CSGL theory is shown to encode a new type of a field-induced 2nd-order quantum phase transition that is simple enough not to obscure the observed universality of FQHE.

We stress again that in order to arrive at a realistic description of FQHE, the CSGL theory has to be extended to include a $1/|\mathbf{x}|$-Coulomb potential and impurities. Both will change the critical behavior we have found in this subsection. In particular, the Coulomb potential will change the Gaussian value $z = 2$ into $z = 1$.

6.4 Random Theory

In Sec. 6.2, we have seen that in the absence of impurities the repulsively interacting bosons undergo a 2nd-order quantum phase transition as function of the chemical potential. As has been pointed out there, this universality class is of relevance to various condensed-matter systems. However, in most of the systems mentioned there, as well in ^{4}He in porous media, impurities play an essential if not decisive role. For example, the two-dimensional superconductor-to-insulator transition investigated by Hebard and Paalanen[63] is driven by impurities. This means that, e.g., the correlation length ξ diverges as $|\Delta_c - \Delta|^{-\nu}$ when the disorder strength Δ characterizing the randomness approaches the critical value Δ_c. Hence, a realistic description of the critical behavior of these systems should include impurities.

To incorporate these, we proceed as before and add the random term (119) to the nonrelativistic $|\phi|^4$-theory (45). The random field $\psi(\mathbf{x})$ has the Gaussian distribution (120). We shall study the theory in the symmetrical state where the bare chemical potential is negative and the global U(1) symmetry unbroken. We, therefore, set $\mu_0 = -r_0$ again, with $r_0 > 0$. We leave the number of space dimensions d unspecified for the moment. As we have remarked before, since $\psi(\mathbf{x})$ depends only on the d spatial dimensions, the impurities it describes should be considered as grains randomly distributed in space. When—as is required for the study of quantum critical phenomena—time is included, the static grains trace out straight worldlines. That is to say, these impurities are linelike in the quantum theory. It has been shown by Dorogovtsev[86] that the critical properties of systems with extended defects must be studied in a double ϵ-expansion, otherwise no IR fixed point is found. The method differs from the usual ϵ-expansion, in that it also includes an expansion in the defect dimensionality ϵ_d. To carry out this program in the present context, where the defect dimensionality is determined by the dimensionality of time, the theory has to be formulated in ϵ_d time dimensions. The case of interest is $\epsilon_d = 1$, while in the opposite limit, $\epsilon_d \to 0$, the random nonrelativistic $|\phi|^4$-theory reduces to the classical spin model with random (pointlike) impurities. Hence, ϵ_d is a parameter with which quantum fluctuations can be suppressed. An expansion in ϵ_d is a way to perturbatively include the effect of quantum fluctuations on the critical behavior. Ultimately, we will be interested in the case $\epsilon_d = 1$.

To calculate the quantum critical properties of the random theory, which have first been studied in Ref.[87], we will not employ the replica method[88], but instead follow Lubensky[89]. In this approach, the averaging over impurities is carried out for each Feynman diagram separately. The upshot is that only those diagrams are to be included which remain connected when Δ_0, the parameter characterizing the Gaussian distribution of the impurities, is set to zero[90]. To obtain the relevant Feynman rules of the random theory we average the interaction term (119) over the distribution (120):

$$\int \mathrm{D}\psi\, P(\psi) \exp\left[i^{\epsilon_d} \int \mathrm{d}^{\epsilon_d} t\, \mathrm{d}^d x\, \psi(\mathbf{x})\, |\phi(x)|^2\right]$$

$$= \exp\left[\frac{1}{4} i^{2\epsilon_d} \Delta_0 \int \mathrm{d}^{\epsilon_d} t \, \mathrm{d}^{\epsilon_d} t' \, \mathrm{d}^d x \, |\phi(t, \mathbf{x})|^2 |\phi(t', \mathbf{x})|^2 \right]. \tag{307}$$

The randomness is seen to result in a quartic interaction term which is nonlocal in time. The factor i^{ϵ_d} appearing in (307) arises from the presence of ϵ_d time dimensions, each of which is accompanied by a factor of i. The Feynman rules of the random theory are now easily obtained

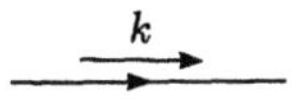

$$= \frac{-i^{-\epsilon_d} \mathrm{e}^{i(\omega_1 + \omega_2 + \cdots + \omega_{\epsilon_d})\eta}}{\omega_1 + \omega_2 + \cdots + \omega_{\epsilon_d} - \mathbf{k}^2 - r_0 + i\eta}$$

$$= -4 i^{\epsilon_d} \lambda_0$$

$$= i^{\epsilon_d} (2\pi)^{\epsilon_d} \delta^{\epsilon_d}(\omega_1 + \omega_2 + \cdots + \omega_{\epsilon_d}) \Delta_0, \tag{308}$$

where we note that the Lagrangian in ϵ_d time dimensions involves instead of just one time derivative, a sum of ϵ_d derivatives: $\partial_t \to \partial_{t_1} + \partial_{t_2} + \cdots + \partial_{t_{\epsilon_d}}$.

Following Weichman and Kim[91], we evaluate the integrals over loop energies assumming that all energies are either positive or negative. This allows us to employ Schwinger's propertime representation of propagators[92], which is based on the integral representation (88) of the Gamma function. The energy integrals we encounter to the one-loop order can be carried out with the help of the equations

$$\int' \frac{\mathrm{d}^{\epsilon_d}\omega}{(2\pi)^{\epsilon_d}} \frac{1}{\omega_1 + \omega_2 + \cdots + \omega_{\epsilon_d} - x \pm i\eta}$$
$$= -\frac{\Gamma(1-\epsilon_d)}{(2\pi)^{\epsilon_d}} \mathrm{sgn}(x) |x|^{\epsilon_d - 1} \left(\mathrm{e}^{\pm i \, \mathrm{sgn}(x) \pi \epsilon_d} + 1 \right), \tag{309}$$

$$\int' \frac{\mathrm{d}^{\epsilon_d}\omega}{(2\pi)^{\epsilon_d}} \frac{\mathrm{e}^{i(\omega_1 + \omega_2 + \cdots + \omega_{\epsilon_d})\eta}}{\omega_1 + \omega_2 + \cdots + \omega_{\epsilon_d} - x + ix\eta}$$
$$= \frac{i\pi}{(2\pi)^{\epsilon_d} \Gamma(\epsilon_d)} (i|x|)^{\epsilon_d - 1} \left[\sin(\tfrac{1}{2}\pi\epsilon_d) - \frac{\mathrm{sgn}(x)}{\sin(\tfrac{1}{2}\pi\epsilon_d)} \right]. \tag{310}$$

The prime on the integrals is to remind the reader that the energy integrals are taken over only two domains with either all energies positive or negative. The energy integrals have been carried out by using again the integral representation (88) of the Gamma function. In doing so, the integrals are regularized and—as is always the case with analytic regularizations—irrelevant divergences suppressed.

By differentiation with respect to x, Eq. (309) can, for example, be employed to calculate integrals involving integrands of the form $1/(\omega_1 + \omega_2 + \cdots + \omega_{\epsilon_d} - x + i\eta)^2$. It is easily checked that in the limit $\epsilon_d \to 1$, where the energy integral can be performed with help of a contour integration, Eqs. (309) and (310) reproduce the right results. When considering the limit of zero time dimensions ($\epsilon_d \to 0$), it should be borne in mind that the energy integrals have been taken over two separate domains with all energies either positive or negative. Each of these domains is contracted to a single point in the limit $\epsilon_d \to 0$, so that one obtains a result which is twice that obtained by simply purging any reference to the time dimensions. The integral (310) contains an additional convergence factor $\exp(i\omega\eta)$ for each of the ϵ_d energy integrals. This factor, which—as we remarked before—is typical for nonrelativistic quantum theories[51], is to be included in self-energy diagrams containing only one ϕ-propagator.

Before studying the random theory, let us briefly return to the repulsively interacting bosons in the absence of impurities. In this case, there is no need for an

ϵ_d-expansion and the formalism outlined above should yield results for arbitrary time dimensions $0 \le \epsilon_d \le 1$, interpolating between the classical and the quantum limit. After the energy integrals have been performed with the help of Eqs. (309) and (310), the standard technique of integrating out a momentum shell can be applied to obtain the renormalization-group equations. For the correlation-length exponent ν we obtain in this way[30]

$$\nu = \frac{1}{2}\left[1 + \frac{\epsilon}{2}\frac{m+1}{(m+4)-(m+3)\epsilon_d}\cos^2\left(\frac{1}{2}\pi\epsilon_d\right)\right]. \tag{311}$$

Here, $\epsilon = 4 - 2\epsilon_d - d$ is the deviation of the *effective* spacetime dimensionality from 4, where it should be noted that in (canonical) nonrelativistic theories, time dimensions have an engineering dimension twice that of space dimensions. (This property is brought out by the Gaussian value $z = 2$ for the dynamic exponent z.) For comparison, we have extended the theory (45) to include m complex ϕ-fields instead of just one field. In the classical limit, Eq. (311) gives the well-known one-loop result for a classical spin model with $2m$ real components,[29]

$$\nu \to \frac{1}{2}\left(1 + \frac{\epsilon}{2}\frac{m+1}{m+4}\right), \tag{312}$$

while in the quantum limit it gives the result $\nu \to 1/2$, as required.

The exponent (311), and also the location of the fixed point, diverges when the number of time dimensions becomes $\epsilon_d \to (m+4)/(m+3)$. Since this value is always larger than one, the singularity is outside the physical domain $0 \le \epsilon_d \le 1$. This simple example illustrates the viability of the formalism developed here to generate results interpolating between the classical and the quantum limit.

We continue with the random theory. After the energy integrals have been carried out, it is again straightforward to derive the renormalization-group equations by integrating out a momentum shell $\Lambda/b < k < \Lambda$, where Λ is a high-momentum cutoff and $b = \exp(l)$, with l infinitesimal. Defining the dimensionless variables

$$\hat{\lambda} = \frac{K_d}{(2\pi)^{\epsilon_d}}\lambda\Lambda^{-\epsilon}; \quad \hat{\Delta} = K_d\Delta\Lambda^{d-4}; \quad \hat{r} = r\Lambda^{-2}, \tag{313}$$

where

$$K_d = \frac{2}{(4\pi)^{d/2}\Gamma(d/2)} \tag{314}$$

is the area of a unit sphere in d spatial dimensions divided by $(2\pi)^d$, we find[30]

$$\begin{aligned}
\frac{d\hat{\lambda}}{dl} &= \epsilon\hat{\lambda} - 8\left[\Gamma(1-\epsilon_d) + (m+3)\Gamma(2-\epsilon_d)\right]\cos(\tfrac{1}{2}\pi\epsilon_d)\hat{\lambda}^2 + 6\hat{\Delta}\hat{\lambda} \\
\frac{d\hat{\Delta}}{dl} &= (\epsilon + 2\epsilon_d)\hat{\Delta} + 4\hat{\Delta}^2 - 16(m+1)\Gamma(2-\epsilon_d)\cos(\tfrac{1}{2}\pi\epsilon_d)\hat{\lambda}\hat{\Delta} \\
\frac{d\hat{r}}{dl} &= 2\hat{r} + 4\pi\frac{m+1}{\Gamma(\epsilon_d)}\frac{\cos^2(\pi\epsilon_d/2)}{\sin(\pi\epsilon_d/2)}\hat{\lambda} - \hat{\Delta}.
\end{aligned} \tag{315}$$

These results are to be trusted only for small values of ϵ_d. The set of equations yields the fixed point

$$\begin{aligned}
\hat{\lambda}^* &= \frac{1}{\left[16\cos(\frac{1}{2}\pi\epsilon_d)\Gamma(1-\epsilon_d)\right]}\frac{\epsilon + 6\epsilon_d}{[2m(1-\epsilon_d)-1]} \\
\hat{\Delta}^* &= \frac{1}{4}\frac{m(1-\epsilon_d)(2\epsilon_d - \epsilon) + 2\epsilon_d(4-3\epsilon_d) + \epsilon(2-\epsilon_d)}{2m(1-\epsilon_d)-1},
\end{aligned} \tag{316}$$

and the critical exponent

$$\nu = \frac{1}{2} + \frac{\epsilon + 2\epsilon_\mathrm{d}}{16} + \frac{m+1}{16} \frac{(6\epsilon_\mathrm{d} + \epsilon)[\epsilon_\mathrm{d} + \cos(\pi\epsilon_\mathrm{d})]}{2m(1-\epsilon_\mathrm{d}) - 1}. \tag{317}$$

The dynamic exponent is given by $z = 2 + \hat{\Delta}^*$. When the equations are expanded to first order in ϵ_d, we recover the IR fixed point found by Weichman and Kim[91] using an high-energy cutoff:

$$\hat{\lambda}^* = \frac{1}{16} \frac{\epsilon + 6\epsilon_\mathrm{d}}{2m-1}; \quad \hat{\Delta}^* = \frac{1}{4} \frac{(2-m)\epsilon + 2(m+4)\epsilon_\mathrm{d}}{2m-1}, \tag{318}$$

with the critical exponent

$$\nu = \frac{1}{2}\left[1 + \frac{1}{8} \frac{3m\epsilon + (5m+2)2\epsilon_\mathrm{d}}{2m-1}\right]. \tag{319}$$

The value of the critical exponent (319) should be compared with that of the classical spin model[86] with $2m$ components in the presence of random impurities of dimension ϵ_d:

$$\nu = \frac{1}{2}\left[1 + \frac{1}{8} \frac{3m\epsilon + (5m+2)\epsilon_\mathrm{d}}{2m-1}\right]. \tag{320}$$

Taking into account that in a nonrelativistic quantum theory, time dimensions count double as compared to space dimensions, we see that both results are equivalent. As to the dynamic exponent, notice that the perturbative result $z = 2 + \hat{\Delta}^*$, with $\hat{\Delta}^*$ given by (318), is far away from the exact value $z = d$ for $\epsilon_\mathrm{d} = 1$ predicted in Ref. 80.

6.5 Experiments

Superconductor-to-Insulator Transition. The first experiments we wish to discuss are those performed by Hebard and Paalanen on superconducting films in the presence of random impurities.[63,85] It has been predicted by Fisher[93] that with increasing applied magnetic field such systems undergo a zero-temperature transition into an insulating state. (For a critical review of the experimental data see Ref. 2.)

Let us restrict ourselves for the moment to the $T\Delta$-plane of the phase diagram by setting the applied magnetic field H to zero. For a given disorder strength Δ, the system then undergoes a Kosterlitz–Thouless transition induced by the unbinding of magnetic vortex pairs at a temperature T_KT well below the bulk transition temperature (see Sec. 4.3). The Kosterlitz–Thouless temperature is gradually suppressed to zero when the disorder strength approaches criticality $\Delta \to \Delta_\mathrm{c}$. The transition temperature scales with the correlation length $\xi \sim |\Delta_\mathrm{c} - \Delta|^{-\nu}$ as $T_\mathrm{KT} \sim \xi^{-z}$.

In the $H\Delta$-plane, i.e., at $T = 0$, the situation is as follows. For a given disorder strength, there is now at some critical value H_c of the applied magnetic field a phase transition from a superconducting state of pinned vortices and condensed Cooper pairs to an insulating state of pinned Cooper pairs and condensed vortices. The condensation of vortices disorder the ordered state as happens in classical, finite temperature superfluid- and superconductor-to-normal phase transitions[34]. When the disorder strength approaches criticality again, H_c is gradually suppressed to zero. The critical field scales with ξ as $H_\mathrm{c} \sim \Phi_0/\xi^2$. In fact, this expresses a more fundamental result, namely that the scaling dimension d_A of A is one,

$$d_A = 1, \tag{321}$$

so that $A \sim \xi^{-1}$. From this, it in turn follows that $E \sim \xi_t^{-1}\xi^{-1} \sim \xi^{-(z+1)}$, and that the scaling dimension d_{A_0} of A_0 is z,

$$d_{A_0} = z, \tag{322}$$

so that $A_0 \sim \xi_t^{-1} \sim \xi^{-z}$. Together, the scaling results for $T_{\rm KT}$ and H_c imply that[93]

$$H_c \sim T_{\rm KT}^{2/z}. \tag{323}$$

This relation, linking the critical field of the zero-temperature transition to the Kosterlitz-Thouless temperature, provides a direct way to measure the dynamic exponent z at the $H = 0$, $T = 0$ transition. This has been first done by Hebard and Paalanen[63,85]. Their experimental determination of $T_{\rm KT}$ and H_c for five different films with varying amounts of impurities has confirmed the relation (323) with $2/z = 2.04 \pm 0.09$. The zero-temperature critical fields have been obtained by plotting $\mathrm{d}\rho_{xx}/\mathrm{d}T|_H$ versus H at the lowest accessible temperature and interpolating to the field where the slope is zero. The resulting value $z = 0.98 \pm .04$ is in accordance with Fisher's prediction[93], $z = 1$, for a random system with a $1/|\mathbf{x}|$-Coulomb potential.

Hebard and Paalanen[63] have also investigated the field-induced zero-temperature transition. The control parameter here is $\delta \propto H - H_c$. When plotted as a function of $|H - H_c|/T^{1/\nu_H z_H}$ they saw their resistivity data collapsing onto two branches; an upper branch tending to infinity for the insulating state, and a lower branch bending down for the superconducting state. The unknown product $\nu_H z_H$ is experimentally determined by looking for which value the best scaling behavior is obtained. Further experiments carried out by Yazdani and Kapitulnik[94] have also determined the product $\nu_H(z_H + 1)$ (see below). The two independent measurements together fix the critical exponents ν_H and z_H separately. From their best data, Yazdani and Kapitulnik have extracted the values[94]

$$z_H = 1.0 \pm 0.1, \quad \nu_H = 1.36 \pm 0.05. \tag{324}$$

Quantum Hall Systems. We continue to discuss the field-induced quantum phase transitions in quantum Hall systems. Since an excellent discussion recently appeared in the literature[4], we shall be brief, referring the reader to that review for a more thorough discussion and additional references.

One can image transitions from one Hall liquid to another Hall liquid with a different (integer or fractional) filling factor, or to the insulating state. Experiments seem to suggest that all the quantum-Hall transitions are in the same universality class. The transitions are probed by measuring the resistivities ρ_{xx} and ρ_{xy}. From the dependence of the conductivity σ on the superfluid mass density, Eq. (297), and the scaling relation (292), it follows that it scales as[95]

$$\sigma \sim \xi^{-(d-2)}. \tag{325}$$

In other words, the scaling dimension of the conductivity and, therefore, that of the resistivity is zero in two space dimensions. On account of the general finite-size scaling form (281), we then have in the limit $|\mathbf{k}| \to 0$:

$$\rho_{xx/y}(k_0, H, T) = \varrho_{xx/y}(k_0/T, |\delta|^{\nu z}/T), \tag{326}$$

where the distance to the zero-temperature critical point is measured by $\delta \propto H - H_{\nu_H}^{\pm} \sim T^{1/\nu z}$. This scaling of the width of the transition regime with temperature has been corroborated by DC or $k_0 = 0$ experiments on various transitions between integer quantum-Hall states which were all found to yield the value[96] $1/\nu z = 0.42 \pm 0.04$.

A second measurement of the critical exponents involves the applied electric field. As we have seen above, it scales as $E \sim \xi^{-(z+1)}$, so that for the DC resistivities we now obtain the scaling form:

$$\rho_{xx/y}(H,T,E) = \varrho_{xx/y}(|\delta|^{\nu z}/T, |\delta|^{\nu(z+1)}/E). \tag{327}$$

The scaling $|\delta| \sim E^{1/\nu(z+1)}$ has again been corroborated by experiment which yields the value[97] $\nu(z+1) \approx 4.6$. Together with the previous result, obtained from the temperature scaling, this gives

$$z \approx 1, \quad \nu \approx 2.3. \tag{328}$$

The value of the dynamic exponent strongly suggests that it is a result of the presence of the $1/|\mathbf{x}|$-Coulomb potential. The correlation length exponent ν is large.

2*d* **Electron Systems.** Recently, silicon MOSFET's at extremely low electron number densities has been studied.[98-101] Earlier experiments at higher densities seem to confirm the general believe, based on the work by Abrahams *et al.*,[95] that such two-dimensional electron systems do not undergo a quantum phase transition. In that influential paper, it has been demonstrated that even weak disorder is sufficient to localize the electrons at the absolute zero of temperature thus excluding conducting behavior. Electron-electron interactions were, however, not included. As we have seen in Sec. 3.2, at low densities, the $1/|\mathbf{x}|$-Coulomb interaction becomes important and the analysis of Abrahams *et al.*[95] no longer applies.

The recent experiments have revealed a zero-temperature conductor-to-insulator transition triggered by a change in the charge carrier density $\bar{n}$. That is, the distance to the critical point is in these systems measured by $\delta \propto \bar{n} - \bar{n}_c$. Like in the quantum-Hall systems, these transitions are probed by measuring the resistivity. It scales with temperature near the transition according to the scaling form (327) with H set to zero. For $\bar{n} < \bar{n}_c$, where the Coulomb interaction is dominant and fluctuations in the charge carrier density are suppressed, the electron system is insulating. On increasing the density, these fluctuations intensify and at the critical value $\bar{n}_c$, the system reverts to a conducting phase. By plotting their conductivity data as a function of $T/|\delta|^{\nu z}$ with $\nu z = 1.6 \pm 0.1$, Popovíc, Fowler, and Washburn[101] saw it collapse onto two branches; the upper branch for the conducting side of the transition, and the lower one for the insulating side. A similar collapse with a slightly different value $1/\nu z = 0.83 \pm 0.08$ has been found in Ref. 99, where also the collapse of the data when plotted as function of $|\delta|/E^{1/\nu(z+1)}$ was obtained. The best collapse resulted for $1/(z+1)\nu = 0.37 \pm 0.01$, leading to

$$z = 0.8 \pm 0.1, \quad \nu = 1.5 \pm 0.1. \tag{329}$$

The value for the dynamic exponent is close to the expected value $z = 1$ for a charged system with a $1/|\mathbf{x}|$-Coulomb interaction, while that of ν is surprisingly close to the value (324) found for the superconductor-to-insulator transition.

A further experimental result for these two-dimensional electron systems worth mentioning is the suppression of the conducting phase by an applied magnetic field found by Simonian, Kravchenko, and Sarachik.[100] They applied the field *parallel* to the plane of the electrons instead of perpendicular as is done in quantum-Hall measurements. In this way, the field presumably couples only to the spin of the electrons and the complications arising from orbital effects do not arise. At a fixed temperature, a rapid initial raise in the resistivity has been found with increasing field. Above a value of about 20 kOe, the resistivity saturates. It was pointed out that both the behavior in a magnetic filed, as well as in a zero field strongly resembles that near the superconductor-to-insulator transition discussed above, suggesting that the conducting phase may in fact be superconducting.

Conclusions. We have seen that general scaling arguments can be employed to understand the scaling behavior observed in various quantum phase transitions. Most of the experiments seem to confirm the expected value $z = 1$ for a random system with a $1/|\mathbf{x}|$-Coulomb interaction. The number of different universality classes present is yet not known. Even if the conductor-to-insulator transition observed in silicon MOSFET's at low electron number densities turns out to be in the same universality class as the superconductor-to-insulator transition, there are still the field-induced transitions in quantum-Hall systems, which have a larger correlation-length exponent.

The paradigm provided by a repulsively interacting Bose gas, seems to be a good starting point to describe the various systems. However, high-precision estimates calculated from this theory with impurities and a $1/|\mathbf{x}|$–Coulomb interaction included are presently lacking.

REFERENCES

1. D. I. Uzunov. "Introduction to the Theory of Critical Phenomena," World Scientific, Singapore (1993).
2. T. Liu, and A. M. Goldman, *Mod. Phys. Lett.* B 8:277 (1994).
3. S. Sachdev, *in*: "Proceedings of the 19th IUPAP International Conference on Statistical Physics," Xiamen, China (1995) edited by B. -L. Hao, World Scientific, Singapore (1995), p. 289 [e-print cond-mat/9508080 (1995)].
4. S. L. Sondhi, S. M. Girvin, J. P. Carini, and D. Shahar, *Rev. Mod. Phys.* 69:315 (1997).
5. C. M. Fraser, *Z. Phys.* C 28:101 (1985); I. J. R. Aitchison and C. M. Fraser, *Phys. Rev.* D 31:2605 (1985).
6. E. P. Gross, *Nuovo Cimento* 20:454 (1961); L. P. Pitaevskii, *Sov. Phys. JETP* 13:451 (1961).
7. N. N. Bogoliubov, *J. Phys. USSR* 11:23 (1947).
8. S. T. Beliaev, *Sov. Phys. JETP* 7:289 (1958).
9. N. M. Hugenholtz, and D. Pines, *Phys. Rev.* 116:489 (1959).
10. J. Bernstein, and S. Dodelson, *Phys. Rev. Lett.* 66:683 (1991); K. M. Benson, J. Bernstein, and S. Dodelson, *Phys. Rev.* D 44:2480 (1991).
11. J. Gavoret and P. Nozières, *Ann. Phys.* (N.Y.), 28:349 (1964).
12. A. M. J. Schakel, *Int. J. Mod. Phys.* B 8:2021 (1994).
13. A. M. J. Schakel, *Mod. Phys. Lett.* B 4:927 (1990).
14. A. M. J. Schakel, *Int J. Mod. Phys.* B 10:999 (1996).
15. P. W. Anderson, *Rev. Mod. Phys.* 38:298 (1966).
16. H. Leutwyler, *Phys. Rev.* D 49:3033 (1994).
17. A. M. J. Schakel, *Nucl. Phys.* B [FS] 453:705 (1995).
18. M. Greiter, F. Wilczek, and E. Witten, *Mod. Phys. Lett.* B 3:903 (1989).
19. J. F. Donoghue, *Phys. Rev.* D 50:3874 (1994); e-print gr-qc/9512024 (1995).
20. T. Y. Cao, and S. S. Schweber, *Synthese* 97:(1993); *in*: "Renormalization," edited by L. M. Brown, Springer-Verlag, Berlin (1993).
21. P. B. Weichman, *Phys. Rev.* B 38:8739 (1988).
22. A. L. Fetter, and J. D. Walecka. "Quantum Theory of Many-Particle Systems," McGraw-Hill, New York (1971).
23. E. H. Lieb, and W. Liniger, *Phys. Rev.* 130:1605 (1963); E. H. Lieb, *Phys. Rev.* 130:1616 (1963).
24. D. I. Uzunov, *Phys. Lett.* A 87:11 (1981).
25. U. C. Täuber, and D. R. Nelson, *Phys. Rep.* 289:157 (1997).
26. O. Penrose, and L. Onsager, *Phys. Rev.* 104:576 (1956).
27. P. Nozières and D. Pines. "The Theory of Quantum Liquids," Addison-Wesley, New York (1990), Vol. II.
28. L. S. Brown. "Quantum Field Theory," Cambridge University Press, Cambridge (1992).
29. S. K. Ma. "Modern Theory of Critical Phenomena," Benjamin, London (1976).
30. A. M. J. Schakel, *Phys. Lett.* A 224:287 (1997).

31. S. Giorgini, L. Pitaevskii, and S. Stringari, *Phys. Rev.* B 49:12938 (1994).
32. K. Huang, and H.-F. Meng, *Phys. Rev. Lett.* 69:644 (1992).
33. A. L. Fetter, *Phys. Rev.* 138:429 (1965).
34. H. Kleinert. "Gauge Fields in Condensed Matter," World Scientific, Singapore (1989), Vol. I.
35. H. Kleinert, *Int. J. Mod. Phys.* A 7:4693 (1992).
36. H. Kleinert, *in*: "Formation and Interactions of Topological Defects," Proceedings NATO ASI, Cambridge (1994), edited by A. C. Davis, and R. Brandenburger Plenum, New York (1995).
37. H. Lamb. "Hydrodynamics," Dover, New York (1945), p. 230.
38. W. Yourgrau, and S. Mandelstam. "Variational Principles in Dynamics and Quantum Theory," Pitman, London (1968).
39. M. V. Berry, *Proc. Roy. Soc.* A 392:45 (1984).
40. F. D. M. Haldane, and Y. -S. Wu, *Phys. Rev. Lett.* 55:2887 (1985).
41. F. Lund, *Physica* A 159:245 (1991).
42. E. J. Yarmchuk, and R. E. Packard, *J. Low Temp. Phys.* 46:479 (1982).
43. V. L. Berezinskii, *Sov. Phys. JETP* 34:610 (1972).
44. J. M. Kosterlitz, and D. J. Thouless, *J. Phys.* C 6:1181 (1973).
45. D. R. Nelson, and J. M. Kosterlitz, *Phys. Rev. Lett.* 39:1201 (1977).
46. S. Schenker, *in*: "Recent Advances in Field Theory and Statistical Mechanics," proceedings of Les Houches summerschool, Session XXXIX, 1982, edited by J. B. Zuber, and R. Stora, Elsevier, Amsterdam (1984).
47. J. Bardeen, L. N. Cooper, and J. R. Schrieffer, *Phys. Rev.* 108:1175 (1957).
48. S. Weinberg, *Prog. Theor. Phys. Suppl.* 86:43 (1986).
49. H. Kleinert, *Fortschr. Phys.* 26:565 (1978).
50. A. J. Leggett, *in* "Modern Trends in the Theory of Condensed Matter," edited by A. Pekalski, and J. Przystawa, Springer-Verlag, Berlin (1980), p. 13.
51. R. D. Mattuck. "A Guide to Feynman Diagrams in the Many-Body Problem," McGraw-Hill, New York (1976).
52. R. Haussmann, *Z. Phys.* B 91:291 (1993).
53. M. Drechsler, and W. Zwerger, *Ann. Phys.* (Germany), 1:15 (1992).
54. C. A. R. Sá de Melo, M. Randeria, and J. R. Engelbrecht, *Phys. Rev. Lett.* 71:3202 (1993).
55. M. Marini, F. Pistolesi, and G. C. Strinati, e-print cond-mat/9703160 (1997).
56. U. Eckern, G. Schön, and V. Ambegaokar, *Phys. Rev.* 30:6419 (1989).
57. M. Randeria, J. -M. Duan, and L. -Y. Shieh, *Phys. Rev.* B 41:327 (1990).
58. F. A. Schaposnik, *Phys. Rev.* D 18:1183 (1978).
59. J. Pearl, *in*: "Low Temperature Physics—LT9," edited by J. G. Danut, D. O. Edwards, F. J. Milford, and M. Yagub, Plenum, New York (1965).
60. P. G. de Gennes. "Superconductivity of Metals and Alloys," Addison-Wesley, New York (1966).
61. M. R. Beasley, J. E. Mooij, and T. P. Orlando, *Phys. Rev. Lett.* 42:1165 (1979).
62. M.-C. Cha, M. P. A. Fisher, S. M. Girvin, M. Wallin, and A. P. Young, Phys. Rev. B 44:6883 (1991); S. M. Girvin, M. Wallin, M.-C. Cha, M. P. A. Fisher, and A. P. Young, *Prog. Theor. Phys. Suppl.* 107:135 (1992).
63. A. F. Hebard, and M. A. Paalanen, *Phys. Rev. Lett.* 65:927 (1990).
64. R. B. Laughlin, *Phys. Rev. Lett.* 50:1395 (1983).
65. V. L. Ginzburg, and L. D. Landau, *Zh. Eksp. Teor. Fiz.* 20:1064 (1950); reprinted *in*: L. D. Landau. "Collected Papers," Pergamon, London (1965), p. 546.
66. L. P. Gorkov, *Sov. Phys. JETP* 9:1364 (1959).
67. S. M. Girvin, *in*: "The Quantum Hall Effect," edited by R. E. Prange and S. M. Girvin, Springer-Verlag, Berlin (1986), Ch. 10; S. M. Girvin, and A. H. MacDonald, *Phys. Rev. Lett.* 58:1252 (1987).
68. S.-C. Zhang, T. H. Hansson, and S. Kivelson, *Phys. Rev. Lett.* 62:82 (1989).
69. S.-C. Zhang, *Int. J. Mod. Phys.* B 6:25 (1992).
70. F. Wilczek. "Fractional Statistics and Anyon Superconductivity," World Scientific, Singapore (1990).
71. D. P. Arovas, J. R. Schrieffer, and F. Wilczek, *Phys. Rev. Lett.* 53:722 (1984).
72. B. Widom, *J. Chem. Phys.* 43:3892 (1965); 43:3898 (1965).
73. M. P. A. Fisher, P. B. Weichman, G. Grinstein, and D. S. Fisher, *Phys. Rev.* B 40:546 (1989).
74. J. A. Hertz, *Phys. Rev.* B 14:1165 (1976).
75. I. Affleck, *Phys. Rev.* B 41:6697 (1990).
76. T. R. Kirkpatrick, and D. Belitz, *Phys. Rev. Lett.* 76:2571 (1996).
77. M. P. A. Fisher, G. Grinstein, and S. M. Girvin, *Phys. Rev. Lett.* 64:587 (1990).

78. O. Bergman, *Phys. Rev.* D 46:5474 (1992).
79. G. G. Batrouni, R. T. Scalettar, and G. T. Zimanyi, *Phys. Rev. Lett.* 65:1765 (1990).
80. D. S. Fisher, and M. P. A. Fisher, *Phys. Rev. Lett.* 61:1847 (1988).
81. S. Coleman, and B. Hill, *Phys. Lett.* B 159:184 (1985).
82. J. D. Lykken, J. Sonneschein, and N. Weiss, *Int. J. Mod. Phys.* A 6:1335 (1991).
83. G. Lozano, *Phys. Lett.* B 283:70 (1992); O. Bergman, and G. Lozano, *Ann. Phys.* (N.Y.) 229:416 (1994).
84. F. D. M. Haldane, *Phys. Rev. Lett.* 51:605 (1983); B. I. Halperin, *Phys. Rev. Lett.* 52:1583 (1984).
85. A. F. Hebard, and M. A. Paalanen, *Helv. Phys. Act.* 65:197 (1992).
86. S. N. Dorogovtsev, *Phys. Lett. A* 76:169 (1980).
87. E. R. Korutcheva, and D. I. Uzunov, *Phys. Lett.* A 106:175 (1984).
88. G. Grinstein, and A. Luther, *Phys. Rev.* B 13:1329 (1976).
89. T. C. Lubensky, *Phys. Rev.* B 11:3573 (1975).
90. For a review see, for example, J. A. Hertz, *Physica Scripta* T 10:1 (1985).
91. P. B. Weichman and K. Kim, *Phys. Rev.* B 40:813 (1989).
92. J. Schwinger, *Phys. Rev.* 82:664 (1951).
93. M. P. A. Fisher, *Phys. Rev. Lett.* 65:923 (1990).
94. A. Yazdani, and A. Kapitulnik, *Phys. Rev. Lett.* 74:3037 (1995).
95. E. Abrahams, P. W. Anderson, D. C. Licciardello, and T. V. Ramakrishnan, *Phys. Rev. Lett.* 42:637 (1979).
96. H. P. Wei, D. C. Tsui, M. A. Paalanen, and A. M. M. Pruisken, *Phys. Rev. Lett.* 61:1294 (1988).
97. H. P. Wei, L. W. Engel, and D. C. Tsui, *Phys. Rev.* B 50:14609 (1994).
98. S. V. Kravchenko, G. V. Kravchenko, J. E. Furneaux, V. M. Pudalov, and M. D'Iorio, *Phys. Rev.* B 50:8039 (1994); S. V. Kravchenko, W. E. Mason, G. E. Bowker, J. E. Furneaux, V. M. Pudalov, and M. D'Iorio, *Phys. Rev.* 51:7038 (1995).
99. S. V. Kravchenko, D. Simonian, M. P. Sarachik, W. E. Mason, and J. E. Furneaux, *Phys. Rev. Lett.* 77:4938 (1996).
100. D. Simonian, S. V. Kravchenko, and M. P. Sarachik, e-print cond-mat/9704071 (1997).
101. D. Popovíc, A. B. Fowler, and S. Washburn, e-print cond-mat/9704249 (1997).

LIST OF CONTRIBUTORS

Aksenov, V. L.,
Frank Laboratory of Neutron Physics,
Joint Institute for Nuclear Research,
141980 Dubna, Moscow region, Russia.

Copley, J. R. D.,
NIST Center for Neutron Research,
National Institute of Standards and Technology,
Gaithersburg, MD 20899, USA.

De Cesare, L.,
Dipartimento di Scienze Fisiche "E. R. Caianiello,"
Università di Salerno, I–84081 Baronissi, Salerno, Italy.

Folk, R.,
Institut für Theoretische Physik, Johannes Kepler Universität Linz,
A–4040 Linz, Austria.

Fomin, I.,
P. L. Kapitza Institute for Physical Problems,
ul. Kosygina 2, 117334 Moscow, Russia.

Holovatch, Yu,
Institute for Condensed Matter Physics,
Ukrainian Academy of Sciences, UA–290011 Lviv, Ukraine.

Michel, K. H.,
Department of Physics, University of Antwerp,
UIA, 2610 Antwerpen, Belgium.

Nemirovskii, S.,
Institute of Thermophysics, Prospect Lavrentyeva 1,
630090, Novosibirsk, Russia.

Nikolaev, A. V.,
Institute of Physical Chemistry of RAS,
117915, Moscow, Leninskii prospect 31, Russia.

Piacentini, M.,
Dipartimento di Energetica, Università di Roma "La Sapienza,"
via Scarpa, 14, 100161 Rome, Italy.

Plakida, N. M.,
Bogolubov Laboratory of Theoretical Physics,
Joint Institute for Nuclear Research, 141980 Dubna, Moscow Region, Russia.

Schakel, A. M. J.,
Institut für Theoretische Physik,
Freie Universität Berlin,
Arnimallee 14, 14195 Berlin.

Shakhmatov, V. S.,
Frank Laboratory of Neutron Physics,
Joint Institute for Nuclear Research,
141980 Dubna, Moscow region, Russia.

Uzunov, D. I.,
CPCM Laboratory, G. Nadjakov Institute of Solid State Physics,
Bulgarian Academy of Sciences, BG–1784 Sofia, Bulgaria.

Zema, N.,
Istituto di Struttura della Materia,
Area di Ricerca del CNR a Tor Vergata, Rome, Italy.

INDEX

MIX
Papier aus verantwortungsvollen Quellen
Paper from responsible sources
FSC® C105338

If you have any concerns about our products,
you can contact us on
ProductSafety@springernature.com

In case Publisher is established outside the EU,
the EU authorized representative is:
Springer Nature Customer Service Center GmbH
Europaplatz 3, 69115 Heidelberg, Germany

Printed by Libri Plureos GmbH
in Hamburg, Germany